The NEW complete book of SELF-SUFFICIENCY

The NEW complete book of SELF-SUFFICIENCY

JOHN SEYMOUR

WITH WILL SUTHERLAND

A DORLING KINDERSLEY BOOK

Contents

London, New York, Munich, Melbourne, and Delhi

Produced for Dorling Kindersley by
Editor **James Harrison**
Designer **Edward Kinsey**

Digital colour enhancement **Simon Roulstone**
Linocut illustrations **Jeremy Sancha**
Line illustrations **David Ashby, Sally Launder, John Woodcock, Kathleen McDougall, Peter Bull Associates, Simon Roulstone**
Assistant Editors **Clare Hill, Carla Masson**

For Dorling Kindersley Ltd
Editor **Gary Werner, Christine Heilman**
DTP designer **Louise Waller**
Production controller **Amanda Inness**

First published in Great Britain in 2002
by Dorling Kindersley Limited
80 Strand
London WC2R 0RL

A Penguin Company

2 4 6 8 10 9 7 5 3

Copyright 1976, 1996, 2002
Dorling Kindersley Limited
Text copyright 1976, 1996, 2002
John Seymour

A CIP catalogue record for this book is available from the British Library.

ISBN 07513 6442 8

Colour reproduction by Colourscan, Singapore
Printed and bound by AGT, Spain

See our complete catalogue at
www.dk.com

Preface

I first met John Seymour nearly 10 years ago. I had just resigned from my job as a senior management consultant in the City (of London) and was editing an alternative political magazine called *Ideas for Tomorrow Today*. I had discovered John's writings and I asked him to a conference I was organizing prior to the U.N. Rio Earth Summit in 1992. John Seymour, who is now 88 years old, had written the original *The Complete Book of Self-Sufficiency* in 1975, a book which has now sold over 600,000 copies and has become a virtual "bible" for those contemplating moves towards the "good life". After the Summit was over, I decided to explore some of the book's philosophy for myself.

I'll never forget that first trip to visit John and his companion Angela Ashe. The blackest of winter nights surrounded my BMW motorbike as it lurched unsteadily onwards through knee-deep Irish slurry. With all the studied objectivity of an ex-Whitehall civil servant, I weighed up the prospect of ending my first trip to Ireland lost in bottomless cow manure. But the fates were kind, the slurry released me and a kind Irish family redirected me to Killowen. Supper was delicious, the stove was warm and I found myself part of a totally different approach to life and living. We had eaten home-reared beef, drunk home-brewed beer and supped extraordinarily delicious home-made ice-cream. Over a whisky after supper John took out his accordion and Angela sang songs.

Ten years on I realize that my journey had not ended at Killowen — no indeed — it had only just begun. I had fallen in love with a new lifestyle, moved house to Ireland and later fell in love with Angela who is now my wife, and the mother of my three children, Liam, Roisin, and Hal. There was much to learn and I am still learning and working with John and Angela. Together we run our smallholding, rear our children and try (through our courses) to show others a different approach to life.

The aim of the courses is to give "students" an insight into, and hands-on experience of, life on our self-sufficient smallholding. The new book has benefitted from our combined experience in giving these courses and has expanded in two ways: catering more for those who live in urban situations by including new sections on the urban garden, and secondly, helping people rediscover how to work together with other people and therefore "helping each other to help ourselves".

The New Complete Book of Self-Sufficiency updates a classic treatise with full-colour artwork, together with many new sections arising from questions and issues raised by students attending our self-sufficient smallholding. Despite his 88 years John, Angela, and I are still able to work together in the garden — and enjoy late night Irish sessions in the pub. It has been a real pleasure for Angela and myself to be able to help John to extend and revise his original "bible" of self-sufficiency.

WILL SUTHERLAND

FOREWORD TO THE FIRST EDITION - Dr. E.E. Schumacher, CBE

We can do things for ourselves or we can pay others to do them for us. These are the two "systems" that support us; we might call them the "self-reliance system" and the "organization system". The former tends to breed self-reliant men and women; the latter tends to produce organization men and women. All existing societies support themselves by a mixture of the two systems; but the proportions vary.

In the modern world, during the last hundred or so years, there has been an enormous and historically unique shift: away from self-reliance and towards organization. As a result people are becoming less self-reliant and more dependent than has ever been seen in history. They may claim to be more highly educated than any generation before them; but the fact remains that they cannot really do anything for themselves. They depend utterly on vastly complex organizations, on fantastic machinery, on larger money incomes. What if there is a hold-up, a breakdown, a strike, or unemployment? Does the state provide all that is needed? In some cases, yes; in other cases, no. Many people fall through the meshes of the safety net; and what then? They suffer; they become dispirited, even despondent. Why can't they help themselves? Generally, the answer is only too obvious: they would not know how to; they have never done it before and would not even know where to begin.

John Seymour can tell us how to help ourselves, and in this book he does tell us. He is one of the great pioneers of self-sufficiency. Pioneers are not for imitation but for learning from. Should we all do what John Seymour has done? Of course not. Total self-sufficiency is as unbalanced and ultimately stultifying as total organization. The pioneers show us what *can* be done, and it is for every one of us to decide what *should* be done, that is to say, what we should do to restore some kind of balance to our existence.

Should I try to grow *all* the food my family and I require? If I tried to do so, I probably could do little else. And what about all the other things we need? Should I try to become a jack-of-all-trades? At most of these trades I would be pretty useless and horribly inefficient. But to grow or make some things by myself; for myself: what fun, what exhilaration, what liberation from any feelings of utter dependence on organizations! What is perhaps even more: what an education of the real person! To be in touch with the *actual processes of creation*. The inborn creativity of people is no mean or accidental thing; neglect or disregard it, and it becomes an inner source of poison. It can destroy you and all your human relationships; on a mass scale, it can — nay, it inevitably will — destroy society.

Contrariwise, nothing can stop the flowering of a society that manages to give free rein to the creativity of its people — *all* its people. This cannot be ordered and organized from the top. We cannot look to government, but only to ourselves, to bring about such a state of affairs. Nor should anyone of us go on "waiting for Godot", because Godot never comes. It is interesting to think of all the "Godots" modern humanity is waiting for: this or that fantastic technical breakthrough or automation so that nobody, or hardly anybody, will have to lift a finger anymore; government policies to solve all problems once and for all: multinational companies to make massive investments in the latest and best technologies; or simply "the next upturn in the economy".

John Seymour has never been found "waiting for Godot". It is the essence of self-reliance that you start now and don't wait for something to turn up; and though the technology behind John Seymour's self-sufficiency is still quite rudimentary, it can of course be improved. The greater the number of self-supporters, the faster will be the rate of improvement, that creation of technologies designed to lead people to self-reliance, work-enjoyment, creativity, and therefore *the good life*. This book is a major step along that road, and I wholeheartedly commend it to you.

Foreword

THE FIRST EDITION (UK)
The cover of the UK 1st edition
published in 1976.

This book first came out over 25 ago, created by Dorling Kindersley, with a little help from me, but published by Faber and Faber. It is now being delivered into the new millennium kicking and screaming! Since I first wrote it the book has certainly got about. I have travelled in at least dozens of countries since I wrote it (to say nothing of four continents) and in every one of them people have come up to me with their copy for me to sign. I have been delighted to find wine stains on the wine-making pages, and good honest dirt on the gardening pages. I have indeed updated it for the new millennium, but have not sacrificed any of the techniques and tips that have stood me well all that time and continue to do so.

Since I first wrote the first version of this book back in 1975 I now think there is a far more urgent reason for it. Very few people today can fail to see that the present course that man- and woman-kind is embarked upon is unsustainable. We are changing our atmosphere with frightening rapidity – we are altering the world's climate – we are burning up the Earth's enormous stores of carbon as quickly as possible. We are utterly dependent on cheap fossil fuels, so any interruption in supply would bring disaster. Imagine trying to feed and service our huge cities if the fuel ran out! It is now urgently necessary to dismantle the whole fabric of world trade and replace it with a far less fuel-hungry, less polluting, less dangerous arrangement.

Most people know all this, but they are afraid that their quality of life will decline if we change course. Well, the purpose of this book is to show that this is not the case. We could live without pillaging our planet – and live very well indeed! We have allowed ourselves to get where we are because of the "blind workings of the market". But we are not blind, so we must now start using our good sense to "break this sorry scheme of things and then remould it to our hearts desire," as old Omar Khayam had it.

To allow ourselves to be dependent on some vast Thing created by the Merchants of Greed is madness. It is time to cut out what we do not need so we can live more simply and happily. Good food, comfortable clothes, serviceable housing and true culture – those are the things that matter. The only way this can happen is by ordinary people, us, boycotting the huge multinational corporations that are destroying our Earth – and creating a new Age – an Age of Healing in place of the current Age of Plunder.

I would like to acknowledge my fellow self-supporters, Will Sutherland and Angela Ashe, for their unfailing help over the last three decades. They have shared the trials and labours as well as the joys of this way of life, and have come into partnership with me to start a school of self-sufficiency in Ireland, to which all honest men, women, and children are welcome, provided they can find the fees!

Of course, I have learned a lot since I wrote it (when you cease learning things they take you away in a box), and I continue to live a pretty self-sufficient life. There are very few processes described in this book that I have not performed myself – albeit, perhaps, some of them ineptly. I have embarked on many an enterprise without the faintest idea of how to do it – but I have always ended up with the thing done and with a great deal more knowledge than I had when I started.

Would I advise other people to follow this lifestyle? I wouldn't advise anybody to do anything. The purpose of this book is not to shape other people's lives but simply to help people to do things if they decide to. This way of life suits me – it has kept me fighting fit and at least partly sane into my 88th year, and it has prevented me from doing too much harm to our poor planet.

I would offer this advice: do not try to do everything at once. This is an organic way of life and organic processes tend to be slow and steady. I would also like to offer this motto: "I am only one. I can only do what one can do. But what one can do, I will do!" Happy grub-grubbing! (Better than money-grubbing any day!)

John Seymour,
Killowen,
New Ross,
County Wexford,
Ireland

Introduction

In the lives we lead today, we take much for granted, and few of us indeed remember why so many "advanced" civilizations of the past simply disappeared. When I left college I went to Africa and roamed for six years. I rode the veld in the Karroo, in South Africa, looking after sheep; I managed a sheep farm in Namibia on the verge of the Namib Desert; I hunted buck and shot lions. I spent a year deep sea fishing and six months working in a copper mine in what is now Zambia. And then I travelled all over central Africa for two years innoculating native cattle.

One of the best friends I made during my time in Africa was a man of the Old Stone Age. White people, unable to get their tongues around his real name, which was a conglomeration of clicks, called him Joseph. He was a bushman of the Namib desert but he had been caught by a white farmer and made to work so he knew Afrikaans. I knew some of this language and so could communicate with him.

I used to go hunting with Joseph. First he would hand over the flock of sheep to his wife, and then we would walk out into the bush in search of gemsbok or oryx. Joseph had what seemed to be an amazing knack for knowing where they were. When he knew me better he asked me to leave my rifle behind, and he used to put his arm into a thorn bush and pull out the head of a spear. It was quite illegal for a "native" to own a spear. He would cut a shaft from a bush and fit the spearhead, and using three dogs we would bring a buck to bay and Joseph would kill it with the spear. Later Joseph took me on an expedition to meet his people. They lived in the most desolate and inhospitable part of Africa, but they still lived well. They hunted food by lying in wait near waterholes and shooting game with small poisoned arrows. They found water by cutting open the stomach of the gemsbok and drinking the contents – as I learned to do myself. Or they would find an insignificant looking creeper, dig down under it and bring up a soggy mass of vegetation as big as a football. They sucked the liquid from this, and very nasty it tasted too, but very welcome when it could keep you alive.

These people did not "work". They could walk 40 miles in a night and had endless patience whilst waiting for game. Life was hard in this fierce climate, but they spent most nights dancing and singing and telling stories in the light of their fires. They were completely at home in the natural world around them and they knew every living being in it. They never felt for one moment that they were in any way special or apart from the rest of nature.

I tell all this because I want to point out the enormous change in lifestyle which took place when humans began to practice agriculture. Suddenly, only 10,000–12,000 years ago, people discovered they could plant crops and, just as important, domesticate animals. But there were only a few places where natural conditions were such as to allow this to be possible year after year (mostly fertile river valleys) and thus permit development of cities. As history shows, very few civilizations developed cultures sufficiently wise and robust to last for more than a thousand years: they simply exhausted their soils or were conquered by more aggressive neighbours.

Now we have had the Industrial, the Technological, and are in the midst of the Information Revolution, which again is bringing about great changes. It is bringing great material prosperity to the few who have their hands on or adjacent to the levers of power.

Elsewhere most of humankind lives in appalling conditions, forced to work in slum cities for starvation wages and sing to the tunes of the big multi-nationals. Farmers and farm workers are either starving or being forced to adopt methods which they know are damaging the land. All over the earth the soil is going, eroded by tractor cultivation and slowly poisoned by chemicals from agri-businesses. And so we have created lifestyles which are simply not sustainable nor pleasant. But there are many simple changes which individuals can make to their lifestyles which could change all this. And, if we are wise, we will not wait for the apocalypse before making some adjustments. I do not ask you to blindly follow my suggestions, but merely consider them as you think about the future.

ENERGY

One day I was invited to attend a public symposium on energy, and a public relations officer for the nuclear power industry was there. He showed us all a frightening graph showing the world's energy consumption from 1800 to the present day. The graph started at virtually zero and went up at increasing speed until it was almost vertical. What he had not noticed was that the line pointed straight at the "EXIT" sign above! Now surely a moment's reflection is enough to tell us that everyone in the world simply cannot live at the energy levels favoured by the 21st-century Western ideal. Of course, for thousands of years muscle energy and the heat from fires were all that humans had to depend on. When I was born in 1914, things began to change radically with the discovery of how to exploit oil, and today we have released so much carbon back into the atmosphere that no one can predict the consequences. But

what difference will it make (you may ask) if I walk up the stairs instead of using the lift, or turn down the heating a couple of pips, or use my bike instead of the car? I get no substantial cash benefits because no-one can really pay me for benefitting the "commons". This is called the tragedy of the commons – no-one pays us to keep the oceans or the air clean.

But if it's true that the only person over whom I have control of actions is myself, then it does matter what I do. It may not matter a jot to the world at large, but it matters to me. And there is by good fortune one important factor that can help us make progress in saving energy. For not only does using our muscles help the planet, but it also keeps us fit and healthy. Of course there are many other sources of benign energy. Solar energy, wind energy, and water power (see the chapter on energy) are three obvious alternatives which are increasingly easy to harness with modern technology. I count planned tree planting and coppicing as one of the best solar energy collecting devices. And let us never forget that energy saved is as good as energy bought. It is often much cheaper to buy energy-saving equipment than to pay for the energy used by less effective arrangements.

TRANSPORT

Unless we come from a race of travelling peoples, we are all pretty much "locals". We live somewhere, and what goes on in the locality of where we live is much more important than what goes on in Paris, London, or Washington, DC. If we could once again run our world on a local scale, with decisions made on a local basis, then many of our problems would be stopped in their tracks. Let me explain "local" by comparing two villages: these are in Crete

but they could be anywhere. One village is high up in the mountains, just to the south of the cave where Zeus was born. It can only be reached by an unpaved road full of potholes and quite unsuitable for buses. The only contact with the outside world that I could see was that a man with a very tough truck would brave the potholes once each week and bring a load of fish from the small fishing port down on the coast. Sheep were exported from the village to bring in the cash for this exchange.

Apart from this exchange, the community on this mountain was self-supporting. There were enough tiny terraced fields to grow wheat, wine, and olives. There was an oil mill for pressing the olives. There were plenty of nut trees as well as lemon groves, fig trees, and many other kinds of fruits. There were beehives, and the sheep provided meat in abundance. The mountain village houses were beautiful, simple, and comfortable in that climate. Clothes were made by the women. There was a loom maker in a neighbouring village, a boot-maker in another, and a knife-maker in yet another. Was there culture, you might ask? Well, there was singing and dancing and music in plenty. There were few books, but if the villagers had wanted them, they could have been available. The villagers paid no taxes and had just one policeman. They knew their own laws and kept to them.

Now the other Cretan village I wish to describe was lower down the mountain and had a "good" road. This gave access to the city but also gave the city access to the countryside. City money came in and bought much of the land, uprooting the old trees and vineyards and planting quick-growing olive trees, so providing an olive crop for cash sale. Now the villagers had to pay for their olive oil and were dragged quickly into the money economy. All sorts of traders had access to the village and a small supermarket opened. Suddenly the villagers found they "needed" all sorts of things they had never needed before. Television arrived and brought with it aspiring visions. Young people in the village no longer sang or danced; they wanted Western pop music and Coca-Cola. Even though their fine road looked like a road to freedom, it was actually a road to sadness, wage slavery, and discontent, and one from which the youngsters could not return.

WORK

I once knew an old lady who lived by herself in the Golfen valley of Herefordshire. She was one of the happiest old women I have met. She described to me all the work she and her mother used to do when she was a child: washing on Monday, butter-making on Tuesday, market on Wednesday, and so on. "It all sounds like a lot of hard work," I said to her. "Yes, but nobody ever told us then," she said in her Herefordshire accent. "Told you what?" "Told us there was anything wrong with work!" Today work has become a dirty word, and most people would do anything to get out of work. To say that an invention is labour saving is the highest praise, but it never seems to occur to anyone that the work might have been enjoyable. I have ploughed all day behind a good set of horses and been sad when the day came to an end!

This book is about changing the way we live, and I am well aware that the subject is fraught with difficulties. The young couple who have mortgaged themselves to buy a house, have to pay huge monthly sums for their season ticket to get to work, have a bank overdraft and debts with the credit card sharks, are in no position to be very choosy about what work they do. But why should we get into such a situation? Why should we all labour to enrich the banks (for that is what we are doing)? There is not necessarily anything wrong with doing things that are profitable. It is when "profit" becomes the dominant motive that the cycle of disaster begins.

In my work with self-sufficiency I have met hundreds of people in many countries and four continents who have withdrawn themselves from conventional work in big cities and moved out into the countryside. Almost all of them have found good, honest, and useful ways of making a living. Some are fairly well off in regard to money; others are poor in that regard but they are all rich in things that really matter. They are the people of the future. If they are not in debt they are happy men and women.

HOME

A true home should be the container for reviving real hospitality, true culture and conviviality, real fun, solid comfort, and above all, real civilization. And the most creative thing that anybody can do in this world is to make a real home. Indeed, the homemaker is as important as the house, and the "housewife" is the most creative, most important job on Earth.

One of the essential characteristics of a good home is "craftsmanship". It seems to me that all human artifacts give off a sort of cultural radiation, depending on how much love and art has gone into their production. A mass-produced article of furniture comes from a high-speed, high-tech factory, using plastics and often working with wood that has been destroyed by "chipping" and glueing. The noise and smell of these places is quite terrifying. And this factory-made rubbish, although it may look fine for a few years, is fit only for the landfill site (you cannot burn it, for it gives off dioxins). Furniture made by a craftsman, on the other hand, is full of care and made with sympathy for the wood. It will last for generations and be a constant beacon of beauty in the home. Of course I am not suggesting that everyone should build their own home or furniture. After all, if houses were well built and the population was stable, everyone should inherit a good house. What I am saying is that if you can build your own furniture, or indeed house, either with your own hands or with the help of a builder, it is a marvellous thing to do.

FOOD

It is true that our friends in the supermarkets have made many "advances" in the complexity of pre-prepared meals. But the sad fact is that our food now travels thousands – yes thousands – of miles between the place it is produced and our mouths. Most people never get an opportunity to taste real local-grown fresh food – they do not know what they are missing. This book is about quality of life, and I submit that if there is no quality in the food we eat, then we must just hope to get through life as quickly as possible. Because the sources of our food are getting further away from our tables and the food goes through more and more industrial processing, the only quality now deemed important is long shelf life. Such food is dead food: all the life has been taken out of it. The best food of all comes from our own garden and our own land. Next best is food from a local farm, or farmer's markets, and then food from a local shop. If we take the trouble to seek out good, "flavour-full", real food we will be benefitting ourselves and, just as important, giving support to those who take the trouble to produce such food.

"We had never had any real conscious drive to self-sufficiency. We had thought, like a lot of other people, that it would be nice to grow our own vegetables. But living here has altered our sense of values. We find that we no longer place the same importance on artifacts and gadgets as other people do. Also — every time we buy some factory-made article we wonder what sort of people made it — if they enjoyed making it or if it was just a bore — what sort of life the maker, or makers, lead. I wonder where all this activity is leading. Is it really leading to a better or richer or simpler life for people? Or not? I wonder about the nature of progress. One can progress in so many different directions. Up a gum tree for example. I know that the modern factory worker is supposed to lead an "easier" life than, say, the peasant. But I wonder if this supposition is correct. And I wonder if, whether "easier" or not, it is a better life? Simpler? Healthier? More spiritually satisfying? Or not?

So far as we can, we import our needs from small and honest craftsmen and tradesmen. We subscribe as little as we can to the tycoons, and the Ad-men, and the boys with their expense accounts. If we could subscribe to nothing at all we would be the better pleased."

JOHN SEYMOUR FAT OF THE LAND 1976

CHAPTER ONE

THE MEANING OF SELF-SUFFICIENCY

The Way to Self-Sufficiency

The first questions we must answer are: What is this book about? What is self-sufficiency, and why do it? Now self-sufficiency is not "going back" to some idealized past in which people grubbed for their food with primitive implements and burned each other for witchcraft. It is going forward to a new and better sort of life, a life which is more fun than the over-specialized round of office or factory, a life that brings challenge and the use of daily initiative back to work, and variety, and occasional great success and occasional abysmal failure. It means the acceptance of complete responsibility for what you do or what you do not do, and one of its greatest rewards is the joy that comes from seeing each job right through – from sowing your own wheat to eating your own bread, from planting a field of pig food to slicing a side of bacon.

Self-sufficiency does not mean "going back" to the acceptance of a lower standard of living. On the contrary, it is the striving for a higher standard of living, for food which is fresh and organically grown and good, for the good life in pleasant surroundings, for the health of body and peace of mind which come with hard, varied work in the open air, and for the satisfaction that comes from doing difficult and intricate jobs well and successfully.

A further preoccupation of the self-sufficient person should be the correct attitude to the land. If it ever comes to pass that we have used up all, or most of, the oil on this planet, we will have to reconsider our attitude to our only real and abiding asset – the land itself. We will one day have to derive our sustenance from what the land, unaided by oil-derived chemicals, can produce. We may not wish in the future to maintain a standard of living that depends entirely on elaborate and expensive equipment and machinery, but we will always want to maintain a high standard of living in the things that really matter – good food, clothing, shelter, health, happiness, and fun with other people. The land can support us, and it can do it without huge applications of artificial chemicals and manures and the use of expensive machinery.

But everyone who owns a piece of land should husband that land as wisely, knowledgeably, and intensively as possible. The so-called "self-supporter" sitting among a riot of docks and thistles talking philosophy ought to go back to town. He is not doing any good at all, and is occupying land which should be occupied by somebody who can really use it.

Other forms of life, too, besides our own, should merit our consideration. Man should be a husbandman, not an exploiter. This planet is not exclusively for our own use. To destroy every form of life except such forms as are obviously directly of use to us is immoral, and ultimately, quite possibly, will contribute to our own destruction. The kind of varied, carefully thought-out husbandry of the self-supporting holding fosters a great variety of life forms, and every self-supporter will wish to leave some areas of true wilderness on his holding, where wild forms of life can continue to flourish undisturbed and in peace.

And then there is the question of our relations with other people. Many people move from the cities back to the land precisely because they find city life, surrounded by people, too lonely. A self-supporter, living alone surrounded by giant commercial farms, may be lonely, too; but if he has other self-supporters near him he will be forced into cooperation with them and find himself, very quickly, part of a living and warm community. There will be shared work in the fields, there will be relief milking and animal feeding duties when other people go on holiday, the sharing of child minding duties, there will be barn-raisings and corn-shuckings and celebrations of all kinds. This kind of social life already happens in those parts of Europe and North America where self-supporting individuals, or communities, are becoming common.

Good relations with the old indigenous population of the countryside are important, too. In my area, the old country people are very sympathetic to the new "drop-ins". They rejoice to see us reviving and preserving the old skills they practised in their youth, and they take pleasure in imparting them to us. They wax eloquent when they see the hams and flitches of bacon hung up in my chimney. "That's real bacon!" they say, "better than the stuff we get in the shops." "My mother used to make that when I was a boy, we grew all our own food then," they might add. "Why don't you grow it now?" I ask. "Ah! times have changed." Well, they are changing again.

Self-sufficiency on a small scale

Self-sufficiency is not only for those who have five acres of their own country. A city dweller who learns how to mend his or her own shoes becomes, to some extent, self-sufficient. They save money and increase their own satisfaction and self-respect too. We were not meant to be a one-job animal. We do not thrive as parts of a machine. We are intended by nature to be diverse, to do diverse things, to have many skills. The city person who buys a sack of wheat from a farmer on a countryside visit and grinds his own flour to make his own bread cuts out the middlemen and furthermore gets better bread. He gets good exercise turning the handle of the grinding machine too. And any suburban gardener can dig up some of that useless lawn and put some of those dreary hardy perennials on the compost heap and grow his own cabbages. An urban garden, an allotment, these are both a sound base for a would-be self-supporter (see pp.26–29), and a good-sized suburban garden can practically keep a family. I knew a woman who grew the finest outdoor tomatoes I ever saw in a window box 12 storeys up in a tower-block. They were too high up to get the blight.

So good luck and long life to all self-supporters! And if every reader of this book learns something useful that he or she did not know before, and could not very easily find out, then I, and the dedicated people who have done the very arduous and difficult work of putting it together, will be happy and feel it has not been in vain.

THE FIRST PRINCIPLES OF SELF-SUFFICIENCY

The only way that the homesteader can farm his or her piece of land as well and intensively as possible is to institute some variant of what was called "High Farming" in Europe over two centuries ago. This was a carefully worked out balance between animals and plants, so that each fed the other: the plants feeding the animals directly, the animals feeding the soil with their manure, and the land feeding the plants. A variety of both animals and plants were rotated about the same land so that each species took what it needed out and put what it had to contribute back, and the needs of the soil were kept uppermost always in the husbandman's mind. Each animal and crop was considered for what beneficial effect it might have on the soil.

If the same crop is grown on a piece of land year after year, the disease organisms that attack that crop will build up in the area until they become uncontrollable. Nature abhors monoculture: any cursory inspection of a natural plant and animal environment will reveal a great variety of species. If one species becomes too predominant, some pest or disease is sure to develop to strike it down. Man has managed to defy this law, to date, by the application of stronger and stronger chemical controls, but the pests (particularly the fast-evolving viruses) adapt very quickly to withstand each new chemical and so far the chemist has managed to keep only a short jump ahead of the disease.

New homesteaders will wish to husband their land in accordance with the principles of High Farming. They will have to substitute the labour of their hands for imported chemicals and sophisticated machinery. They will have to use their brains and their cunning to save the work of their hands. For instance, if you can get your animals to go out into your field and consume their share of your crops there, then you will save yourself the work of harvesting the crops for them and carrying them in. In other words, take the animals to the crops, not the crops to the animals. So, also, if you can get the animals to deposit their dung on your land, then this will save you the labour of carrying the dung out yourself. Thus the keeping of animals on limited free range will appeal to you: sheep can be "folded" on arable land (folding means penning animals on a small area of some fodder crop and moving the pen from time to time), chickens can be housed in arks that can be moved over the land so as to distribute the hens' manure while allowing the hens to graze fresh grass, and pigs can be kept behind electric fences which can also be easily moved. Thus the pigs harvest their food for themselves and also distribute their own manure. (To say nothing of the fact that pigs are the finest free cultivators that were ever invented! They will clear your land, and plough it, and dung it, and harrow it, and leave it nearly ready for you to put your seed in, with no more labour to you than the occasional shifting of an electric fence.)

In planning the layout of the smallholding the homesteader will take careful account of natural shelter, considering especially the effects of the prevailing wind. Trees will be planted to create a barrier on the north and east (in the northern hemisphere) and permanent thorn hedges established to divide the holding into sensible stockproof fields. Existing water and streams will be carefully assessed for possible use in irrigation, water power, or suitable ponds for duck and geese. Care will be taken to make good advantage of all natural features of the site. Walls will be constructed to create a south-facing shelter suitable for excellent fruit trees. The buildings will be sited where they are most convenient both to each other and to the productive areas of the smallholding.

Now "husbanding" homesteaders will not keep the same species of animal on a piece of land too long, just as they will not grow the same crop year after year in the same place. They will follow their young calves with their older cattle, their cattle with sheep, their sheep with horses, while geese and other poultry either run free or are progressively moved over their grassland and arable land (by arable I mean land that gets ploughed and planted with crops as opposed to land that is grass all the time).

All animals suffer from parasites, and if you keep one species on one piece of land for too long, there will be a build up of parasites and disease organisms. As a rule, the parasites of one animal do not affect another and therefore following one species with another over the land will eliminate parasites.

Also, "husbanding" homesteaders will find that every enterprise on their holding, if it is correctly planned, will interact beneficially with every other. If they keep cows, their dung will manure the land which will provide food, not only for the cows but for the humans and pigs also. The by-products of the milk of the cows (skimmed milk from butter-making and whey from cheese-making) are a marvellous whole food for pigs and poultry. The dung from the pigs and poultry helps grow the food for the cows. Chickens will scratch about in the dung of other animals and salvage any undigested grain.

All crop residues help to feed the appropriate animals – and such residues as not even the pigs can eat, they will tread into the ground, and activate with their manure, and turn into the finest in situ compost without the husbandman lifting a spade. All residues from slaughtered birds or animals go either to feed the pigs or the sheepdogs, or to activate the compost heap. Nothing is wasted. Nothing is an expensive embarrassment to be taken away to pollute the environment. There should be no need for a dustman on the self-sufficient holding. Even old newspapers can make litter for pigs, or be composted. Anything that has to be burnt makes good potash for the land. Nothing is wasted – there is no "rubbish".

But before the potential self-supporter embarks on the pursuit of true "husbandry", he or she should acquaint themselves with some of the basic laws of nature, so that they can better understand why certain things will happen on their holding and why other things will not.

Humans & their Environment

True homesteaders will seek to husband their land, not exploit it.

They will wish to improve and maintain the "heart" of this land, its fertility.

They will learn by observing nature that growing one crop only, or keeping one species of animal only, on the same piece of land is not in the natural order of things.

They will therefore wish to nurture the animals and plants on their land to ensure the survival of the widest possible variety of natural forms.

They will understand and encourage the interaction between them.

They will even leave some areas of wilderness on their land, where wild forms of life can flourish.

Where they cultivate, they will always keep in mind the needs of the soil, considering each animal and each plant for what beneficial effect it might have on the land.

Above all, they will realize that if they interfere with the chain of life (of which they are a part) they do so at their peril, for they cannot avoid disturbing a natural balance and natural cycle.

THE FOOD CHAIN

Life on this planet has been likened to a pyramid: a pyramid with an unbelievably wide base and a small apex.

All life needs nitrogen, for it is one of the most essential constituents of living matter, but most creatures cannot use the free, uncombined nitrogen which makes up a great part of our atmosphere. The base of our biotic pyramid, therefore, is made up of the bacteria that live in the soil, sometimes in symbiosis with higher plants, and have the power of fixing nitrogen from the air. The number of these organisms in the soil is unimaginably great: suffice it to say that there are millions in a speck of soil as big as a pin-head.

On these, the basic and most essential of all forms of life, lives a vast host of microscopic animals. As we work up the pyramid, or the food chain, whichever way we like to consider it, we find that each superimposed layer is far less in number than the layer it preys upon.

On the higher plants graze the herbivores. Every antelope, for example, must have millions of grass plants to support it. On the herbivores "graze" the carnivores. And every lion must have hundreds of antelopes to support it. The true carnivores are right at the apex of the biotic pyramid. Humans are somewhere near the top but not at the top because they are omnivores. They are one of those lucky animals that can subsist on a wide range of food: vegetable and animal.

INTER-RELATIONSHIPS

Up and down the chain, or up and down between the layers of the pyramid, there is a vast complexity of inter-relationships. There are, for example, purely carnivorous micro-organisms. There are all kinds of parasitic and saprophitic organisms: the former live on their hosts and sap their strength, the latter live in symbiosis, or in friendly cooperation, with other organisms, animal or vegetable. We have said that the carnivores are at the apex of the food chain. Where in it stands a flea on a lion's back? Or a parasite in a lion's gut?

And what about the bacterium that is specialized (and you can bet there is one) to live inside the body of the lion flea? A system of such gargantuan complexity can best, perhaps, be understood by the utter simplification of the famous verse:

Little bugs have lesser bugs upon their backs to bite 'em,
And lesser bugs have lesser bugs and so ad infinitum!

This refers to parasitism alone, of course, but it is noteworthy that all up and down the pyramid everything is consumed, eventually, by something else. And that includes us, unless we break the chain of life by the purely destructive process of cremation.

Now humans, the thinking monkeys, have to interfere with this system (of which we should never forget that we are a part), but we do so at our peril.

If we eliminate many carnivores among the larger mammals, the herbivores on which these carnivores preyed become overcrowded, overgraze, and create deserts. If, on the other hand, we eliminate too many herbivores, the herbage grows rank and out of control and good pasture goes back to scrub and cannot, unless it is cleared, support many herbivores. If we eliminate every species of herbivore except one, the grazing is less efficiently grazed.

Thus sheep graze very close to the ground (they bite the grass off with their front teeth), while cows like long grass (they rip grass up by wrapping their tongues around it). The hills produce more and better sheep if cattle graze on them too. It is up to us as husbandmen and women to consider very carefully, and act very wisely, before we use our powers to interfere with the rest of the biotic pyramid.

Plants, as well as animals, exist in great variety in natural environments and for very good reasons. Different plants take different things out of the soil, and put different things back. Members of the pea-bean-and-clover family, for example, have nitrogen-fixing bacteria in nodules on their roots. Thus they can fix their own nitrogen. But you can wipe the clovers out of a pasture by applying artificial nitrogen. It is not that the clovers do not like the artificial nitrogen, but that you remove the "unfair advantage" that they had over the grasses (which are not nitrogen-fixing) by supplying the latter with plenty of free nitrogen and, being naturally more vigorous than the clovers, they smother them out.

It is obvious from observing nature that monoculture is not in the natural order of things. We can only sustain a one-crop-only system by adding the elements that the crop needs from the fertilizer bag and destroying all the crop's rivals and enemies with chemicals. If we wish to farm more in accordance with the laws and customs of nature, we must diversify as much as we can, both with plants and animals.

Ultimately, it all comes back to the first rule in becoming self-sufficient: that is, to understand the "Natural Cycle": namely, the soil feeds the plants, the plants feed the animals, the animals manure the land, the manure feeds the soil, the soil feeds the plants (*see pp.22–23*). True "husbanding" homesteaders will wish to maintain this natural cycle, but they have to become part of the cycle themselves; as plant-eaters and carnivores they are liable to break the chain unless they observe at all times the "Law of Return". The Law of Return means that all residues (animal, vegetable, and human) should be returned to the soil, either by way of the compost heap, or the guts of an animal, or the plough, or by being trodden into the ground by livestock. Whatever cannot be usefully returned to the soil, or usefully used in some other way, should be burned; this will make potash for the land. Nothing should be wasted on the self-sufficient holding and I believe this applies as much to a modest allotment as it does to a holding of several acres.

THE SOIL

Because soil derives from many kinds of rock, there are many varieties of soil. As we cannot always get exactly the kind of soil that we require, the husbandman must learn to make the best of the soil available. Depending on the size of their particles soils are classified as light or heavy, with an infinite range of gradations in between. Light means composed of large particles. Heavy means composed of small particles. Gravel can hardly be called soil but sand can, and pure sand is the lightest soil you can get. The kind of clay which is made of the very smallest particles is the heaviest. The terms "light" and "heavy" in this context have nothing to do with weight but with the ease of working of the soil. You can dig sand, or otherwise work with it, no matter how wet it is, and do it no harm. Heavy clay is very hard to dig or plough, gets very "puddingy" and sticky, and is easily damaged by working it when it is wet.

What we call soil generally has a thickness to be measured in inches rather than feet. It merges below with the subsoil, which is generally pretty humus-free, but may be rich in mineral foods needed by plants. Deep-rooting plants such as some trees, lucerne or alfalfa, comfrey, and many herbs, send their roots right down into the subsoil, and extract these nutriments (nourishing food) from it.

The nature of the subsoil is very important because of its influence on drainage. If it is heavy clay, for example, then the drainage will be bad and the field will be wet. If it is sand, gravel, decayed chalk or limestone, then the field will probably be dry. Below the subsoil lies rock, and rock goes on down to the centre of the Earth. The rock, too, can affect drainage: chalk, limestone, sandstone and other pervious rocks make for good drainage: clay (geologists consider this a rock, too), slate, mudstone, some shales, granite, and other igneous rocks generally make for poor drainage. Badly drained soils can always be drained – provided enough expenditure of labour and capital is put into doing it.

Types of soil

Let us now consider various types of soil:

Heavy clay This, if it can be drained and worked with great care and knowledge, can be very fertile soil, at least for many crops. Wheat, oak trees, field beans, potatoes, and other crops do superbly on well-farmed clay. Farmers often refer to it as strong land. But great experience is needed to farm it effectively. This is because of the propensity of clay to "flocculate"– that is, the microscopic particles which make up clay gather together in larger particles. When this happens, the clay is more easily worked, drains better, allows air to get down into it, and allows the roots of plants to penetrate it more easily. In other words, it becomes good soil. When it does the opposite of flocculate, it "puddles" – it forms a sticky mass, such as the potter uses to make his pots, becomes almost impossible to cultivate, and gets as hard as brick when it dries out. The land forms big cracks and is useless.

Factors which cause clay to flocculate are alkalinity rather than acidity, exposure to air and frost, incorporation of humus, and good drainage. Acidity causes it to puddle; so does working it while wet. Heavy machines tend to puddle it. Clay must be ploughed or dug when in exactly the right condition of humidity, and left strictly alone when wet.

Clay can always be improved by the addition of humus (compost, "muck" or farmyard manure, leaf-mould, green manuring: any vegetable or animal residue), by drainage, by ploughing it up at the right time and letting the air and frost get to it (frost separates the particles by forcing them apart), by liming if acid, even, in extreme cases, by incorporating sand with the clay. Clay soil is "late" soil, which means it will not produce crops early in the year. It is difficult soil. It is not "hungry" soil – that is, if you put humus in it the humus will last a long time. It tends to be rich in potash and is often naturally alkaline, in which case it does not need liming.

Loam This is intermediate between clay and sand, and has many gradations of heaviness or lightness. A medium loam is perhaps the perfect soil for most kinds of farming. Most loam is a mixture of clay and sand, although some loams probably have particles all of the same size. If loam (or any other soil) lies on a limestone or chalk rock, it will probably be alkaline and will not need liming, although this is not always the case: there are limestone soils which, surprisingly, do need liming. Loam, like every other kind of soil, will always benefit from humus addition.

Sand Sandy soil, or the lighter end of the spectrum of heavy light soils, is generally well-drained, often acid (in which case it will need liming), and often deficient in potash and phosphates. It is "early" soil; that is, it warms up very quickly after the winter and produces crops early in the year. It is also "hungry" soil; when you put humus into it, the humus does not last long. In fact, to make sandy soil productive you must put large quantities of organic manure into it, and inorganic manure gets quickly washed away from it. Sandy soils are favoured for market gardening, being early and easy to work and very responsive to heavy dressings of manure. They are good soils for such techniques as folding sheep or pigs on the land. They are good for wintering cattle on because they do not "poach" (turn into a quagmire when trodden) like heavy soils do. They recover quickly from treading when under grass. But they won't grow as heavy crops of grass or other crops as heavier land. They dry out very quickly and suffer from drought more than clay soils do.

Peat Peat soils are in a class of their own, but unfortunately are fairly rare. Peat is formed of vegetable matter which has been compressed in anaerobic conditions (underwater) and has not rotted away. Sour wet peatland is not much good for farming, although such soil, if drained, will grow potatoes, oats, celery, and certain other crops. But naturally drained peatlands are, quite simply, the best soils in the world. They will grow anything, and grow it better than any other soil. They don't need manure, they are manure!

The Natural Cycle

One of the most important maxims you may remember from your school day science lessons is that matter can be neither created nor destroyed. Nowhere is the principle more critical than in understanding the process by which the fertility of the land can be increased. The major life processes on Earth require large quantities of certain basic elements: carbon, oxygen, and nitrogen in particular. And the energy which makes all this life shake, rattle, and roll is provided by sugars which are essentially made by photosynthesis driven by the sun. Whilst carbon and oxygen are both common and reactive, nitrogen is not quite so easy for our life processes to manage. Even though there is a huge quantity of nitrogen gas in our atmosphere, plants cannot absorb nitrogen directly. Apart from a small amount of nitrogen which is made water soluble by the action of thunderstorms, we depend upon bacteria to fix nitrogen and change it into forms which can be used by plants. The animals that run about have to get their nitrogen secondhand by courtesy of the plant kingdom. The bacteria we need are present by the millionfold in healthy soil.

MUCH-NEEDED NITROGEN
Huge quantities of usable nitrogen are present in human, animal, and plant wastes. It is much easier for the soil to recycle these than for soil bacteria to "fix" nitrogen from scratch. This is why all effective husbandry places such emphasis on the importance of composting and healthy soil. It is easy to understand the cycle from this simple diagram.

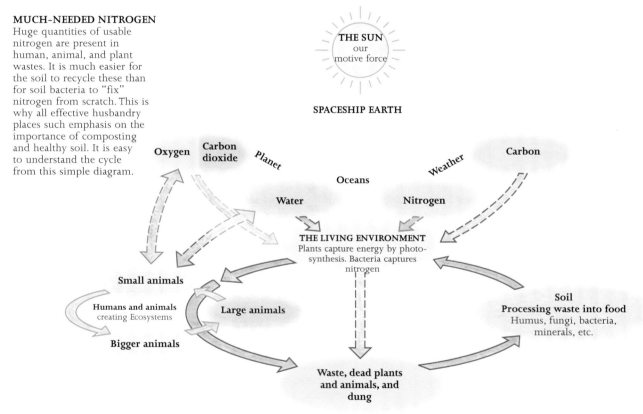

THE ECOLOGICALLY SOUND HOLDING
One of the chief features of the High Farming era of 18th-century England was the famous "Norfolk Four Course Rotation". It was an ecologically sound system of husbandry – a model for the productive growing of a variety of crops in both large and small-scale farming. It worked like this:
I One-Year Ley A ley is grass and clover sown for a temporary period. The grass and clover was grazed off by stock and the purpose of it was to increase the fertility of the land by the nitrogen fixed in the root nodules of the clover, by the dung of the grazing animals, and ultimately by the mass of vegetation ploughed into the land when the ley was ploughed up.
2 Root Break The crops in the Root Break might have been turnips or swedes to be fed to cattle, sheep, or pigs, potatoes to be fed mostly to humans, mangolds for cattle, and various kinds of kale – the latter not actually "roots," of course, but taking the same place in the Root Break. The effect of the Root Break was to increase the fertility of the soil, because nearly all the farmyard manure produced on the farm was applied to the root crop, and to "clean" the land – that is, make the earth weed-free. Root crops are "cleaning-crops" because, by being planted in rows, they have to be hoed several times. The third effect of the Root Break was to produce crops which stored the summer's growth for winter feeding.
3 Winter Cereal Break This was wheat, beans, barley, oats, or rye sown in the autumn. It "cashed" the fertility put into the land by the Ley and the Roots, benefitted from the cleanliness of the land after Roots, and was the farmer's chief "cash crop" – the crop from which he made his money. The beans, however, were for feeding to livestock.

4 Spring Cereal Break This was possibly spring-sown wheat but it was more likely to be barley. After the barley had been drilled, grass and clover seed was undersown – that is, broadcast on the ground along with the cereal seed. As the barley grew, the grass and clover grew, and when the barley was harvested, a good growth of grass and clover was left to be grazed off next spring and summer, or to be cut for hay and grazed the following winter too. The barley went principally to feed stock, but the best of it went to be malted for beer. The oats and barley straw were fed to the cattle, the wheat straw went under their feet to provide all that vast tonnage of farmyard manure (the best compost that ever was invented), rye straw was used for thatching, the roots were mostly fed to the cattle or to the sheep, and wheat, malting barley, beef, and wool went off to be sold to city folk.

Historically, land properly managed in this way often grew two tons of wheat to the acre and this with no input of oil-derived chemicals whatever. There weren't any. Now we can emulate this ecologically sound system, changing it to suit our different needs. We may not wish to live primarily on the bread, beef, and beer of the 18th-century Englishman. We may need more dairy products – butter, cheese, and milk – more vegetables, and a greater variety of food altogether. Also, we have new techniques: new crops such as Jerusalem artichokes, fodder radish, fodder beet, maize in northern climates, and devices such as the electric fence, which widen our possible courses of action. So, whether our would-be self-supporters have nothing more than a back garden, or perhaps a city allotment, or perhaps a hundred-acre farm, or whether they are part of a community owning a thousand acres, the principles to follow are the same. We should try to work with Nature, not against it, and we should, as far as we can while still serving our own ends, emulate Nature in our methods.

EIGHT POINTERS TO A HEALTHY PLOT
Thus you are to improve and maintain the heart of your land, you should remember these eight essential pointers:
1 Monoculture, or the growing of the same crop on land year after year, should be avoided. Disease organisms which attack any particular crop always build up in land on which that crop is grown year after year. Also, each crop has different requirements from the soil and its residues return different materials to the soil.
2 The keeping of one species of animal on the soil and one only should also be avoided. The old High Farming practitioners in England used to say: "A full bullock yard makes a full stack yard." In other words, the dung from the animals is good for the soil. Mixed stocking is always better than mono-stocking, and rotational grazing is the best of all: the penning or folding of a species of animal over the land so that the animals leave their droppings (and inevitable parasite eggs) behind and so break the lifecycle of the parasites. Following one species with another in such a rotation should be practised wherever possible.

3 To grow "leys", graze them; ultimately plough them in.
4 To practise "green manuring". That is, if you don't want to grow some crop to graze off or feed off to animals, grow the crop anyway and then plough it in, or, better still, work it in with discs or other instruments.
5 To avoid ploughing too much or too deep. To bury the topsoil and bring the subsoil to the surface is not good. On the other hand, chisel ploughing (the cutting of furrows in the soil by dragging knives through it) does not invert the soil, helps drainage, breaks "pans" (hard layers under the surface), and can only do good.
6 To suffer not the land to remain bare and exposed to the weather and elements more than is absolutely necessary. When it is covered with vegetation, even with weeds, it will not erode or deteriorate. If left bare, it will. A growing crop will take up and store the nitrogen and other elements of the soil and release them when it rots down. In bare soil many soluble plant-foods are "leached out" (washed away).
7 To attend to drainage. Waterlogged soil is no-good soil and will deteriorate unless, of course, you are growing rice, or keeping water buffalo.
8 To observe, at all times, the Law of Return (*see* p. 20). All crop and animal residues should be returned to the soil. If you sell anything off the holding then you should import something of equal manurial value back on to it. (The Law of Return should apply to human excrement, too.) Now if the Law of Return is properly observed it is theoretically possible to maintain, if not increase, the fertility of a piece of land without animals at all. Careful composting of vegetable residue is necessary, but it is noteworthy that on holdings where no animals are kept, but a high standard of fertility is maintained, almost always vegetable matter is brought in from outside the holding, and very often other high-energy substances, such as compost activator, too. Seaweed, leaf mould from woods, dead leaves from city street cleaning services, waste vegetables from greengrocers, straw or spoiled hay, nettles or bracken mown on public or neighbours' land, or waste ground: all such inputs of vegetable residues are possible, and will keep up the fertility of land which has no animals.

It is difficult to see why putting vegetable matter into animals and then returning it to the land as manure should be better than putting it direct on to the land, but it is demonstrably so. There is no doubt about it, as any self-supporter with any experience knows, there is some potent magic that transmutes vegetable residues into manure of extraordinary value by putting it through the guts of an animal. But when it is realized that animals and plants have evolved together on this planet, perhaps this is not surprising. Nature does not seem to show any examples of an animal-free vegetable environment. Even the gases inhaled and exhaled by these two different orders of life seem to be complementary: plants inhale carbon dioxide and exhale oxygen, animals do the opposite.

The Seasons

EARLY SPRING

Plough and rotavate your land when the winter's frosts have broken up the soil. Rotavate beds several times in dry weather at two-week intervals to knock out the weeds. Harrow fields for sowing with discs or spikes. Spread compost or well-rotted animal manure on the land before cultivation. Start ordering or buying the seeds you will need when the weather warms up. Make the best of the last of the shooting season. Plant up your onion sets.

LATE SPRING

Sow seeds when the ground is warm and mark rows carefully for future hoeing. Sow batches every two weeks and make sure you take every chance to hoe and give the weeds a pasting when the weather is dry and breezy. Put in your early potatoes and use cloches to protect melons and other squashes from late frosts. Harvest your delicious sprouting broccoli. Turn out animals on to grass. Make sure you have plenty of home brew on the go to prepare you for the thirsty work of shearing and haymaking later on. Buy in young piglets for fattening over the summer. Expect to start milking again as soon as your cow has calved.

EARLY SUMMER

Shear your sheep and look forward to the fact that wool from five of them will clothe a large family. With the summer flush of grass, your cow(s) will pour out milk and you should make butter nearly every day. Make a good cheese each week for storing for the winter. In mid-summer comes the backbreaking but satisfying job of haymaking. Make sure your equipment is in good fettle and line up friends to give you a hand. Check again that you have ample supplies of home brew. Prune plum trees now whilst there is still vigorous growth. Keep weeding the vegetable beds. Plant out tender plants such as tomatoes and sweetcorn.

LATE SUMMER

Harvest your wheat and barley when it is good and ripe. The harvest is the crown of the year, and you will need help from friends. Top fruit (from large trees) in the orchard will start being ready to pick; soft fruits must be picked and processed. The early potatoes will be lifted: lovely fresh ones for lunch each day. Wild berries can be gathered for jams and bottling. Wine-making continues throughout the late summer period. Many parts of the garden will now need harvesting daily: peas and beans to be processed for the deep freeze, lettuce and tomatoes to be eaten fresh. By now you will have beaten the weeds and your plants are pretty well strong enough to stand up for themselves.

AUTUMN

This is the time to harvest root crops and get them into clamps for winter storage. Now you will be in full production of cider, both from your own apples and also those you can pick up from friends. Mushrooms of all shapes and sizes will be a tasty addition to your rations and if there are any extra, they will go into the deep freeze. Take some time now to sort out your firewood for the winter. Bring wood in to cut and split before it gets too wet. Begin spinning of flax and wool in preparation for winter weaving. Pickle the remainder of your beetroot and dry out and string up your onions. Rotavate and tidy up all parts of the garden which have been harvested.

WINTER

This is the time to rebuild hedges and plant new ones. Press on now with repair work on gates and fences. Tools and implements must be sharpened and repaired. Bring in grazing animals if the weather turns nasty. Depending where you live and the livestock laws there, when the weather is cold, expect to slaughter livestock and butcher meat in the appropriate environment. Kill pigs for ham and bacon. Prepare areas for your winter tree planting programme. Consider racking off your demi-johns of summer wine. Above all, take time out to enjoy the fruits of your labours with friends.

The Urban Garden

It is amazing what can be packed into an urban garden. Even the smallest space can be made productive and what could be more attractive than the sight of well tended succulent fruit and vegetables right outside your back door. If you have space then think about a small greenhouse (*see* p.82) which will extend your growing season and offer a chance for more exotic produce. And don't forget that a beehive (*see* p.126) will send your little foragers out to gather nectar from all your neighbours' flowers – pounds of lovely honey and it's robbery that is all perfectly legal!

I remember meeting a fascinating man when I was in California who made his living creating "easy-run" urban vegetable gardens for the old and infirm. These were all on raised beds made of either brick or treated wood constructions. These beds raised the soil level to a comfortable height for weeding and picking while, at the same time, giving the plants more light and providing the garden with a pleasing 3-D effect. This sort of garden layout is expensive, but when land is scarce in a city it provides maximum planting areas.

The raised gardens I saw in California were of the "deep-bed" type. They were raised containers created by using old sleepers, brick or block constructions, and then filled with extremely good quality top soil to a depth of at least 18 inches (45 cm) deep. This allows for very dense planting, high output, and vigorous, drought-resistant rooting. The smaller your plot, the more intensively you will be able to cultivate.

At the other extreme from the raised bed easy-run urban garden we have the traditional allotment. We have all seen these along the side of railways with their little wooden potting sheds, staked runner beans, and well-guarded winter brassicas. Most towns will have allotment societies of one kind or another, so check with the local council or in the local library. You will not be building a raised bed system in your allotment, although you may very well wish to develop deep-bed plantings or even a small polytunnel.

Planning your urban garden

It's a lovely occupation and a great luxury to be dreaming about your future garden in the depths of winter, but finally crunch time comes: what exactly are you going to put where? The first essential is to think about the orientation of the site: where is the sun and where is the shade? You will not want to have tall plants on the south side of your garden (in the northern hemisphere at any rate); equally, you will want to use any south-facing fence or wall for sun-loving plants – for example, espaliered fruit trees. Ideally, you will want to break up the cultivated area into smaller blocks by using perennials, such as the soft fruit bushes or artichokes.

Next, you have to decide what kind of crops you want to grow. In part this is a matter of personal preference, but also it is a question of what the site will support. Some plants –

for example, marrows or blackberries – are very large and aggressive. These plants "with attitude" will not make comfortable plantings for your small urban garden. Equally, potatoes take up a great deal of space for very little real gain, you can easily obtain first-rate spuds from other sources. As a compromise you might want to plant just a few earlies. The green salad crops are wonderful to grow because they taste so much better pulled up fresh from the garden or allotment. Mange tout or sugar snap peas are also especially delicious when fresh. Runner beans are very exciting to grow. They are a real 3-D plant that produces a mass of food from a very small space. A few soft fruit bushes and raspberries will pay their way well in terms of labour and space, as will espaliered fruit trees. You might also consider rhubarb, which is a nice early "fruit" crop, but asparagus would be a real space guzzler, as would sweet corn.

Carrots are an ideal vegetable to grow, too, and taste absolutely delicious when pulled fresh from the soil. Tomatoes can also be a brilliantly productive plant in a tight space. Strawberries grow well in specially designed pots, but make sure the pot you buy is frost-proof because many of the cheaper ones are not and will crack, causing you to lose your lovely crop. As well as growing strawberries in pots, you can also grow them in little crevices made in stone walls. In the urban garden I think this is the best place for them.

The best piece of advice I can give you when planning your urban garden is to remember that the more use you can make of all three dimensions, the better your garden – and produce – will be. It is a notable feature of the natural world that pests such as aphids or rabbits breed extremely fast. Once they get going, their numbers soon go exponential and your vegetable plot is fast consumed. So there is vital value in having predators ready and waiting right at the start of the breeding season. One or two bushy perennials will provide shelter for these predators over the winter – for example, blackcurrants or raspberries. You may even like the idea of a little box hedge as this makes a neat traditional border which will also help deter cats and dogs.

The Central Deep Bed
This is the workhorse of your garden. Keep fertility and soil condition high with regular applications of rotted manure and compost. Make sure you vary the choice of crop each year to avoid disease.

AN URBAN MICRO-GARDEN
Make use of all your urban garden space by
using all 3 dimensions. Paving and bricks cut
out mud and eliminate weeds.

Wire Supporters
Use treated timbers and galvanized
wire for supports. Even better, find
stainless steel wire which is less
abrasive on bark.

Fruit Espalier Hedges
Apples and plums can be
trained to make attractive
and productive "hedges".
These should face south
or west.

Productive Climbers
Raspberries and runner
beans provide a 6 foot
"wall" of productivity.

Larger Plants
Use tomatoes, artichokes, rhubarb, or
even courgettes as a lower layer of
robust productivity.

The Beehive
Keep a clear space around your hive to allow the
bees access and ensure that no tall plants block
getting to the hive. The bees need a good few
yards of clear run to their landing board.

Composting
An enclosed bin
keeps out rats and
flies. A perforated
metal sheet
underneath will let in
worms but keep out
rats and mice.

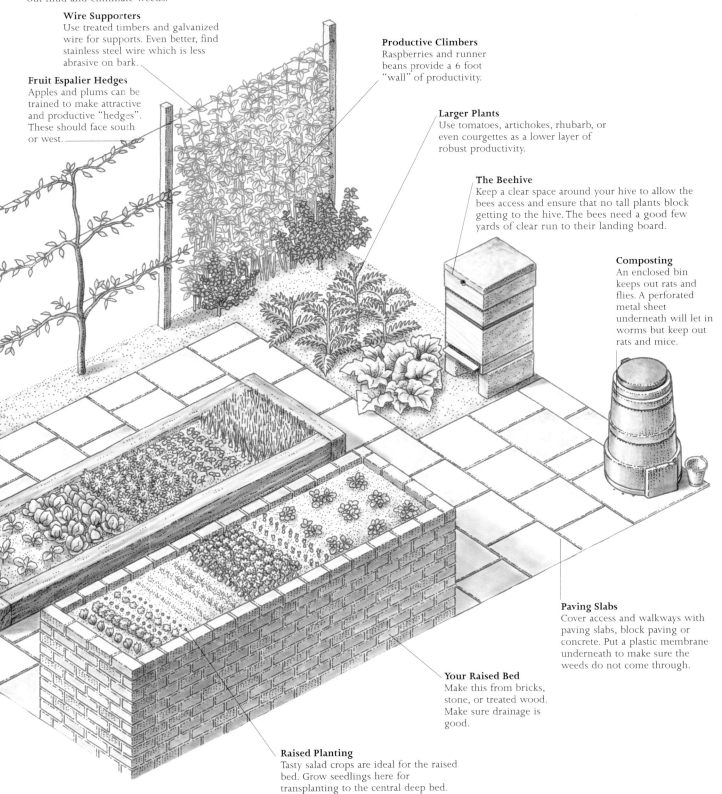

Paving Slabs
Cover access and walkways with
paving slabs, block paving or
concrete. Put a plastic membrane
underneath to make sure the
weeds do not come through.

Your Raised Bed
Make this from bricks,
stone, or treated wood.
Make sure drainage is
good.

Raised Planting
Tasty salad crops are ideal for the raised
bed. Grow seedlings here for
transplanting to the central deep bed.

The Allotment

Your allotment will, I hope, become part of a long-term relationship between your family and the soil. So think about how it needs to develop over a number of years. Plants like soft fruit bushes and fruit trees will grow more than you imagine – at least that has been my experience, as I always seem to end up having to thin out my blackcurrant and gooseberry bushes after five or six years.

The layout of paths, hedges, and perennial plantings needs to be thought about carefully. I always tell people to leave enough space to take a small tractor into the heart of the garden. If you do not do this you will almost certainly find frustration a few years later if you need to take building materials or farmyard manure into the garden. Once hedges and trees have grown it is very difficult to move them! And do not forget that if you want to use a rotavator you will need to leave enough room in your allotment for turning at the end of rows.

THE URBAN ALLOTMENT
Check with your local library, post office or local council to find out about getting an allotment. Make contact with your local allotment society.

Training Espaliers
Be brutal when training espaliers, pinch out shoots, and prune off branches you do not want.

Row Crops
Grow row crops like lettuce, onions, and peas in six-year rotation.

Espaliered Fruit Trees
Buy new trees and train them to make a productive hedge, preferably facing south or west. This is an attractive and effective way to grow top fruit.

Central Bed
Grow beetroot for a heavy yield and a great winter salad substitute when pickled in cider vinegar.

Cultivated Beds
Make these long but not too wide so you can use a rotavator to make life easy. 5 feet (1.5 m) is maximum for a deep-bed garden; use 6–12 feet (1.8–3.5 m) for conventional row planting.

Permanent Posts
Make sure you use either concrete or treated posts for your espalier fencing. Much cheaper to do a good job in the first place than to replace 10 years down the line.

Brassicas
Well spaced mature brassicas are a great crop for winter greens.

Raspberry Supports
Always use treated wooden posts to make secure supports for raspberry canes which can be re-used year after year.

Beech Hedges
These are an ideal way to shelter your garden from blustery winds. They are easily controlled, good looking, and very durable. Keep the height down to about 4 feet (1.2 m) to prevent shade.

Seedbeds
Plant brassicas and salad crops in seedbeds for transplanting later in the year when there is more room in the garden.

Runner Bean
These are a brilliantly productive "aerial" crop. Slugs love the young shoots, and don't forget they grow at least 8 feet (2.5 m) tall, casting lots of shade.

Compost Bins
Use a two-bin compost system, layering the material carefully in square, not loose heaps. Add farmyard manure if you can get it. And make the family use a piss bucket if decorum allows.

Paving
Well-laid paving helps restrain weeds and generally makes working the garden easier.

Soft Fruits
Use soft fruits and other large perennials to break up the garden and provide shelter.

close to the earth as your skill allows. Rake off the cut vegetation with a garden fork. Press the fork down firmly as you pull it towards you to get the surface weeds pulled out. When the surface clearing is complete, get to work with your fork again. By using a fork you will avoid cutting up the deep roots of perennial weeds such as docks or the dreaded spear grass. You will not be able to dig if it is too wet, and you will not be able to dig if it is too dry. A nice, cool breezy day in winter is best of all – especially if the sun is shining to keep your spirits high. Fork the land over in large lumps, pulling out deep roots and couch strands by hand. Don't worry too much about surface-rooting annual weeds, as you are going to give them a hard time anyway in the weeks to come. Just turn over the biggest square sods that you can with the green side down. Keep your wheelbarrow close by and take the weeds and roots straight to your compost heap. Spread them out in layers to make a stack and not a city folk's "heap".

When you are digging it is essential to pace yourself and develop a sort of meditational rhythm. Let your mind wander as your body steadily does the work. It may take several hours or even days to dig the plot, but that's fine. We cannot hurry nature, and your next meal will taste much better after a good session. When the job is finished you will have a brown and very lumpy plot spread out in front of you. Now you must leave things as they are for at least three or four weeks. This will allow the weather, the wind, and the sun to work on your soil.

The ideal next step is to find yourself a good rotavator. This should not go down too deep – 9 inches (23 cm) is about right. Try to choose a bright sunny morning, really as warm and windy as possible. Take a quick walk over the plot before you start to rotavate and pull out any remaining deep-rooted perennials. Now you can start your rotavator and make at least two passes over the plot. On the last pass try not to leave any footmarks. You want the soil to be left as fluffed up as possible so that weed roots will dry out and die. You are now going to leave your plot again for a couple of weeks, although at this stage you may want to put in some of the larger plantings, for example, rhubarb, soft fruits, artichokes, and fruit trees.

Caring for compost

Choose a shady place for your composting heaps – preferably with a bit of shelter from the prevailing wind so that they do not get too soaked with rain. And always remember that the compost heap is the foundation of a successful garden (see pp.43, 234–235). Nowadays you will find many foolish folk who will readily part with their animal manure – whether local horse owners or local farmers. Their loss is your gain as well-rotted farm manure is going to make your garden the best on the block. Manure added to the compost heap also greatly assists in turning weeds and garden rubbish into good humus for your soil.

Avoid the temptation to get everything done overnight and you will happily make haste in clearing a weedy allotment! If you leave any perennial deep-rooted weeds (nettles, thistles and docks, for example) lurking in your soil, it will be an impossible task trying to keep them at bay later.

To start clearing an allotment, first remove all the surface vegetation and set up a composting area. You will probably need a space about 5 square feet (4.5 sq m) for this. A scythe is the perfect tool for cutting the weeds as

The One-Acre Holding

Everyone will have an entirely different approach to husbanding his or her land, and it is unlikely that any two smallholders with one acre each will adopt the same plan or methods. Some people like cows, other people are afraid of them. Some people like goats, other people cannot keep them out of the garden. (I never could and I don't know many people who can.) Some people will not kill animals and have to sell their surplus stock off to people who will kill them; others will not sell surplus stock off at all because they know that the animals will be killed. Some people are happy to keep more stock than their land can support and to buy in fodder from outside, while other people regard this as contrary to the principles of self-sufficiency.

Myself, if I had an acre of good well-drained land, I think I would keep a cow and a goat, a few pigs, and maybe a dozen hens. The goat would provide me with milk when the cow was dry. I might keep two or more goats, in fact. I would have the cow (a Jersey) to provide me and the pigs with milk, but more important I would keep her to provide me with heaps and heaps of lovely manure. For if I was to derive any sort of living from that one acre, without the application of a lot of artificial fertilizer, it would have to be heavily manured.

Now the acre would only just support the cow and do nothing else, so I would, quite shamelessly, buy in most of my food for the cow from outside. I would buy all my hay, plenty of straw (unless I could cut bracken on a nearby common), all my barley meal and some wheat meal, and maybe some high protein in the form of bean meal or fish meal (although I would aim to grow beans).

It will be argued by many that it is ridiculous to say you are self-supporting when you have to buy in all this food. True, you would grow much of the food for cows, pigs, and poultry: fodder beet, mangolds, kale, "chat" (small potatoes), comfrey, lucerne, or alfalfa, and all garden produce not actually eaten by people. But you would still have to buy, say, a ton or a ton and a half of hay a year and, say, a ton a year of grain of different sorts including your own bread wheat, and a ton or two of straw. For I would not envisage growing wheat or barley on such a small area as an acre, preferring to concentrate on dearer things than cereals, and produce that it was more important to have fresh. Also, the growing of cereals on very small acreages is often impossible because of excessive damage by marauding birds, although I have to say I have grown wheat successfully on a garden scale.

A cow or not a cow?
The big question here is – a cow or no cow? The pros and cons are many and various. In favour of having a cow is the fact that nothing keeps the health of a family, and a holding, at a high level better than a cow. If you and your children have ample good, fresh, unpasteurized, unadulterated milk, butter, buttermilk, soft cheese, hard cheese, yoghurt, sour milk, and whey, you will simply be a healthy family, and that is the end of it. A cow will give you the complete basis of good health. If your pigs and poultry also get their share of the milk by-products, they too will be healthy and will thrive. If your garden gets plenty of cow manure, that too will be healthy and thrive. This cow will be the mainspring of all your health and well-being.

On the other hand, the food that you buy in for this cow will cost you in the hundreds per year. Against this you can set whatever money you would pay for dairy produce in that year for yourself and your family (and if you work that out, you will find it to be quite substantial), plus the increased value of the eggs, poultry-meat, and pig-meat that you will get (you can probably say that, in value, a quarter of your pig meat will be creditable to the cow), plus the ever-growing fertility of your land. But a serious contra consideration is that you will have to milk the cow.

Twice a day for at least 10 months of the year you will have to milk the cow. It doesn't take very long to milk a cow (perhaps eight minutes), it is very pleasant when you really know how to do it and if she is a quiet nice cow, but you will have to do it. So the buying of a cow is a very important step, and you shouldn't do it unless you do not intend to go away very much, or you can make arrangements for somebody else to relieve you with milking. (Of course, even if you only have a budgerigar somebody has got to feed it.)

So let us plan our one-acre holding on the assumption that we are going to keep a cow.

ONE-ACRE HOLDING WITH A COW
Half the land will be put down to grass, leaving half an acre arable. (I am not allowing for the land on which the house and buildings stand.) Now the grass half could remain permanent pasture and never be ploughed up at all, or it could be rotated by ploughing it up, say, every four years. If the latter is done it is better done in strips of a quarter of the half acre each, so each year you grass down an eighth of an acre of your land. Thus there is some freshly-sown pasture every year, some two-year-old ley, some three-year-old ley, and some four-year-old ley. The holding will be more productive if you rotate your pasture thus every four years.

The holding may break naturally in half: for example, an easily worked half acre of garden, and a half acre of roughish pasture. You will begin then by ploughing up or pigging (allowing pigs to root it up behind an electric fence) or rotavating half of your holding. This land you will put down to a grass-and-clover-and-herb mixture.

If you sow the seed in the autumn you can winter your cow indoors on bought hay and hope for grazing next spring. If your timetable favours your sowing in the spring, and if you live in a moist enough climate to do so, then you will be able to do a little light grazing that summer. It is better not to cut hay the first summer after spring-sowing of grass, so just graze it lightly with your little cow.

Grazing

At the first sign of "poaching" (destruction of grass by treading) take the cow away. Better still, tether your cow, or strip-graze behind an electric fence. Just allow the cow to have, say, a sixth part of the grass at one time, leave her on that for perhaps a week, then move her to the next strip. The length of time she stays on one strip must be left to your common sense (which you must develop if you are to become a self-supporter).

The point about strip-grazing is that grass grows better and produces more if it is allowed to grow for as long as possible before being grazed or cut, then grazed or cut right down, then rested again. If it is grazed down all the time it never really has a chance to develop its root system. In such super-intensive husbandry as we are envisaging now, it is essential to graze as carefully as possible.

Tether-grazing, on such a small area, might well be better than electric fencing. A little Jersey quickly gets used to being tethered and this was, indeed, the system that they were developed for on the island of Jersey, where they were first bred. I unequivocally recommend a Jersey to the one-acre man, incidentally, because I am convinced that for this sort of purpose she is without any peer. I have tried Dexters, with complete lack of success, but if you really know of a Dexter that gives anything like a decent amount of milk (my two gave less than a goat), is quiet and amenable, then go ahead and get a Dexter and good luck to you. But remember, a well-bred Jersey gives plenty of milk (quite simply the richest in butter fat of any milk in the world), is small, so docile that you will have trouble resisting taking her into the house with you, moderate in her eating demands, lovable, healthy, and very hardy.

Now your half acre of grass, once established, should provide your cow with nearly all the food she needs for the summer months. You are unlikely to get any hay off it as well, but if you did find that the grass grew away from the cow, then you could cut some of it for hay.

A highly intensive garden

The remaining half of your holding – the arable half – will then be farmed as a highly intensive garden. It will be divided, ideally, into four plots, around which all the annual crops that you want to grow will follow each other in strict rotation. (I discuss this rotation more on page 40.) The only difference that you will have to make in this rotation is that every year you will have to grass a quarter down, and every year plough a quarter of your grassland up. I suggest that your potatoes come after the newly ploughed bit.

An ideal rotation might go like this:

1 *grass (for four years)*
2 *potatoes*
3 *pea-and-bean family*
4 *brassica (cabbage family)*
5 *roots*
6 *grass again (for four years)*

To sow autumn-sown grass after your roots, you will have to lift them early. In a temperate climate it would be quite practicable to do this; however, in countries with more severe winters it might be necessary to wait until the following spring. In areas with dry summers, unless you have irrigation, it would probably be better to sow in the autumn. In some climates (dry summers and cold winters) it might be found best to sow your grass in the late summer after the pea-and-bean break instead of after the root break, for the peas and beans are off the ground earlier than the roots. It might then pay you to follow the grass with potatoes, and your succession could be like this:

1 *grass (for four years)*
2 *potatoes*
3 *brassica (cabbage family)*
4 *roots*
5 *pea-and-bean family*
6 *grass (for four years)*

A disadvantage of this might be that the brassica, following main-crop potatoes, might have to wait until the summer following the autumn in which the potatoes were lifted before they could be planted. When brassica are planted after pea-and-bean family they can go in immediately, because the brassica plants have been reared in a nursery-bed and it is not too late in the summer to transplant them after the peas and beans have been cleared. But potatoes cannot be lifted (main crop can't anyway) until the autumn, when it is too late to plant brassica. Actually, with this regime, you will be able to plant some of your brassica that first summer, after early potatoes. Or if you grow only earlies, you may get the lot in.

One possibility would be to follow the potatoes immediately with brassica (thus saving a year) by lifting some earlies very early and planting immediately with the earliest brassica, then following each lifting of potatoes with more brassica, ending with spring cabbages after the main crop have come out. This would only be possible in fairly temperate climates, though.

All this sounds complicated: but it is easier to understand when you do it than when you talk about it. And consider the advantages of this sort of rotation. It means that a quarter of your arable land is newly ploughed-up four-year-ley every year: intensely fertile because of the stored-up fertility of all that grass, clover, and herbs that have just been ploughed in to rot, plus the dung of your cow for four summers. It means that because your cow is in-wintered, on bought-in hay, and treading and dunging on bought-in straw, you will have an enormous quantity of marvellous muck to put on your arable land. It means that all the crop residues that you cannot consume go to help feed the cow, or the pigs or poultry, and I would be very surprised if, after following this regime for a few years, you did not find that your acre of land increased enormously in fertility, and that it was producing more food, for humans, than many a 10-acre farms run on ordinary commercial lines.

DIVIDING UP YOUR ONE ACRE

If you had one acre of good, well-drained land, you might choose to use all of it to grow fruit and vegetables. Myself, I would divide it in half and put half an acre down to grass, on which I would graze a cow, and perhaps a goat to give milk during the short periods when the cow would be dry, a sow for breeding, and a dozen chickens. I would admittedly have to buy in food from outside to feed these animals through the winter, but this is preferable to buying in dairy products and meat, which would be the alternative.

My remaining half acre I would divide into four plots for intensive vegetable production, devoting a plot each to potatoes, pulses (peas and beans), brassica (cabbage family), and roots. I would divide the grass half acre into four plots as well and rotate the whole holding every year. This means I would be planting a grass plot every year and it would stay grass until I ploughed it up four years later. I would not have enough grass to keep the cow outdoors all year. I would have a greenhouse for tomatoes and hives for bees, and I would plant a vegetable patch with extra household vegetables, herbs, and soft fruit.

Peas and beans

Grow at least three kinds of beans, say, French, runners, and broad, and plenty of peas. Plant brassica on this plot next year.

Brassica

On your brassica plot grow a variety of cabbages, cauliflower, broccoli, and sprouts for yourself. Grow kale, turnips, and swedes, which are roots but also brassica, to feed to your animals. Next year this plot should be planted with roots.

A half acre of grass

Your half acre of grass will feed your cow all through the summer. Let your hens run on it and give them a movable chicken ark. When you want to plough up your annual eighth of an acre, put the pigs on it and let them do the work for you.

Movable pig sty

Tethered cow

Hay

Cabbage

Kale

Broccoli

Peas

French be

Swedes

French be

Broad beans

Cauliflower

Sprouts

Potatoes

Each year plant your potatoes in the plot which has just been ploughed up from grass.

Vegetable patch
In your home vegetable patch plant extra vegetables for your own consumption. Spinach, carrots, lettuce, celery, leeks and onions, when added to your brassica, pulses, and potatoes should give you a varied diet. Plant a herb garden near the kitchen and sunflowers so you can press your own oil.

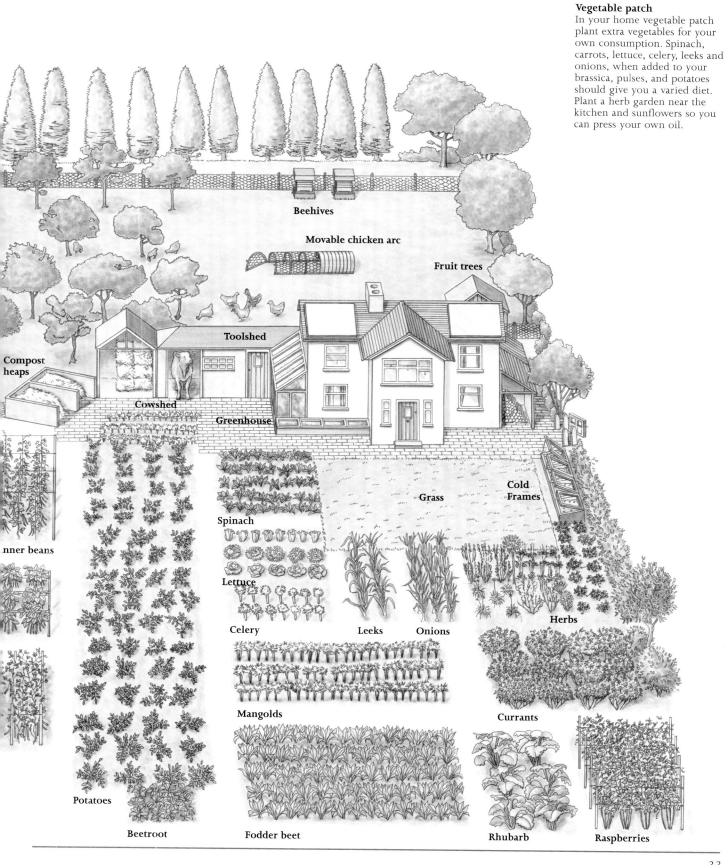

Beehives

Movable chicken arc

Fruit trees

Toolshed

Compost heaps

Cowshed

Greenhouse

Grass

Cold Frames

Spinach

nner beans

Lettuce

Herbs

Celery   Leeks   Onions

Mangolds

Currants

Potatoes

Beetroot   Fodder beet   Rhubarb   Raspberries

Half-acre rotation

You may complain that by having half your acre down to grass you thus confine your gardening activities to a mere half acre. But actually half an acre is quite a lot, and if you garden it really well it will grow more food for you than if you "scratch" over a whole acre. And the effect of being under grass, and grazed and dunged, for half its life, will enormously increase the fertility of it. I believe you will grow more actual vegetables than you would on the whole acre if you had no cow, or grass break.

Small-acreage tips

Here are a few pointers to bear in mind:

Cows First, the cow will not be able to be out of doors all the year. On such a small acreage she would poach it horribly. She should spend most of the winter indoors, only being turned out during the daytime in dry weather to get a little exercise and fresh air. Cows do not really benefit from being out in all weathers in the winter time, although they put up with it. They are better for the most part kept in, where they make lovely manure for you, and your cow will have plenty of greenstuffs and roots that you will grow for her in the garden. In the summer you will let her out, night and day, for as long as you find the pasture stands up to it. You could keep the cow on "deep litter": that is, straw which she would dung on and turn into good manure, and you would put more clean straw on it every day. I have milked a cow for years like this and the milk was perfect, made good butter and cheese, and kept well.

Or you could keep the cow on a concrete floor (insulated if possible), giving her a good bed of straw every day and removing the soiled straw, and putting it carefully on the muck-heap – that fount of fertility for everything on your acre – every day. You would probably find that your cow did not need hay at all during the summer, but she would be entirely dependent on it right throughout the winter, and you could reckon on having to buy her at least a ton. If you wished to rear her yearly calf until he reached some value, you would need perhaps half a ton more hay, too.

Pigs You would have to be prepared to confine pigs in a house for at least part of the year (and provide straw for them). This is because on a one-acre holding you are unlikely to have enough fresh land to keep them healthy. The best thing would be a movable house with a strong movable fence outside it, or you could have a permanent pig-sty as well. But the pigs would have a lot of outdoor work to do: they would spend part of their time ploughing up your eighth of an acre of grassland; they could run over your potato land after you had lifted the crop; they could clear up after you had lifted your roots, or after you had lifted any crop. But they could only do this if you had time to let them do it. Sometimes you would be in too much of a hurry to get the next crop in. As for food, you would have to buy in some corn, barley, or maize. This, supplemented with the skimmed milk and whey you would have from your cow, plus a share of the garden produce and such specially grown fodder crops as you could spare the land for, would keep them excellently.

If you could find a neighbour who would let you use his boar, I would recommend that you kept a sow and bred from her. She might well give you 20 piglets a year. Two or three of these you would keep to fatten for your bacon and ham supply, the rest you would sell as "weaners" (piglets 8–12 weeks old, depending on the requirements of your particular market), and they would probably fetch enough money to pay for every scrap of food you had to buy for them, the poultry, and the cow too. If you could not get the service of a boar, you would probably buy weaners yourself – just enough for your own use – and fatten them.

Poultry Poultry could be kept on the Balfour method (*described on p.120*), in which case they would stay for years in the same corner of your garden. Or better, in my opinion, they could be kept in movable arks on the land. They could then be moved over the grassland, where by their scratching and dunging they would do it good. I would not recommend keeping very many. A dozen hens should give you enough eggs for a small family, with a few occasionally to sell or give away in the summer time. You would have to buy a little corn for them, and in the winter some protein supplement unless you could grow enough beans. You might try growing sunflowers, buck-wheat, or other food specially for them. You might consider confining them in a small permanent house, with two outdoor runs à la Balfour system, during the worst months of the winter, with electric light on in the evenings to fool them that it was the time of the year to lay and thus get enough winter eggs.

Goats If you decided to keep goats instead of a cow (and who am I to say this would not be a sensible decision?), you could manage things in much the same way. You would only get a small fraction of the manure from goats, but on the other hand you would not have to buy anything like so much hay and straw – indeed, perhaps not any. You would have nothing like so much whey and skimmed milk to rear pigs and poultry on, and you would not build up the fertility of your land as quickly as you would with a cow.

Crops Crops would be all the ordinary garden crops, plus as much land as you could spare for fodder crops for the animals. But bear in mind that practically any garden crop that you grew for yourself would be good for the animals too, so everything surplus to your requirements would go to them. You would not have a "compost heap": your animals would be your compost heap.

Half-an-acre If you kept no animals at all, or maybe only some poultry perhaps, you might well try farming half an acre as garden, and growing wheat in the other half acre. You would then rotate your land as described above but substituting wheat for the grass-and-clover ley. And if you were a vegetarian this might be quite a good solution. But you could not hope to increase the fertility, and productiveness, of your land as much as with animals.

The Five-Acre Holding

The basic principles I have described for running a one-acre holding will also broadly apply to larger acreages. The main difference would be that if you had, say, five acres of medium to good land in a temperate climate, and the knowledge, you could grow all the food necessary for a large family, except such things as tea and coffee, which can only be grown in the tropics. And you could, of course, do without such things. You could grow wheat for bread, barley for beer, every kind of vegetable, every kind of meat, eggs, and honey.

Just as every person in the world is different, so is every five-acre plot, but here is a possible pattern, assuming one acre was set aside for house and buildings, orchard, and kitchen garden, the remainder could be divided up into eight half-acre plots. It would be necessary to fence them permanently: electric fencing would do. Or, if you are a tetherer, you might tether your cows, and your pigs, and your goats if you have any, and not have any fencing at all. I tried tethering a sheep once, but the poor thing died of a broken heart, so I wouldn't recommend it.

The rotation could be something like this:

1 *grass (for three years)*
2 *wheat*
3 *roots*
4 *potatoes*
5 *peas and beans*
6 *barley, under-sown with grass-and-clover-grass (for three years)*

This would only leave you, of course, one and a half acres of grassland, but it would be very productive grassland, and in a good year it could be supplemented with something like: a ton of wheat; 20 tons of roots; four tons of potatoes; half a ton of peas or beans; three quarters of a ton of barley. You might well manage to get two tons of hay off your grassland, and then have enough "aftermath" (grass which grows after you have cut the hay) to give grazing to your cows until well into the autumn.

Flexibility: the essence of good husbandry

There are a thousand possible variations of this plan, of course. Flexibility is the essence of good husbandry. You could, for example, take potatoes after your ploughed-up grassland, and follow that with wheat. You could grow oats as well as barley, or oats as well as wheat. You could grow some rye: very useful if you have dry light land, or want good thatching straw, or like rye bread. You could grow less peas and beans. You could try to grow all your arable crops in four half-acre plots instead of five and thus leave two acres for grassland instead of one and a half. You might find you had some grassland to spare in your "home acre", in your orchard, for example, if your trees were standards and therefore too high to be damaged by the stock. Of course if you were in maize-growing country you would grow maize, certainly instead of barley, maybe instead of roots or potatoes. A good tip is to seek out farming neighbours, and ask them which crops grow best in your area.

Stock As for stock, you might well consider keeping a horse to help you do all that cultivating, or you might have a small garden tractor instead. Your ploughing could be done with pigs. With five acres you might well consider keeping enough sows to justify a boar. Four is probably the minimum; we kept six sows and a boar for many years and they were astonishingly profitable. Indeed, in good years and bad, they paid all our bills for us: the Irish call the pig "the gentleman who pays the rent" and one can see why. But pigs won't pay you very well unless you can grow a great deal of their food for them. You could look upon your pig herd, whether large or small, as your pioneers: they would plough up your half-acre of grass every year for you, plough your stubbles after corn, clean up your potato and root land after harvest, and generally act as rooters-up and scavengers.

Poultry Poultry, too, would be rotated about the holding as much as possible. Put on wheat or barley stubble, they will feed themselves for some time on spilled grain, besides doing great good scrapping out leatherjackets and wireworm. Following the pigs after the latter have rooted up a piece of land they will also do good by eating pests and will do themselves good too. Ducks, geese, turkeys, tame rabbits, pigeons: your five acres will provide enough food and space for them all, and they will vary your diet.

Cows I would recommend keeping two cows, so you would have ample milk all the year. You would have enough milk to make decent hard cheese during the summer to last you through the winter, and enough whey and skimmed milk to supplement pig and poultry feed. If you reared one calf a year, and kept him 18 months or two years, and then slaughtered him, you would have enough beef for family use. That is, if you had a deep freeze. If you did not, then you could sell your bullock and use that money for buying beef from the butcher, or, much better, you could make an arrangement with several smallholder neighbours that you each took turns to slaughter a beast, then divided the meat up amongst you so it could all be eaten before it went bad. In a cold winter you can keep beef at least a month without a freezer.

Sheep On such a small acreage, sheep are a more doubtful proposition because they need very good fencing, and also it is uneconomic to keep a ram for less than, say, six sheep. But you could keep some pet ewes, get them mated with a neighbouring ram, rear the lambs, and keep yourself in mutton and wool.

The above is only an introductory outline of how a prospective self-supporter might organize a five-acre holding. Each person will wish to adapt according to his or her circumstances, the size of his or her family or community, and the nature of the land. But the main body of this book is aimed at providing you with as much practical help as possible in selecting and managing your acreage, your crops, and your livestock, and in making them the productive agents in your search for self-sufficiency and the good life.

DIVIDING YOUR FIVE ACRES

If you had five acres of good well-drained land, you could support a family of, say, six people and have occasional surpluses to sell. Of course, no two five-acre plots are ever the same, but in an ideal situation I would set aside one of my acres for the house, farm buildings, kitchen garden, and orchard and the other four acres I would divide into eight half-acre plots. Three of them I would put down to grass every year, and there I would run: two cows for dairy produce; four sows, a boar, some sheep and some geese for meat; and some chickens for eggs. As well as these animals I would keep ducks, rabbits, pigeons, and bees wherever I could fit them in.

Now, in the five remaining plots I would sow wheat, roots, Jerusalem artichokes or potatoes, peas and beans, oats, and barley undersown with grass and clover. I would rotate all eight plots every year so no plot ever grew the same crop two years running, unless it was grass. A grass plot would stay grass for three years before being ploughed.

Pasture

Your pasture can cover one and a half acres. Here you can graze cows, sheep, geese, and chickens; and when you want to plough up some of your grassland, you can bring pigs back from the woods and fold them on small areas at a time. The top end of the field has not yet been cut for hay.

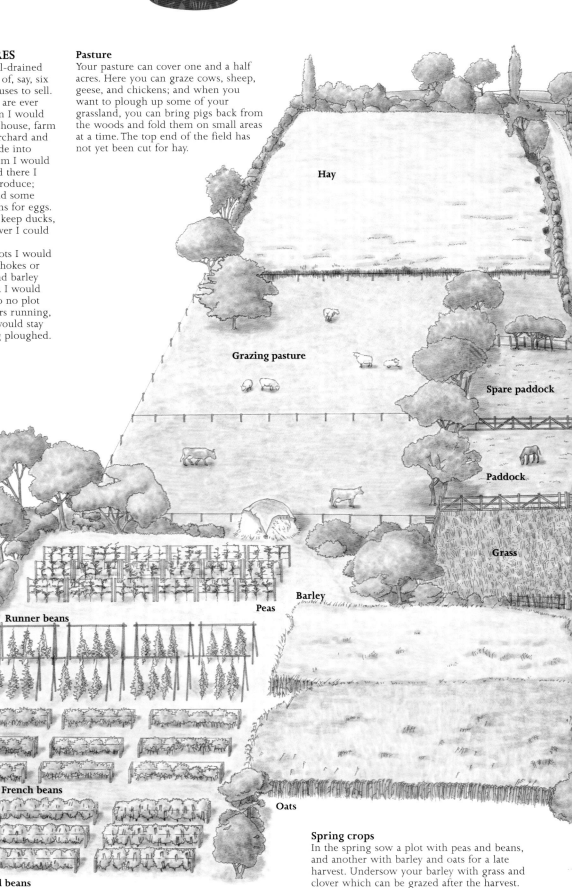

Hay

Grazing pasture

Spare paddock

Paddock

Grass

Runner beans

Peas

Barley

French beans

Broad beans

Oats

Spring crops

In the spring sow a plot with peas and beans, and another with barley and oats for a late harvest. Undersow your barley with grass and clover which can be grazed after the harvest.

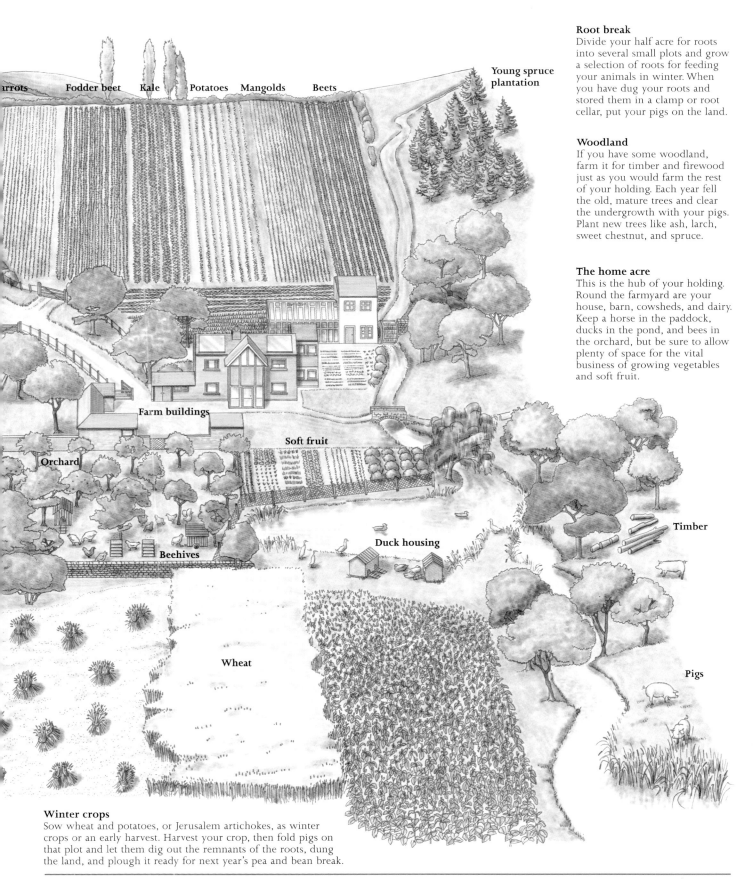

Carrots Fodder beet Kale Potatoes Mangolds Beets Young spruce plantation

Farm buildings

Orchard

Soft fruit

Beehives

Duck housing

Timber

Wheat

Pigs

Root break
Divide your half acre for roots into several small plots and grow a selection of roots for feeding your animals in winter. When you have dug your roots and stored them in a clamp or root cellar, put your pigs on the land.

Woodland
If you have some woodland, farm it for timber and firewood just as you would farm the rest of your holding. Each year fell the old, mature trees and clear the undergrowth with your pigs. Plant new trees like ash, larch, sweet chestnut, and spruce.

The home acre
This is the hub of your holding. Round the farmyard are your house, barn, cowsheds, and dairy. Keep a horse in the paddock, ducks in the pond, and bees in the orchard, but be sure to allow plenty of space for the vital business of growing vegetables and soft fruit.

Winter crops
Sow wheat and potatoes, or Jerusalem artichokes, as winter crops or an early harvest. Harvest your crop, then fold pigs on that plot and let them dig out the remnants of the roots, dung the land, and plough it ready for next year's pea and bean break.

"We don't bother to do a lot of things in our garden. We let things take their chance, and every year some crops are good and others are bad; but at least there is always enough to eat and we always get a taste of everything. If we did all the spraying and sprinkling and dusting and fumigating that one is told to do in the books, we would spend a fortune on chemicals and have no time left over for anything else. In fact, growing a big variety of crops and never the same crop two years on the same ground, and heavily manuring with the dung of a variety of animals seems to give our crops the strength to resist most pests and diseases. Only sometimes do we come a cropper...

The failure to use artificials is not crankiness. It is simply this: our aim is to grow our food for nothing. If we spend money on buying artificial manures we are not doing this. Also we realize now that food tastes a lot nicer if it has been grown with natural and not artificial manures."

JOHN SEYMOUR FAT OF THE LAND 1976

CHAPTER TWO
FOOD
FROM THE
GARDEN

The Food-Producing Garden

The country garden of my childhood was a mixture of vegetables, flowers, soft fruit, tree fruit (oh, those greengages!) and very often tame rabbits, almost certainly a hen run, often pigeons, and often ferrets. It was a very beautiful place indeed. Now, alas, it has disappeared under a useless velvety lawn and a lot of silly bedding plants and hardy perennials, but of course the owner feels compelled to keep up with the people next door.

However limited the space available, you only need the determination to abandon your space-wasting lawn and flowerbeds in exchange for a programme of planned crop rotation for every inch of your garden to become a productive unit. You will save money, your end products will be fresh, and your garden will be a fine example of a dying breed: the cottage garden of yesteryear. But how can we best reproduce the old cottage garden, which was one of the most productive places on Earth? Well, divide the garden area into six parts: seven if you want a small lawn-and-flower area for the sublime fragrance of flowers.

The clever thing to do is to use perennial food plants as "hedges" to divide up the garden into plots. These are plants that go on from year to year – thus providing valuable overwintering homes for beneficial insects, as well as shelter from the wind and weather. Plants for this purpose are those such as asparagus, globe artichoke, horseradish, rhubarb, and many of the soft fruits, including raspberries. You want to avoid large open areas that encourage spread of disease. At the same time you do not want a garden so claustrophobic that you cannot manoeuvre the rotavator effectively for the spring cultivations. Larger trees like the top fruits are best kept separate in the orchard, as they really do shade and sterilize a large area of soil. Of course, neat espaliered fruit trees can and do make excellent and very productive "hedges" dividing up your beds.

Our six parts can then be used in a six-part rotation which will keep potatoes and brassicas from being grown too frequently in the same position. Each yearly crop is called a "break". The six yearly crops I use are:

Spuds (which means potatoes and is a very good word!)
Pea and bean family
Brassicas (cabbages, broccoli, turnips, and so on)
Greedy plants like sweetcorn, pumpkins and cucumbers
Salad and catch crops, onions, shallots
Roots (carrots, beetroot, parsnips, beets, celery, and so on)

Liming

If your land is acid it will need lime. You can test for this with a very simple device bought from any garden shop – or by asking a neighbour. You should lime before the pea and bean break. The peas and beans like the lime and the cabbage tribe that follows them likes what is left of it. Lime has more time to combat the dreaded club-root disease, which is carried by brassica, if it is in the soil for a few months before the brassica are planted.

Mucking & mulching

If you have muck – farmyard manure – and I hope you have, or if you have compost, concentrate this on your potato break. The potatoes benefit enormously by it. In fact, you won't grow very many without it. It is better not to put it on the root break because some roots, carrots and turnips in particular, are apt to "fork" if they have too much fresh muck. It is better not to put muck on the pea and bean break, because you lime that, and lime and muck don't go very well together in the same year.

It is quite advantageous to put a mulch, a covering of some dead greenstuff, on the surface of the soil between the cabbage-tribe plants, but only after you have hoed them two or three times to suppress the weeds. If you mulch on top of weeds, the weeds will simply grow through the mulch and the mulch will then impede the hoe.

Organic gardening

The aim of the organic gardener should be to get as much humus into his land as possible. Muck, compost, seaweed, leaf-mould, spoiled hay, nettles, roadside cuttings, anything of vegetable or animal origin: compost it (*see pp.234–237*) and put it on the land or just put it on the land. If you dig it in well, you dig it in. If you just leave it on top, the worms will dig it in for you.

PERCENTAGE VALUES OF ORGANIC FERTILIZERS

	Nitrogen	Phosphorus	Potash	Calcium
Average farmyard manure	0.64	0.23	0.32	nil
Pure pig dung	0.48	0.58	0.36	nil
Pure cow dung	0.44	0.12	0.04	nil
Compost	0.50	0.27	0.81	nil*
Deep litter on peat	4.40	1.90	1.90	2.20
Deep litter on straw	0.80	0.55	0.48	nil
Fresh poultry dung	1.66	0.91	0.48	nil
Pigeon dung	5.84	2.10	1.77	nil

*Unless lime has been added

Unless you keep animals on your garden, you will have to bring organic matter, or inorganic matter if you are not "organically minded", in from outside if you want a really productive garden. I subsidize my garden with manure made by animals that eat grass, hay, and crops grown on the rest of the farm. There is much wild talk by would-be organic gardeners who think a garden will produce enough compost material to provide for itself. Well, let them try it. Let them take a rood of land, grow the bulkiest compost-making crop they can on it, compost it, and then see how far the compost it has made goes. It will not go very far.

True, deep-rooting plants, such as comfrey and lucerne (alfalfa), can do great work in bringing up minerals, and phosphates and potash as well, from the subsoil to add to your soil. Trees do an even better job. But the land that is devoted to growing the comfrey or the trees is out of use for growing food crops.

Of course if your own sewage goes back, in one form or another, into the soil of your garden, one big leakage of plant nutriments is stemmed. The old cottage gardens of the past had all their sewage returned, because the sewage system was a bucket and the contents of that were buried in the garden. Provided the ground in which they were buried was left undisturbed for a time, any pathogens in the sewage would die a natural death. These country gardens owed their phenomenal fertility to the fact that the inhabitants were importing food from outside all the time, as well as eating their garden's own produce, and lots of matter from both sources ended up in the soil. But if you annually extract large amounts of produce from a piece of soil, and either export it or eat it and export the resulting sewage, and don't import any manure or fertilizer, the laws of nature are such that you will ultimately exhaust that soil.

It is vital that your garden be well drained, and it is an advantage if the land beneath it is not too heavy.

GARDEN TOOLS

1 Dutch hoe, for pushing, good for using backwards, leaving the ground free of footprints. **2** Draw hoe, for pulling, much faster, goes deeper and tackles tougher weeds. **3** Mattock, excellent for cutting through tree roots. **4** Spade, essential for inverting soil and digging in manure; keep clean. **5** Fork, for loosening up soil quickly without inverting it, for incorporating compost or manure in first few inches and forking out roots of creeping weeds, a must for digging up spuds. **6** Knapsack (pressure) sprayer, essential for large areas. **7** Secateurs, quicker and kinder for pruning than a knife, also for severing chickens' neckbones when gutting them. **8** Trowel, for setting out plants. **9** Pruning knife. **10** Dibber; make one by cutting down a broken spade or fork handle, for setting out seedlings. **11** Garden reel, wooden or iron for winding up line. **12** Garden line, use light cord that does not get tangled up like string. **13** Watering can, preferably big and galvanized. **14** Rake; use a large steel one for fine seed beds and covering seed. **15** Precision drills, which pick seeds up one at a time and drop them in exactly the right intervals, save seed and save work-thinning later for the market gardener.

A well-drained medium loam is most desirable, but sandy soil, provided you muck it well, is very good, too. Heavy clay is difficult to manage, but will grow good brassica crops. Whatever your soil is, you can scarcely give it too much muck, or other humus or humus-forming material.

Green manuring

Green manuring is the process of growing a crop and then digging or ploughing it in to the soil, or else just cutting or pulling the crop, and throwing it down on top of the soil. This latter form of green manuring is "mulching". Ultimately, the green matter will rot and the earthworms will drag it down into the soil in their indefatigable manner. If you dig in green manure crops you should do it at least three weeks before you sow the next crop on top of them. The only way round this is to add plenty of available nitrogen to help the green manure to rot down without robbing the soil.

Green manuring improves the quality of the soil because the vegetable matter rots down into humus. The amount of humus added by an apparently heavy crop of green manure is smaller than you might think, but the great value of such crops is that they take up the free nitrogen in the soil. Bare soil would lose this nitrogen to the air, whereas the green crop retains nitrogen and only releases it when it has rotted, by which time the subsequent crop should be ready to use it.

It should be the aim of the organic gardener to keep as much of his land as possible covered with plants. Bare soil should be anathema unless for a very good reason it has to be bare temporarily. The old gardener's idea of "turning up land rough in the autumn to let the frosts get in it in the winter" should not stop us encouraging a green covering of vegetation after harvesting – a green covering which can be readily scythed off for the compost heap before it seeds and before the soil is broken up. The key here is to leave large lumps (12-inch/30-cm square) that will weather well without leading to erosion. Soil dug over like this in January will be perfectly ready to rotavate down into a seedbed in late March. If you can manage two or three passes in dry weather with at least a week between each you will ensure your seed bed is quite free of weeds.

Using weeds

Even weeds can be a green manure crop. Many annual weeds will pop out of nowhere after your crops have been harvested. Tolerate these with good heart for they make an excellent addition to the compost heap when scythed off tight to the ground before they seed. Their roots will help bind the ground against winter rains. But whatever you do, don't let them seed. For one thing, "one year's seeding is seven years' weeding", and for another, all green manure crops should be cut or pulled at the flowering stage, or earlier, when their growth is young, succulent, and high in protein. They then have enough nitrogen in them to provide for their own rotting down.

So look upon annual weeds as friends, provided you can keep them under control. Perennial weeds (weeds that go on from year to year) should not be tolerated at any cost. They will do you nothing but harm, except perhaps nettles and bracken. If you grow these two crops on otherwise waste land you can cut them and add them to the compost heap. They will do great good, as they are both deep rooting and thus full of material they bring up from below.

Planting green manure

Green manure crops can be divided into winter and summer crops, and legumes and non-legumes. People with small gardens will find winter crops more useful than summer ones, for the simple reason that they will need every inch of space in summer for growing food crops. Legumes make better green manure than non-legumes, because they have bacteria at their roots which take nitrogen from the air, and this is added to the soil when they rot.

Grazing rye

Of winter green manuring crops, grazing rye is probably the best. It can be broadcast at a rate of 2 ounces of seed per square yard (70 g per sq m) after early potatoes have been lifted. Rake the seed in, leave it to grow all winter, and then dig it in during spring. You can plant grazing rye as late as October, although you won't get such a heavy crop.

Comfrey

Comfrey is a perennial to grow for either magnificent green manure or good compost. Plant root cuttings from existing plants 2 feet (0.6 m) apart in really weed-clear land in spring and just let it grow. The roots will go down into the soil as far as there is soil for them to penetrate, and they will live for a decade giving heavy yields of highly nitrogenous material, rich in potash, phosphate, and other minerals, too.

Other green manure and compost crops

Tares are legumes and winter crops and so are doubly valuable. They can be sown from August to October, and dug in next spring. As a summer crop, they can be sown any time in spring and dug in when in flower. Sow the much-used mustard after early potatoes are lifted. Give the dug-over ground a good raking, broadcast the seed tightly and rake it in. Dig the crop into the ground as soon as the first flowers appear. Red clover seeds are expensive but it is a fine, bulky nitrogen-rich legume to sow after early potatoes and dug in in the autumn. Lupins are a large legume. Put the seeds in at 6-inch (15-cm) intervals both ways in the spring or early summer. *Tagetes minuta* is a kind of giant marigold, and is an interesting crop to plant for compost material. It grows 10 feet (3 m) high and has two marvellous effects. It kills eelworm and it wipes out ground elder and bindweed. Sunflowers make bulky compost material. Plant the seed ½ inch (1 cm) deep and 12 inches (30 cm) apart both ways in the spring, and cut when it is in flower.

The Deep Bed

When we face a semi-circle of keen students raring to get their teeth into something tough and physical, our favourite question is: "How would you like to dig a deep bed?" General puzzlement all around is the usual response, followed by, "What is a deep bed?"

A deep bed is a highly intensive and effective method of producing vegetables in a small space – especially if you are looking for drought resistance and vigorous growth. The idea seems to have originated in California – at least that is where I first saw it. The technique is ideal for those with limited space.

The first essential is obviously to understand how to create the bed. Then we have to learn to get the best out of it. The principles of the deep-bed system follow these three simple steps:

1 Create a highly fertile, deep, and well-drained block of topsoil, working in plenty of good compost so it will give you vigorous root growth as well as drought resistance.

2 Work your plot from the edges, without the need for walking on the soil. A deep bed should not be any more than 5 feet (1.5 m) wide – and it might need to be narrower for those who have short arms! This way you can do your planting and weeding from the sides.

3 Plant your vegetables in a close pattern that will reduce the need for weeding (much of which you will have to do by hand).

Marking out

Begin by marking out the area you have chosen for your bed. This may well be in established grassland turf – or it may be an area of already cultivated soil convenient for the purpose (perhaps a specifically created area within an urban garden).

The area may be 5 feet (1.5 m) wide by 20 feet (6 m) long. Note that the deep bed is an excellent way to bring old grassland into vegetable production – the turf is turned over well below two "spits" of earth (a "spit" is the countryman's term for the depth of one good spadeful of earth, that is about 9 inches /22 cm) where it will rot.

If your bed is 5 feet wide you will begin by marking off, say, six widths using a garden line or straight batten to guide you. Before you start it is useful to find a large piece of old plastic or even some old pieces of plywood. This will provide a convenient place to dump the diggings from the first row. After cutting the turf, dig out the first row.

Next you are going to go back to where you started and dig out the next spit of soil. Do not worry if the second spit of soil contains some subsoil. By bringing some subsoil to the top you make it available to mix with compost. And as your deep bed matures over the years you will be regularly turning over the top 18 inches (45 cm) or so of soil, creating a massively expanded layer of fertile topsoil for your plants.

After finishing your first trench you can fork in a 2–3-inch (5–8-cm) layer of good manure into the bottom before turning over the next sod.

If you don't have manure, don't worry; your bed will work well anyway – but the manure forms a superb reservoir of moisture for plants in a dry period. Give the turves a few chops with your spade if they are too lumpy and uneven to fit the trench.

Continue the process until you have dug the final trench. At this point you take in the turf and then the soil from the very first trench (from your plastic of plywood sheet). Bingo: your deep bed is dug.

The process can be tough work, especially if you are working with land that has not been cultivated before. In our Irish garden we have many huge stones – well hidden underground, for the most part. You will need to be prepared for this kind of thing. Have your wheelbarrow ready, a pickaxe, and a good, strong iron bar.

Be prepared, too, to find tree roots many yards away from large trees. When dealing with roots or a stone, always hold the spade with your two hands grasping the shaft. This way you can use the momentum of the spade's own weight to cut roots or loosen stones *without* hurting your hands. It is a useful tip, believe me.

Don't be surprised when you find that your deep bed is now raised 9 inches (22 cm) above the surrounding garden! This is quite normal; indeed, the "heaped" effect is beneficial for the use of the bed. But, of course, this is what happens to all soil when disturbed and there will be considerable subsidence with weather and rain.

Making the edges

One feature that makes a deep bed more effective is to add substantial edging of some kind once the bed has been created. I like to use large paving slabs, but you can use old sleepers or treated wood or anything similar. Put the slabs on to sheets of building plastic (you can easily buy these from the local builder's merchant) and this will stop weeds growing through the cracks. Not only does the solid edging allow you to cultivate and harvest the bed more comfortably, it also lets you rest heavy equipment on it without disturbing the bed. There are several other beneficial effects:

1 Rainfall from both sides of the deep bed is gathered from the edges, almost doubling the effective summer rainfall into the cultivated area.

2 It prevents the constant spread of creeping weeds from the edges.

3 The warm concrete acts as a major deterrent for the slugs (especially if you sprinkle on a little salt or ash from time to time), which are always wandering around in the garden looking for their next meal.

4 Surprisingly, the sheer capacity of slabs to soak up heat during the day helps to keep off frost and maintain a better temperature through the period of darkness. I have used this feature very effectively to keep frost away from tender young trees like walnuts – you can use a couple of big 9-inch (22-cm) concrete blocks, either side of each young tree. It prevents accidental damaging by kids and beasts!

THE DEEP BED

DIGGING A DEEP BED

1 First mark out the boundaries of your bed. Use all your weight on a sharp spade to cut the turf in two parallel lines right down the length of the first row of the plot before you start to dig.

2 The cut lines should be about 12 inches (30 cm) apart, allowing you now to dig out square lumps of turf 6 inches (15 cm) thick. Pull out deep rooted weeds like docks or thistles.

3 Put these turves (the first "spit") in a row on the plastic or board which you have put close to the edge of your plot. Put the row at the far side of the board so there is enough space for the soil which comes out next.

4 When you have taken out the whole of the top layer of turf (the top "spit") then continue to dig the trench by taking out the next "spit" of soil. Put this on to the board next to the turf.

Put the earth from under the turf on the first trench into a heaped row beside the turf.

Put the turf from the second row grass-side down into the bottom of trench one.

Put the turf from the first trench on to the far edge of your board or plastic sheeting.

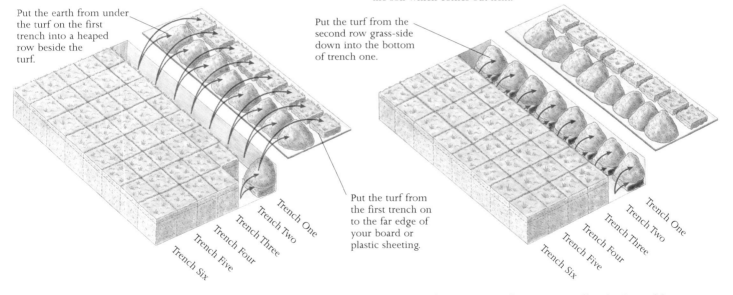

Trench One / Trench Two / Trench Three / Trench Four / Trench Five / Trench Six

5 When you have dug out all the soil from the first row, you begin the second trench by cutting squares of turf as for the first row; but these go (grass face-down) into the trench you have already dug.

6 The second trench is now completed by digging out the next "spit" of soil and turning it on top of the upturned turf in the first row. Chuck any large stones (you may need a pickaxe) into the wheelbarrow.

7 Continue the sequence with each row going upside down into the trench. Add rotted manure under each turf for the ideal result.

8 Finally, take the turf from your first trench and put it face down into the last trench. When you cover this with the soil from the first trench, your bed is complete.

Using the deep bed

Let your new bed settle for at least a week before you get it ready for seeding (depending, of course, on the time of year). Fork over the top layer and smooth it about with your fork or a strong rake. Ideally, you will want to work some good compost into the top layer if it contains a high proportion of subsoil. If you don't do this your germination may be poor. You can avoid this risk if you transplant seedlings from elsewhere and this, in my view, is the best approach. You will find that crops such as onions, lettuce, sweet corn, cucumbers, marrows, and strawberries simply race away, as will the brassicas. However, plants like carrots and beetroot must obviously start from seed.

Sowing & Planting

Some people are said to have "green fingers", meaning that when they plant a seed or a tree, it grows. I suspect that this mysterious power is merely common sense and sympathy – sympathy for the new life that you are helping to nurture. After all, what does a seed want? Moisture, warmth, and soil friable enough for its shoots to grow upwards and its roots downwards. This soil should be in close contact with the seed, and there shouldn't be too much soil between the seed and the light, because the plant's growth depends on the energy collected from the sun by photosynthesis in its green leaves. This energy takes over from the energy stored in the seed when that is exhausted, and helps to protect the plant from its enemies.

Plants vary in their requirements, of course, but broadly speaking, there are two ways to establish vegetables. One is by sowing the seed direct into the ground where it is going to stay. The other is by sowing it somewhere else, and then transplanting in due course. And there are even occasions when we transplant the plants from where we sowed the seed into another bed, leave them there to grow for a while, and then transplant them again into their final bed.

There are two quite sensible reasons for this seemingly laborious and time-consuming procedure. First, by crowding the seed in a seed bed, we release the land that the plants will ultimately take up, and can use it for another, earlier crop. So nearly all our brassicas (cabbage tribe), leeks, and those other plants that will grow through the autumn and possibly part of the winter occupy very little ground for the first half of the summer. Then we put them in ground vacated by earlier crops, such as early spuds or peas, and so we get two crops off the land in one year. The second reason for transplanting is to give seeds a good start.

This is done by sowing seeds in a seed bed, but under glass, plastic, or some other covering. This way, we who live in temperate climates can start them earlier and give them an initial boost, so that they will come to harvest during our short summer. After all, many of our vegetable crops evolved in warmer climes than the ones we grow them in.

Peat pots
There are certain crops which respond far better to being grown in peat pots before they are transplanted, rather than in flats or seed boxes. These are crops which don't like having their roots interfered with. When you plant the peat pot direct in the ground, the roots will simply drive their way through the wet peat and the plant won't suffer. Maize, melons, squashes, and many other semi-hardy plants benefit from this treatment.

Soil for seed boxes
The sort of soil you put in your flats, or seed boxes, or pots, or whatever, is very important. If you just put in ordinary topsoil, it will tend to crack, and dry out, and it will have insects and disease organisms in it that may flourish in the hot air of the greenhouse. This won't give you very good results.

If you can get and afford prepared potting composts, then get them. The expense is justified by results. These composts are carefully blended and well sterilized. If you can't or don't want to buy them, you will have to manufacture potting composts of your own. The fundamental ingredients of the prepared composts are loam, peat, and sand.

SOWING

1 Fork over the ground. Mark out the rows, and stretch a garden line along each one. Drive a drill with a draw hoe at a suitable depth.

2 Sprinkle tiny seeds thinly. Large seeds like peas and beans should be planted at regular intervals, usually recommended by the supplier. Water them gently.

3 When sowing is finished, rake the bed all over, so the entire surface becomes a fine tilth. This top layer of crumbly soil is the most important feature.

4 After you have raked the soil, step it down firmly with your feet or with the base of the rake. This ensures that the seeds are in close contact with the earth.

You can make the loam by digging top quality meadow turves and stacking them, grass-side down, with a sprinkling of good compost or farmyard manure in between each layer of turves. Stack them in six-foot layers, and leave them for six months to a year. The loam should be sterilized. This is best done by passing steam through it: put the loam in any container with holes in the bottom and place it over a vessel of boiling water.

Peat can be bought in bales, or it can be dug from a peat bog, and then sterilized by simply boiling it in water.

Seed compost, for putting in flats or seed boxes, is, by volume: 2 parts sterilized loam; 1 part sterilized peat; 1 part coarse sand. To each bushel (25.5 kg) of the above add 1½ oz of (40 g) superphosphate of lime and ¾ oz (20 g) of ground chalk or ground limestone.

Potting compost is by volume: 7 parts sterilized loam; 3 parts sterilized peat; 2 parts coarse sand. To each bushel (25.4 kg) add ¼ lb (115 g) of base fertilizer and ¾ oz (20 g) ground chalk or limestone.

And base fertilizer is by weight: 2 parts hoof and horn meal; 2 parts superphosphate of lime; 1 part sulphate of potash.

Transplanting

The same qualities are needed for transplanting a growing plant successfully as are needed for sowing seed: sympathy and common sense. Consider what a trauma transplantation must be for a plant, which is a life form evolved for growing all its life in one place. It is wrenched out of the ground, and most of the friendly earth is shaken from its tender roots which themselves are probably severely damaged. Then it is shoved roughly into some alien soil, possibly with much of its root system not in contact with the soil at all and the rest jammed together into a matted ball. It is quite amazing that transplanted seedlings ever survive, let alone grow into mature plants.

So dig plants out gently and be sure that as much soil adheres to their roots as possible. Transplant them gently into friable soil with their roots spread naturally as they were before. Make sure the soil is well firmed, but not roughly trampled so as to break off tender roots. Then water them well. "Puddling" transplants, which means completely saturating them, is nearly always a good idea. It is drying out that kills most transplants. Of course, if we have masses of brassicas to plant, we can't be too particular. Sheer pressure means banging them in pretty quickly, but even then it is surprising how one person will have 100% success with his plantings, while another has many failures.

Putting the plant in

Plant when it is raining, or rain is forecast. Put large plants in with a trowel and smaller ones with a dibber (basically a pointed stick). Farm labourers transplanting thousands of brassicas go along at a slow walking pace, jabbing the dibber in beside the plant and moving the dibber over towards it to jam the earth tight around its roots. If a moderate tug on the plant doesn't pull it out, it will be all right.

With larger or more delicate plants, such as tomatoes or sweet corn, keep a ball of soil on the roots and very carefully place them in a hole dug with a trowel. Then firm the soil around them. If you have grown them in pots, carry them in the pots to their planting station. Water them well, then take them gently out of the pots immediately before you place them in the ground.

PLANTING

   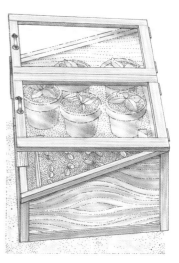

1 Crowd the seeds into a seed box, so that the land where they will eventually grow will be able to bear another early crop in the meantime.

2 Or you can plant your seeds in pots. As the seedlings grow, thin them out to allow the strongest seedlings more room for their roots to develop.

3 When the seedlings of your first seed box look over-crowded, prick them out. This means thinning and removing to a box or bed for room to grow.

4 Give your seeds a good start by putting them in pots or seed boxes under glass. They will grow and thrive earlier than they would if left in the open air.

Growing under Cover

You can get a greenhouse which has an interior like a space module about to make a landing on the moon, with thermostats, propagators, electric fumigators, and hell knows what. But if you buy this sort of equipment you are spending the money that would buy you out-of-season vegetables at the nearest greengrocers for many decades. Consider: is it really worthwhile going to great trouble and expense in order to have some vegetable or fruit ready a fortnight earlier than you would otherwise? If you are growing for market, the answer is yes.

Greenhouse production – or intensive cloche production – for sale is a very sensible and valid way of making the small amount of money that every self-supporter must have to conduct their limited trade with the rest of the world. I write books, one of my neighbours gives piano lessons, another makes wooden articles. If anyone wants to make under-glass cultivation their money-spinning enterprise, then they must get some good specialist books on the subject, which is a very complex one, and requires a great deal of knowledge to make the difference between success and complete, expensive failure.

But unless the self-supporter intends to make greenhouse production a main item of foreign exchange or money earning, only the simplest of greenhouses is justified, with maybe some cold frames, "hot frames", or a variety of cloches. You can buy your greenhouse ready-made or build it yourself. Buying the frames with glass in them is often the best thing, for you can then build your own lean-to greenhouse (*see pp.82–83*).

Cold frames

If you make four low walls and put a pane of glass on them, sloping to face the sun, you have a frame. The walls can be made of wood, bricks, concrete blocks, rammed earth, what you will. The glass must be set in wooden frames so that it can be raised or lowered. Frames are fine for forcing on early lettuces and cabbages, for growing cucumbers later in the summer, or for melons and all sorts of other things. Most of them are too low for tomatoes.

Hot frames

These are much used by the skilful French market gardeners, and are a fine and economical way to force on early plants, but they do need skill. You make a "hot bed". This is a pile of partly-rotted farmyard manure or compost.

The best is stable manure: horse manure mixed with straw, mixed with an equal part of leaves or other composting material so it won't be too hot. Turn this a couple of times until the first intense heat of fermenting has gone off along with the strong smell of ammonia, then lay it down in your frame with a shallower layer of earth on top.

You should manure 2½ feet (75 cm) deep with 12 inches (30 cm) of soil over it. The seed should be put in when the temperature falls to about 80°F (27°C). You can transplant plants into the bed. You would need to do

this in the late winter or early spring, so that as the hot bed cools, the spring advances, and the heat of the sun replaces the heat of the manure, which will be gone by the time you no longer need it. You will then have lovely, well-rotted horse manure. Growing in a hot frame is not as easy as it sounds, but if you get the procedure right it is highly effective. It is sad that it is not used more often. Maybe as heating greenhouses with oil and electricity becomes more expensive, it will be. Of course, you must first find your horse. A well-made compost heap with some activator will work, too.

Cloches

The first cloches were bell-shaped glass bowls, and were much used in France. They were simply inverted over the plants to be forced. These were replaced by continuous cloches, which are tent or bar-shaped glass sheds placed end to end to form long tunnels. These are much cheaper,

HOT FRAME
Enough heat to last from late winter right through spring comes from a thick layer of decomposing manure or compost. Cover with a layer of soil.

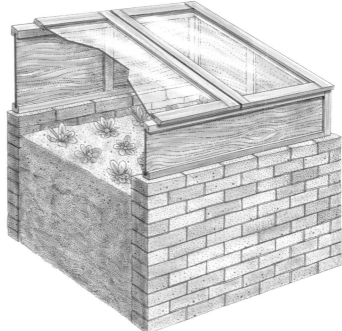

CARDBOARD BOX
A cardboard box painted black absorbs the sun's heat and aids germination.

PLASTIC SHEETING
A transparent plastic sheet will help germination and force on early vegetables.

which is a good thing, because if you are half as clumsy as I am your cloche-managing career will be incessantly punctuated by the merry tinkling sound of breaking glass. If I just look at a glass cloche it falls to pieces, so when you reflect that you have to hoe round crops, hand-weed, water (very necessary for crops under cloches as they do not get the rain), thin, inspect and harvest, and that the cloches have to keep coming off and going back on every time you do one of these operations, you will realize that cloche mortality can be very high.

Polythene tunnels supported by inverted U-shaped wires were the next development. They don't shatter, but can very easily be blown away in a gale, and blown to pieces. However, they do work, and many people use them now; many market gardeners have them on a large scale. Getting them on and off enormously increases the labour involved in growing a crop, but harvesting a fortnight

early may well make the difference between profit and loss. PVC, by the way, retains the heat more efficiently than polythene, but is more expensive. And don't neglect the humble jam-jar. One of these inverted over an early-sown seed or plant of some tender species will protect it as well as any cloche. A sheet of any transparent plastic spread on the ground and weighted down on the edges with earth is fine for forcing on early potatoes and so on. When you do this sort of thing you must be careful to "harden off" the plants sensibly and gradually.

Propagators
You can use a propagator to get very early seeds going. This is an enclosed glass box with soil in it and under-soil electric heating. It produces the condition known as "warm feet but cold head," which many plants like. Tomato seeds can be germinated in one of these in January in a temperate climate, but the air above it must be kept at 45°F (7°C) at least, as well as the soil being warm. A propagator is probably a worthwhile investment if you have electricity – and the time and skill to grow your own tomato plants from seed.

CLOCHES AND A COLD FRAME
A cold frame is 4 walls with glass across them (far right, top). Cloches are portable and there are many types: (left to right) hard plastic cloche; glass barn cloche; soft plastic tunnel cloche; simple corrugated plastic cloche; glass tent cloche. A weighted-down plastic "fleece" (top centre) will also do.

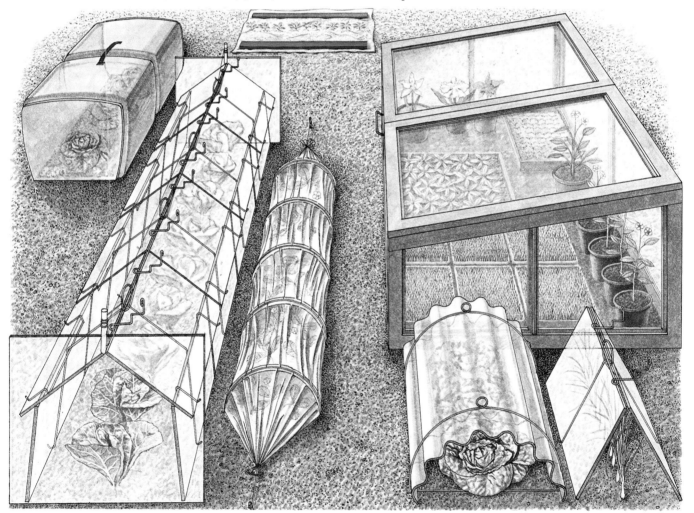

Protecting from Pests

The weeds that grow so merrily in our gardens, in defiance of all our efforts to wipe them out, are tough organisms, and well-adapted to protect themselves from most enemies and diseases. They wouldn't be there otherwise. But our crops have evolved gradually through artificial selection so as to be succulent, good to eat, and productive of high yields. As a result, their natural toughness and immunity against pests and diseases have often been sacrificed to other qualities. We must therefore protect them instead; but avoiding attack by pests and diseases is not so easy. You will always get some losses, but they should not reach serious proportions. An organic farmer I know who farms a thousand acres with never an ounce of chemicals, and whose yields for every crop he grows are well above the national average, says that in his wheat he can show you examples of every wheat disease there is, but never enough of any one for it to make the slightest difference to his yield.

If you observe the principles of good husbandry, by putting plenty of animal manure or compost on the land, and by keeping to strict crop rotations (never grow the same annual crop on a piece of land two years running, and always leave the longest possible gap between two crops of the same plant), you will avoid many troubles. It's also important to encourage a highly diversified floral and faunal environment for balance between species: plenty of predators of various sorts kill the pests before they get out of hand. Destroy all forms of life with poisonous chemicals and you destroy all the predators, too. So when

WORK WITH NATURE, NOT AGAINST IT

Nasturtiums repel cucumber beetle and Mexican bean beetle.

Toads will eat nasties such as slugs, aphids, and mosquitoes.

Thrushes eat snails which would otherwise damage your plants.

Hedgehogs eat pests, including millipedes, that like potatoes.

Mint with its smell keeps white fly off beans.

Lacewings and their larvae destroy aphids.

Centipedes eat slugs' eggs and are the gardener's friend.

Ladybirds aren't just pretty. They consume aphids by the thousand.

you do get a plague of some pest there will be no natural control, and you will be forced to use chemicals again. Still, no matter how organically you farm, there are times when some pest or disease gets the upper hand and something must be done if you are not to lose the crop.

Chemical pest control

Orthodox gardeners will say use poison. You can indeed use some poison, and maybe sometimes you will have to; but surely it is far better and more skilful gardening practice to save your crops without using poison? Any fool can keep disease at bay simply by dousing his crop with chemicals, but what of the effect on other, benign, forms of life? If a chemical is poisonous to one thing you can be certain it

SIMPLE METHODS OF PROTECTION

Young plants and bushes need protection from birds. Four sticks and some soft netting can cage in a growing bush. Cover seedlings with wire netting stretched across hoops, or with a mesh of string wound on wooden pegs.

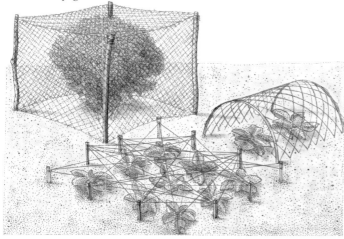

Intercropping works wonders. Carrots and onions, for example, repel each other's enemies.

Strips of sand soaked in paraffin between rows of onions will deter onion fly.

A piece of rhubarb under a brassica seedling frightens off club-root.

Slugs like beer. Trap them by sinking a saucerful in the earth.

will be poisonous to other forms of life too — including human life — and will do damage even if it doesn't kill.

The only chemicals I use are Bordeaux mixture against blight in potatoes, various poisoned baits against slugs, and derris or pyrethrum or a mixture of both against caterpillars and green or black fly. Derris and pyrethrum are both derived from plants, are non-persistent, and are harmless to any non-insect. I have tried calamine (mercuric chloride) against club-root, but it was ineffective.

Biological pest control

Comparatively little research has been done into natural, or biological, means of defence, simply because there is no money to be made out of doing such research. No big company will look into ways of controlling pests and diseases which aren't going to make it any profit (and which will even operate against the profits it already makes by selling poisonous chemicals).

Here are some tips recommended from research into natural controls conducted by Lawrence D. Hills of the Henry Doubleday Association. Many of the findings are merely confirmations of old and tried methods that have been used by countrymen for centuries. The Association also sells a little book called *Organic Pest and Disease Management*, which is very useful (*see also p.301*).
- Tie sacking strips or corrugated cardboard round fruit trees in late summer, and then burn complete with weevils, codling moth grubs, and other nasties.
- Put a very good old-fashioned grease-band around tree boles to catch nasties coming up. Most predators fly.
- Cut off all dead wood from stone fruits in early summer and burn as a guard against silverleaf and die-back.
- Spray winter (tar) wash on fruit trees in winter — only if needed, for it kills useful predators as well as nasties.
- Use plenty of potash to prevent chocolate-spot in beans.
- Grow winter-sown broad beans instead of spring-sown to avoid blackfly. Pick out (and cook and eat!) the tips of broad-bean plants at the first sign of aphid attack.
- Avoid carrot fly by interplanting with onions. The smell of the one is said to "jam" the smell of the other; thus you avoid both carrot and onion fly. Putting sand soaked with paraffin around carrot or onion rows may be more effective.
- Rigorously get rid of every brassica weed (charlock and shepherd's purse) so as not to harbour club-root.
- Drop bits of rhubarb down each hole before you plant out brassica seedlings, or better still, water seed beds and seedlings with rhubarb-water. The smell of the rhubarb is said to deter the club-root organism. This is an old remedy.
- Sink basins full of beer into the ground to trap slugs. Or save the beer and use milk and water instead.

My own experience is that, except for potato blight, if you don't spray occasional plagues of caterpillars on brassicas, and occasional aphid or greenfly and blackfly attacks, there is no need to worry as long as you work with nature, not against it. A few pests on healthy crops do very little harm — certainly not enough to worry about.

Pests, Fungus, & Disease

The mere mention of pests, fungus, and disease can seem like an X-rated exposure to topics best left out of the upbeat miasma of many expositions on organic gardening. Most members of the human race are fixed on the idea that there should be "magic bullets" to cure all ills. You will indeed find many promises of such in most garden centres. In fact, much of their brightly coloured, heavily promoted, and very nasty products are simply offshoots from chemical warfare. Most selective weedkillers, defoliants, insecticides, and fungicides are deadly poison. And it is no accident that statistics show that the use of garden chemicals is closely correlated with increases in rates of serious cancers – especially for children who frolic so often in and around our gardens. Suffice it to say we do not want to pollute our own small havens of the natural world with by-products of chemical warfare or, at the very least, chemicals whose effects we know very little about. So what is our answer?

First, it must be clear that in the natural world there is an enormous and very beautiful variety of healthy plants with all their attendant bugs and fungi. Nature itself does have a multitude of checks and balances which militate against epidemics and vegetative wipeouts. We can try to mirror some of these stratagems in our planting and management.

We want variety. We want to avoid large areas of single-species planting. We want to mix perennials with annuals. We want areas of wilderness close by. We want birds and insects in profusion. We want invertebrates, dragonflies, frogs, and toads. We want a vibrant and healthy soil. We want to avoid planting the same plant in the same place year in, year out. We want to dispose of garden rubbish and keep our boxes and tools clean. We want to limit damage from domestic animals and children. We want to keep out birds and animals much encouraged by human activities elsewhere such as pigeons, deer, and rabbits.

By incorporating all these features into the design and management of our land we can give ourselves a head start in maintaining a good natural balance. And where there is a natural balance, there are no epidemics, just minor damage. Good soil, wise garden layout, effective garden hygiene, and sensible precautions against ineffective composting – these are the number-one priority. Making regular garden bonfires is the number-two priority, and having proper sheds and buildings with easily cleaned concrete floors is number three. Be ruthless about tearing up and composting or burning diseased material, always use good seed, and never try to store anything other than top-quality produce.

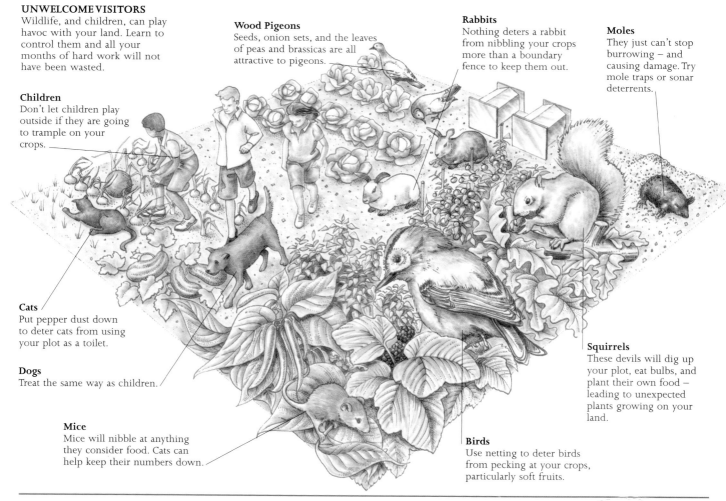

UNWELCOME VISITORS
Wildlife, and children, can play havoc with your land. Learn to control them and all your months of hard work will not have been wasted.

Children
Don't let children play outside if they are going to trample on your crops.

Wood Pigeons
Seeds, onion sets, and the leaves of peas and brassicas are all attractive to pigeons.

Rabbits
Nothing deters a rabbit from nibbling your crops more than a boundary fence to keep them out.

Moles
They just can't stop burrowing – and causing damage. Try mole traps or sonar deterrents.

Cats
Put pepper dust down to deter cats from using your plot as a toilet.

Dogs
Treat the same way as children.

Mice
Mice will nibble at anything they consider food. Cats can help keep their numbers down.

Birds
Use netting to deter birds from pecking at your crops, particularly soft fruits.

Squirrels
These devils will dig up your plot, eat bulbs, and plant their own food – leading to unexpected plants growing on your land.

GOOD PRACTICE

My preference is for a six-year rotation in the vegetable garden. Pests such as nematodes (worms like wireworm and eelworm and many more) can wreak havoc if crops are grown too frequently in the same plot.

Companion planting

The smells and chemicals given off by one plant can effectively mask those given off by another. In certain cases this can be used very effectively to confuse pests – for example, onions and herbs such as rosemary will confuse the carrot fly if planted nearby.

Sacrifice cropping

One way to minimize damage on plants you want to harvest is to plant others that the pests may favour but which you do not want to harvest. For example, slugs love Chinese cabbage in preference to fresh spring lettuce. I even make a dummy seed bed some years to attract the local cats, while I put rabbit wire around my real seed beds on a temporary basis.

FRIENDS TO THE GARDEN

Bees

Not only do bees supply us with copious quantities of honey, but they also provide a vital service in pollinating our garden crops. Make sure you have a healthy bee population in and around your garden.

Birds

Although some birds cause us trouble by robbing fruit and nibbling vegetables, there are many which eat insects, so we would be wise to encourage these. Blackbirds, robins, tits, starlings, and thrushes constantly search out insects, caterpillars, and slugs. You can encourage birds by feeding in winter and by providing water in summer.

Centipede

These are fast-moving little hunters who must not be confused with the millipede, which is slow and has more legs. Centipedes eat mites, bugs, and slugs.

Frogs and Toads

These require water for breeding and are extremely sensitive to poisons. Both eat insects in large quantities. They will travel all over your garden, provided you have areas of long grass or shade for them to lurk in.

Ground Beetles

There is a large family of beetles, mostly large black ones, that live on the ground and eat other insect pests, including the root fly larvae. These beetles need cover to survive the winter, hence the importance of perennial crops and hedges.

Hedgehogs

This little fellow is the great consumer of slugs. Encourage him if you can with a bowl of bread and milk. He will need rough hedges and leaves to make his winter hibernation bed.

Hover Fly

Looking like small wasps, these pretty insects have larvae with a voracious appetite for aphids.

The larvae are nasty-looking little creatures with a pair of nippers to hold their aphid prey. Encourage hover flies by growing flowers in your vegetable garden – they seem particularly fond of yellow flowers.

Lacewing

These delicate insects so beloved by trout are encouraged by water. Their larvae, which are flattish with a brown body, have a great appetite for aphids.

Ladybirds

The famous ladybird is the best-known predator for aphids. The adult and larvae both eat aphids. The larvae are grey in colour and are very active killers.

Worms

Earthworms and manure worms are the two species of worm that we most want to see in our gardens. Earthworms are the great soil builders, constantly aerating the soil and bringing new soil to the surface as worm castes and dragging organic material down. If you have ever wondered why Roman remains are buried under several feet of earth, you only need to watch what happens on your garden lawn. I have taken soil plugs from my lawn where you can see clearly the depth of various dressings I have applied over the years. It is as plain as a pikestaff that these layers get deeper by about one inch (2.5 cm) every 10 years. They get deeper as the worms push out new soil above them – so 1,000 years of worm activity would represent about 8 feet (2.5 m) of earth! Yes, it really does happen – worms are constantly digging under those old ruins and pushing up new soil all around them.

Manure worms, on the other hand, are thinner and redder than earth worms. They rapidly turn organic material into compost and are the essential basis of worm composting. Buy them from fishing shops as "brandlings" and you can build your own worm composter very easily. Manure worms like warm, wet places and will avoid material that is drying out.

ANIMAL PESTS

Birds

Many birds will dig up seeds, pull out onion sets, eat emerging leaves of peas and brassicas, and steal your soft fruits. A cat or two may well help keep damage down, but the cat itself will do damage of its own kind. A tethered hawk, or falcon is much more effective – especially if flown regularly over the garden. Birds definitely stay well clear of anywhere frequented by birds of prey – this really does work. I used to fly a falcon regularly over my vineyards and never a bird was to be seen – even though the bird only flew for 20 minutes each day. Physical protection is an effective remedy against bird damage.

You can use cloches over seed beds or, even easier, make yourself a few small mesh wire frames about 6 feet by 1 foot long (2 m x 30 cm). It is easy to do this using roofing battens and rabbit wire. Not only will this protect your seed bed from birds, but it will also keep off cats, children, and white butterflies (if the mesh is small enough). Such wire frames are a godsend. They are cheap, effective, and easy to move to the next seed bed once your seedlings are established.

For peas I always use rabbit wire as a support for the vines because this also seriously discourages birds, who do not like their flight paths and escape routes being interfered with.

Fruit cages will bring complete protection to your soft fruits. In some bad areas this may be necessary, but I have always found that we have ample crops to afford a bit extra to feed the birds.

Cats

Cats simply adore a freshly prepared seed bed. Rolling, scratching, and pooping are the order of the day. A rudimentary temporary rabbit wire fence will keep them off, as will the wire frames mentioned earlier. Sometimes a sacrificial seed bed (without the seeds!) may be one way to distract them from the real thing. Having your own dog is the best way to keep neighbours' cats away. If your dog does not do the trick, I have tried various other remedies – even resorting to scattering red pepper about the place on one occasion. In the end I have found nothing works better than a good jet of cold water. You can either keep a water pistol handy or simply use an old washing up bottle full of water. The more sudden the shock, the greater the effect – you can always make up with your pet later and no bones will be broken!

Deer

Deer are becoming an increasing nuisance. They are very agile and hard to stop with fences, although they will avoid fences if there is a choice. Deer particularly love to eat and destroy young trees – cedar trees seem to be a particular favourite. Plastic tree protectors will be the best option. Note that some young trees, particularly poplar, are very prone to being killed by mice gnawing the bark in a hard winter. Even plastic tree protectors may not prevent this. Cats and traps are your only hope.

Dogs

Do not allow dogs to get into the habit of playing in your garden. A garden hose or catapult can keep unwanted dogs at bay. You will have to train your family and guests not to play with your own dog around the vegetables. If dogs feel comfortable in your vegetable garden they will dig, scratch, and poop. Keep them out if you can.

Children

Kids love the garden and they love working with grown ups. Even the simplest of simple garden jobs will keep them happy and out of mischief – and that applies to other people's kids as well as your own. Small wheelbarrows and tools are great for this. Fast-germinating seeds like radish and sunflower will almost keep pace with a child's eagerness to see quick results. And for hunting and destroying caterpillars there can be no better weapon than a well motivated three-year-old. Children particularly love harvesting. And there is always the woodpile to be sorted and tidied up if the worst comes to the worst. Involve your kids in the gardening work and you will find they love it and soak up knowledge and experience in the most remarkable way. Chickens and other farmyard animals are the child's best friend – providing all sorts of duties (not least collecting the eggs) – but warn them about potentially aggressive cocks and geese.

And somehow children seem to understand the circle of nature, including the butchering of domestic animals, without a great wad of sentimentality. They apparently can accept their place at the top of the food chain without problem.

Mice

Not only do mice destroy stored food, they also rob pea and bean seeds very effectively. Keep your mice population down with cats or, alternatively, use traps. Soaking the seeds in paraffin or pre-germinating them before planting will also help.

Moles

Moles love good soil (and live off all the fat worms that go with it). If you get moles in your garden, then set traps immediately. Setting a mole trap is a strange black art. The trap must be left outside in the soil for at least a month before you use it. Some experts say you should rub the traps onto the skins of any other mole you have caught. But the vital thing is that the trap should not be touched by human hand or contaminated by oil or grease. The mole lives through his nose and will immediately be wary of a strange smell. Dig the trap into his run and hope you catch him.

Rabbits and Hares

Keep rabbits and hares out of your garden by making sure you have a good boundary fence or an active dog. A good boundary fence that is rabbit-proof is in any case a good investment – it will keep your chickens, kids, and dogs in, too!

Squirrels

These agile pests can do a great deal of damage. Netting is the only cure, although cats, dogs, and hawks will obviously help.

INSECT PESTS

Ants

These little eager beavers of the insect world can be a nuisance in the greenhouse and where the soil is light and dry. Their workings can disturb the soil around your plants – and don't forget they are also experts at cultivating aphids to harvest the sap of your vegetables and fruit. Unless you are unlucky or there is a very hot summer, ant infestations are unlikely to cause you much trouble.

We have most trouble with the red ants, which get into your shoes and bite like hell. My advice: wear long boots when scything or working in ant areas. Ants can be killed by boiling water or, if necessary, by a solution of equal parts of borax and icing sugar.

Aphids

Aphids come in all shorts of shapes, colours and sizes and affect a huge range of plants. Aphid damage will be limited if your plants are strong and healthy. Believe it or not, these plants will almost instantly create their own chemical defences to limit the attacks.

If your plants are weak and your garden has few natural aphid predators, then you will have to resort to other control methods. Soft soap mixed with warm water and then sprayed on will reduce small infestations. Alternatively, you can brew up your own homemade deterrents, such as boiling up rhubarb leaves or wormwood with water.

The most lethal solution can be made from the deadly poison nicotine, which you can extract from old cigarette ends immersed in hot water.

Beetles

There are a range of beetles that are harmful to the garden – mostly through the appetites of their larvae. The most common are the chafer beetle and the flea beetle. Chafer beetles feed mainly on the leaves, flowers, and fruits of plants, whilst the grub destroys the roots, making the leaves wilt and the plant die. Continuous cultivation and good weed control are the main preventive measures. As you cultivate you may see and be able to gather up the grubs you find near affected plants.

Flea beetles eat small, round holes in the leaves of plants, while their grubs tunnel into stems and eat roots. Control the beetle by making sure the soil is kept clear of debris in the winter to deprive them of wintering quarters. Keep seedlings growing fast in dry weather by watering. Dust seedlings with calcified seaweed, especially when the dew is still on the leaves in the early morning.

Big Bud Mite

This small mite attacks blackcurrants and sometimes red or white currants. As the name implies it creates "big buds" in early spring – these contain the mites and can be broken off and burned. You can spray with a solution of lime sulphur, but only when the buds have opened at the time when the plants are just showing full leaf. Use 1 pint (0.5 l) of lime sulphur to 7 gallons (4.5 l) of water and spray every 3 weeks.

Capsid Bugs

Similar to aphids but smaller and more virulent, capsid bugs cause small brown patches to appear on leaves, making them distorted and causing them to fall off. Keep the bottoms of your fruit trees well cleaned out during winter and replace mulches to prevent overwintering. Winter tar washes are helpful but the only effective summer control is by spraying with nicotine.

Caterpillars

The white butterfly is the great creator of caterpillars – and is a pretty but unwelcome guest in our summer gardens. The main control is to net seedlings with lightweight wire frames. Then, when plants are too large for the frames, remove the caterpillars by hand each day and destroy. If plants are really badly affected, they are better destroyed than simply abandoned.

Gooseberry Sawfly

This common pest will eat all the gooseberry leaves during and after fruiting. You can limit the damage by spraying with a soft soap solution but, in practice, the yield of healthy plants will be scarcely affected. Remember to keep the bases of plants clean in winter.

Pea Moth

The larvae tunnel into the peas in their pods and ruin them. Early or late varieties of pea are less affected. Once an attack has started there is no cure. The best control is good cultivation. Hoeing the ground after the peas have been lifted will encourage birds to eat the cocooned caterpillars.

Cutworm

Cutworms are the caterpillars of any of several moths, which feed at night on the stems of young plants, cutting them off at the ground. They attack the stems of seedlings at ground level. The damage is similar to slugs but without the slime. Keep weeds down to discourage cutworms. You can collect them by hand and then destroy if attacks are serious.

Eelworms

These nematode worms are very small, even microscopic, but they breed in immense numbers and can do considerable damage to a range of crops. Prevention by good rotation and hygiene is the best remedy. Always buy your plants from reputable suppliers. Affected leaves may show brown discolorations between the veins. There is no really effective cure, although affected leaves and stems may be removed and burned.

Potato eelworm is a major problem if good rotation is not practised. Leaves turn yellow before dying off, and when lifted the tubers will be marble-sized. The roots are attacked, creating small white or yellow cysts that can be seen with a magnifying glass. Each crop of potatoes should be at least 6 years apart on the same piece of land. Affected plants should be burned.

Root knot eelworm affects a range of plants, causing deformities in the roots and small galls, which may run together and cause general swelling. The affected plants die and there is no cure – plants must be burned and the same crop not planted in that area for at least 6 years.

Rootfly

There are a number of important plants which are attacked by their own particular rootfly. Cabbage and carrot are the most common. Affected plants will discolour and wilt before they die. Cure is by prevention. Attack is most likely after cabbage seedlings have been transplanted. Make sure these are well-watered and earthed up.

In carrots, the attacks are encouraged by even the minutest damage to plants when thinning or weeding. Great care should be taken to remove all thinnings immediately – thin when the ground is damp and make sure the remaining plants are well covered with soil at ground level.

If your carrots are continually affected, cover the whole bed with a proprietary fine-mesh netting, burying the edges carefully to prevent the flies from gaining access. Once affected, there is no cure. Be careful to avoid storing any carrots that may have rootfly damage and do not leave any damaged carrots lying about for the flies to multiply.

Red Spider Mites

These little mites look (under a magnifying glass) as their name suggests, and the main types attack greenhouse plants and fruit trees. Mites and larvae live on sap from leaves, which first have yellow spots and then turn brown.

Keep the greenhouse plants cool and moist to avoid attack. Do this by spraying with water twice daily. There are biological controls available which are used commercially in greenhouses. With fruit trees control is difficult – remove and burn affected leaves.

Slugs

These grey, black, and orange plant munchers are the gardener's constant companions. Keep weeds and grass controlled to minimize their habitat — good garden hygiene plays a major role in prevention. Slugs can devastate your vegetables and have particular preferences, such as runner beans and basil. You can use piles of old leaves as hiding places which you can pick up and compost — old newspaper or planking placed on grass will also attract slugs. You can prepare slug traps using jamjars set into the earth and filled with sweetened beer and water. Or, finally, as a last resort, you can go out at night with torches and collect the little blighters.

Dressing your soil with calcified seaweed or fresh wood ash will also limit attacks, as slugs dislike travelling over sharp dry material. Soot also works well but it will damage plants if it touches them.

Wasps

These are another of the gardener's love-hate insects. In early summer wasps hunt a large range of insects and their larvae consume thousands of aphids and so help the garden. But, as the summer matures, the wasps develop a longing for sweet food as they begin to transform into queens for their winter hibernation. The only effective control is to find and destroy nests but, on the whole, I believe the wasp is an insect we have to learn to live with. In most years damage to fruit is not too bad — just watch out when you are picking plums so that you do not get stung into the bargain.

Weevils

The larvae of weevil beetles attack a wide range of plants — particularly apple blossom, peas, beans, and turnip. In the orchard the pest can be controlled by catching the pupating larvae in bands of sacking around the trunks in June — then burning later. Keeping the soil clean, free from debris, and full of fertility will also reduce damage, as will good crop rotation.

Wireworms

These destructive worms inhabit grassland and can be very destructive when it is first cultivated. They are the larvae of the click beetle and live as brown wiry worms about 1 inch (2.5 cm) long for 4 or 5 years before pupating. They are particularly destructive in potatoes in late summer. Newly cultivated grassland may be affected for up to 3 years — so you can plant resistant crops such as peas and beans. You can catch wireworms and destroy them by punching holes in a tin can, filling it with potato peelings, and burying it in the earth. Lift the can every 3 or 4 days and then tip out the peelings into a bucket of water to drown the wireworms. A green manure of mustard dug into the soil can help reduce the pest, but there are no effective sprays.

DISEASE
Blight

This is the most serious fungal disease of potatoes (and tomatoes). Once affected, there is no cure. Blight needs warm humid weather with little wind in order to develop — in some countries you will hear blight warnings on the radio. Dark patches develop on leaves.

Underneath a white mould develops. Prevention by spraying with Bordeaux mixture is the only remedy. Spray both top and bottom of leaves and do so every 2 weeks, starting in late June.

Canker

The bark on pear and apple trees may shrink and crack and a deep, swollen wound develops. In the winter, small red growths cover the affected area. In the worst cases, whole trees may be destroyed. The fungus attacks trees where drainage is poor. All affected branches must be cut off and burned. Make sure that cuts are clean, with no damage to surrounding bark that could lead to re-infection.

Club root

This is the most serious of the diseases affecting brassicas (Brussels sprouts and the cabbage family). Roots become swollen (and smell foul) and the plants wilt. The disease organisms lie in the soil — sometimes for over 20 years. They can easily be imported into your garden on bought-in seedlings, so try to grow your own plants from seed. There is no cure so good crop rotation is a must. All affected plants must be burned not composted.

Damping Off

This fungus affects seedlings where the stems turn black and rot. Over-watering and overcrowding, combined with a lack of ventilation, can cause this problem. Also, be careful not to touch stems or roots as you transplant small seedlings — hold them by the seed leaves only.

Fire Blight

This is a deadly disease of pears and sometimes apple trees. Affected leaves turn dark brown but do not drop off — whole branches may be affected. In some countries you must notify government authorities if this disease is contracted and control must be carried out under their supervision. Trees must be uprooted and burned.

Mildews

There are many fungal mildews — some occur in damp and overcrowded conditions (downy mildew), others in hot, dry, and overcrowded conditions on light soils (powdery mildews). Essentially, these are different moulds. Bordeaux mixture will offer protection but avoiding poor conditions is the priority. Seriously affected plants should be burned.

Rusts

There are a huge range of fungal rusts affecting plants. In most cases strong plants will be no more than discoloured with brown or yellow rusty spots. Bordeaux mixture will control but not cure. Badly affected plants should be burned.

Scab

Common scab affects potatoes, beetroot, radish, and swedes. It is not a very serious disease, simply causing "scabs" on the surface of these vegetables. Scab is most common on newly dug soils, where soils have been heavily limed, or where there is a dry summer. There are resistant varieties of potato that will reduce the damage.

More important, you can reduce the alkalinity of the soil (you can buy a simple test kit for testing your soil pH values). The prime method for achieving this alkalinity is to add more organic material — you can even dig trenches and grow the potatoes directly on a bed of compost, peat or well-rotted manure.

PLANNING FOR PREVENTION

If we do get diseases, pests, or fungi then it pays us to learn from other experience in finding ways to deal with them. Always remember that every garden and every location will have different characteristics, and measures that work in one place will not necessarily work in another. Use your own imagination and intuition in harness with the knowledge and experience of others.

At the beginning, you will either be laying out a garden or smallholding from scratch or you will be inheriting an existing layout. Two factors are immediately relevant. First, you may be working with land where there have been heavy applications of poisons in the past. If this is the case it can take quite a few years for nature to settle down again — and you will have to be patient. Second, you must review the design and layout of your land. We have already touched upon this and some of the basic essentials in other parts of the book, but the main points bear repeating again:

Variety — we do not want great big plots with uniform cropping, which lends itself to the spread of pests or disease. Smaller fields or plots will help minimize the dangers.

Mix Perennials and Annuals — we want to arrange our plantings so there is a mix of perennials and annuals. Perennials provide excellent winter shelter for valuable black beetles and other insect predators. I like to think we have no annuals in the garden which are more than 15 feet (4.5 m) from a perennial.

Build up your Soil — A healthy soil with plenty of compost added annually is the best guarantee of strong, healthy plants. Regular cultivation of your soil, especially by rotavating, will also discourage the breeding of many pests. Good organic soil conditioners such as seaweed, dried blood, and bonemeal will also help.

Fencing and Netting — A good boundary fence or wall is an invaluable defence against four-legged pests. Use netting to protect against bird and butterfly damage.

Encourage Helpful Beasts — Birds, beetles, bugs, bees, and many other insects and animals can be your greatest allies. Even chickens and ducks will help keep pests at bay in the orchard. Your own well-behaved dog can keep away the neighbours' marauding cats, whose scraping, defecating, and rolling will devastate your seed beds. (Make sure no one is allowed to throw sticks for your dog in the garden — smashed plants and mangled seed beds are sure to follow.)

As we have seen, pests and diseases can provide a constant challenge but, despite all the above advice, the best effects are produced if you ensure GGH — Good Garden Hygiene.

Vegetables

If you grow just a few of the vegetables listed below, you can eat your own fresh produce from early spring to late autumn. And, if you grow the right things and store them properly (*see* p.196), or if you set yourself up with a greenhouse (*see* p.82), you can have your own vegetables year-round and need never again suffer a flabby lettuce or a tasteless tomato.

ARTICHOKES
Globe artichokes
Use Globe artichokes are perennials and so are a long-term proposition. I would not recommend them as the crop to feed a hungry world, but the object of the self-supporter should be to live a rich and varied life, and part of this must be a rich and varied diet. Basically, globe artichokes are huge thistles, and we eat their flower heads – and not even all of these; just the little bit at the base of each prickly petal and the heart, which lies under the tuft of prickles that are immature petals. Globe artichokes are delicious beyond description. Just boil the whole flower head, pull off the petals, and eat with just warmed butter or an oil and vinegar mix.

Sowing In spring, plant the artichoke suckers from an existing plant, each with a piece of heel of the old plant attached. Plant them 4 inches (10 cm) deep in good well-manured, well-drained soil at 3 feet (1 m) intervals.
Aftercare Keep them well hoed.
Harvesting Spare the artichokes the first year but pluck the heads the second and each ensuing year. After 5 or 6 years, dig out and plant a new row somewhere else. If you plant a new row every year and scrap an old one, you will never have a year without artichokes to eat. Muck well every year and cover in winter with a thick mulch of straw.

Cardoons
Use Cardoons are exactly like globe artichokes, except that they are annuals and they have slightly smaller flowers. The flowers can be eaten like globe artichoke flowers, but cardoons are really grown for earthing up in the spring so that you can eat the stems like celery.

Planting Plant from seed, which is generally sown in a greenhouse in spring, and then prick out a fortnight after the last frost.

Jerusalem artichokes
Use These are a useful standby in winter as a substitute for potatoes. They can be lightly boiled or fried in slices. Despite their name, they have absolutely nothing to do with globe artichokes, or Jerusalem.
Planting Just like potatoes, the Jerusalem artichoke grows from tubers planted in early spring. They are very easy to grow, and need only a little extra lime if your soil requires it. They are rarely attacked by pests.
Aftercare Hoe until the foliage is dense enough to suppress weeds.
Harvesting Dig them up as you need them. They can be left in the ground throughout the winter. Save a few mature tubers to plant next year.

ASPARAGUS
Use Asparagus are perennial vegetables, so once they are planted they can't be moved each year. They take three years to get established, but it is well worth waiting. They come very early – just when you need them – and are delicious and nutritious, perhaps one of the most valuable crops you can grow. Do not be put off by any puritanical ideas that as a luxury crop they are somehow sinful. They are delicious, nourishing, and come just when you don't have anything else.

Soil They like a deep, light loamy fertile soil, which must, above all, be well drained. They will grow on sand as long as it has plenty of muck.

Make absolutely sure there are no perennial weeds in your future asparagus bed: couch grass or ground-elder can ruin a bed because they cannot be eliminated once the asparagus begins to grow. The roots will get inextricably intertwined. People always used to have raised beds for asparagus, but nowadays some people plant in single or double rows. It doesn't really matter. I like a raised bed with three rows of plants, and as the years go on the bed tends to get higher because I put so much stuff on it. It is a good plan to cover it thickly with seaweed in the autumn. If it hasn't rotted down by spring, take it off and compost it.
Sowing Muck really heavily in the autumn; in spring buy or beg three-year-old plants and plant them 18 inches (45 cm) apart, measuring from their middles. The plants look like large spiders. Don't let them dry out before you plant them, and most importantly, pile a few inches of soil on top of them. Make sure the soil does not dry out, and keep weeding the bed. Don't let any weed live.
Treatment Don't cut any asparagus the first year – not a single stick. In late autumn, cut the ferns down to the ground and muck well again. The following spring you can feed well with fish meal, mature chicken dung, seaweed, or salt (yes salt – asparagus is a seashore-plant), and weed again. That second year you may just have a feed or two, but delay cutting until June. Muck again in the late autumn, and feed again in the spring.
Harvesting The third year when the shoots look like asparagus shoots, cut them just below the ground. You can cut away fresh asparagus ready to eat every two or three days. They soon shoot up again, and you can go on cutting until the third week in June and then stop. No more cutting. By then you will have plenty of other green things to eat, anyway. Let the tall ferns come up again, cut them down in the late autumn to confound the asparagus beetle by destroying their eggs. Then, muck them or feed them, or both, for these are lime- and phosphate-hungry plants and like plenty of humus.

AUBERGINE OR EGGPLANT
Use A delightful Mediterranean food, aubergines have an exotic taste and can be used for stews, ratatouille, and Greek moussaka, or grilled on their own.
Sowing Sow aubergine seeds indoors in early spring. Sow them in compost and try to keep the

temperature close to 60°F (16°C). Pot out into peat pots or soil blocks about a month later.

Planting Plant them out in the open in early summer. Protect them with cloches if you live in a cool climate. When you plant them, pinch out the growing points to make them branch. Or you can sow seeds out in your garden under cloches in late spring and you will get a late crop.
Harvesting Pick them when they are deep purple and glossy. Pick before frost sets in.

BEETROOT
Use Beetroot is a very rich source of betain, which is one of the B vitamins. Beetroot therefore keeps you healthy, particularly if you grate it and eat it raw, but it tastes a hell of a lot better cooked, although tiny immature beet are good raw.

Soil Beetroot likes light, deep loam, but most soil will do.
Treatment It doesn't like freshly manured land, and wants a good fine seed bed.
Sowing In early summer sow your beetroot maincrop very thinly; you only need a couple of seeds every 6 inches (15 cm). The seeds are multiple ones and you will have to thin anyway. Sow 1 inch (2.5 cm) deep in rows 1 foot (30 cm) apart.
Aftercare Thin and hoe. You can eat the thinnings raw in salads.

Harvesting You can leave the beetroot in the ground until they are needed or until the heavy frosts set in. Alternatively, you can lift them in autumn. Twist (don't cut) the tops off, and do not cut too near the roots or they will bleed. Clamp or store in sand in a cool cellar.

BROAD BEANS
Use You can pick off the tops of autumn-sown broad beans and cook them. You can eat the seeds when they are green, which is their main use, or you can dry them for winter. It is best to rub the skins off winter-dried beans to make them more tender.

Soil They will grow in most soil.
Treatment Treat the same as you would peas (see below). Lime well and use plenty of mulch.
Sowing I like to sow broad beans in late autumn, but then the climate here is fairly moderate. If the weather got intensely cold it might nip them off, but it never has yet, although once or twice in severe frost they have looked pretty sick. They perk up again, though. If your winter is too severe, or if you haven't been able to get seed in the autumn, you can plant in early spring on light, well-drained soil. The later you sow, the more trouble you are likely to have from black-fly. Sow 3 inches (8 cm) deep, each seed 8 inches (20 cm) from the next, in two rows that are 8 inches (20 cm) apart. Common sense will tell you to stagger the seed in the rows. Each pair of rows should be at least 30 inches (75 cm) away from the next pair.
Aftercare In the spring, just as soon as the black-fly attack (as they inevitably will), pick the tender tops off and eat them. Hoe, of course.

Harvesting Pick them as they are ready. Go on picking as hard as you can, and dry any that are left after the summer.

BROCCOLI
Hearting broccoli or winter cauliflower
Use Hearting broccoli are like cauliflower. They are a damned good winter and early spring standby, and you can have heads from late summer one year until early summer the next if you plant successionally and use a number of different varieties.

Soil They like good, heavy, firm soil, but will grow in most soils as long as it is well manured.
Treatment Like all brassica, broccoli needs lime and doesn't like acid soil. It likes deeply cultivated, but very firm soil.
Sowing Start sowing in seed beds in late spring and go on for 4 or 5 weeks.
Planting Plant out as soon as the plants are ready and you have the ground. Seedlings are ready when they are a few inches high and have made at least four leaves. Plant 2 feet (60 cm) apart in rows 30 inches (75 cm) apart.
Aftercare Hoe regularly until the weeds stop growing in the autumn.
Harvesting Autumn varieties can be cut in September and October; winter varieties from January to March; spring varieties up to April. To get late heads, protect the curds (the white cauliflower heads) by bending leaves over them. Always cut when ripe and don't boil, just steam lightly. (Don't "steam launder" any brassica, as hospital kitchens and schools do. This boils the life out of them.) Steam lightly until soft but still firm.

Sprouting broccoli, purple or green.
Use These are quite different from hearting broccoli. Purple sprouting broccoli is very hardy and therefore the great standby in late winter and early spring when there is not much else about. Green broccoli or calabrese is a delicious vegetable for autumn use.
Treatment This is the same as for hearting broccoli (see above) except that green broccoli is planted in mid-summer. You pick and eat the purple or green shoots when they appear. Don't pick the leaves until the very last moment, and then eat them, too.

BRUSSELS SPROUTS
Use Brussels sprouts are the most useful and delicious winter green vegetable. You simply cannot have too many of them.

Soil Sprouts like deeply worked rich loam, but they will give a crop when planted in most soils as long as it is deeply worked and made very firm.
Treatment Put on compost or muck in the previous autumn, or plant after a well-mucked crop. If your soil is lime deficient, plant after a limed crop.
Sowing Sow out in the open in seed beds during early spring. If you want late sprouts, sow again in a few weeks' time.
Planting Plant them out in early summer in rows that are 3 feet (90 cm) apart. It is useful, especially in windy places, to give each plant a stake so that it can be supported and kept straight as it grows taller.

Aftercare Hoe when required. "Intercrop" (plant in alternate rows), if you like, with lettuce or another quick-growing catch-crop, because the spaces are wide. Keep free of slugs and caterpillars. If you didn't stake the plants in spring, in autumn earth up the stems to give support and to encourage the growth of new roots.
Harvesting Early sprouts are ready in late summer but look on them, that is, if you live in a reasonably temperate climate, as a winter standby. Christmas dinner without sprouts is a travesty and they should keep you going until spring. Pick off the leaves only after they have gone yellow. Use the tops of the plants after you have picked the sprouts.

CABBAGE
Use Cabbage is the most reliable of all the brassicas. It is not fussy about soil and treatment, yields a heavy crop per acre and some varieties can be stored in clamp, cellar, or sauerkraut vat. What we would do without cabbages I cannot think. There are three sorts: spring, summer and autumn, and winter.

Spring cabbage
Soil Light soil is ideal.
Treatment They like fertile soil which is not acid, and it needn't be particularly firm.
Sowing Sow during the summer in a seed bed.
Planting Plant spring cabbage in autumn, 12 inches (30 cm) apart in rows 18 inches (45 cm) apart.
Aftercare Hoe regularly, and top-dress with nitrogen if you think they need it.
Harvesting Use as spring greens in the hungry gap – early spring – or leave a few to heart for eating in late spring and early summer.

Summer and autumn cabbage
Soil They are not very fussy.
Treatment *See* Spring cabbage.
Sowing Sow in late winter in a cold frame or outdoors in spring.
Planting Plant a few where there is room in early summer.
Aftercare *See* Spring cabbage.
Harvesting You don't need many cabbages in summer, but pick when you feel like a change.

Winter cabbage
Soil They like a heavy loam.
Treatment *See* Spring cabbage.
Sowing Sow in seed beds in April and May.
Planting Plant 2 feet (60 cm) apart in rows also 2 feet (60 cm) apart in mid-summer.
Aftercare Hoe regularly. Don't bother to top-dress.
Harvesting Where winters are not too severe, leave them in the ground until you want them. Where there's lots of snow and ice, cut in autumn and clamp or make sauerkraut.

Red cabbage
Treat the same as winter cabbage. Pickle or cook in oil and vinegar with spices. Cook it for some time because it is tough stuff.

CARROTS
Use Carrots have more vitamin A than anything else we are likely to grow. During World War II, it was put about that the uncanny success of British night fighter pilots was due to their huge consumption of carrots, which helped them see in the dark. In fact, it was all due to radar, which the Germans knew nothing about.

Carrots store well through the winter and are a most useful source of good food for the self-supporter. They can be eaten raw in salads or cooked with absolutely anything.

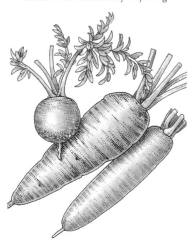

Soil Carrots like a deep, well-cultivated sandy loam. They grow well in very light soil, almost sand in fact.
Treatment Like most roots they fork if planted in soil which has recently been heavily manured with muck or compost, although well-matured compost doesn't seem to affect them so much. Shakespeare compared Man to a forked carrot. So don't plant them after fresh muck. They don't like sour ground (a pH of about 6 is fine). The land must have been deeply dug and then worked down into a fine tilth.
Sowing There is no point in sowing carrots until the ground is dry and warm, say, in the late spring. Sow very shallowly, as thinly as you can, and tamp down rows with the back of the rake afterwards. Some people sow a few radishes in with them to show where the rows are before the slower carrots emerge. Then they pull the radishes for eating when they are ready. Some people intercrop with onions, in the belief (and I support this) that the carrot fly is put off by the onions, and the onion fly is put off by the carrots.
Aftercare If you sow in dry weather, it is good to water the rows to start germination. Hoe frequently and carefully so as not to damage the carrots, and hand-weed as well. Suffer no weeds to exist in your carrot rows. To get a heavy crop, thin to about 3 inches (8 cm) apart, then harvest every other carrot so as to leave them 6 inches (15 cm) apart. This is best for big, tough carrots for winter storing. But, for summer and autumn use, don't bother to thin at all. When you do thin, try to do it when it is raining (to thwart the carrot fly), or if it's not raining, sprinkle derris dust around the plants. After thinning, draw the soil around the plants and then tamp down so the scent of bruised carrots will not attract that beastly carrot fly.
Harvesting Pull them young and tender whenever you feel like it. Lift the main crop with a fork before the first severe frost of winter, and store in sand in a cool place such as a root cellar. You can clamp (*see* p.198) them but they sometimes go rotten in the clamp. Washed carrots won't keep at all whatever you do. They rot almost immediately.

CAULIFLOWER
Use Eat your cauliflower in summer and autumn, as hearting broccoli are apt to take over during the winter. Cauliflowers yield well but in order to grow them successfully, you need skill and good land. They are not a beginner's crop.

Soil Cauliflowers want deep, well-drained, well-cultivated soil that has been well manured, and given ample water. They won't grow on bad land or under bad conditions.

Treatment They must have non-acid conditions, like all brassicas, which means you must lime if necessary. A fortnight before planting, fork on or harrow in a good dressing of fish manure or the like. They also need some potash.
Sowing Sow under cold glass in September or sow in a warm greenhouse in January or February. Sow outdoors in late spring. Plant 24 inches (60 cm) apart in rows 30 inches (75 cm) apart.
Planting Autumn- and winter-sown plants go out in spring, spring sown ones in summer.
Aftercare Hoe, of course. Ensure there is always plenty of moisture, as they can't withstand drought. Top-dress with nitrogen if you have any. Keep them moving – in other words, don't let them stop growing.
Harvesting Cut them when they are ready, early in the morning if possible. Don't boil them to death. They are nice boiled and then dipped in batter, fried, and eaten cold.

CELERIAC
Use You can grate the big swollen roots and eat them raw. Or you can peel and boil, or boil and then fry.

Sowing Sow, prick out, and plant out just like white celery.
Aftercare When you hoe, draw the

soil away from the plants instead of earthing them up as you would for celery.
Harvesting Begin harvesting in autumn. Earth them up in the middle of November for protection against winter frosts.

CELERY
White celery
Use It is said that celery is best after the first frost has been on it. If you are lucky it will keep going until a few weeks after Christmas, provided that you ridge it well. Celery is a most delicious and useful winter vegetable, whether eaten raw, as the blanched stems should be, or cooked in stews as the tops should be.

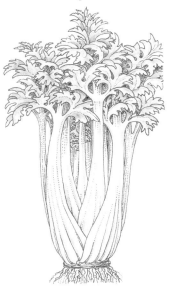

Soil It loves deep, fertile soil that is very moist but not swampy. The best celery is grown in soil that is high in organic matter and retains moisture. Don't let the soil dry out.
Treatment Celery prefers acid to alkaline conditions, so never give it lime. It needs plenty of humus so dig in muck or well-rotted compost where it is going to grow.
Sowing Sow under glass at a temperature of between 60°F and 65°F (16°C and 19°C) in spring, or buy plants from a nursery. The seedlings must be kept moist. Spray them with water at least twice a day.
Planting Early summer is the usual time to plant. Plant very carefully 12 inches (30 cm) apart in trenches with muck underneath. Soak the plants well with water.
Aftercare You can grow catch crops such as lettuce or radish on ridges between the furrows. When these catch crops have been harvested, earth up the celery. Cut off the side shoots. Then hold the plants in a tight bunch and earth them up.

Do this so that only the tops of the leaves are above the new ridges. Always keep ground moist. Never let it dry out. To prevent leaf blight, spray with Bordeaux mixture (*see* p.164) once or twice, as you would spuds. If you want to extend the eating season, in the winter protect the plants with straw, bracken, cloches, or what you will. You can also heel them in dry ground in a protected position if you fear very hard frost, but it makes harvesting difficult.
Harvesting Dig them out whenever you want them and eat them fresh.

Self-blanching celery
You can grow this on the flat in the same conditions as ordinary or white celery (*see above*) but you don't need to earth it up. It gets used before the white celery and must be finished before hard frosts begin as it is not frost-hardy. It is not as good to eat as white celery but is a good standby in the autumn before white celery is ready.

CHICORY
Use Chicory makes a good winter salad ingredient.

Sowing Sow the Witloof type in early summer in a fine tilth and thin to 12 inches (30 cm) apart in rows 18 inches (45 cm) apart.
Aftercare Cut down to just above the crown in November. Lift and plant in pots and keep in the dark at 50°F (10°C) or thereabouts. They will then shoot.
Harvesting Break the shoots off just before you need them. They should grow again every 4 weeks or so. Keep picking.

CORN SALAD
Use If you like eating salad in the winter this is an ideal crop for you.

It produces leaves that taste like tender young lettuce leaves.
Sowing Sow in drills 12 inches (30 cm) apart in late summer.
Harvesting Cut it when the plant is short and has just three or four leaves. Don't let it get too lanky.

CUCUMBERS AND GHERKINS
Use You can grow ridge cucumbers and gherkins, both of which are fine pickled, out of doors. Frame cucumbers, which are better looking and better tasting when fresh, are grown in frames or under cloches. A heated greenhouse is even better because you will get your cucumbers earlier.

Soil Cucumbers will grow on light soil if it has plenty of manure in it. They must have plenty of moisture, and they don't like acid soil.
Treatment Dig plenty of mulch in during the previous autumn.
Sowing Frame cucumbers can be sown under cover in early spring. It is ideal if you can start them off in a heated greenhouse, keeping the temperature at about 60°F (16°C). Outdoor types can't be sown until early summer unless they are protected for the first month. In wet climates plant six seeds of an outdoor variety on a small hill, 4 inches (10 cm) high, and later thin out to the three best plants on each hill. In dry climates use the same method, but plant in a small depression that has had plenty of muck or compost dug below it the previous autumn.
Planting Outdoor cucumbers just continue to grow where you plant them. Frame cucumbers can be hardened off in early summer. If you grow cucumbers in a greenhouse, pot them in peat pots as they grow big enough to handle, then plant them, pot and all, in the greenhouse soil when they look

about to outgrow the pot. Always water them with warm water. Make sure you keep the greenhouse humid and well aired.
Aftercare They must have plenty of water and never be allowed to get dry. It helps to soak muck in the water. Ridge cucumbers should have all the male flowers pinched off them so that the female flowers that produce the cucumbers don't get fertilized. If they do, the fruit will be bitter. Ridge cucumbers should also have the growing points nipped out when the plant has seven true leaves.
Harvesting Pick them regularly while they are young and they will go on cropping. Pickle the last lot before the first frosts.

ENDIVES
Use A plant related to chicory, the crisp, curly, or broad leaves are used in salads during winter instead of lettuce.

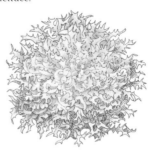

Sowing Sow in mid-summer and put cloches over in late summer. Whitewash the cloches to keep out the light and the endives will blanch and make good winter salading. Blanching also helps to reduce the bitter flavour. For summer endives, sow out in the open from spring onwards and eat in salads.

FRENCH BEANS AND DRIED BEANS
Use Haricots are ripe French beans that have been dried for winter use. Butter beans and Lima beans are

specifically for drying and using in the winter. For vegetarians, such dried beans are really necessary, because they are about the only source of protein readily available to them in winter time. French beans can be eaten green, pods and all, just like runner beans.
Soil They all like lightish, well-drained and warm soil. It's no good trying to grow them in heavy clay or on sour land.
Treatment Like all the legumes, they grow best after a heavily mucked crop. Lime the soil well if necessary.
Sowing In temperate climes, sow in early summer. They are all very frost-tender, and won't thrive if sown in cold, damp ground. Sow in a wide drill, about 2 inches (5 cm) deep, in two staggered rows, so the beans are about 6 inches (15 cm) apart.
Aftercare Hoe well, and draw the soil around the plants. Dwarf varieties don't need sticking but high varieties do. Any arrangement of sticks, or wire and string supported on poles, will do.
Harvesting If the beans are for drying for the winter, let them get quite ripe and then pull the plants intact. Hang them upside down from the roof of an airy shed. Thresh them as required. If you are eating them green, pull them and pull again. The secret of having plenty of them, young and fresh, is to keep on picking.

KALE
Use Kale is very hardy and so is an excellent winter green standby. It will grow in both cold and wet climates where there is little other greenstuff in winter and early spring. In the Highlands of Scotland the "kale-yard" has often been the only source of greenstuff in the harsh winters.

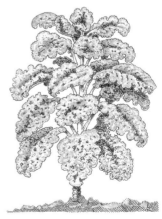

Soil Kale is not at all fussy but the richer the soil, the better the crop.
Treatment *See* Spring cabbage.

Sowing Sow kale during late April and early May in colder climes and in early April in warmer ones.

Planting It is a good idea to sow the seed in situ, and not transplant it but thin it instead. But you can transplant it if you need the land.

Aftercare *See* Spring cabbage.

Harvesting Leave kale until you really need it, that is, after the Brussels sprouts have rotted, the cabbages are finished, the slugs have had the rest of the celery, and the ground is 2 feet (60 cm) deep in snow and only your kale plants stand above it like ship-wrecked schooners.

LEEK

Use In cold, wet areas this is one of the most useful plants, for it stands the winter and provides good food and vitamins in the months when perhaps little else has survived except kale. Onions are hard to grow and to keep but leeks are an easy substitute. The Welsh are very sensible to have this excellent plant as an emblem and not some silly inedible flower or a damned thistle.

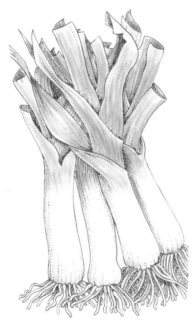

Soil Leeks grow on pretty well any soil as long as it is not waterlogged.

Treatment Heavy manuring is advantageous. Most people plant leeks out on land from which early potatoes have been harvested and which has been heavily mucked for that purpose. However, if you can't lift your earlies before mid-summer, this is too late. You must plant on other ground, which should be well dug and manured.

Sowing Sow the seed in the general seed bed 1 inch (2 cm) deep and sow in rows 12 inches (30 cm) apart in spring.

Planting The traditional way to plant leeks is to chop the bottoms off the roots and the tops off the leaves of the little plants and just drop the plants in small holes and leave them. If you do this they grow and make leeks, but I have come to believe that this is a silly idea and it is better not to mutilate the plants and also better to plant them properly. Why not try both methods and compare them? Draw drills 3 inches (8 cm) deep with a hand-hoe or a wheel-hoe and plant the leeks 5 inches (13 cm) apart in the furrows. Make a biggish hole for each leek and plant carefully, making sure the little roots are not doubled up. Don't press down as you would onions. Just water them in and this will wash a little loose earth into the hole round the roots.

Aftercare Hoe them of course and ridge them, raising the ridges from time to time so as to blanch the lower parts of the stems.

Harvesting Leave them until you really need them and then towards the end of the winter, dig them out and "heel them in" on another small piece of ground. Heeling in means opening a slot in the ground with a spade, putting the leeks in quite thickly, and heeling the earth back on their roots. They won't grow any more like this but they will keep alive and fresh until you need them. They are very hardy and don't mind frost.

LETTUCE

Use Lettuce is the firm base of salads throughout all the fair months of the year, and with a little glass protection we can even have it through the winter if we feel we must. It is not a brassica so we needn't worry about clubroot. Try growing different types of lettuce – some are much crisper than others.

Soil It likes good soil but will grow on most soil, especially if it is richly manured. Lettuce likes it cool and will stand shade, but will not grow well near trees. It likes a moist climate.

Treatment Dig in well-rotted muck or compost for summer lettuce, but not for winter, as winter lettuce doesn't like too much fresh manure: it gets botrytis. Work down to a fine seed bed.

WINTER LETTUCE

Sowing and planting Sow about 1 inch (2 cm) deep in late summer and then expect to protect them with cloches or something over the winter. Of course, in very cold climates, winter lettuce is out. You can sow winter lettuce in seed beds with the intention of planting them out in early spring to get an early crop. And of course you can get lettuce all winter in a heated greenhouse.

SUMMER LETTUCE

Sowing and planting Sow thinly starting in the spring with about 18 inches (45 cm) between rows. Thin the plants out to over a foot apart and transplant the thinning elsewhere because they transplant easily. Don't sow too much at one time; instead try to keep on sowing throughout the summer.

Aftercare Hoe and hoe and water whenever necessary. Keep eating.

MARROWS, SQUASHES, AND PUMPKINS

Use They can be kept for the winter and are rich in vitamins and highly nutritious.

Soil Nothing is better for these than to grow them on an old muck heap, and that is what we often do. They love a heavy soil.

Treatment If you don't plant on a muck heap, dig in plenty of muck or compost in the autumn.

Sowing Sow seeds in situ in late spring under cloches or better still, under upturned jam jars.

Otherwise sow in soil blocks or peat pots under glass. Harden plants off gradually in early summer by propping the jam jars up in the day, for example, and putting them down again at night. Remove the glass, or plant the potted plants out in the open a few weeks later. Plant three seeds to a station and have the stations about 6 feet (1.8 m) apart because these things like to straggle.

Aftercare Hoe of course, water when necessary, mulch if you can, and beware of slugs.

Harvesting Keep cutting them when they are young and tender and you will get more. Young marrows, or courgettes, are particularly good. In late summer leave some to ripen and store them out of the frost in a cool place, preferably hung up in a net. In southern Africa, where you don't get too much frost, pumpkins are thrown up on corrugated iron roofs and left there all winter. They dry out in the winter sun, become delicious and form the chief winter vegetable of that part of the world.

MELONS

Use Melons grow outdoors in warm climates and can be grown outdoors in cool climates as long as you start them off under cloches after the last frost. But they are best grown under protective frames in cooler climates.

Treatment Treat them exactly like cucumbers but don't remove the male flowers. Plant them on small hills 6 feet (1.8 m) apart.

ONIONS

Use Good food is inconceivable without onions.

Soil Onions are demanding: they like medium loam well drained, deeply dug and richly composted.

Treatment The soil must not be acid, so if necessary lime it in the autumn. Dig deeply in the autumn months and incorporate manure or compost. Get it down to a fine tilth in the spring and then get it really firm because firm soil is a necessity for onions to grow well.

Sowing You can sow in mid-summer and leave in the seed bed until spring. Or you can sow in early spring, or as early as the ground is dry enough to walk on without it caking. Sow very shallowly, very thinly in rows of about 10 inches (25 cm) apart, if you intend to thin the onions and grow them in situ. But you can have the rows much closer together if you intend to plant them all out. Rake the seed in very lightly and firm the soil with the head of the rake.

Planting Plant very firmly in firm soil, but don't plant too deeply. Plant summer-sown seedlings in early spring – whenever the soil is dry enough – Inter-rowing with carrot is said to help against onion fly, and I believe inter-rowing with parsley is even better.

Aftercare Growing onions means a fight against weeds, which seem to love onions, and the onions have no defence against them from broad, shading leaves as many crops have. Now I know that some people say onions will grow well in a mass of weeds, but my experience is that you must keep them free of weeds in the early months of their growth. It is true that if large annual weeds grow among them for, say, the last month, they may still grow into good onions. I like to keep them weed-free and mulch them well with pulled-out weeds in their later stages. If you are growing onions in situ in the seed bed, single them in about 4 inches (10 cm) apart. If you have sown very thinly you might like to try not thinning at all.

You will get smaller onions but they will keep better.

Harvesting When the tops begin to droop, bend them all over to the ground. This is said to start the onions ripening, and possibly it also stops them from growing up and going to seed. After a few days pull the onions and lay them down on bare soil or better still on a wire netting frame to keep them clear of the ground. Turn them occasionally. The more hot sun that falls on them the better. Before the autumn string them and hang them up, or hang them in net bags, or lay them on wire netting in a cool and draughty place. The air must be able to get between them. They don't mind some frost, but can't stand lack of ventilation.

Shallots

Sow the bulbs in late winter and you get lots of little onions that grow around the first bulb next summer. You can then go on picking until autumn. Keep some of the best bulbs to plant next year.

Tree onions

These onions are perennial so once you have planted them they will grow year after year. Each year, when the plant grows, little onions will form at the tips of the stems. When this happens you must support the weight of the plant on sticks. Plant 6 inches (15 cm) apart in rows 18 inches (45 cm) apart. You can use the onions that form underground as well as those on the leaf tips.

Pickling onions

These like poor soil. Broadcast the seed in spring and lightly rake it in. Hand-weed but don't thin. Pull and pickle when ready.

Salad onions

Sow these like ordinary onions in late summer and again if you like in early spring. Don't thin, and pick to eat as required.

Onion sets

These are the lazy man's way of planting onions. Sets are immature bulbs, with their growth arrested by heat treatment. Plant them early in the spring very firmly, and replant any the birds pull out. Then treat them like ordinary onions. They are much easier to grow.

PARSNIPS

Use Parsnips make the best of the root wines, and, properly cooked and not just boiled to death, are a magnificent vegetable, very rich in vitamins A, B, and C.

Soil Parsnips will grow on any soil, provided it is deep and not too stony. As with all root vegetables, don't use fresh manure.

Treatment Parsnips like potash, and the ground must be deeply dug. If you want to grow really big ones, make a hole with a fold pritch, or steel bar, and fill the hole with peat and compost, or a potting compost, and sow on this.

Sowing Drill to 1 inch (2.5 cm) deep and 14 inches (36 cm) apart in early spring or as soon as the land is open and dry enough. They take a long time to grow, so sow some radishes with them, as these declare themselves first and enable you to side-hoe.

Aftercare You can intercrop with lettuces for one lettuce crop. Then hoe and keep clean.

Harvesting Leave them in the ground as long as you like. They are far better after they have been frosted. If you want them during hard frost, when it would be difficult to dig them out, pull them before the frost and leave them in a heap outside or in a shed. You can boil them in stews, but they are far better baked around a joint in fat, or partly boiled and then fried in slices. There are old boys in Worcestershire who devote half their gardens to rhubarb and half to parsnips. And the whole lot of both crops goes to make wine!

PEAS

Use Eaten green, peas are delicious and extremely nutritious. Allowed to dry, they can be kept through the winter and cooked like lentils. It is better to have fresh green peas in their season, and only then, so that you come to them every year with a fresh and unjaded palate. Freezing them is a bore.

Soil They like a medium loam but will grow on most soils. Like all legumes (and brassica) they don't like acid ground. They like to be kept moist.

Treatment If you want a bumper crop, dig a trench in the autumn, fill it with muck, compost, or any old thing so long as it's organic, and bury it. Lime the soil well. Plant in what is left of the trench in the spring. But this is very laborious. Put your peas in after your spuds and your land should be well mucked already.

Sowing I personally sow peas thick in a little trench dug about 3 inches (8 cm) deep with a hoe. And I eat a hell of a lot of peas. Plant each pea 2 or 3 inches (5–8 cm) from its neighbour. Cover and firm the soil over the peas. It helps a lot to have soaked the peas for two or three days first, to get them germinating so they sprout early. Also swill your seed in paraffin to deter mice. Birds, too, are a menace; wire pea-guards are an answer, and so is a good cat.

You can sow some round-seeded peas in November in mild climates, and some more in February. For this the land must be light and dry. Of course if you cloche them it helps. You will thus get very early pickings but for most of your crop sow from mid-March onwards, in successional sowings right into July. Finally, for your last sowings use "early" varieties (paradoxically). They will ripen quickly before the frosts cut them down.

Aftercare Hoe until the pea vines themselves smother the weeds. And mulch does wonders with peas, for it keeps the ground cool and moist, which is just what peas like.

Harvesting Pick them young to eat raw in salads, and then when the pods are tighter packed, pick for cooking. Keep picking as hard as you like, and if you have more than you can eat green, let them ripen on the vines and harvest properly. In other words, pull the vines when they are dead ripe (but before the autumn) and hang them up in the breeze but out of the rain. Thresh them in due course, stow them in jars, and eat them in soup.

PEPPERS

Use Peppers are either round and mild, or thin and long, and very much hotter! Peppers and chillis (the capsicums) spice up the blandest of meals and can be grown successfully under polytunnels.

Sowing and planting Sow seed indoors in early spring, and plant out in the garden on well-mucked ground at least a fortnight after the last possible frost; under cloches if you have them in cold climates. Plant 2 feet (60 cm) apart in rows 3 feet (90 cm) apart. After the ground has really warmed up, mulch. Peppers need moisture but not too much or they will die off. So in a wet climate plant them on the tops of ridges.

Harvesting Harvest peppers when they are green, or you can leave them to turn red.

POTATOES

Use Quite simply you can live on them. They are one of the best storable sources of energy we can grow, and are our chief source of vitamin C during the winter.

Soil Potatoes like good, strong soil. They will grow in clayey loam, love peat, and are one of the few crops that not only tolerate but like acid soil. If you lime before planting they will get scabby. They want plenty of muck.

Treatment Better to dig deeply in the autumn and dig again in the spring, this time making ridges and furrows. They don't want a fine tilth but they want a deep one. Throw as much muck or compost as you can spare into the furrows before planting. Plant the spuds straight on top of it.

Planting Put your first earlies in when other people in your locality do, or a fortnight earlier under cloches or transparent plastic. The slightest touch of frost on the leaves blasts them and they have to start growing all over again. If you want early potatoes, chit your seed potatoes, that is, lay your early seed in trays, on shelves, or in old egg cartons, in the light and not in the frost: 40°F–50°F (5°C–10°C) is right. When you plant them, be careful not to knock off all the shoots. Leave two on each tuber. Don't chit the main crop. Bung them straight in in the late spring, but not before. Plant earlies only about 3 inches (8 cm) deep, a foot (30 cm) apart in rows 18 inches (45 cm) apart. Plant main crop 18 inches (45 cm) apart in rows also 18 inches (45 cm) apart, but plant them about 5 inches (13 cm) deep.

Aftercare As soon as leaves show, earth up. In other words, band earth lightly over the potatoes. Three weeks later, earth up some more, and, with the main crop, earth up again in another couple of weeks. Hoe between the rows. Spray with Bordeaux mixture (see p.164) when the weather gets warm and muggy to prevent blight.

Harvesting If you have plenty of early potatoes in, don't deny yourself a meal or two when they are quite tiny: why should you? Then go on digging earlies until they are finished. If you have second earlies, go on to them. Your main crop will then take over for immediate eating, but don't lift the bulk of the main crop until the haulms (tops) have completely withered away. Then fork them out carefully and let them lie on the ground for a day and a half to set their skins (more than two days might start them going green, in which case they tend to become poisonous). Then clamp them, or put them in a root store in the dark. Potatoes must never be allowed to be affected by frost or they will go bad, which is a waste.

RADISHES

Use Radishes grow just about anywhere. Add to salads for extra flavour, crunchiness, and colour.

Sowing Sow the large seeds in drills and pick them when they are ripe, after about 6 weeks. They are brassicas, but grow so quickly that they don't get, or perpetuate, club-root. Put in successional sowings all through the spring and summer so as to have a constant supply of tender young ones. Don't let them get old and go to seed.

RHUBARB

Use Rhubarb is a perennial, and once you have planted it, or inherited it, you have it for good.

Soil Pretty well any soil is fine.
Treatment Put on plenty of muck.
Planting Buy or cadge crowns and plant them in late autumn Leave about 3 feet (90 cm) between plants and 4 feet (1.2 m) between rows, and put some nitrogenous fertilizer on top to turn it into self-activating compost heap. Put upturned pots or buckets over some of the plants in spring to force them on early.
Aftercare Cover the beds with deep straw in autumn.

Harvesting Pull what you want when the stems are thick and tall. Leave what common sense will suggest, so as not to rob the plants too much.

RUNNER BEANS

Use These come later than the drying beans described earlier. They yield very heavily, are tougher and have a coarser, and I think better, flavour. They need more care in planting and must have tall sticks. Salted they are a great standby for the winter.

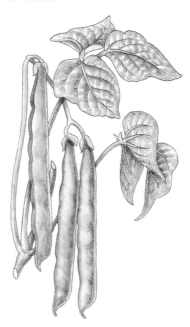

Soil They like good, rich, deep soil.
Treatment Double dig a deep ditch in early spring and incorporate plenty of compost or muck in the bottom of it. If you have comfrey leaves, dig them in, because they are rich in potash, which all beans like. As they come in your bean break, you will already have limed the ground, if you had to, the previous autumn.

Sowing Sow them in the early summer in a wide but shallow trench 2 inches (5 m) deep, in two staggered rows with the seeds about 9 inches (23 cm) apart. Leave at least 5 feet (1.5 m) between stands of beans. Put in tall sticks early enough for the beans to get a good start.

Otherwise, you can pinch the growing tops out and let the vines straggle on the ground, but you won't get much of a crop and in my opinion it's a poor way of growing these magnificent climbing plants, which can be about the most beautiful and productive things in your garden.

Aftercare Hoe and keep well watered in dry seasons. When they start to flower, make sure they have plenty of water. Mulch with compost if you can, and spray the flowers with water occasionally, because this "sets" the flowers in the absence of rain.

Harvesting Just keep on picking. If you can't cope with the supply, and you probably can't because they crop like hell, just pick anyway. String the beans, slice them (you can buy a small gadget for this), and store them in salt (*see p.208*). Pick them and give them to the pigs rather than let them get old and tough. Keep some though to get ripe for seed for next year.

SOYA BEANS

Use Soya beans have been grown in Asia for centuries. They came to the West less than 200 years ago – and are now proving to be a very worthwhile crop to grow in warm areas because of their high protein value. They do need a long, warm growing season though – at least 100 days. They can be eaten green like peas or the beans can be left to ripen and then dried for use all through the winter. The beans can be ground into flour.

Preparation Dig the ground in autumn and add plenty of lime.
Sowing Sow in the late spring about 1 inch (2.5 cm) deep, and 3 inches (8 cm) apart in rows 2 feet (60 cm) away from each other.
Harvesting Pick the beans for eating green when they are young, certainly before they turn yellow. It is easier to remove the beans from the pods if they are boiled for a few minutes first. If the beans are for drying or for flour, leave them on the plants to ripen, but they must be picked before the pods burst and release the beans. Judge this carefully but be guided by the colour of the stems on the plant – they should still be green.

SPINACH
Use There are several kinds of spinach – New Zealand spinach, spinach beet, perpetual spinach, and seakale beet – but treat them all as one.
Soil Like nearly everything else spinach likes a good, rich loam, so give it as much muck as you can. It will do well on clay, but is apt to run to seed on sandy land unless you give it plenty of muck.

Sowing Sow 1 inch (2.5 cm) deep in drills 12 inches (30 cm) apart. Later thin the plants in their rows to 6 inches (15 cm) apart.
Aftercare Hoe, mulch and water during the summer.
Harvesting Pick the leaves when they are young and green, raking only a few from each plant, leaving the smaller ones to grow bigger. Don't boil spinach. Wash it in water and put the wet leaves in a saucepan and heat over a fire. When you harvest seakale beet, pull off the stems as well as the leaves. Eat the stems like asparagus.

SWEDES AND TURNIPS
Use Swedes and turnips can be eaten young and tender in the summer and autumn and clamped for winter use. Turnips will survive in the ground until severe frosts begin, maybe till Christmas in temperate climates. Swedes are much hardier and will live in the ground all winter. All the same, it is handier to pull them and clamp them so you have them where you need them. They are cruciferous, which means they are subject to club-root, and should therefore be part of the brassica break so that this disease is not perpetuated. You want to leave the longest possible gap between crops that are prone to club-root. Kohl-rabi, or cabbage turnip, is much like turnip and is grown in the same way.
Soil Light fertile loam is best. Keep it well-drained but not too dry. But turnips, particularly your main crop for storing, will grow on most soils.

Treatment In heavy rainfall areas, say over 35 inches (90 cm) a year, it is a good thing to grow turnips and swedes on the tops of ridges to aid drainage. So ridge up your land with a ridging plough, or on a small scale with a spade, and drill on the ridges. If you want to grow them on the flat, just treat the land as you would for spring cabbages (*see p.57*).
Sowing Very early sowing can be done in the early spring or a week or two before the last probable frosts, but you can sow turnips and swedes right up until August. Sow the seed shallowly in drills about 9 inches (23 cm) apart. Cover and press down.
Aftercare Beware the flea-beetle. These are little jumpers that nibble tiny holes in the leaves. You can get rid of them by dusting with an insecticide, or you can trap them with a special little two-wheeled arrangement. The sticky underside of a board goes along just over the plants and a wire brushes the plants. The beetles jump and get stuck to the board. It sounds silly but it works. Thin to 4 inches (10 cm) apart in the rows while they are still quite small. Hoe at least twice again afterwards.
Harvesting Eat swedes and turnips when they are ready (after two months), or leave until early winter and pull, top, and clamp them.

SWEET CORN
Use Sweet corn is maize that has not been allowed to get ripe. The seeds are still fairly soft and slightly milky, and the carbohydrate is mostly in the form of sugar, which is soluble and can therefore move about the growing plant. As the cobs mature or when they are picked, the sugar changes to starch. It will grow in the hottest climates, and in temperate climates if you grow hardy varieties.

Soil Sweet corn will grow in most good well-drained soils, but it is a greedy feeder, likes plenty of muck and a pH of about 6.5.
Sowing A long growing season is essential but sweet corn can't stand frost, so if we plant it a week or two before the last likely frost under upturned jam jars, or little tents of plastic, or cloches, so much the better. In warm climates you can sow it straight out in the open in early summer, but if your summers are a long time coming you would do better to sow it in peat pots indoors in late spring and then plant it out. Sow the seed about 1 inch (2.5 cm) deep, 15 inches (28 cm) apart in rows 30 inches (75 cm) apart. And try to sow in blocks – nothing narrower than four rows for example – because maize is wind-pollinated, and if it is sown in long thin lines many plants will not get pollinated.
Planting If you have grown it in pots, plant it out very carefully, because it doesn't like being disturbed anyway. Plant out when it is about 5 inches (13 cm) high, and preferably plant peat pot and all. Water well after planting. It is a lot better if you can sow them in their final position.

Aftercare Hoe and top dress with nitrogen about a month after planting sweet corn if your soil is not as rich as it should be to keep the plants growing. Apparently the Amerindians used to bury a dead fish under each plant. This is a very good idea, for the nitrogen would become available last when it was needed. I know a vet who gave all the dogs he "put down" to a fruit farmer who buried them under his newly planted apple trees for the same reason.

Harvesting Break the cobs off in the milky stage after the tassel has begun to wither and turn brown. To test, pull the leaves off part of a cob and press your thumbnail into the grain. It should be milky. They say you can walk down the garden to pick your corn, but you must run back to cook it; it must be absolutely fresh. This is because the sugar starts turning into starch as soon as you pick it, and it loses flavour. If you have too many cobs, you can dry them (see p.198-199 for harvesting and storing).

The straw makes good feed for cows, litter for pigs, or material for the compost heap, and it is a valuable crop for this reason alone.

SWEET POTATOES

Use Sweet potatoes can be your staple food in a dry, warm climate, but you won't get much of a crop in a damp, cool environment. They are very frost-tender.

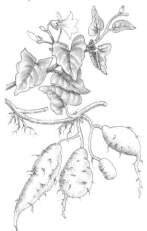

Soil They grow in sand, or sandy loam, and they don't like rich soil.
Treatment Just dig deeply. You needn't add anything.
Planting Plant tubers just like potatoes (if you are sure they haven't been sprayed with a growth-inhibitor). Plant them 16 inches (40 cm) apart in rows 2½ feet (75 cm) apart. Don't plant them anywhere in the world until two weeks after the last frost.

Aftercare Just hoe.
Harvesting Dig them up very carefully at least a fortnight before the first frost. Cure them by laying them carefully on hay and leaving them out in the sun for about ten days. They don't go green because they are no relation of real potatoes. Turn them from time to time. If there isn't enough sun keep them somewhere with 90 percent humidity between 80°F (27°C) and 90°F (32°C) for ten days. Store them packed lightly in straw in an airy place at not less than 50°F (10°C).

TOMATOES
Outdoor tomatoes

Use Outdoor tomatoes are a dicey business in any cold, wet climate. What they need is a warm, dry ripening season in late summer, and that is what, where I live, they don't get. But green tomatoes make excellent chutney and if you store them well they sometimes get ripe in store, although they never taste like sun-warmed fruit picked off the vine and eaten straightaway. But if you can grow them, they are an enormously valuable crop for bottling to keep your family healthy during the dark days of winter. They really are preserved sunshine.

Soil The soil must be well-drained, and in a sunny but sheltered position in cold climates.
Treatment I ridge the land in the autumn, put well-rotted compost or muck in the trenches in early spring, split the ridges over it and then plant the tomatoes on the new ridges.
Sowing The most luxurious tomatoes I ever saw growing were on the overspill of a sewage works, which leads one to think that it would be better to eat the seed before we plant it. But failing such extreme measures, sow thinly under glass in any seed compost, including the kind you make

yourself. If you sow in the late spring in a temperate climate, the plants will grow even if you have no heat in your greenhouse, but if you do have a little heat, so much the better. If you have no heat put thick newspapers over the seedlings at night to keep them warm: 55°F (12°C) is right. Water diligently with lukewarm water but not too much. Don't drown them. Or you can sow direct, in situ, a week or two later, under cloches in warmer climates, or just outdoors in hot climates.

Planting Most people plant twice. First, when they have three to four true tomato-type leaves, they plant in either soil blocks or peat pots or in compost in small flowerpots. These pots can be put into cold frames and the plants gradually hardened off. Then plant out in the first fine warm weather in early summer. Plant very carefully, retaining as much of the compost on the roots as you can, and plant a little deeper than they were before. Plant on the mucked ridges described above. Give each plant a tall stake for support as it grows bigger and heavier.

Aftercare Hoeing and mulching, within reason, do help, and with low fruiting varieties it is common sense to put clean straw on the ground to protect the fruit. Pick out all side-shoots. These are little shoots that grow between the fruiting branches and the main stem, rather as if you had another little arm growing out of your armpit.

You cannot pamper tomatoes too much. Water them whenever they need it. Many gardeners soak muck in the water so that they feed the plants as they water them. As they grow taller, tie them carefully to the stakes with raffia or string. Spray them with Bordeaux mixture (see p.164) to protect them from potato blight. (The tomato is so closely related to the potato that it is almost the same plant.) Don't touch them with tobacco-stained hands, because you can convey tobacco virus disease to them. (The poisonous tobacco plant is also closely related to the tomato.) Allow the plants to set about four trusses. To ripen tomatoes in dull climates it is often advantageous to lay them down flat on clean straw and place clothes over them. Some people pull leaves off "to let the sun get to the fruit"; I don't think this is worth it.

Harvesting Home-grown tomatoes are so good to eat (immeasurably better than tomatoes bred for a long "shelf life" and not for flavour) that you will not be able to stop eating them as they ripen.

But try to bottle as many as you can. We wallow in vitamins in the summer: it is for the winter and the hungry gap that we need them.

Indoor tomatoes

Use If your greenhouse is heated, you can sow seed in early winter and get ripe tomatoes in spring. If you don't want to eat them all, you can sell them at a good price.
Sowing If you have an adequately heated greenhouse, sow tomato seed in November at a temperature of 70°F (21°C). Never let it fall below 60°F (16°C) during the winter. If you can't raise a temperature of 70°F (21°C), sow seed in February and keep the temperature at 60°F (16°C). Sow in compost made of two parts sifted loam, one part leaf mould and a little sand. Cover with glass to prevent evaporation. Keep them moist.
Planting When the plants have formed two rough leaves, pot the plants singly in 5-inch (13-cm) pots. Use the same compost as before but add some well-rotted muck. When the first truss of flowers is formed, move the plants into much larger pots (about 12 inches or 30 cm in diameter) or into the greenhouse soil.
Aftercare Treat your greenhouse tomatoes in the same way as outdoor tomatoes, but you can let them set up to ten trusses.
Harvesting Begin picking the tomatoes as soon as they are red. This will be much earlier in the year than outdoor tomatoes.

WATERCRESS

Use Watercress is one of the richest sources of vitamin C likely to come your way. It makes a superb salad, or it can be cooked.

Sowing Sow seed or rooted cuttings in a damp, shady spot in late spring or midsummer. Dig the soil deeply and work in some peat if you can get it. Rake the bed, flood it, and sow thickly when the water has drained away. You can grow it in an unpolluted stream.

Herbs

Herbs are a very cheap and easy way of improving the flavour of food; they also make it more digestible and do you good at the same time. In ancient times they were valued as much for their healing properties as for their culinary ones. The coming of the Industrial society saw a decline in the use of herbs, and up to a time only parsley, mint and – in enlightened circles – horseradish were being much used in the North American or British kitchen. Now, the revival of a flourishing international cuisine has once again made people eager to experiment with a variety of new tastes. Consequently, growing fresh herbs to add natural enhancement to food is becoming an increasingly attractive proposition for everyone. Even people without gardens can grow them in pots.

A drift of borage or a sea of thyme look splendid from the kitchen window. There is really no reason why herbs should not take the place of inedible flowers in beds near the house instead of being relegated to an inaccessible patch at the back of the garden. But unless you are planning to become a herbalist, it is better to concentrate on a few herbs that will have a culinary or medicinal use to you, rather than cultivate scores of varieties, most of which you will neglect.

Herbs divide fairly straightforwardly into two groups: perennial and annual, with just the odd biennial to complicate matters. Most herbs prefer a light, well-drained soil and plenty of sun, although a few prefer the shade. All respond to constant picking.

DRYING HERBS

You dry herbs in order to keep the colour and aroma of a fresh herb in a dried one. It is a delicate operation, because it requires both speed and care, but most herbs can be dried and stored.

As a general rule, harvest the leaves and stems just before the plant flowers, on the morning of a fine, hot day after the dew is gone. If you are going to preserve the herbs, take them to a drying rack immediately. Do not over-handle them. They bruise easily, and every minute you waste means the loss of more volatile oils. These are what give herbs their flavour and quality.

Tie the herbs in small bunches and hang them in an airy place. Ideally you dry them at a temperature between 70°F and 80°F (21°C–27°C), in the strongest possible draught of air. You can leave them hanging up indefinitely, but they will collect dust. A better thing to do is to rub leaves off the stem when they are quite dry and brittle (but you hope still green), crumble them up and store in sealed glass or pottery jars in the dark. If the air is too damp to get them dry, lay them in a cool oven at 110°F (44°C) on sheets of paper overnight. Or you can hang them in a solar drier (see p.244) which is ideal for drying herbs, but in that case watch the temperature by using a thermometer.

Below, and over the next few pages, I describe many of the herbs that the self-supporter might find most useful for flavouring food or fortifying the spirit or even banishing ailments.

ANGELICA
Angelica archangelica
Biennial

Use Once thought to cure the plague, angelica-scented leaves make a fine tisane. The roots and stems can be candied or they can be crystallized.
Soil Angelica needs a rich, moist soil and a shady position.
Sowing Seeds must be fresh or they won't germinate. Plant in mid-summer as soon as they are ripe, in drills one inch (2.5 cm) deep.

Planting Transplant seedlings or young plants in the autumn and thin to 6 inches 15 cm) the first year, 2 feet (60 cm) the following year. In the third year distance them 5 feet (1.5 m) apart. They grow very tall and their leaves are spreading.
Harvesting Leaves should be cut in early summer while still a good colour. Pick stalks and leaves in late spring or they become too hard for candying. Roots should be dug up in the first year in autumn before they really get too woody. Wash thoroughly, then plait and dry as quickly as possible.

ANISE
Pimpinella anisum
Annual

Use Anise has valuable digestive properties. The fragrant seeds can be used to impart a slight licorice flavour to breads, cheeses and puddings.
Soil A moderately rich and fairly dry soil is best.
Sowing Sow in situ in late spring, and thin later on to 8 inches (20 cm) apart. Take care when thinning, as the herb is fragile and easily upset.
Harvesting The seeds will mature the first year after 120 days, as long as they are exposed to full sun. Harvest when the seedheads turn grey-brown and thresh them when they have dried out thoroughly.

BALM
Melissa officinalis
Perennial
Use The leaves impart a fresh lemony flavour to soups and summer drinks.
Soil Balm likes a fairly rich, moist soil in a sunny, sheltered spot. If it is too shady the aroma will be stifled; if too dry, the leaves turn yellow.
Sowing Grows easily from seed which it self-sows profusely. Sow in spring or early summer in a cold frame. It should germinate in 3–4 weeks. Pick out and plant in the garden when 4 inches (10 cm) high, or sow the seed in your

garden in mid-summer and plant seedlings out in the early summer of the following year.
Planting Keep 10 inches (25 cm) between the rows and 12 inches (30 cm) between the plants. Balm is susceptible to frost, so protect your plants by earthing them up or giving them a light cover of manure, peat or leaf mould.
Harvesting Don't expect too much the first year. Harvest just before the buds flower and then again in the autumn. Balm bruises easily, so hands off it as much as possible. Dry in the dark with plenty of ventilation, then store in stoppered jars in the dark. The temperature should never go above 100°F (38°C) or it will lose its flavour.

BASIL
Ocimum basilicum
Annual

Use A fine pungent herb, basil is superb in sausages, spaghetti, and stuffed tomatoes.
Soil Basil needs dry, light, well-drained soil and a sunny, sheltered position.
Sowing A hardy perennial in hot countries, basil is a delicate plant in colder climes, where it has to be grown annually from seed. Sow indoors in early summer.
Planting Seedlings should not be planted until the soil is warm. Plant 8 inches (20 cm) apart in rows 12 inches (30 cm) apart.
Harvesting Basil needs plenty of water to keep the leaves succulent. The leaves can be picked off as soon as they unfurl. Cut down for drying in late summer or early autumn. Basil needs a longer drying time than most herbs; it is also very sensitive to light and heat, and it bruises easily, so handle it as little as possible.

BAY
Laurus nobilis
Evergreen

Use Once used to crown poets in ancient Greece, bay leaves are now more often used in casseroles.
Soil Bay is amenable to any reasonable soil. Give it shelter from harsh winds; it will grow in the shade, though it likes the sun.

Intense frost will kill it; in colder climates bay is almost always grown in tubs so that it can be moved indoors in winter.
Planting It propagates rapidly from hardwood cuttings of half-ripened shoots. Don't let it dry out; feed manure occasionally.
Harvesting The leaves can either be dried (at a low temperature, which helps retain their natural colour) or picked fresh all year.

BORAGE
Borago officinalis
Annual

Use Tradition has it that borage will stimulate the mind and fortify the spirit. Add a sprig or two to your wine and you will certainly notice a difference. The blue flowers can be used raw to garnish salads and the leaves can be chopped into soups and stews.
Soil Borage needs sun and a well-drained loamy or sandy soil.
Sowing Seed is best sown in spring in drills about 1 inch (2.5 cm) deep, 3 feet (90 cm) apart, 3 seeds to a station. Later, thin to one plant per station. Seeds will germinate early and thereafter sow themselves and need only to be kept weeded.
Harvesting Leaves are ready for use in approximately 8 weeks; only the young leaves should be picked. The herb is ready for harvesting as soon as it flowers but it needs quick-drying at a low temperature.

SALAD BURNET
Poterium sanguisorba
Perennial
Uses Young tender salad burnet leaves lend a cucumber flavour to iced drinks or salads. They provide the perfect accompaniment to cream or cottage cheese. The dried leaves make a good burnet vinegar.

Soil It grows well in dry, light, well-limed soil.
Sowing Sow from seed in early spring and thin to 12 inches (30 cm) apart. You can also grow burnet from cuttings. Full sun is essential; seed should be sown annually if a constant supply of fresh leaves is required.
Harvesting The plant is hardy in most climates. Pick young leaves frequently for salads or for drying.

CARAWAY
Carum carvi
Biennial

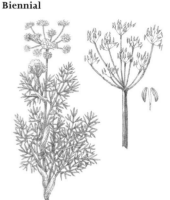

Use As well as using caraway seed for cakes and breads, sprinkle the ground seeds on liver or roast pork, or cook them with goulash and sauerkraut. Leaves can go into salads, and the roots make a good vegetable if you boil them and serve them like parsnips.
Soil Caraway likes a fertile clay loam and a sheltered position. It is winter-hardy, and thrives in cool temperate climates.
Sowing Sow from seed in mid-summer, and it will flower and seed the following year. Protect flower stalks from the wind to prevent the seedheads from shattering before the seed is ripe.
Harvesting Cut off the flower heads as the seed turns brown, and dry the seed in an airy place before threshing.

CHAMOMILE
Matricaria chamomilla
Annual
Use Sometimes used in flower borders, this herb is grown chiefly for medicinal purposes. Chamomile tea is a cleansing aid to digestion, and an infusion of two teaspoons of flowers to a cup of boiling water makes a splendid gargle, or a soothing cure for a toothache.
Soil Any good garden soil with full sun suits chamomile admirably.
Sowing Sow the very fine seeds mixed with sand or wood-ash on a humid day in early spring.

Thin later to 9 inches (23 cm) apart. The seeds self-sow easily. Watering is advisable during germination.

Harvesting Flowers appear and are ready for picking eight weeks after sowing. Pick often, but only on sunny days, when the oil content of the flowers is highest. Try not to touch the flowers too much.

CHERVIL
Anthriscus cerefolium
Biennial

Use Chervil is famed for the flavouring it imparts to soups and sauces. It is well worth growing. Use it as a garnish, or make that classic dish – chervil soup.
Soil Chervil will grow in most soils but it will not thrive in a heavy, badly drained soil.
Sowing Sow from seed in early spring out of doors and in the greenhouse at over 45°F (7°C) all winter. Sow in drills 12 inches (30 cm) apart. After that it will self-sow easily. Chervil does not transplant well so sow where you want it to grow. Seedlings should be thinned out when 2–3 inches (5–8 cm) high. Keep beds weeded and moist.
Harvesting You can eat chervil 6–8 weeks after sowing. Always pick leaves from the outside to allow it to go on growing from the centre. Don't allow it to flower – it takes away the flavour. Chervil is a difficult herb to dry, needing a constant low temperature, but as it is available fresh all year this should be no problem.

CHIVES
Allium schoenoprasum
Perennial

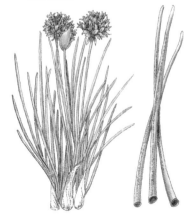

Use Chives add an onion flavour with a green, fresh difference to salads, soups, or any savoury dish. Snip into scrambled eggs and cream cheese. The bulbs can be picked like small onions.
Soil Chives like a warm, shady position, and will grow in almost any soil, but they must have humidity. So plant them near a pond or water tank if you can.
Sowing Sow from seed in spring in drills, 12 inches (30 cm) apart. Chives will thrive on doses of strong humus, and then need careful, frequent watering.
Harvesting Chives are ready for cutting about 5 weeks after spring planting. Plants that are sown in a greenhouse in winter at 80°F (27°C) will be ready in 2 weeks. Cut close to the ground.

CORIANDER
Coriandrum sativum
Annual

Use An important ingredient in Indian cooking, coriander can be grown successfully in cold countries. Use the seeds crushed or whole in curried meats or stuffed vegetables; add some to marmalade to make an exotic change. Seeds are sometimes sugar-coated and eaten as sweets.
Soil Coriander needs a sunny, well-drained site in fairly rich soil.
Sowing Sow in late spring in drills 12 inches (30 cm) apart, and thin seedlings to 6 inches (15 cm). They will grow rapidly to about 2 feet (60 cm).
Harvesting Cut the seedheads when the pods are ripe, and allow the seeds to dry thoroughly before using, as they will taste bitter if they are still green. Thresh and store in the usual way.

DILL
Anethum graveolens
Annual

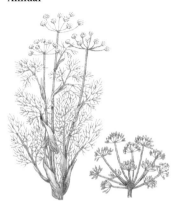

Use The name comes from the Norse "dilla" meaning to lull to sleep, and the seeds were once called "meeting house" seeds, for they were taken to church to be nibbled during endless sermons. While dill seed is the soporific ingredient in gripewater, the leaves can enliven your cooking. It is good with fish, roast chicken, vegetables, and chopped up raw into salads and sauces.
Soil Dill needs a well-drained medium soil in a sunny spot.
Sowing Sow consecutively through late spring and early summer in rows 12 inches (30 cm) apart and later thin to 9 inches (23 cm). Keep plants well watered.
Harvesting Leaves can be used from 6–8 weeks after planting. Cut dill for drying when 12 inches (30 cm) high, before the plant flowers. For pickling seed, cut when flower and seed are on the head at the same time. If seeds are wanted for sowing or flavouring, leave longer, until they turn brown. Seedheads should be dried and then shaken or threshed. Never dry the leaves in a temperature higher than blood heat or you will cook them and they will lose their strong flavour.

FENNEL
Foeniculum vulgare
Perennial

Use Fennel's sharp-sweet flavour is specially suited to the oilier sea fish. Chop the leaves in sauces, salad dressings, and marinades. The broad base can be sliced into salads or cooked whole with a cheese sauce. The seeds can be put into sausages, bread, or apple pie.
Soil Fennel needs sun, a rich, chalky soil, and plenty of sunshine.
Sowing Seeds should be sown in spring in stations of three to four seeds 18 inches (45 cm) apart. If you want to get seed you will have to sow earlier under glass and in heat. If propagated by division, lift the roots in spring, divide and replant 12 inches (30 cm) apart in rows 15 inches (38 cm) apart.
Harvesting Leaves can be used through the summer months; seed heads are ready for drying in the autumn. Harvest the seeds when they are still light green and dry, in a very low temperature, never in direct sunlight. Lay in thin layers and move often as they sweat. Harvest the whole fennel when it takes on a grey-brown hue.

GARLIC
Allium sativum
Perennial
Use Garlic is the basis of good health and good cooking. Unhappy are the nations who have to do without it. Use it liberally and use it often. Take no notice of foolish injunctions to "rub a suspicion" around the salad bowl. Chop a clove or two and put it in the salad.
Soil Garlic needs a rich soil, plenty of sun, and a reasonable amount of moisture. If your soil is light, enrich it with manure.
Planting Plant individual cloves in spring just like onion sets to a depth of 2 inches (5 cm), 6 inches (15 cm) apart.

They will be ready for eating in the autumn. Plant again then, and you will have garlic all year round.

Harvesting When the leaves have died down, lift the crop. Allow to dry in the sun for a few days, then plait and hang in bunches under cover in a dry, airy room.

HORSERADISH
Cochlearia armoracia
Perennial

Use Shred roots finely and use as it is or mix into a paste either with oil and a little vinegar, or grated apples and cream. Horseradish sauce is delicious with roast beef; it is also good with smoked trout and ham.
Soil It needs a rich, moist soil and a fairly shady position.
Sowing The horseradish plant grows furiously and spreads large tap roots with equal abandon. So give it maximum space. Plant the roots in early spring. Dig trenches 2–3 feet (61–90 cm) deep, throw about 15 inches (38 cm) of topsoil in the bottom, dig in a layer of good compost on top of this, and fill with the rest of the soil. Take 3-inch (8-cm) pieces of root, plant roughly 12 inches (30 cm) apart. And keep it weeded. Seed can also be sown in early spring and plants thinned to 12 inches (30 cm) apart.
Harvesting Roots are ready for eating 9 months after planting. Use the larger ones in your kitchen and the smaller roots for replanting.

HYSSOP
Hyssopus officinalis
Perennial

Use Mentioned in the Bible for its purgative properties, monks now use hyssop to make green Chartreuse. You can use sprigs of it in salads, or chop it into soups and stews. Its slightly minty flavour is pleasant in fruit pies. I like it with fat mackerel. Use it sparingly.
Soil Hyssop prefers light well-limed soil and a sunny plot.
Sowing Hyssop grows easily from seed and often self-sows. It can also be propagated by division, from cuttings taken either in the spring before flowering or in the autumn after it. Sow from seed in drills ¼ inch (0.5 cm) deep and plant out seedlings 2 feet (60 cm) apart when 6 inches (15 cm) high.
Harvesting Cut back the tops of the plants often to keep leaves young and tender. Cut for drying just before flowering.

MARJORAM (POT)
Origanum onites
Perennial

Use Pot marjoram has less flavour than sweet marjoram; use it in sausages and stuffings.
Soil It prefers a dry, light soil, and it needs sun.
Sowing Grow it from seed in spring in shallow ½-inch (1-cm) drills, 8 inches (20 cm) apart. When the seedlings are big enough to handle, transplant to 12 inches (30 cm) apart. Alternatively, grow it under glass from cuttings taken in

the early summer and plant out later, allowing 2 feet (60 cm) between plants and between rows.
Harvesting Harvest as for sweet marjoram. For marjoram seeds ripen in late summer or early autumn. Cultivated pot marjoram can last for years.

MARJORAM (SWEET)
Origanum majorana
Annual

Use Sweet marjoram lends a spicy flavour to sausages, and to game and poultry stuffings.
Soil It needs a medium-rich soil, plenty of compost and a warm, sheltered spot.
Sowing Sow sweet marjoram in pots under glass in early spring and then plant out in early summer 12 inches (30 cm) apart.
Harvesting Leaves and flowers are best collected just before the bud opens towards the end of summer. Dry in thin layers, at temperatures not over 100°F (38°C).

MARJORAM (WILD)
Origanum vulgare
Perennial

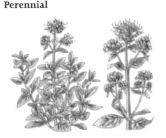

Use Wild marjoram (oregano) turns up in many spicy dishes which incorporate its overpowering flavour with ease. In delicate dishes use it in moderation.
Soil It needs a warm, dry place to grow, and prefers a chalky or gravelly soil.
Sowing Sow from seed in early spring. The distance between plants should be as much as 20 inches (50 cm); if you sow in drills you should thin to 8–12 inches (20–30 cm). Like pot marjoram, it can be grown from cuttings.
Harvesting Harvest as for sweet marjoram. Seeds ripen in early autumn.

MINT
Mentha species
Perennial

Use There are several kinds of mint, with different properties and flavours, but they can be treated similarly. For mint sauce use Bowles mint rather than garden mint if you want a stronger flavour. A few sprigs of peppermint make a fine tisane. Mint added to any fruit dish or drink peps it up.
Soil Mint has a rampant root system and is best planted away from all other herbs. Grow mint in the sun and it will have a fuller flavour, but it needs a moist, rich soil and plenty of water.
Sowing Plant mint in autumn or spring from roots or runners. Lay horizontally in drills 2 inches (5 cm) deep, 1 foot (30 cm) apart. Hoe frequently during the first weeks and compost liberally.
Harvesting Mint for drying should be harvested at the beginning of the flowering season (mid-summer), but fresh leaves can be cut at any time. Frequent cutting helps the plant to grow. Don't cut for drying in damp, rainy weather, for the leaves will only blacken and go mouldy. Keep peppermint leaves whole when drying for tea. Rub them and they will have a totally different taste.

NASTURTIUM
Tropaeolum major or *minus*
Annual

Use The round, hot-flavoured leaves are delicious tossed in rice salads. They are a healthy alternative to pepper for people who like spicy food. The flowers are good with cream cheese. The young green

seeds can be pickled and used like capers. They are excellent with roast mutton.
Soil Given a light, sandy soil and plenty of sun, nasturtiums will grow almost anywhere. Plants grown for leaves need a ground rich in compost.
Sowing Sow the seeds in situ in late spring. If they are planted near other plants, they are said to protect them from pests.
Harvesting The highest vitamin content is found in the leaves before they flower in mid-summer, so harvest then. Chop or dry, then rub or shred. The leaves dry well, but the flowers should always be eaten fresh.

PARSLEY
Petroselinum crispum
Biennial

Use There are several varieties of parsley and all are rich in vitamin C, iron, and organic salts. Chop it up into tiny pieces and use lavishly as a garnish, as well as in cooking; it's especially complementary with fish.
Soil Parsley needs rich soil with a fine tilth.
Sowing Sow parsley fresh every year as it runs to seed. Sow in early spring and later in mid-summer at a distance of 8–12 inches (20–30 cm) in drills 2 inches (1 cm) deep. Cover thinly and water well, especially during the 5- to 8-week germination period. When the seedlings are 1 inch (2.5 cm) high, thin to 3 inches (8 cm) and finally to 8 inches (20 cm) when mature. Keep it well watered.
Curly parsley can often be sown three times a year: sow in a border in early spring, on open ground in early summer, and in a sheltered spot in mid-summer.
Harvesting Pick a few leaves of parsley at a time. Bunches should not be picked until the stem is 8 inches (20 cm) high. Pick for drying during the summer and dry quickly. Plain parsley is the only herb requiring a high drying temperature; it must be crisp and brittle before you start rubbing it.

ROSEMARY
Rosmarinus officinalis
Perennial

Use This evergreen shrub was brought by the Greeks to stimulate the mind. We use it to stimulate meat, fish and game dishes.
Soil Rosemary can grow to well over 5 feet (1.5 m). It likes a light, dry soil in a sheltered position, and it needs plenty of lime.
Sowing Sow seeds in early spring in shallow drills 6 inches (15 cm) apart. Transplant seedlings to a nursery bed when they are a few inches high, keeping 6 inches (15 cm) distance between plants, and finally plant out 3 feet (90 cm) apart. Cut in mid-summer so shoots have a chance to harden off before winter sets in. Then cover the soil over the roots with leaf mould and sacking for the winter.
Harvesting Leaves can be picked from the second year on, at any time of the year, although late summer is the best time for drying purposes. Rosemary flowers should be picked just before they are in full bloom.

SAGE
Salvia officinalis
Perennial

Use Although now better known for its presence in stuffings, sage was for centuries regarded as one of the most versatile of all healing remedies. Narrow-leaved sage is better for cooking, while broad-leaved sage is much more suitable for drying.

Soil Sage grows to around 2 feet (60 cm) and needs a light, dry chalky soil. It makes a good border plant and loves the sun.
Sowing
Narrow-leaved sage Sow seed in late spring in humid soil and cover lightly. Germination then takes 10–14 days. Transplant seedlings 15–20 inches (38–50 cm) apart in the early summer.
Broad-leaved sage is always propagated from cuttings taken in very late spring. When rooted plant out 15–20 inches (38–50 cm) apart in rows 2 feet (60 cm) apart.
Harvesting Second-year plants are richer in oils and give a better harvest. Broad-leaved sage is best cut in mid-summer and again a month later to prevent it from becoming too woody. Don't expect it ever to flower in a temperate climate. Cut narrow-leaved sage in early autumn. Sage leaves are rough and need a longer drying time than most herbs.

SAVORY (SUMMER)
Satureja hortensia

Use Summer savory is known as the "bean-herb" and brings out the innate taste of all beans.
Soil Being a bushy plant, growing about 12 inches (30 cm) high, it flourishes best in a fairly rich, humid soil, without compost.
Sowing Sow in late spring or early summer, in rows 12 inches (30 cm) apart. Thin seedlings to 6 inches (15 cm). You will get two cuts from this sowing, one in mid-summer and another smaller one in autumn.
Harvesting Cut shoots for drying shortly before flowering occurs (from mid-summer through to autumn). Harvest seeds as soon as they are brown.

SAVORY (WINTER)
Satureja montana
Perennial

Use Winter savory has a strong flavour and goes well with sausages, baked fish, or lamb.
Soil Winter savory makes an ideal herb garden hedge, prefers a chalky, well-drained soil and plenty of sun.
Sowing Winter savory is germinated by light, so don't cover the seed. Sow in late summer in drills 12–15 inches (30–38 cm) apart, and propagate by cuttings in spring, planted out 2 feet (60 cm) apart. Plants will continue to grow healthily year after year in the same place.
Harvesting Cut shoots and tips from early summer of the second year onwards. Cut before flowering to get oils at their peak.

SORREL
Rumex acetosa
Perennial

Use Pick young leaves and eat them raw or cook like spinach. Sorrel's acid taste combines well with rich stews and fish. Sorrel soup is a speciality of France.
Soil Sorrel needs a light, rich soil in a sheltered, sunny spot.
Planting The herb is best propagated by the division of roots in spring or autumn. Plant out about 15 inches (38 cm) apart. When the plant flowers in early summer, cut it back to prevent it from going to seed.
Harvesting Pick 3–4 months after planting when it has for or five leaves. Harvest shoots and tips for drying in the early summer before flowering starts.

TARRAGON
Artemisia dracunculus
Perennial

Use An important cooking herb, tarragon is a classic for shellfish and is also delicious with chicken and buttered vegetables (especially courgettes). The young leaves are fine in salads.
Soil Drainage is important if you are to grow tarragon well. Slightly sloping, sunny ground is ideal.
Planting Tarragon is another sun-loving herb, and the roots will

spread out about 4 feet (120 cm), so give it growing room. The best way to establish is to buy plants from a nursery and plant out 2 feet (60 cm) apart after the last frost of winter. Pull the underground runners away from the main plant for propagation in late spring. Transplant cuttings in either spring or autumn.

Harvesting Fresh leaves can be picked continually all summer long and this will encourage new ones to grow. Harvest the leaves for drying at the beginning of the flowering period.

THYME
Thymus vulgaris
Perennial

Use Garden thyme is a good herb to put in the pan with any roast meat, or to use in soups and stuffings. It should not be used too freely, as it can drown other tastes.
Soil Thyme thrives in a dry, well-drained position, with light soil.
Planting Seeds can be sown in late spring ¼-inch (0.5-cm) drills 2 feet (60 cm) apart, but the herb is generally grown from cuttings taken in early summer. Side shoots can be layered in spring. Transplant the rooted cuttings or layers 12 inches (30 cm) apart in rows 2 feet (60 cm) apart. Keep beds well watered and free from weeds.
Harvesting In the first year only one cutting should be made. Two cuttings can be taken from the second year on, the first in early summer, just before flowering, the second in mid-summer. Do not cut stems from the base of the plant; cut shoots about 6 inches (15 cm) long. Trim the plant after flowering to prevent it from growing leggy.

Vegetables through the Year

Exactly the same principle of crop rotation applies to the garden as to field crops, but in the garden there are two main factors to consider: you want the biggest possible gap (at least 3 years) between brassica crops to prevent club-root disease from building up, and the biggest possible gap between potato crops to guard against eelworm. You should also take into account that potatoes don't like freshly-limed ground, which makes them scabby, whereas beans and peas do. Brassica prefer limed ground, but after the lime has been in it a few months. The root crops don't like land too freshly mucked or dunged.

You can pander to the needs of all these plants if you adopt a 4-year rotation something like this: Manure the land heavily and sow potatoes. After the potatoes are lifted, lime the land heavily and the next year sow peas and beans. Once the peas and beans are lifted, set out brassica immediately from their seed bed or their "holding bed" (see below). The brassica will all have been eaten by the next spring and it will be time to put in what I call mixed crops. These will be onions, tomatoes, lettuce, radishes, sweet corn, and all the gourd tribe (marrows, squash, courgettes, pumpkins, and cucumbers). Follow these with root crops such as carrots, parsnips, beet, and celery. (Both mixed crops and root crops can be very interchangeable.) Don't include turnips or swedes which suffer from club-root and therefore must go in the brassica rotation if you aren't already growing them on a field scale, which suits them better. Then back to spuds again, which is where we started.

This suggested rotation will suit a garden in a temperate climate with a fairly open winter. (Snow doesn't hurt unless it is extremely deep, but intense frost stops you from having anything growing outside in the winter at all.) Probably no one would stick to this (or any) rotation slavishly. I know that there are idiosyncrasies in it, but I also know that it works. For example, I cram the brassica break in after the peas and beans, and clear the land of brassica the subsequent spring: this may be crowding things a bit, but two main crops are being produced in one year. To do this (and personally I find it a very good thing to do) you must sow your brassica seed in a seed bed, preferably not on any of your four main growing plots at all but in a fifth plot, which is for other things such as perennials.

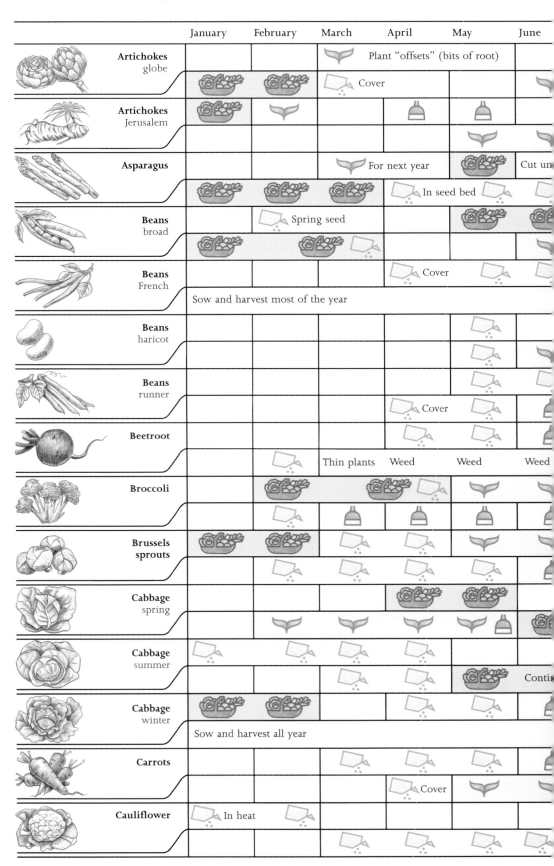

	January	February	March	April	May	June
Artichokes globe			Plant "offsets" (bits of root)			
			Cover			
Artichokes Jerusalem						
Asparagus			For next year		In seed bed	Cut un...
Beans broad		Spring seed				
Beans French		Cover				
	Sow and harvest most of the year					
Beans haricot						
Beans runner				Cover		
Beetroot			Thin plants	Weed	Weed	Weed
Broccoli						
Brussels sprouts						
Cabbage spring						
Cabbage summer					Conti...	
Cabbage winter						
	Sow and harvest all year					
Carrots				Cover		
Cauliflower	In heat					

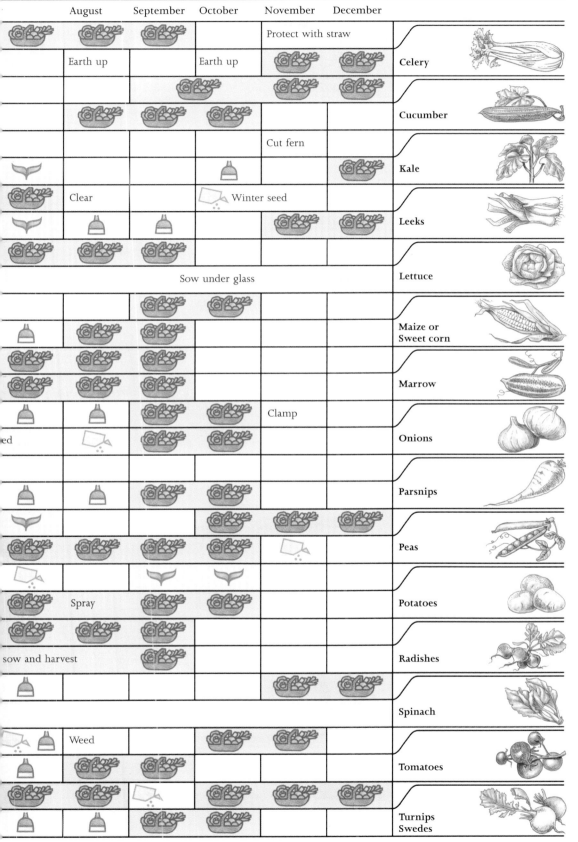

Then you must plant out the little plants from your crowded seed bed to a "holding bed" (a piece of clean, good land, in which these small brassica plants can find room to grow and develop). It will be late in the summer before many of them can go in after the just-harvested peas and beans, and it would be fatal to leave them crammed in their original seed bed until that late. So cramming 5 main crops into 4 years requires a holding bed as well as a seed bed. We can then look upon such quick growing things as lettuces, radishes, and early peas (which are actually best sown late) as catch-crops, ready to be dropped in wherever there is a spare bit of ground. Perhaps you think that radishes are brassica and therefore should only go in the brassica break? Well, we pull and eat ours so young that they don't have time to get and perpetuate club-root. But don't leave them in to get too old and go to seed, or they will spread this rather nasty disease. Provided you keep brassica crops 3 years away from each other, you won't go far wrong.

Climate is all-important, and for the seasonal plans on the following pages I have taken as the norm a temperate climate, which will support brassica outdoors all winter but will not allow us to grow subtropical or Mediterranean plants outdoors. In a climate with no winter frosts we could get three or four crops a year, provided we had enough rain or enough water for irrigation. In climates too cold for winter greens outside, we would have to devote the summer to one plot of brassica for storage during the winter.

A vegetable calendar

The chart shows when to sow, plant out, hoe, and harvest for vegetables you might grow in a temperate climate. As a tip, check what your neighbours are growing; the climate where you live could make as much as a month's difference.

Sow

Plant out

Hoe

Harvest

Winter

Winter is a time for building and repairing, for felling timber and converting it, for laying hedges, digging drains and ditches, building fences and stone walls. If the soil in the garden is heavy clay, it is best to keep off it as much as possible, because digging or working such soil in winter only does harm to it. On lighter land the same inhibition does not apply. In cold climates the land may be deep under snow anyway, and all the crops that have been harvested will be safe in clamp (*see* p.198) or root cellar, or stored away in jars or bottles, crocks, and barrels. Good husbandmen and women should start the winter feeling that their labours have secured them a store of good and varied food, to keep them through the dark months, and to provide hospitality for their friends. So winter is also a time for feasting.

Greenhouse and perennials

In the greenhouse, it is time to clear winter lettuce, and the enriched soil that grew tomatoes last year goes out to the garden, with fresh soil barrowed in and mixed with compost. Tomato and cucumber seed are sown in the heat of the greenhouse. "Hot beds" can be built up in the cold frames. Mature compost is emptied on to the land intended for potatoes. Any remaining compost then goes into an empty bin to aerate it, and a new compost heap is begun. Perennial plants protected from the winter cold by straw and seaweed are resting, preparing for their spurt of growth in the spring.

Plot A

This plot will have been very heavily mucked after the potatoes were lifted last autumn. A small proportion of it may well be winter-sown broad beans this year if the winter is mild. The rest will go under winter rye or another winter green crop, which will stop the loss of nitrogen, and keep it ready to dig in as early as the land is dry enough to work in the spring. The plot was limed last autumn after the spuds had been lifted, and this will benefit the peas and beans that are to follow and also the brassica crop which will come after them. A small part of this bed will have been planted with spring cabbage plants last autumn, and there will be a bed of leeks, which will be ready to pull.

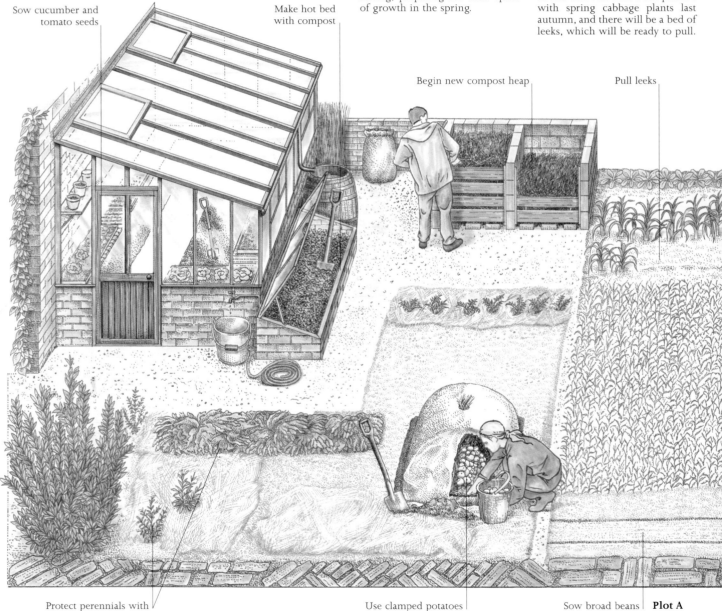

Sow cucumber and tomato seeds

Make hot bed with compost

Begin new compost heap

Pull leeks

Protect perennials with straw and seaweed

Use clamped potatoes

Sow broad beans | **Plot A**

Plot B

This plot should be full of big brassica plants: Brussels sprouts, hearting broccoli (also called "winter cauliflowers"), big hard-hearted winter cabbage, kale, red cabbage, and any other brassica plant that can weather the winter. There may well be a few rows of swedes if these haven't already been harvested and clamped. Turnips must be in the clamp or root cellar by now, for they can't stand the winter as swedes can. This plot will provide most of the greenstuff during the winter, helped by the leeks in plot A. In temperate climates, this helps to avoid much complicated canning and bottling. Shallots can be planted out, as this plot becomes the "miscellaneous" break next summer.

Plot C

This plot is under green manure such as rye or some other winter crop. Last year it bore the miscellaneous, short-lived crops. As soon as the land is dry enough to work, the green manure can be forked into the ground, so that it can begin to rot down. There is no hurry because this plot is going to be "roots" this year, and most of these will not have to be planted out very early.

Plot D

This plot is fallow, or else under green manure, although if the roots were harvested late last year, there may not have been time to sow any green manure. It is time to barrow out compost or muck for the future crop of spuds. If barrowing is done in heavy frost, it is easier to push the barrow. It also does the ground less damage. There may be a row of celery left undug, and this can be remedied as the winter progresses.

Fruit plot

Fruit trees only need spraying with a winter wash if pests have afflicted them badly. 2½ pounds (1.1 kg) of caustic soda dissolved in 10 gallons (45 litres) of water was the old-fashioned remedy, but most people now buy commercial winter washes. After the middle of February fruit trees, gooseberries, and other bushes are pruned. The blackcurrants may well have been pruned in the autumn. Muck or compost is barrowed and dumped around trees and bushes, and the ground between soft fruit bushes is forked lightly. All prunings should be burned.

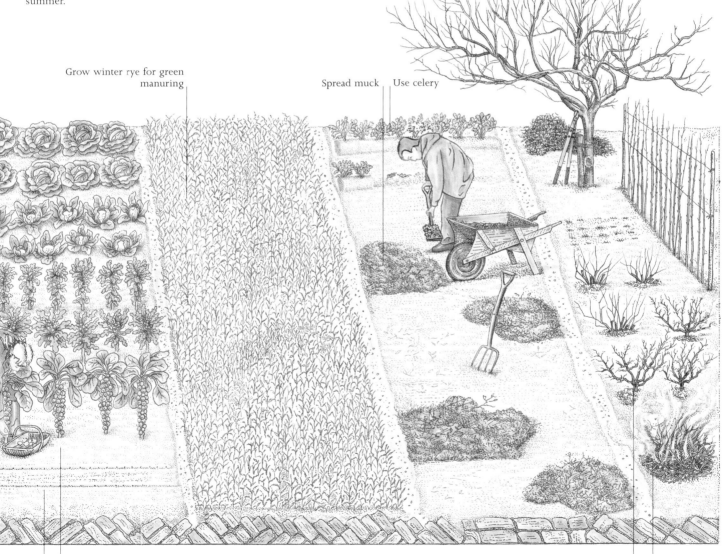

Grow winter rye for green manuring

Spray and prune fruit trees if necessary

Spread muck   Use celery

Plot B   Use brassicas   Plant shallots   Plot C   Plot D   Prune fruit bushes late in the season   Burn prunings   Fruit Plot

Spring

There is so much to do in spring that it is difficult to get going fast enough. For a start, the green manure crops are turned in (a rotavator is helpful), the seed beds prepared, and seed sown. But it is no good being in too much of a hurry to sow seeds, because they can't grow in half-frozen ground, and wet ground is cold ground. It is better to sow a week or two later, in dry, warm soil, than earlier, in wet, cold soil. Some things, like parsnips, need a very long growing season, and can be put in early. Some others are best started off early, but under glass. Cloches are a great help at this time of year, to warm the soil for early sowing. In March I have a big sheet of transparent plastic over February-planted early potatoes. The soil under it feels warm to the touch, while the soil outside is freezing.

Greenhouse and perennials

In a heated greenhouse, sweet corn is sown in peat pots and green peppers in seed boxes. As the tomato and cucumber plants become big enough they can be planted out in pots or greenhouse soil. Cucumbers can be sown in the hot bed. In the herb garden it is time to lift, divide, and replant perennial herbs such as mint, sage and thyme if they need it. The seaweed covering should be removed from the asparagus bed, and the seaweed put in the compost heap. Rhubarb is forced under dark cover. Globe artichokes should be progressing well. Seeds are sown in the seed bed ready for planting out later: onion, all the broccoli, cabbage, kale, cauliflower, Brussels sprouts, and leeks. Lettuce can also be sown in the seed bed if space is wanting in the main garden and lettuce plants are wanted ready for replanting later when there is room.

Plot A

The leeks are cleared and eaten as the spring advances. The winter-sown broad beans will be growing well, but if there aren't enough of them spring-sown varieties may be planted early in the season, when early peas will also go in. After this, peas will be sown in succession as the year advances. However many are grown, there will still never be enough! Early turnips, soya bean, and swedes should go in this plot, which will be brassica next winter. The row of spring cabbage will do for fresh greens, and will get eaten as spring advances.

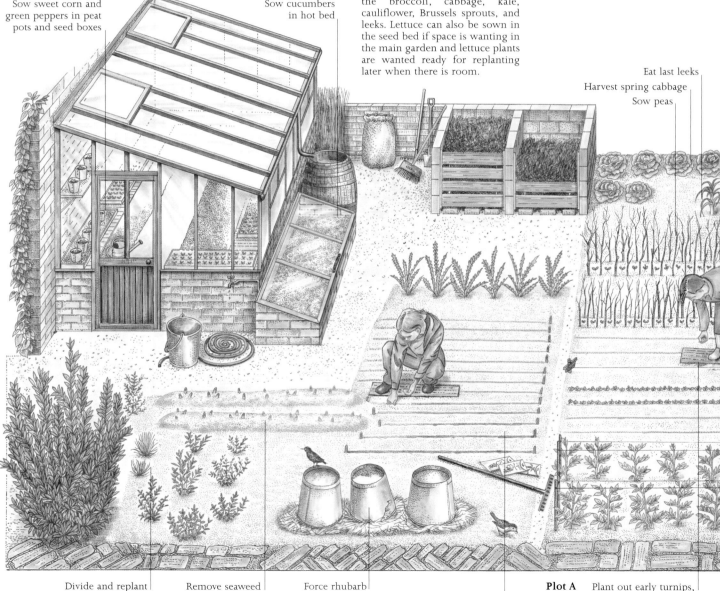

Sow sweet corn and green peppers in peat pots and seed boxes

Sow cucumbers in hot bed

Eat last leeks
Harvest spring cabbage
Sow peas

Divide and replant perennial herbs

Remove seaweed from asparagus

Force rhubarb
Sow brassicas and onions in seed bed

Plot A Plant out early turnips, swedes, and soya beans

Plot B

Spring is more of a hungry-gap than winter, but the late hardy brassica, together with the leeks, tide things over. The brassica are nearly finished, but maybe a few Brussels sprouts are still standing, with some kale, some sprouting broccoli, and perhaps a few hearting broccoli. As the plants are finished, they must be pulled out, the stems smashed with an axe and then put on the compost heap. The shallots should be growing well.

Plot C

By now the winter rye sown last year as a green manure crop should be dug in, to make way for the roots to be sown later in the year. The only root crop sown early on is parsnip, but as spring progresses onion seed and carrots are sown in the bed. It is time to plant out onion sets and autumn-sown onions. If there is no garlic in the herb garden, it must be put out here early in spring. As spring turns into early summer more and more crops go into this root break bed.

Plot D

A row of early potatoes could be growing under cloches or transparent plastic. These will have been planted towards the end of February in mild climates, or late March in severer ones. The main crop won't go in until mid-April. The earlies get planted shallowly but the main crop go in deep furrows, both with ample muck compost. They are ridged up as they grow.

Fruit plot

Prune gooseberries early in the season. Some people set out strawberry plants in March or April. The ground around soft fruit such as the blackcurrants, gooseberries, and raspberries should be kept hoed and cultivated to prevent grass from growing. Insect pests are to be avoided and something must be done about them if they attack. Grease-bands put around fruit trees will catch crawlies climbing up. It is important not to spray insecticides on flowering trees, as they kill the beneficent bee.

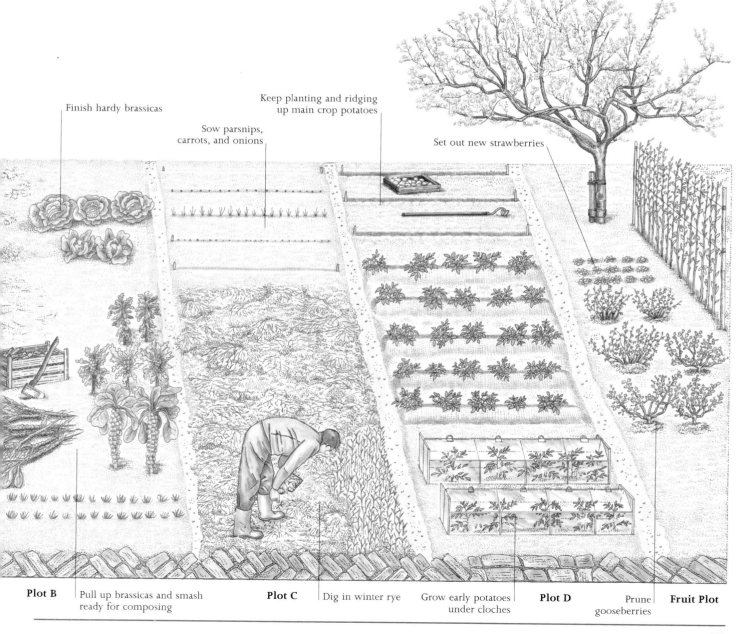

Finish hardy brassicas

Sow parsnips, carrots, and onions

Keep planting and ridging up main crop potatoes

Set out new strawberries

Plot B | Pull up brassicas and smash ready for composing | **Plot C** | Dig in winter rye | Grow early potatoes under cloches | **Plot D** | Prune gooseberries | **Fruit Plot**

Early Summer

Successional planting must go on unabated with many crops during April, May, and June. A constant supply of fresh peas, lettuces, radishes, and French beans can be maintained by planting these short-lived plants little and often. Fresh young turnips should be available all summer too. Hoeing should never be neglected during the early summer, as this is the time when the weeds are raring to get a foothold along with everything else. If they are allowed to get away with it, the crops will be miserable or non-existent. Onions and carrots must be meticulously hand-weeded. If some radish seed is sown along with the parsnips, radishes will be clearly visible before the slow-growing parsnips have declared themselves, and can be side-hoed with safety.

Greenhouse and perennials

Asparagus can be cut and eaten until the end of June, when it must be abandoned and allowed to grow. Herbs will thrive on frequent pickings. Artichokes are growing fast. The seed bed is kept weeded. If flea beetle appear on brassica seedlings, they can be dusted with derris or pyrethrum dust. Ventilation in the greenhouse must be carefully adjusted. A good airing is vital during the day, but cold air must be kept out at night. The air is kept humid by spraying the floor and plants. The top glass should be lightly shaded with whitewash. Tomato plants are fed with water in which muck has been soaked, and as small cucumbers begin to develop they, too, are fed. Brassica plants are pricked out into a holding bed. The lids on cucumber frames should be propped open. Forcing of rhubarb continues.

Plot A

Peas are sown in succession and given sticks to twine around as they need it. More turnips and swedes can be sown. May, or June in later districts, is the ideal time to sow out French and runner beans on previously prepared well-composted beds. These need regular weeding and watering; all these legumes want frequent watering in a dry season. It is time to harvest broad beans, and if there had been any signs of black-fly earlier, the tops of the broad beans should have been snapped off immediately and cooked. As soon as they are finished, cut them down and sow French beans in their place.

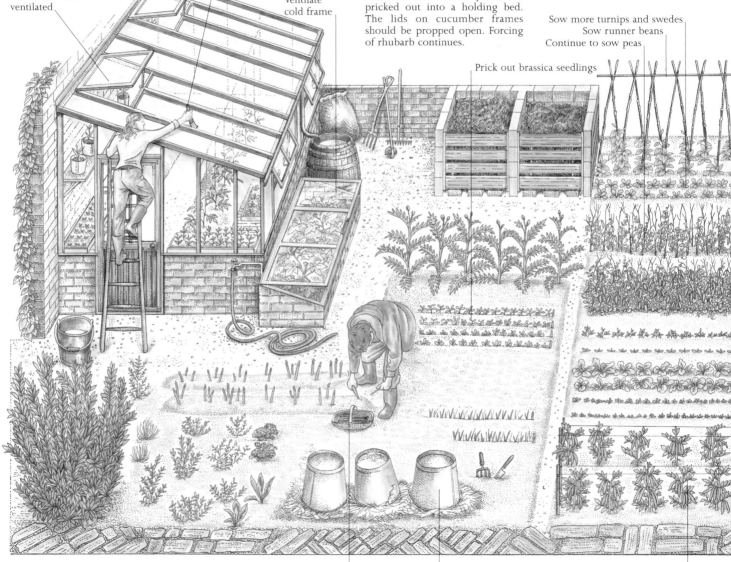

Keep greenhouse humid and well ventilated

Whitewash greenhouse roof

Ventilate cold frame

Sow more turnips and swedes
Sow runner beans
Continue to sow peas

Prick out brassica seedlings

Cut asparagus until the end of June

Continue to force rhubarb

Plot A   Harvest broad beans

Plot B

Now cleared of last winter's brassica, this plot becomes the new miscellaneous bed, for outdoor tomatoes, courgettes, melons, marrows, pumpkins, squashes, radishes, lettuce, ridge cucumbers, spinach, and sweet corn. As all these things – some of them reared in the greenhouse or cold frame – become ready, and the weather is warm enough, they are planted out, and should be watered and tended. A good mulch of well-rotted muck or compost, if it can be spared, will do them all good. It revives the soil, and shouldn't make next year's roots "fork" too much, if put on well in advance.

Plot C

Onions in the root break plot should be growing well and will need weeding and thinning. The carrots should be thinned if they are wanted for winter storing but not if intended for summer eating. The wily carrot fly must be avoided. Carrots should only be thinned when it is raining; otherwise paraffin, or some other strong-smelling stuff, must be sprinkled on the row after thinning. Parsnips are thinned and weeded. Endive and beet are sown. Celery should be planted out before the end of May in a previously prepared celery trench, and never allowed to dry out at all.

Plot D

The potatoes already planted should be earthed up as they grow. Very early morning, or late evening, is the best time to do this, because the leaves lie down and sprawl during the day and make earthing up difficult. Turnips can be sown to come up in the brassica break when the early potatoes are out. The trick of planting leeks after spuds have been lifted can only be done if the spuds are early ones. Earlies are being eaten by June, so leeks can be transplanted into the ground when it is clear.

Fruit plot

Nets go over strawberries, straw underneath them, and birds must be kept off other soft fruit, too. Soft fruit such as gooseberries can now be picked, starting with the hard ones for cooking, so as to give the younger ones a chance. Insects and various blights must be kept at bay. The ground between soft fruit bushes is hoed, and a mulch of compost or anything else put on. It is vital on light land.

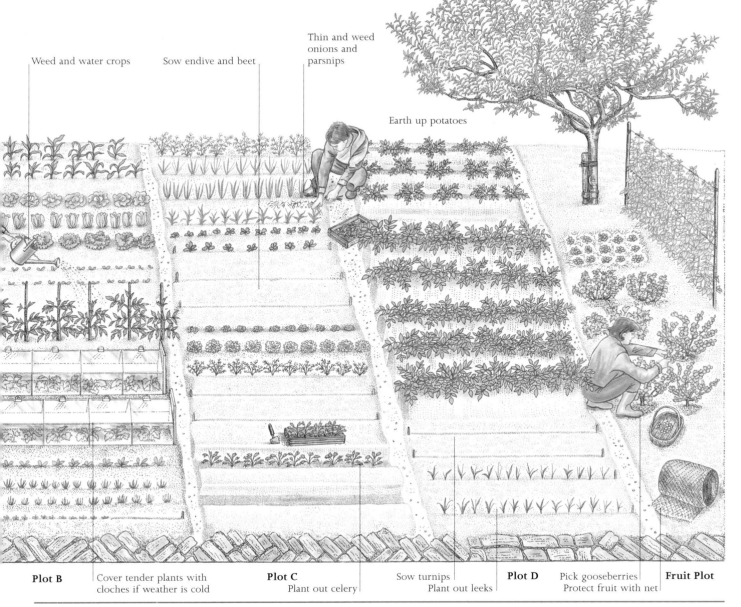

Weed and water crops

Sow endive and beet

Thin and weed onions and parsnips

Earth up potatoes

Plot B | Cover tender plants with cloches if weather is cold

Plot C | Plant out celery

Sow turnips
Plant out leeks

Plot D | Pick gooseberries
Protect fruit with net | **Fruit Plot**

Late Summer

Earlier labour will now start bearing fruit in earnest. There is almost an *embarras de richesse* of harvest, and it is time to think of giving away, or trading, the surplus of many crops. The surplus of French or runner beans can be stored in salt, and the haricot beans and peas prepared for dry storage. As fast as peas and beans are harvested and cleared, the space is filled with well-grown brassica plants. Fitting the main brassica crop in as a catch crop after the peas and beans have been cleared is made possible by the use of the "holding bed", which comes into its own this season. Brassica seems to benefit from the twice planting out. Hand-weeding must go on incessantly, for weeds that are too big to hoe must be pulled out before they have time to seed: one year's seeding is seven years' weeding.

Greenhouse and perennials

With the lid now taken off the cold frames, the cucumbers will run riot. Tomatoes, cucumbers, and peppers in the greenhouse will be bearing, and will want watering and feeding. They now need plenty of ventilation. In the seed bed early spring cabbage seed can be sown. The herb and asparagus beds are kept weeded and the rhubarb needs to be regularly pulled. Soon the flowers of globe artichokes will be eaten; they should not be neglected, because uncut plants will not produce any more. However, it is fun to leave a few to burst out in brilliant blue flowers and add to the scenery.

Plot A

The peas and beans are watered if they need it, and the flowers of runner beans sprayed with water every evening, to help the flowers set. Peas, French beans, and runner beans are now ready to be picked. So are the turnips. As each row passes its best, it must be ruthlessly cleared out of the way, and the space planted up with well-grown brassica plants from the holding bed. When the runner beans begin to yield, they must be picked and picked again and never allowed to get old and tough. A great many are salted for the dark days of the winter. The true countryman always bears the winter in mind; it is easy enough to get plenty to eat in July.

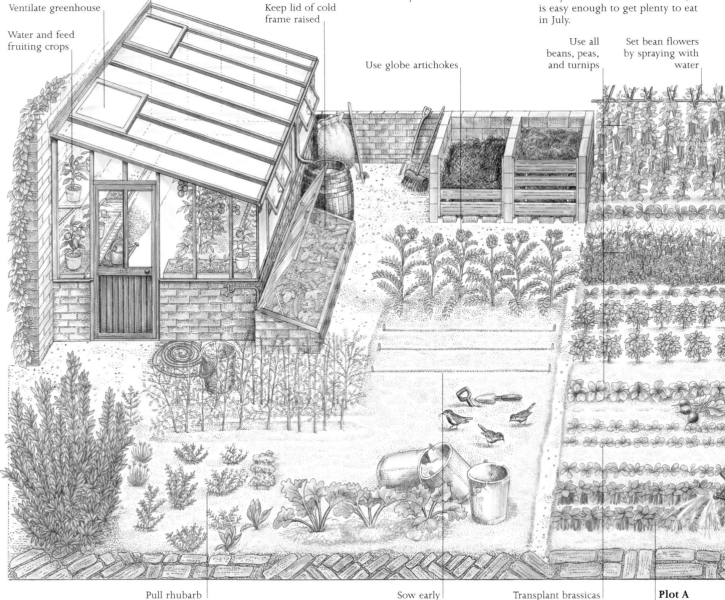

Ventilate greenhouse

Water and feed fruiting crops

Keep lid of cold frame raised

Use globe artichokes

Use all beans, peas, and turnips

Set bean flowers by spraying with water

Pull rhubarb

Sow early spring cabbage

Transplant brassicas

Water French beans

Plot A

Plot B

Any straggling vines of melons, pumpkins, and squashes must be cropped. The tomatoes must be staked, side-shoots picked out, and the plants stopped when they have four trusses. They must be well watered if it is dry. In damp climates, it is a good idea to lay the tomatoes down at the end of August and put cloches over them so that more of them may ripen. Outdoor cucumbers are stopped before they get out of hand, and they must be picked hard and continuously so as not to get too big and bitter. All male flowers must be picked off. Lettuce should be eaten when ready, and should not be allowed to go to seed. Successional plantings of both lettuce and radishes continue. Sweet corn is now high, in a block to facilitate wind pollination. The shallots can now be harvested.

Plot C

There is little to do now but hoe all the root crops, keep the weeds down, and kill any slugs lurking in the vegetables. In fact, this is by far and away the best time of the year to clear all the weeds in the garden. Celery can be earthed up and sprayed with a Bordeaux mixture in preference to leaf-spot. Start harvesting the onion tribe now.

Plot D

By now the early potatoes are gradually being eaten, and the second lot started on if there are any. The main crop must not be lifted yet, but can be sprayed twice with Bordeaux mixture, if blight is feared. Warm, muggy weather is the enemy. The main crop must be well earthed up, but when the tops meet across the furrows it won't be possible or necessary to hoe any more, though the big annual weeds should still be hauled out. Turnips and leeks should be establishing themselves.

Fruit plot

Cut any superfluous suckers from the base of raspberry plants. Thin immature apples where they are too thick on the tree (the "June drop" may have done this naturally) and summer-prune fruit trees, particularly cordons and trained trees. Eat plums and soft fruit while birds are eating the cherries. I think August is the time to plant a new strawberry bed, so root the strawberry runners in small sunken pots. Carry on hoeing between soft fruits to keep the grass down and give the birds a chance to eat creepies.

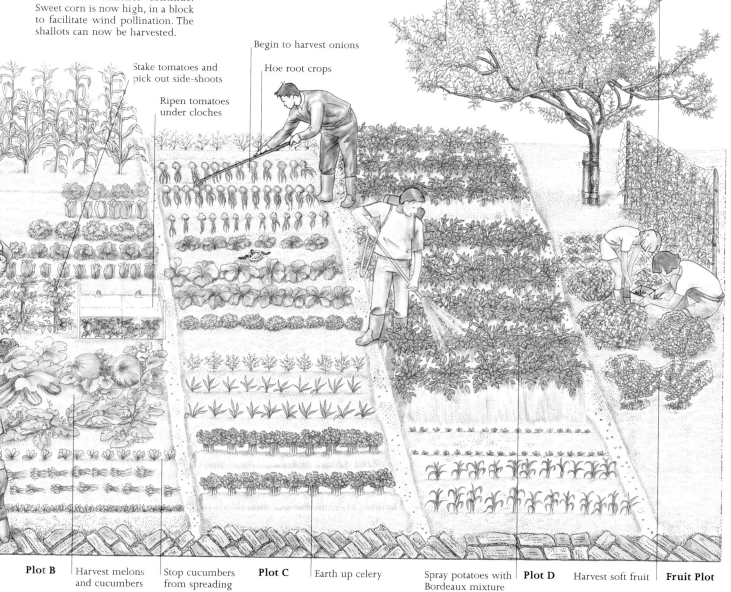

Remove raspberry suckers

Thin apple crop

Begin to harvest onions

Stake tomatoes and pick out side-shoots

Hoe root crops

Ripen tomatoes under cloches

| **Plot B** | Harvest melons and cucumbers | Stop cucumbers from spreading | **Plot C** | Earth up celery | Spray potatoes with Bordeaux mixture | **Plot D** | Harvest soft fruit | **Fruit Plot** |

Autumn

Autumn is the season of mists and mellow fruitfulness according to Keats. It is also the real harvest time, when all the main crops have to be gathered in and stored for the winter. The good gardener will try to broadcast green manure seed wherever beds are left empty, although on very heavy soil old-fashioned gardeners are fond of leaving it "turned up rough" after digging so that frost can get at it. Use annual weeds as a green manure if you like.

After the first frost has touched celery and parsnips, it is time to start eating them, and time to think of parsnip wine for Christmas (or the Christmas after next, as purists would have it).

Greenhouse and perennials

The frames and greenhouse can be sown with winter lettuce, spring cabbage, and summer cauliflower. The last two will be planted out next spring. Asparagus ferns are cut down and composted, thus defying the asparagus beetle. Potatoes may well be clamped near the house or put in the root cellar (or anywhere cold, dark, and frost-proof). Globe artichokes are cut as long as there are any left. Then they are abandoned, except for a covering of straw, as they die down, to protect them against frost. It is a good idea now to cover the asparagus bed with seaweed, or manure, or both. All perennial crops want lavish manuring.

Plot A

Now it is time to clear away all the peas and beans, even the haricot beans, soya beans, and any others intended for harvesting and drying for the winter. This bed will hold winter and spring brassica, planted late perhaps, but none the worse for that, as they have been growing away happily in their holding bed. The cabbages will benefit from the residual lime left by the peas and beans and the residue of the heavy manuring given to the previous spuds. When all weeds are suppressed, it is a good idea to mulch the brassica with compost, but slugs must be kept down.

Sow next year's brassicas in greenhouse and frames

Plant winter and spring brassicas

Cut and compost asparagus ferns | Manure perennials | Clamp potatoes | Mulch brassicas well | **Plot A**

Plot B

All the plants in this bed (which are plants with a short growing season) will have been harvested. After the bed has been cleared, it should be lightly forked over, and winter rye planted for green manure. Unfortunately, it is not much good trying a clover for this, as it is too late in the year; only a winter-growing crop such as rye will work.

Plot C

Parsnips can stay in the ground indefinitely. Once earthed up, celery will also survive much of the winter. The rest of the roots are lifted in September and put safely in store. Red beet needs lifting carefully, as its roots bleed when damaged. As the land is cleared rye can be sown in it at least up until the end of September. This bed will be spuds next year, and manuring can now begin.

Plot D

Lift the main potato crop quite late, just before the first frosts are expected. This way the tubers will harden in the ground and keep better, and if blight is present there is less chance of spores being on the ground's surface to infect the tubers when they are lifted. The spuds should lie drying out on the surface for a day or two, while their skins set. Then they are clamped or stored away. The leeks are earthed up to be a great winter standby. As this plot will be the pea and bean break next year, broad beans are planted in October, or September if you have hard winters.

Fruit plot

Runners are cut away from the strawberries, the ground cleared and given a good top-dressing of muck or compost. All fruit is harvested as it becomes ripe, then apples and pears are stored in a cool (not a frosty) place, so that they don't touch each other. The old fruiting canes of raspberries are cut out, leaving the young wood, and blackcurrants are pruned in November or December. New fruit trees can be planted in November if the ground is not too wet. As tree leaves fall, take them up and compost them, because they harbour troublesome pests.

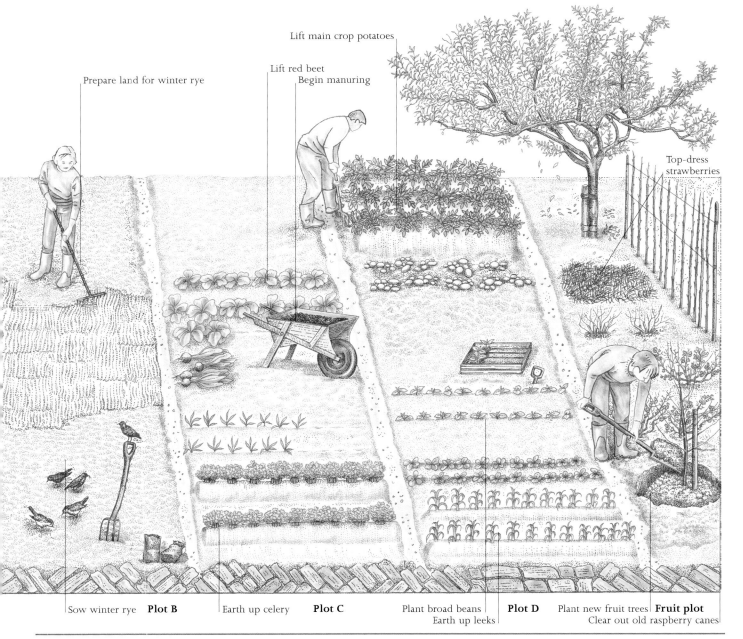

Prepare land for winter rye

Lift main crop potatoes

Lift red beet
Begin manuring

Top-dress strawberries

Sow winter rye **Plot B**   Earth up celery   **Plot C**   Plant broad beans   **Plot D**   Plant new fruit trees **Fruit plot**
Earth up leeks   Clear out old raspberry canes

The Greenhouse

A greenhouse can be a very basic thing; it can consist of a 3-foot (90-cm) high foundation of brick, concrete or stone, a wooden framework containing the glass (heavy glass is best), a door, and four ventilators (two at each end of the building, one high up and another low down). Inside you need staging for standing seed boxes on, and you should be able to remove this so that in the summer you can plant tomatoes in its place.

Unheated greenhouses

In countries where grapes and tomatoes will grow reliably out of doors I personally would not bother to have a greenhouse, but would spend the money on other things. But in cooler climates even an unheated greenhouse is enormously useful for starting off things like celery seed, sweet corn, early summer cabbage, and anything else you wish to get off to a flying start out of doors as soon as the frosts are over. You can also use it during the summer for growing that magnificent plant, the tomato. Tomatoes are a most desirable crop for the self-supporter. They are expensive to buy, but easy to grow; they bottle well, and having a store of them makes all the difference between some possibly pretty dull food in the winter and *la dolce vita*. A couple of dozen large kilner jars filled with fine red tomatoes on the shelves come autumn are a fine sight and give us hope for the future.

And in summer your cold greenhouse may nurture such luxury crops as aubergines, melons, green peppers (which turn into red ones if you leave them long enough), and of course cucumbers. The cucumbers you grow inside a greenhouse taste much better than frame or ridge cucumbers grown out of doors. And you can have lettuce nearly all the year round if you grow it in a greenhouse. In spite of this, a cold greenhouse will not help you much in the winter time, except by bringing along some early cabbage or some winter lettuce, or something that is pretty hardy anyway, because the temperature inside the greenhouse, when there is no winter sun, may go well below freezing point. So do not expect marvels. Remember the limitations.

Heated greenhouses

If you can just manage – by hook or by crook, by oil or electricity, or wood-burning or coal – to keep the temperature of the air in your greenhouse above freezing all winter, and your greenhouse is big enough, you can have peaches, pears, nectarines, grapes, and most Mediterranean climate fruits every year in any climate.

If you want to heat your greenhouse, you can have water pipes running through it. The pipes should slope gently up from the boiler as far as they go, since the hot water will tend to rise and the cold to sink back to the boiler. At the highest point of the pipes there must be a bleeder valve to let out air or steam that may collect. If the masonry inside the greenhouse is painted black, heat is absorbed during the day and let out during the night to allay the frost.

The self-supporter will like the idea of heating his greenhouse without buying fuel. This can be done possibly by harnessing an all-purpose furnace (*see* p.287) or by water or wind-generated electricity (*see* pp.242–245). Solar heating, properly used, has always been adequate to heat greenhouses in the warmer months of the year.

Greenhouse temperatures

In the winter the temperature should be about 40°F (4°C) at night. The sun should bring this to about 50°F (10°C) on bright days. The day temperature should not be allowed to get too high, but it must not be kept down by admitting freezing air into the place, as this will inevitably kill tender plants. So cool the air by letting the boiler fire go out, but get it going again in the afternoon.

A LEAN-TO GREENHOUSE
This greenhouse is a practical way of getting such things as melons, peppers, grapes, and tomatoes off to a flying start in a temperate climate.

In this way the temperature can be kept up at night. During the daytime in winter, have the leeward (that is, the side away from the wind) top ventilator open. Then as spring gets into its stride, open both top ventilators a little more. Eventually, open one of the bottom ones as well, but arrange for the cold air coming in through this to go over the hot pipes. In spring and summer sprinkle water on the floor occasionally to keep the air humid. It helps if you can arrange for the water from the roof to go into an adjoining water butt.

Greenhouse soil

Greenhouse space, whether heated or not, is expensive, and it is therefore not practical to fill your greenhouse with any old soil. The better the soil in the greenhouse, the better use you will be making of this expensive space. If you mix very good compost, good topsoil, and sharp sand in equal parts, and add a scattering of ground rock phosphate and a little lime, you will have a very good soil for your greenhouse. You can put this soil in raised beds, or straight on to the existing soil of the greenhouse. The more you rotate crops inside the greenhouse, the better, but if you are driven to growing the same crop year after year then you may have to remove the old, or spent, soil bodily and replace it with new soil. Tomatoes particularly can suffer from disease if grown too many years on the trot on the same soil.

Greenhouse crops

As to what to grow in the greenhouse, we are all guided in this by what we can grow and what we want. A cold greenhouse enables you to grow a slightly greater range, more reliably than you could outdoors. A hothouse enables you to grow practically anything that can be grown on earth. For my part the main uses of the greenhouse are winter lettuce and other saladings; seed sowing in flats or seed boxes in the early spring of celery, tomatoes, peppers, melons, aubergines, sweet corn, and cucumbers; and my greenhouse crop is tomatoes, which go on all throughout the summer. I know you are supposed to be able to grow tomatoes outdoors in a temperate climate, but you can't really. A tiny greenhouse, however, will produce a really impressive tonnage of ripe red tomatoes that can be eaten fresh until you are fed up with tomatoes. Then they can be bottled to provide marvellous food and flavouring right through the year. You simply cannot have too many tomatoes. As for cucumbers, they can be grown out of doors (the ridge and frame varieties), but there is no reason why you should not grow a few in the tomato greenhouse too. The conditions are not ideal for them though: the true cucumber house is much hotter and more humid than the good tomato house. My advice is to keep your house to suit tomatoes and let the cucumbers take pot luck and do the best they can.

And then there is no harm, when you live too far north to grow grapes reliably out of doors, in having a big old vine growing up the back (north) wall of the greenhouse, trained under the roof so as to get the benefit of the sun without shading the precious tomato plants. A fan-trained peach tree, too, is a pleasing luxury in a fairly large greenhouse. And in countries with very cold winters it is quite useful to sow the seeds of temperate things like brassicas in the greenhouse in the very early spring.

Whatever you do, don't overcrowd your greenhouse. It is far better to grow plenty of one really useful crop, like tomatoes, in the summer, and another really useful crop, like lettuces, in the winter than to fill your greenhouse with innumerable exotic fruits and vegetables. Make all the use you can of hot beds under cold frames, cloches, jam jars and sheets of transparent plastic and the like, out of doors (*see* pp.48–49).

Upper ventilator

Black wall to absorb heat during the day and release it at night

Water kept at greenhouse temperature

Lower ventilator

Soft Fruit

It takes courage to plant top fruit (that is, fruit grown on large trees), knowing that you have many years to wait before you harvest any fruit, but unless you have one foot actually in the grave there is no excuse for not planting soft fruit. The fruit comes into bearing quickly enough: strawberries planted one summer will give you a big yield the next, and bush fruit does not take much longer. And soft fruit will give you, besides a lot of pleasure, a source of vitamins, easily storable, which will ensure the good health of you and your family. By far the best soft fruit to plant, for my money, is blackcurrants. They are hardy, prolific, extremely nourishing – about the richest source of vitamin C and other vitamins you can grow – and easily preserved. With blackcurrants you can be sure of an ample source of delicious fruit right through the winter (and hunger) gap. Bottled they taste nearly as good as fresh, and they very

seldom seem to have a crop failure: in fact, in 20 years of growing them I have never known one. White currants and redcurrants are not nearly as heavy yielding as blackcurrants. One might grow a few for the novelty, and for variety, but they won't really make much difference to whether you starve to death or not during the winter months. Raspberries are a good grow – they can be very prolific and are fine for jam. They are also hardy and thrive in wet and cold latitudes. Raspberries are far easier to grow than strawberries and really just as good to eat. They have a long picking season and children can be turned out to graze on them. Blueberries and the many small berries of that ilk are grown by people who are hooked on their flavour. They are so laborious to pick in any quantity that they must be looked upon as a luxury. They are useful, though, in cold climates where lusher fruit will not grow.

BLACKBERRIES OR BRAMBLE FRUIT

Use I live in countryside where brambles are a blasted nuisance, and as we pick bushels of wild fruit from them, I wouldn't dream of planting blackberries. But cultivated brambles provide a heavier crop of bigger, sweeter fruit, and are very hardy. They also make good prickly hedges, although you may prefer a thornless variety.

Planting If you want a hedge of tame blackberries, make sure the ground is completely clear of

perennial weeds such as couch grass. Dig in muck or some phosphatic manure, or both. Then plant small plants every 6 feet (1.8 m). Each plant must have a bit of stem and a bit of root and each stem and root should have been shortened to about half its length. Provide them with a wire fence and they will climb along it. In fact, they will spread at an amazing rate, so keep a close eye on where they are growing.

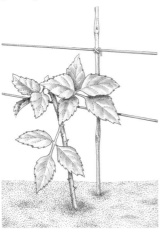

Pruning If you inherit wild brambles, and want to improve them for fruiting, cut the big patches into blocks by clearing rides, or paths, through them. Cut a lot of the dead wood from the bushes, clip the long straggling runners and fling in some phosphatic manure if you really want to make a meal of it.
Aftercare Keep the rides clear, and you will greatly improve both the yield of that bramble patch and the ease of picking the fruit. Do not forget to watch for stray shoots growing up nearby.

BLACKCURRANTS

Use Blackcurrants are by far the most important soft fruit you can grow. They are the richest in vitamin C, and make the best wine of all the fruits.

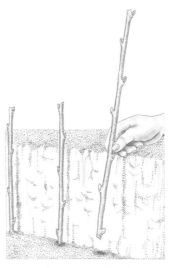

Soil They thrive on a cool and rather heavy soil, even on clay. The land should be limed the previous autumn if it is under about pH6. Get rid of any perennial weeds and dig in plenty of muck.
Planting Take your cuttings from existing bushes in late autumn. Do this in the ordinary course of pruning and remove the tops and bottoms with a very sharp knife. Cuttings should be about 10 inches (25 cm) long. The lower cut should be just below a joint. Make a slot in the soil with a spade, put a little sand in the bottom, and if you are a perfectionist, stick the cuttings into

it with about 12 inches (30 cm) between each. Cover them with leaves or compost as protection against the frost heaving the soil up during the first winter. Nurserymen in cold climates make the cuttings from prunings in November, tie them in bundles and heel them in until March. Then they plant them as described above. Next November lift the young rooted plants carefully and plant them 12 inches (30 cm) apart in rows 18 inches (45 cm) apart. At the end of the second year, lift them and transfer them to their permanent quarters, 6 feet (1.8 m) apart. Don't plant them too deep.
Pruning Blackcurrants, unlike red or white currants, fruit on new wood, so, if you can, cut out all the wood that was fruited on last year. But you will often find that you are faced with a long, old branch with a new branch growing on the end of it, so you will end up retaining some of the old wood. Do not worry if that is the case.

Aftercare Give them plenty of muck every winter and keep the ground clear of grass and weeds.
Pests "Big-bud" is the worst pest. This is caused by a mite, and causes swollen buds. Pick of all such buds and burn them. Another disease they can get is "reversion" when the leaves go a weird shape like nettle leaves. Pull these bushes right out and burn them, so the disease doesn't spread to other plants.
Harvesting Some very lazy people commit the awful atrocity of cutting the fruiting branches off, taking them indoors, and stripping the berries off there! Well of course, it is easier to sit at the kitchen table and pick berries off a branch than to stoop or kneel out in the garden, and you will kill two birds with one stone because you should prune out those already fruited branches that winter anyway. I know people who do it, and it seems to work. But I have never been able to bring myself to do it, because I know that there is still a lot of "nourishment" in that green branch which will go down to the roots as winter comes on, and I feel it is a crime to cut it off before this happens.

BLUEBERRIES
Use Blueberries aren't much good in warm climates, but people living in cold northern regions should consider them very seriously, for they are basically mountain fruit.

Planting Blueberries prefer acid soil to alkaline, so don't put lime on them. They stand up to intense cold and like a rather shallow water table so their roots are near the water. They can't grow in a swamp unless on a hummock. They will grow well on mountain peatland and prefer a pH value of about 4.5, which is very acid. Propagate from cuttings, or buy 3-year-old plants, and plant them 6 feet (1.8 m) away from one another.

Before pruning

After pruning

Pruning When the plant is 4 years old (the first year after planting 3-year-olds), cut out most of the flower clusters and cut away the suckers, the shoots that come up from the roots. Do this for 2 years. Then limit the suckers to two or three for each bush. From then on cut away old wood from time to time. Don't pick the berries until they come off very easily, or they will have little taste.

CRANBERRIES
Use These fruits are most commonly used to make cranberry sauce, which is traditionally eaten with turkey and goes well with game. They will only grow under carefully controlled conditions; for this reason they are rarely grown in gardens.
Soil Cranberries grow in very acid soil. They must be well-drained yet well watered in summer, and then flooded in winter.
Planting Cuttings can be planted in spring in a 3-inch (8-cm) layer of sand on top of peat.

Harvesting After at least 3 years of weeding, watering, and protecting the plants may begin to fruit. The fruits are hand-picked.

GOOSEBERRIES
Use Gooseberries are a very useful source of winter vitamins and they bottle and cook well. You can't have too many of them, and for my part these and blackcurrants and raspberries are the only bush fruit really worth bothering about.

Soil They like a good, deep loam, but you can improve clay for them by digging sand in, and you can improve sand by digging clay in, and you can improve all soils by heavy mucking.
Propagation As for blackcurrants (see opposite), except that you rub out, with your fingers, all the lower buds on the cutting, leaving only four at the top. They also layer well – peg a low branch to the ground and it will root. Cut it off and plant.

Pruning Prune hard the first year or two to achieve a cup-shaped bush (open in the middle, but with no branches straggling down). Then shorten the stems to 3 or 4 inches (8–10 cm) every winter, cutting out all old branches that don't fruit any more. Always keep the middle open

so you can get your hand in to pick the fruit. However, never prune gooseberries in frosty weather.
Aftercare Muck or compost mulch every year. Bullfinches will destroy every bud during the winter if they can, so build a fruit cage if you have to. Leave the cage open in the summer until the fruit can form in order to let friendly birds in to eat the pests, but close it in the winter to keep baddy birds from eating the buds. The bullfinch plague in England and other places is due to gamekeepers. They have destroyed all the predators, like owls and hawks, and small birds have now become a pest.
Diseases A horrible aphid sometimes lives inside gooseberry leaves and makes them curl up. Pick the curled leaves off and burn them. American gooseberry mildew can be sprayed with 2 ounces (55 g) of potassium sulphide dissolved in 5 gallons (23 litres) of water. You can recognize it by a white felt-like growth over leaves and fruit.
Harvesting Just pick them when they are ready. You will find them good for bottling or for wine.

GRAPES
Use Grapes don't mind how cold the winter is, provided the summer is warm enough and there is enough sunshine. They will grow as far north as Suffolk, England. I grew ninety outdoor vines there and got plenty of grapes. The pheasants ate all the grapes but I ate all the pheasants, so that was all right.

Soil Grapes need a very well-drained, warm soil, rich in humus, and they want plenty of sun and air. A south-facing hillside is fine. A pH of 6 is good, so you may have to lime. They can also be grown in a greenhouse and left to climb all over the place.

Propagation They grow well from soft fruit cuttings. Plant rooted cuttings out in lines 6 feet (1.8 m) apart in cold climates, and maybe more in warm. Grapes will fruit better in a cold climate if you keep the vines small and near the ground.

Pruning Have two horizontal wires, one 12 inches (30 cm) from the ground and the other 30 inches (75 cm). Vines fruit on this year's wood, so you can always prune last year's off, provided you leave two or three buds which will produce this year's shoots. In cold, damp climates, don't be too ambitious; leave three shoots to grow. One is a spare in case something happens to one of the others and you cut it off when the other two are established.

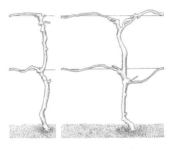

Train the two you leave in the same direction along the two wires by tying them. In warmer climes, leave five shoots. Train four along the wires, two each way, and keep one spare. Prune in late winter. Cut the shoots off after they have made about six buds.

Aftercare Mulch heavily every year with compost. Keep down weeds, and spray with Bordeaux mixture (*see* p.164) in June.

Harvesting Cut the bunches off with secateurs. Never tear them off roughly.

RASPBERRIES AND LOGANBERRIES
Use Both taste excellent with fresh cream, and store well as jam.

Soil These soft fruit like a heavy, moist soil and will thrive in cold northern regions better than most soft fruit. They tolerate shade and a northern aspect. Get rid of all perennial weeds and muck very heavily. They are greedy feeders of muck and will thrive if you give them plenty.

Propagation Buy these young berry plants from a nursery, and then raise them from layers, or just dig them out from the ground near existing raspberries.

Planting Plant them quite shallow, 2 feet (60 cm) apart, in rows 5 feet (1.5 m) apart. Establish a fence for them to climb up, or to contain them. I just have three pairs of horizontal wires and make sure the canes grow between these, but

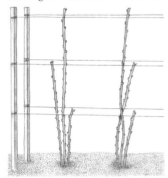

some people tie them to the wires to give the canes extra support and to keep them closer.

Pruning Let them grow but don't let the first shoots flower – cut them down before they do that. The second generation of canes will fruit. Cut the canes out after they have fruited, and just keep three new canes to fruit next year. Cut out all the weak canes. As the years go on, leave more canes to grow, up to about a dozen. Suppress suckers, or dig them up to plant elsewhere. Cut the tips at different levels because they fruit at the tips and you want fruit at all levels on the plant.

RED AND WHITE CURRANTS
Use These are nothing like as useful as blackcurrants, but I grow them for fun. Red and white currants are good for making jelly.

Planting Propagate from cuttings just the same as you would with blackcurrants (*see* p.84).

Pruning They fruit, not on the leaders like blackcurrants, but on spurs like apples. So cut back the

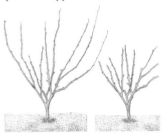

first leader, or new shoots, in half their length the first winter. Then cut all the main leaders back to half their length, and cut out all subsidiary leaders to within ½ inch (1 cm) of where they spring. Fruiting spurs will form at these points. In fact, keep as much older fruiting wood as possible, while cutting out much of the new wood.

Aftercare Otherwise treat them just the same as blackcurrants. They don't get "big-bud" or "reversion".

STRAWBERRIES
Use This fruit is a very good source

of vitamin C, rashes in some small children, and income for home-steaders. Strawberries are very labour-intensive, but they yield about the highest income per acre of anything you can grow. If you grow different strains, you can have strawberries all summer.

Soil They are a woodland plant so they need tons of muck and slightly acid soil: no lime.

Propagation Strawberries make runners which root, and you can dig these out of the ground. Or you can make the runners root in little buried flower pots with compost in them. Then when they are rooted you can cut the runners, remove the pots, and plant them out.

Planting Put little plants in during August and then transplant them 12 inches (30 cm) apart in rows 18 inches (45 cm) apart. Don't

plant them deep and spread the roots out shallow.

Aftercare Hoe and weed constantly or your bed will become a mess, and mulch heavily with peat, if you have it, or compost. Beware of slugs. If you haven't any peat, put straw below the plants to keep the berries clean. If you get botrytis (grey mould), dust with flowers of sulphur.

Harvesting It is best to let them fruit for 3 years, then scalp them. Establish a new bed every year for a constant supply.

Tree Fruit

Happy is the homesteader who inherits a holding that has plenty of established fruit trees. Unfortunately, farms that have previously been tenanted almost never have any. Why should a tenant plant trees on someone else's land? So it often happens that the newcomer finds no fruit at all, and has to wait several years before picking even a solitary apple. The only thing you can do is plant fruit as soon as possible.

Plant some standard or half-standard hard fruit trees; some espalier or cordon hard fruit trees (especially if your space is limited), or else dwarfed trees, which generally yield a much heavier crop much more quickly than full-size trees; and some soft fruit. The latter will give you fruit in three years, or less if you plant 2-year-old bushes. The big standards, or half-standards, will eventually give you a great bulk of fruit, possibly for the rest of your life, and provide you with enough apples for cidermaking. So, to reap your reward quickly, put them in as soon as you possibly can.

However, if you have a very small garden, you had better not grow too much top fruit (that is, fruit grown on large trees), because trees take up a lot of space and sterilize the ground underneath them for some distance by drying out the soil, extracting nutriments from it, and shading it from the sun.

When you site your orchard you must consider drainage. This is very important because no fruit trees will thrive with their feet in water. Air drainage is vital too. Frost runs downhill and therefore a basin is a "frost pocket". Thus you don't want a hedge below a sloping orchard because this will impede the flow of cold air downhill, creating frost, which will impair the quality of your soil. Apples, pears, plums, and the rest need good soil.

APPLES

Use Apples are quite simply the most useful fruit of all for cool and temperate climates. By having both early and late varieties, and long-keeping varieties, you can have prime apples nearly all the year, with maybe a little gap in summer when you have plenty of soft fruit anyway. A raw apple a day can be one of the most valuable items of your diet.

Soil They like good, deep loam but will grow in most soils with plenty of muck. They don't do well on acid soil so you might have to lime. Land must be well-drained and not in a frost pocket.
Preparation Cultivate well and get rid of all perennial weeds. Dig holes bigger than the tree roots are likely to be, and if you can get it, throw some lime builder's rubble in the bottom of the hole (and plant a dead dog down there, too, if you have one).
Planting If you buy trees from a nursery, get them to prune them before they send them. Three-year trees are usual, but if you take immense care with the planting you can have an almost ready-made orchard by planting even 7-year-old trees. Planting trees is covered in detail on p. 90.

Varieties There are at least a thousand varieties in Europe alone so I cannot begin to deal with them here. Get local advice on the best varieties to grow in your area, and make absolutely sure that varieties that need other varieties to pollinate them have their mates nearby. Otherwise they will remain fruitless old maids.
Pruning Pruning is of vital importance if you want large fruit, but don't prune until the middle of February (to guard against rot spores). If your apples are "tip-bearers" – and you must find this out from the nursery – the only pruning you should do is cut out some main branches, and in fact cut out the odd complete branch to keep the tree open and not too densely crowded.

But most other apples you will have to prune more scientifically. Cut all "leaders" (leaders are the long shoots which you want to leave to form new branches) to a third of their length, and cut to about ½ inch (1 cm) beyond an outward-facing bud. This is because the last bud you leave will turn into a branch next year and you want the branches to grow outwards away from the centre of the tree. Try to aim for a cup-shaped tree, open in the middle, with four or five nicely-shaped main branches growing out from the trunk at about 45°. Don't let it get too crowded with minor branches. So for the first year or two remove all young shoots that are not required for leaders to create the final shape of the tree. Cut them off ½ inch (1 cm) from where they join the trunk.

Then the aim should be to encourage fruiting spurs and discourage too much non-fruiting wood. If you cut off young shoots within ½ inch (1 cm) of their base a fruiting spur will probably grow in their place. So, on each small branch, cut the middle, or main, leader back to half its length but cut all the subsidiaries back to within ½ inch (1 cm) of their bases so that they will form additional fruiting spurs. Prune lightly in mid-summer, too. Don't prune leaders, but cut all subsidiary shoots that have grown that year down to within about 4 inches (10 cm) of their base. In later years you may have too many fruiting spurs. In this case you must cut some out. And if a tree bears a lot of small fruit one year and none the next, thin the flowers out. If, during a good year, your tree appears to be supporting an excessive amount of fruit, thin out some of the tiny apples to make sure that the fruit left grows to a good size.

Pruning is very complicated and I would recommend that you find yourself an experienced adviser.
Aftercare Try to keep the ground directly around your trees free of weeds and grass; annual heavy mulching with whatever waste vegetable matter you have, plus muck or compost, helps this. However, don't put too much highly nitrogenous compost on. Pigeon and chicken dung are not good for apples: they cause too much rank growth and not enough apples. Grass the space in between the trees if you don't want to intercrop with something else, but above all keep the grass close-mown all summer and don't remove the cut grass. Leave it there to rot and for the worms to pull down.

Don't start spraying until you are hurt. If you obey all the books you read, you will swamp your trees with deadly poisons (some growers spray a dozen times a year, drenching trees, fruit, and soil with persistent toxins) and you will kill all the predators, the insects and arachnids that feed on your insect enemies, as well as your enemies themselves. If you don't spray at all you will probably be all right.

If you get "canker" (rotting patches on the branches), cut out all dead wood and paint the affected areas with white lead paint. If you get "scab" (brownish scabs on the apples), carefully collect all fallen leaves, pruned branches and so on and burn them every year. Spray with Bordeaux mixture, but add half as much again of water as you would for potatoes. Spray just before the flower petals look as if they are going to open and again just after the flower petals drop. If you get "apple sawfly" (maggots that bore into the apples) spray with quassia, which kills the maggots but not the predators.

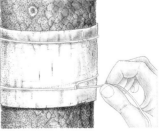

Grease-banding really is a good old-fashioned but effective safeguard against many horrid things.

Just stick bands of greasy material around the tree above the ground. Many nasty things try to climb up the trunk and get stuck in the grease. I believe keeping a few hens under fruit trees is good because the birds scrap out a lot of wicked insects. Planting buckwheat near fruit trees is said to be good, for it attracts beneficial hover-flies. But you may well find you need do nothing to protect your trees, and you still get good apples.

CITRUS FRUIT

Use If I could only grow one citrus fruit tree – in other words if I only had room for one tree in a greenhouse – I would grow a lemon, because you could not hope to produce a significant amount of oranges off one tree, whereas one lemon tree would keep a family in lemons, and without lemons a good cook is lost. You can, of course, grow oranges or lemons in tubs, kept indoors in the winter and put outside in the sunshine in summer, but you will get very little fruit like this.

Soil and climate Citrus fruit will grow well outdoors in subtropical climates. Lemons are slightly more frost tender than oranges: 30°F (-1°C) will kill the young fruit and 26°F (-3°C) may kill the tree. Oranges will put up with a degree or two colder. The best soil is sandy loam, pH between 5.5 and 6.2, and good drainage is essential.
Planting Plant like any other top fruit trees (see p.87).
Aftercare Keep the ground constantly moist for several weeks after planting. After the second year, if you are using irrigation, they should have at least 20 gallons (90 litres) of water a month. They don't need much pruning, except for rootstock suckers and diseased or injured wood. They like plenty of compost mulch but keep it from touching the trunk's foot – if you do not, foot-rot may result.
Harvesting Citrus fruits are harvested during the winter and can be left on the trees quite safely

for many months. So obviously it is best to leave them on the tree until you want them and then pick them while fresh.

CHERRIES

Use Two distinct species of cherry (*Prunus avium* and *Prunus cerasus*) have given rise to the many varieties now cultivated. The former are sweet, the latter sour, but hybrid breeds are common. The vitamin content of the fruit is high and cherry juice has been used to help relieve sufferers from acute arthritis.

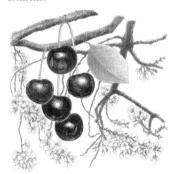

Soil and climate Successful cultivation of cherries depends more on a favourable climate than on any other single factor. An unexpected late frost will kill the crop without fail. Good water and air drainage is crucial. On well-drained soil trees can send their roots down as far as 6 feet (1.8m) at which depth they are not in danger of suddenly drying out. Sweet cherries like a dry, friable loam; sour cherries prefer a clay soil that is more retentive of moisture.
Planting Cherries are best planted in autumn, and the first buds will appear early in spring. A thick mulch applied soon after planting protects the tree.
Aftercare Cherry trees bear their crop early in the season, so if a good mulch is maintained, moisture other than natural spring and winter rain will not be necessary. A young cherry tree should be trained in a way that creates a central trunk with branches coming from it all

the way up, not an open cup-shaped tree, which will bear less fruit. The birds will get all your cherries if you just grow them in the open, so the answer is to grow them up a wall where they can be protected by a hanging net. If they get "dieback" (branches that are dying from the tips), prune the dead wood away and burn it. If they get "leafcurl", spray with Burgundy mixture (see p.164) before the leaves open in spring.
Harvesting Picking cherries with their stems is not simple, for it is easy to damage the fruiting twigs. The smallholder with the single tree may find it easier to pick cherries without stems, although the fruit must then be used at once, before any bacteria have time to enter through the break in the skin. The longer the fruit is allowed to hang when ripe, the sweeter the juice will become.

FIGS

Use The ancient Greeks called figs the Fruit of the Philosophers, and all one can add to this is that the philosophers must have had very good taste, for fresh figs, sunwarmed, are a unique experience.
Soil and climate They are truly a Mediterranean fruit but will bear

fruit out of doors in cooler climates, including many parts of the northern USA. In such climates the Brown Turkey fig is the only one to grow. They are best grown against a south facing wall, and in rainy and fertile land their roots should be confined in some way. A box a cubic yard in size is ideal. The walls should be concrete and the floor should be soil with broken stone on it. The reason for this is that figs grown unconfined in moist and fertile places put on too much leaf and branch growth and not enough fruit. An eccentric parson of my acquaintance confines the roots of his fig trees with tomb stones of the long deceased.
The fig will thrive in most soils but a light or sandy loam is held to be best. In fact, the fig is very much a fruit of poor soils.

Planting Figs grow well from cuttings. Take 2- to 3-year-old wood of under an inch (2.5 cm) in diameter in winter, cut to 10-inch (25-cm) lengths, plant almost completely buried in the soil, and keep moist. In places where figs grow well a fig tree can do with about 20 feet (6 m) of space. In colder climates, train a fan-shape up against a wall.
Aftercare Figs need little pruning unless fan-trained. If their roots are not enclosed and they do not fruit, root-prune them severely. An interesting thing about some figs, particularly the Smyrna fig, is that they can only be fertilized by a certain very slim wasp (*Blastophaga psenes*) that can crawl into the fig's neck. The fig is not a fruit, but a piece of hollow stem that has both male and female fruits inside it. When the Smyrna fig was taken to America, it was not understood why it would not fruit until it was discovered the fig wasp was needed, and these were imported in a certain wild fig called the Caprifig. The Brown Turkey fig, which is the one to grow in northern climates, does not need *Blastophaga* to fertilize it.
Figs can be dried, and make a very nutritious and easily stored food for the winter.

OLIVES

Use Where you can grow it, the olive is the most valuable tree imaginable, for it produces quite simply the best edible oil in the world besides the most delicious and nutritious fruit. In fact, one could live on good bread, olives, and wine, and many people have done so. Olives and the locust bean tree are among the most desirable trees you can grow, for they draw their sustenance from the deep subsoil, and allow intercropping with grass or other smaller herbage. This is true three-dimensional farming, which may well be the subsistence farming of the future.

Soil and climate Olives suffer damage at 18°F (-8°C) and very severe damage at 10°F (-12°C) so they are not suitable for cold climates. But they don't worry about late frosts above these temperatures because their flowers don't come until late spring or early summer when such frosts will not occur. They will not grow at altitudes over 2,600 feet (800 m) unless they are very near the sea but near the sea they suffer from "fumagine" (a sooty mould disease). If you can match these conditions, grow them on a slope if you can, because they cannot stand having their roots in stagnant water. On the credit side, they will put up with practically any soil at all. If on sandy soil in a semi-desert climate, they will survive with as little as 8 inches (20 cm) of rain annually. In clay soils further north they will need 20 inches (50 cm) or over. The best soil of all for olives is sandy soil interspersed with clay layers. They need rain in the summer period and if there is none you must irrigate profusely and regularly.

Propagation If you take cuttings in late summer and plant them in a mist propagator, you can grow trees from these. There are three ways you can do it: you can plant small cuttings of ¾–1½ inches (2–4 cm) diameter and 10–12 inches (25–30 cm) long vertically in the ground; you can plant larger cuttings 1½ inches by 10 inches (4 cm by 25 cm) below the ground horizontally; or you can plant root cuttings (taken from a tree growing on its own roots, of course, not from one grafted on a wild olive root stock) either in a bed or in the position you want your new tree.

Professionals grow trees from seed, then graft them on wild olive stocks, but this is a very tricky business. If you are going to grow olives on any scale you should plant about 250 trees to the acre. Plant them any time between late autumn and early spring. Trees may begin to produce at 5 or 6 years old, be producing heavily at 10–15 years and go on for a hundred. Mature trees will give 90–150 lb (40–70 kg) of fruit and some 18 pints (10 litres) of olive oil.

Aftercare Olives must be heavily pruned, but this is a complicated job which really must be learnt from someone with experience. Alternatively, you can find a professional who will come and prune your trees.

Harvesting You can harvest from the end of November right on through the winter. If you are going to eat your olives you must carefully pick them by hand. If you want them for oil you should shake them down into a sheet.

PEACHES AND APRICOTS

Use Peaches and apricots are perhaps most appreciated in temperate climates where they are not so easy to grow. Increasingly they are found frozen or tinned, so it is worth growing them fresh.

Soil and climate Paradoxically, peaches and apricots need both heat and cold. If they don't get cold in winter, say 40°F (4°C) or below, they don't have their winter sleep and exhaust themselves. On the other hand, one late frost after flowering will wipe the crop out, and they need real heat and sunshine in summer. Most of the people who were going to make a fortune growing outdoor peaches in England after the war have given it up. They like light soil, sandy or gravelly loam.

Planting The fruit is best planted in spring except in climates where the winters are exceptionally mild.

Aftercare Prune right back when you transplant the tree. Prune sensibly in the early stages to shape the tree, and nip out half the fruit if it is too crowded. If they get leaf curl disease, spray with Burgundy

mixture (see p.164) in early spring, just before the buds swell.

Harvesting Peaches and apricots are ripe when all the green in the skin gives way to yellow. Be careful not to bruise the fruit when you are picking it, because once bruised they degenerate very rapidly. You can store them for up to 2 weeks.

Treat almost exactly the same as

PEARS

apples. Pears like a more sheltered spot than apples and are not quite so hardy. Plant a succession of varieties. Give copious top-dressings of manure, but see that it doesn't touch the stem, or roots will grow out from the scion instead of the stock. Incidentally, if you graft pear scions on wild hawthorn bushes they will grow and produce pears! And remember, pears don't keep as well as apples do.

PLUMS

Use A number of very different species are all known as plums. They range from sweet dessert plums to tart damsons exclusively used for jam. Prunes are varieties of plum which have so much natural sugar that they do not ferment while drying out with the pit still inside the fruit.

Varieties Plums are not always self-

pollinating, so you must make sure that the varieties you plant are capable of pollinating each other or you won't get any fruit. If you only want to plant one tree, find out if any of your neighbours have plum trees and choose a variety which can be pollinated by any of them.

Pruning Don't prune plums for the first three years you have them, and then don't prune them until early summer or disease might get into them. Then take out any over-crowded branches, and if the tree is too luxurious, shorten leaders to

12 inches (30 cm) and side-shoots to 6 inches (15 cm). This will slow them down and make them fruit. Always in early summer cut out any "die-back" (branches that are dying from the tips) and paint the wound with paint. Never prune plums in

winter.

Aftercare "Silverleaf disease" is a bad disease of plums. If you get it, the leaves will turn silver and the insides of the twigs brown. Cut off the twigs branches until you get into clean wood, and (this is an old

remedy) slit the bark with a knife right from the cut you have made

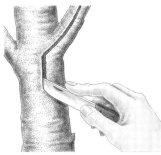

down the trunk to the ground. Of course, burn all affected parts to prevent the disease from spreading.

Harvesting Plums for preserving can be picked as soon as a bloom appears on the skin, but if they're to be eaten fresh they should be left to hang for longer. Their flavour is best when they look and feel over-soft.

Caring for Fruit Trees

Planting

All fruit trees are planted in the same way. If your climate allows, it is best to plant during the winter months when the sap is not moving around the tree. Normally you would buy 3-year-old trees to plant from a nursery, but get the nursery to prune them before they deliver them.

If you take enormous care with planting, you can have an almost "instant orchard" by planting even 7-year-old trees. But these trees would be much more expensive, and you really need to know what you are doing when you plant them. You would have to put a bag around the root ball to keep the soil in, dig right below and all around the roots, plant with immense care, and keep watered for a month. So I would recommend anyone inexperienced in orchard growing to buy 3-year-old trees.

Grafting

If you buy trees from a nursery they will already be grafted: cuttings from the fruit tree that you think you are buying will have been grafted on to another kind of tree. The latter will be some hardy, near-wild variety: for example, it will be a crab if the fruit tree is an apple. Thus you have the advantage of a hardy variety for the all-important root and trunk, and a highly-bred, high-yielding variety for the fruit. Very few amateur gardeners do much grafting, but there is no reason why they shouldn't, as it is easy enough.

It is no good grafting on to an old diseased tree, or one that is prone to, or has had, canker (rot in the bark or wood). A very useful exercise is the top-grafting of old established fruit trees, which are of a poor variety, or are neglected, badly pruned, and in need of reviving. The growing tree you graft on to is called the stock, and the tree you graft on top of it is called the scion. Scions can be made from winter cuttings. Heel in the cuttings (plant them in a cool place) after you have cut them off an existing healthy young tree of the type you fancy, just as if they were ordinary cuttings. Then, in spring, cut all the branches of the old tree you wish to revive down to about a foot from their point of union with the trunk, for top-grafting. Trim the edges of the saw cut with a super-sharp knife, and go about grafting your scions on to each branch.

There are several methods of grafting, according to what sort of branch you are grafting on, but the principle is always the same, and involves bringing the cambium (under-bark) layers of stock and scion into close contact. It is in this layer just under the bark that growth and union of tissue start.

Apples and pears are easily grafted; grafting plums is much more difficult as grafting lets in silverleaf disease. So don't graft plums unless you have to, and then only with great care. You can, by the way, graft pears on to whitethorn or may trees (hawthorn), and get pears! If we had the time and energy we could do this all along our hedges.

PLANTING A TREE

When planting out a tree or a bush put yourself in its place. Consider the shock to the roots, accept that the tree is delicate, and treat it accordingly. Start by digging a hole much bigger than the root ball of the tree.

Drive a stake into the bottom of the hole before you put the tree in. You train the tree up the stake. Then put the tree in and prune off any broken roots or very long ones.

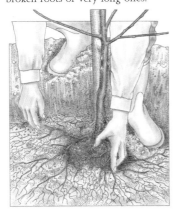

You should only transplant the tree when it is dormant; but even so you can minimize the shock. Put a heap of rich loam in the middle of the hole and spread the roots round it. Make sure that you plant the tree at the same depth as it was before. Sift in more loam round the roots with your fingers, and rub the soil gently into them. Continue filling the hole until the roots are in close contact with the soil.

As the tree grows it will need a good supply of nutriments below it. So the soil under the tree and all round it must be firm; if the soil caves away under the roots and leaves a cavity, the tree will die. You should firm each layer of soil as you plant, making sure it is broken up finely. When you have installed the roots to your satisfaction (and the tree's), throw more soil in on top and stamp gently but firmly.

Do not stamp heavily as this will tear delicate roots. When the hole is completely filled in, and the soil heaped up a little, you can stamp harder. The stake ensures that no movement disturbs the roots of the tree once growth begins.

A tree must have moisture after it has been replanted. So water it well, and then put a good thick mulch of organic matter on the soil around the tree to conserve the moisture.

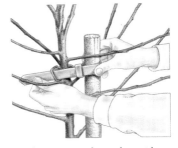

Tie the tree to the stake with a plastic strap and buckle. Do not over-tighten. Regularly check the strap and loosen it as the trunk grows and thickens.

TREE SHAPES

Train your young fruit trees into a variety of decorative shapes. This can save space, and in some cases can considerably increase yields.

Fans

Train a "maiden" (which is a single-stemmed, 1-year-old tree) along a wall or fence, with the help of canes tied to wires 6 inches (15 cm) apart (*as above*).

Cordons

Train a young fruit tree up a fence at an acute angle, and limit it to one stem and no long laterals (*right*).

WHIP GRAFTING

This is a form of grafting which is used when the stock and the scion are approximately the same size. The stock is the branch on to which you graft the scion; the scion is a shoot that you have cut in winter, and then "heeled in" to a cool place

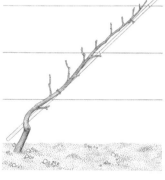

until needed for grafting. Prepare the scion by making a cut just behind a bud at the lower end of the scion so that it slopes away to nothing at the base. The cut might be 2 inches (5 cm) long. Near the top of this cut make another small one upwards, without removing any wood, so that a small tongue is formed. Cut the tip off the scion leaving from three to five buds. Now make cuts on top of the stock branch to correspond with those you have made in the scion.

Fit scion to stock, slipping one tongue down behind the other. The two cambium layers must be in contact with each other.

Tie the two parts together with raffia (cotton will do) and cover the whole joint with grafting wax.

Espaliers

Stretch horizontal wires 12 inches (30 cm) apart between posts. Train the central stem vertically upwards and the lateral shoots at 90°, tying them to canes fastened to the wires.

Dwarf pyramids

The advantages of dwarf trees are that they take up less space than full-size stock, but their fruit yields are as heavy. Restrict the growth of a young tree to 7 feet (2 m). Keep sideshoots short. Dwarf trees fruit earlier, but do not live as long as full-size stock.

BUDDING

This method of grafting is much used by rose-growers, although it can also be done with fruit trees. In summer select a strong healthy scion about 12 inches (30 cm) long and put it in water.

Cut a T-shaped slit along the back of your stock.

Peel back the two flaps of bark formed by the cut.

Take your scion out of water and slice out a shield-shaped piece of bark which contains a bud within a leaf axil.

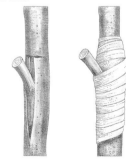

Insert the shield into the T-shaped cut. Remove any of the shield sticking out above the T-shaped cut; put back the flaps on each side. Bind with raffia or tape after insertion. As the bud grows you can cut off any stock above the bud-graft.

"People say to us sometimes 'why do you have cross-breeds and mixed-up strains in your animals and poultry?' ... For us the all-round animal [is one that is] not too highly specialized, not too developed away from the wild creature, not too finicky and highly-strung, not too productive.

People come and stay with us and sometimes express horror at me when I kill an animal. 'How could you do it?' I always ask them what they had for dinner the day before, and if they say 'meat' I know I can treat their scruples with contempt. I do not like killing animals; but having decided, after a great deal of thought, that it is right to kill animals, I do it without worrying myself about it [and] aiming at least for a professional standard in paunching and plucking and cleaning and butchering, and at doing the thing in a workmanlike manner. To connive at the killing of animals while being too lily-livered to kill them yourself is despicable. We could not have eggs unless we kept birds, and if we did not kill off the surplus birds we would very soon be overrun."

JOHN SEYMOUR FAT OF THE LAND 1976

CHAPTER THREE

FOOD
FROM
ANIMALS

The Living Farmyard

Just as I advise against monoculture when it comes to planting crops, so I would urge you not to specialize in one animal, but to keep a wide variety. This is the only way to make the best use of your land's resources and to take advantage of the natural ways in which your animals help each other. Your cows will eat your long grass, and then your horses, sheep, and geese can crop your short grass. After that your pigs will eat the roots and at the same time plough the field ready for sowing with grain crop, which all the animals, especially the hens, can eat. Pigs, of course, thrive on whey, and skimmed milk from butter-making, and cow's milk can nourish orphan foals or lambs. Your different animals will protect each other from disease as well, for organisms which cause disease in one species die when eaten by another.

COWS, HORSES, SHEEP, AND GOATS

Ruminants – cows, sheep, and goats – are the animals best equipped to turn that basic substance, grass, into food in the form of meat and milk. Horses, I hope, will not be exploited for meat or milk, but for their unique ability to turn greenstuff into power. Amongst themselves these animals divide up the available food very efficiently, and in so doing they work to each other's benefit. A lot of horses kept on a pasture with no other animals will not thrive; with other animals they will. It has been said that if you can keep twenty cows on a given piece of pasture, you can keep twenty cows and twenty sheep just as well. Cows will eat the long, coarser grass and the sheep and horses will clean up after them by nibbling what they leave close to the ground. Goats, which are browsers rather than grazers, fill a useful niche because they will eat bark, leaves, brambles and bushes, and if you want your land deforested they will do that as well.

GEESE

Geese will compete directly with the ruminants for grass, and, like the ruminants, though less efficiently, they will turn grass into food in the form of meat. At the same time they will improve your pasture. It is well worth keeping a few, for the more you diversify your grazing species, the better.

DUCKS

Your ducks will eat a lot of food that will otherwise yield you no benefit: water plants and aquatic creatures. And they will patrol your garden and eat that unmitigated nuisance, the slug.

PIGS

The pig is a magnificent animal and really the pioneer on your holding. He will eat anything, and in his efforts to find food he will plough land, clear undergrowth, and devour all surpluses, even your washing-up water. As William Cobbett wrote in his *Cottage Economy*: "In short, without hogs farming could not go on; and it never has gone on in any country in the world. The hogs are the great stay of the whole concern. They are much in small space; they make no show, as flocks and herds do; but without them cultivation of the land would be a poor, a miserably barren concern."

CHICKENS

Chickens are essentially graminivorous, or at least they prefer to live on seed. They will pick up all spilled grain in your harvest fields, eat the "tail corn", the grains which are too small to grind, and eat weed seeds. They will follow the pig, when that splendid animal is rooting away in the ground, and snap up any worms, wireworms, leather-jackets, or whatever else he turns up. And they will keep the rats away by pecking up spilled seed from corn ricks.

The Cow

There are four classes of cows: dairy cows, beef cows, dual purpose cows, and just cows. Fifty years ago there were some magnificent dual purpose breeds in Europe and America, but economic circumstances caused them to die out as the pure dairy and pure beef animals took over. This is sad, from the point of view of the self-supporter, and from every other point of view, in fact, because beef should be a by-product of the dairy herd and not an end in itself.

One cow produces one calf every year, and she's got to do that if she is to give milk. If she is a beef cow, all she can do is suckle this one calf – unless she has twins of course – until it is old enough to be weaned. But if she is a dual purpose cow she can have a good beef calf, provide him with enough milk, and provide you with enough milk as well. And if she is a pure dairy cow she will have a calf each year and provide enough milk for it and more than enough for you; however, the calf won't be a very good one for beef. Now, half the calves you have are likely to be male, and, of the other half that are female, only half will be needed as replacements for the dairy herd. Therefore, ultimately you will be forced, whatever you do, either to sell the three quarters that are not wanted for beef, or to kill them for beef yourself. Otherwise your cattle population will build up on your piece of land until there is no room for you or anybody else.

What breed?

But we can and do eat dairy cattle. Most of Britain's beef comes from the Dutch Friesian, called the Holstein in the US. These cattle are large, bony and give an awful lot of poor milk, meaning milk low in butter-fat. They are hardy and their calves, although from dairy animals, give good beef.

Then at the opposite end of the scale there are the Channel Island breeds. Of these the Jersey makes the best house-cow. They are small and don't give as much milk as

the Friesian, but it is the richest milk there is. Their calves make poor beef and are practically unsaleable, but they are good to eat all the same. I've had them. If you can get a good old-fashioned dual purpose breed I would say get it every time. The Danish Red is fine, the old Red Poll is too, or the dual-purpose Shorthorn. From these you will get good milk and good beef. But there is no cow more tractable, or indeed loveable, than the little Jersey, and if you want a friend as well as a milk supply, I strongly recommend her.

Buying your cow

How do you get a cow in the first place? Nothing can be more difficult. You see, if a person who owns a herd of cows wishes to sell one of them you can be sure of one thing: it is the worst cow in his herd. It may just be that a farmer wants to sell a cow or two because he has too many, and wishes to reduce his herd, but even if he does it for this reason, and his herd is a good one, you can still be sure that the ones he culls will be the worst in the herd. Generally, if a farmer sells a cow, it is because there is something wrong with her.

However, there are a few let-out clauses. One is the farmer who really is just selling up (as many are). His whole herd will be put up for auction and the good cows sold along with the bad. Another is the person (maybe a fellow self-supporter?) who has reared one or more heifer calves with the object of selling them "down calving". This means the cow has just calved and is therefore (another cowman's term) "in full profit", in other words, giving a lot of milk. In this case you can buy a cow which has just calved for the first time, and she is just as likely to be a good cow as any other cow you might buy. The person selling her is not selling her because she is bad, but because he has bred her up especially to sell. Apart from carefully examining the heifer in such cases, also examine the mum. (I remember an old saying about checking out a

BROWN SWISS
Noted for her longevity and the ability to produce large amounts of milk, which is admired by US cheesemakers because of its high percentage of protein and fat.

THE FRIESIAN
The archetypal dairy cow. Big, heavy-yielding, hardy, and given the right bull will produce fine beef calves.

THE JERSEY
The classic house-cow. Affectionate, hardy, with the richest milk in the world, but not so good for beef.

possible future mother-in-law before proposing!) You get an idea of how your beloved is likely to turn out in the maturer years. The same applies to heifers.

There is one disadvantage to buying a first-calver – at least there is if you are new to cows. You will both be learners. She will he nervous and flighty, and may kick. You will be the same. She will have tiny teats, and you may not get on very well together. So if you are a learner, maybe you had better get some dear old cow with teats like champagne bottles, kind eyes, and a placid nature. What does it matter if she is not the world champion milker? What does it matter if she only has three good "quarters"? (Each teat draws milk from one quarter of the udder.) Provided her owner is honest, and says she is healthy, and the cow doesn't kick, she will probably do for you.

Here are a few points to consider when you buy a cow, apart from having to be aware nowadays of the existence of BSE (so make sure the cow has been grass-fed). The rest is mostly down to look and feel:

1 Feel the udder very carefully and if there are any hard lumps in it, don't buy her, because it probably means that she has, or has had and is likely to get again, mastitis very badly. (Mastitis is an extremely common complaint. One or more teats gets blocked, and the milk you get is useless.) But if you are getting a cow very cheaply because she has a "blind quarter" (one quarter of her udder with no milk in it), then that is a different matter, as long as you know about it.

2 Make sure she is TT (tuberculosis tested) and has been tested, and found free from, a disease called brucellosis. In many countries both these tests are obligatory, and it is illegal to sell milk from cows found infected with either.

3 If the cow is "in milk"– in other words has any milk in her udders – try milking her. Try each teat fairly carefully (see p.98). Make sure she doesn't kick when you milk her and handle her, although of course common sense will tell you that, because you are a stranger, and because maybe she has been carted to some strange market or sales place, she will be more than usually nervous. Also make sure she has some milk in each quarter. If you are buying her from her home ask the vendor if you can milk her right out (saving the vendor the trouble!) and then you will really know how much milk she gives.

4 Look at her teeth – they will tell you how old she is (for more guidance see p.101).

5 See that she is quiet and tame enough to let you put your arms round her neck and stroke her behind the ears.

6 Make sure she has that indefinable but nevertheless very real "bloom of health" about her.

7 If you are a real beginner, ask some kindly cow-knowing neighbour to go with you and advise you. If the neighbour is really into cows, do what he or she says.

Now, having got your cow, get her home and make her comfortable. Spoil her a little. Tie her up in your cowshed, give her some good hay, some oats or barley meal, or cow-cake: give her time to calm down. Then, in the evening, milk her.

Feeding

"Agriculturalists" work out the feeding of cows by saying that a cow needs a certain amount of food for her "maintenance ration" and a certain amount more for her "production ration". That is, we work out what it will take to maintain the cow healthily if she is giving no milk and then add some according to how much milk she is giving. Twenty pounds (9 kg) a day of good hay will maintain a large cow during the winter. Twelve pounds (5.5 kg) would maintain a small cow like a Jersey. You can take it that, if you feed a cow nothing else but fair quality hay for her maintenance ration she will need 33 cwt (1,675 kg) of hay for the winter if she is a big girl like a Friesian, or 21 cwt (1,050 kg) of hay if she is a Jersey. Now, if you want to feed other things besides hay for maintenance, the hay equivalents are given on the following page.

THE RED POLL
A good old-fashioned dual-purpose cow, producing plenty of good milk and good beef, too.

THE HEREFORD
Recognized all over the world as a fine beef animal, good for crossing with other breeds for beef.

A ton of fairly good hay is equivalent to:

¾ ton of very good hay
4 tons of kale or other greens
5 tons of mangolds
3 tons of fodder beet

Production ration

Now to give "maintenance plus one gallon" the winter's ration should increase to: 45 cwt (2,285 kg) for a Friesian and 33 cwt (1,675 kg) for a Jersey. The daily ration therefore goes up to 27 lb (13 kg) hay or its equivalent for a Friesian and 20 lb (9 kg) for a Jersey.

Now you can reckon that if you feed 3½ lb (1.5 kg) of a mixed "concentrate" ration for each gallon (4.5 litres) that the cow produces over the first gallon, that will do. And the concentrate could be:

2 parts barley (rolled)
1 part oats (rolled)
1 part beans (broken or kibbled)

To each ton of concentrates you can add if required:

20 lb (9 kg) limestone
20 lb (9 kg) steamed bone flour
20 lb (9 kg) salt

Cows kept naturally on organically farmed pastures and hay are unlikely to suffer from mineral deficiencies, but if you ever did get "grass-staggers" or hypo-magnasemia, or any other diseases that your vet told you were due to mineral deficiency, then you would have to add the missing element either to the diet of the cows or to the land. Seaweed meal is an excellent source of all minerals. You can just dump some seaweed on your pasture from time to time, and allow the cows to nibble it and lick it. They will not suffer from mineral deficiency if they do this.

All this means that if, in an average winter situation, you feed your cows, say 30 lb (13.5 kg) of kale or other greens a day and 12 lb (5.5 kg) of hay for her maintenance plus one gallon, and, say, 3½ lb (1.5 kg) of the above suggested "concentrate" for her production ration for every gallon over the first, you will not go far wrong. But use what food you've got. Use common sense and watch the milk bucket. If the milk goes down, feed more and you should be all right.

Summer feeding

In the summer, if you have plenty of good grass, grass alone should give maintenance plus up to 4 gallons (18 litres). A cow yielding over this would have to have 3½ lb (1.5 kg) concentrate per gallon over 4 gallons, but I hope you wouldn't try to keep such a cow. You don't want a ridiculous lot of milk, and very high yielders need a vet in almost constant attendance and have to be cared for like invalids. But grass varies enormously in value, and, if you suddenly find her milk yield dropping off, give her a little concentrate (even a pound or two) and see what happens.

Stockmanship is a matter of constant keen observation and common sense. Look at your animals: learn what the "bloom of health" means. Watch their condition. Are they getting fatter or thinner? Watch their milk yield; watch how hungry they seem to be. The "stockman's eye" may not be given to everyone, but it can usually be acquired.

Milking

Milk your cow twice a day, ideally at 12-hour intervals. Wash the cow's udder and teats in warmish water. Thoroughly wash your own hands. Then dry well with a towel. The more you massage the udder in doing so, the better. Clean the rear end of the cow generally, so that no dung or dirt will fall into your pail. Give the cow something tasty to eat. Sit down beside her on a stool and grasp the two front teats in your hands. Or, if you are an absolute beginner, grasp one teat with one hand. Common sense will tell you how hard to squeeze. Next, keep holding on with your thumb and finger, and squeeze the teat successively bringing in your other fingers, so as to expel the milk downwards out of the teat. Release the teat and repeat the operation (*see below*). It sounds easy, but actually it is difficult, and it takes a week to learn to milk.

MILKING A COW
Sit down on the right side of the cow (**1**), grip the bucket at an angle between your knees, and grasp the two front teats in your hands. To milk, squeeze high up the top of the teat with the thumb and forefinger (**2**) to stop the milk from going back up to the "bag", as farmers call the udder. Then, bringing in your second (**3**), third (**4**) and little fingers (**5**), squeeze progressively downwards to expel the milk. If you can, practise on a dummy teat to get the necessary rhythmic motion. It is also much better to begin on an old cow who is used to being milked and who won't mind if you fumble.

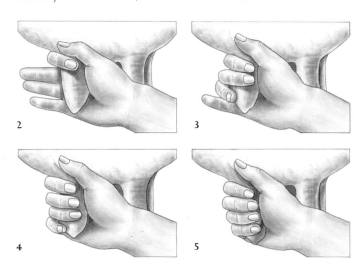

Housing

The modern milking shed is a concrete-floored building arranged so that the cow can be tied up to a vertical post. This has a ring on it that can slide up and down so that the cow can lie down if she likes. She is supposed to dung in the "dunging passage" and you are supposed to clean it every day. But in my view all this is bowing down too far to the great god "hygiene". If you keep your cow in a house for all or part of the time and throw in plenty of dry straw, bracken or other bedding, every day, the dung will slowly build up and you will have the most magnificent muck. Your local Dairying Officer will excommunicate you if you milk cows in such a house and you would certainly not be allowed to sell milk or any milk product in any so-called civilized country; but in fact the milk you get from a cow milked on "deep litter" will be as clean as any milk in the world, provided you observe the other rules of hygiene. We milked our cows like this for 8 years, cleaning the muck out about once a year, and the cows were indoors at night in the winter. Our milk, butter, and cheese were perfect.

A refinement is to have a house for milking the cow or cows and another house for them to sleep, eat and rest in. This can be cleaned out regularly, or it can be strawed every day and the dung left for months. It will be warmer and pleasanter for the cows than the milking stall, and they can be left free.

As to when to house: our cows only come into the cow shed to be milked, and to finish their hay. Winter and summer, night and day, they are out in the field except for an hour morning and evening when they come in to be milked and to eat. In the summer they don't want to come in. They want to stay out and eat grass. We don't feed them in the summer – except maybe a pound or two of rolled barley to make them contented. If there is plenty of grass, grass is enough for them. We don't go in for very high yielders, nor should any self-supporter. In the winter they come in eagerly because the grass is of little value and they are hungry. In bad weather I would prefer to leave them in all night and only turn them out in the daytime, and then, when the weather is really foul, keep them in some days too. But to do this you would need an awful lot of straw in reserve. In really cold climates cows have to be housed all winter. Do what your neighbours do. But don't keep them in half the winter and then turn them out. When you do turn them out in the spring, be careful about it. Wait for good weather, and only push them out for short periods at first. Too much grass may upset stomachs not used to it, and sudden exposure to weather can give cattle a chill. And if you can keep cows out all winter – that is if any of your neighbours do – it is far less trouble.

Mating

A small cow might be mated at 15 months: a large cow at 20 months. A cow will only take the bull, or the artificial inseminator, when she is "bulling" or on heat. This condition occurs at intervals of 21 days in a non-pregnant heifer or cow, and continues each time for about 18 hours. You must watch for signs of bulling and then get the cow to a bull or have her artificially inseminated immediately, certainly by the next day at the latest.

Signs of bulling are: she mounts other cattle or other cattle mount her; she stands about mooing and looking amorous; she swells slightly at the vulva; she will let you put all your weight on her back and apparently enjoy it. It may be a good thing to miss the first bulling period after calving, so as to give the cow a rest, but make sure you don't miss the second one. If there is a bull running with the cows, there is no trouble. He knows what to do and when to do it. But if there is no bull you will just have to

TEACHING A CALF TO SUCK

If you try to milk a cow in competition with a calf, the calf will win and you will lose. Many people therefore milk the cow themselves and feed the calf out of a bucket. To teach a calf to suck, put two fingers, slightly apart, in the calf's mouth. Get him to suck them, then lower your fingers and his nose down into the milk. He may start to choke, or he may find he likes it and start to suck avidly. If he sucks, after a time gently remove your fingers and he will just suck the milk. You will need great patience.

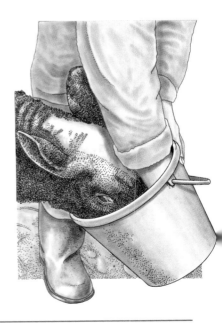

use your eyes, that is all. Then you may take her to a neighbour's bull, or bring him to her, or, in countries which have "AI" (artificial insemination) start by telephoning the relevant authority.

Calving

Leave her alone, out of doors, and you will probably find her one day licking a little calf that has just been born quite easily and naturally. Outdoor cows very seldom have trouble with calving. Watch the two for a few hours until you are quite sure that the calf has got up and sucked. If the calf doesn't suck within an hour, be worried. Get it up and hold it to the teat and make it suck – if necessary tie up the cow.

Cows calve up in the bare mountains in winter and rear their calves perfectly happily in the snow. As long as the calf runs out with the cow, and the cow has enough to eat, and the weather isn't too awful, he is all right. But as soon as you bring him indoors – and if you want the cow's milk you will have to – you have upset the ordinary workings of nature and you must keep the calf warm and out of draughts. He must suck from his mother for at least three days, because this first milk has colostrum, or beestings, a mixture of chemicals, organisms and antibodies which are essential for the health – nay survival – of the calf.

Then you can either take the calf right away from the cow, and keep him out of earshot if possible, or at least in a separate building, or you can keep him close to the cow's head so that she can see him, while you milk the cow. The easiest way for you to get your milk is to take the calf right away, milk his mother yourself, and she, after perhaps a night of bellowing for her calf, accepts the situation. She is not a human, and has a very short memory, and very soon accepts you as her calf substitute. I have ended up with a wet shirt many a time because I could not stop an old cow from licking me while I milked her!

Meanwhile, what do you feed the calf on? Well, if you have enough milk, feed the calf on the following mixture: the mother's milk at a rate of 10 percent the weight of the calf – that is, if the calf weighs 50 pounds (23 kg) give him 5 pounds (2 kg) of milk – plus added warm water (one part water to three parts milk). Feed the calf this mixture twice a day out of a bucket (*see illustrations opposite*) and serve it at blood heat. After about a week or two, see that the calf has got some top-quality hay to nibble if he feels like it. After a month try hull with "cake" rolled barley or other concentrates, and always have clean water available for him. After 4 months wean him off milk altogether and feed him 4 pounds (1.8 kg) hay and 3 pounds (1.5 kg) concentrates, or let him out on a little clean grazing. By 6 months he should be getting 6 pounds (2.7 kg) hay and 3 pounds (1.5 kg) concentrates, unless by then he is on good, clean grazing.

The whole subject of cows may seem to the beginner to be very complicated. Well, it is complicated, and there is absolutely no substitute for making friends with knowledgeable neighbours and asking for their help and

A COW'S TEETH REVEAL HER AGE

A mature cow has 32 teeth of which eight are incisors. All eight are on the lower jaw and munch against a hard layer of palate called the dental pad. Within a month of birth a calf has grown eight temporary incisors. These will slowly be replaced, until at age five the mature cow has eight permanent incisors. Age can therefore be determined by the number of temporary and permanent incisors present in a cow's jaw. The age of cows over five can be gauged by the wear on their teeth (by the age of 12 only stumps remain), and by the roughness of their horns. But beware cattle dealers who sandpaper cows' horns to make them look young.

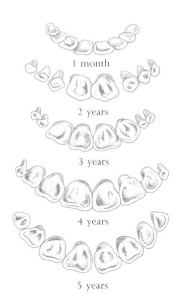

advice. As for diseases of cows, these are many and various – mastitis, already mentioned, lameness, parasites, worm infestation, foot-and-mouth disease, bloat through overfeeding, milk fever, grass staggers – and you must check with a vet to find out which vaccines are recommended for calves and cows in your area. But cows that are not unnaturally heavy yielders, and those that are kept as naturally as possible (not over-fed but not under-fed, and allowed out as much as climate permits) will very seldom get sick. For example, "Husk" is a killer disease, but a calf who has grown up outdoors with his mother will be naturally immune. However, before you put a calf reared indoors out to graze you must inject him with an anti-husk serum obtained from the vet, and feed him very well, on hay and concentrates as well as grass, until he has built up an immunity to husk. Don't turn motherless calves out in the winter time anyway, whatever you do.

There are other ways of rearing small calves and still getting some milk for yourself. If you are very lucky you can find another cow who has her own calf at heel and fool her into thinking she has had twins. Then you can turn her, her calf, and your new calf all out together to pasture. This is the ideal solution because the new calf is absolutely no more trouble and will thrive exceedingly. But such cows are rare. So another solution is to select a foster mother, tie her up twice a day and force her to give milk to the new calf as well as her own. Many cows are pretty complacent about this; others fight like hell. If she fights you may have to tie her legs up to stop her hurting the calf. To get a young calf to suck a cow, put two fingers in his mouth and gently lead him up to the teat. Don't try to heave the young calf forward from behind – that's as good as useless. Another way of getting milk for yourself and keeping the calf happy is to take some milk from the mother and let the calf suck the rest. Or, you can let the calf run with his mum in the day, keep him away at night, and milk the cow the next morning. There are many ways you can play this game.

Beef

Before you kill an ox make sure you have plans for storing and using all the meat. This means emptying and tidying your deep freezes to ensure you have enough space: even the meat from a small bullock or heifer – less than half a ton in weight – will fill up an entire average-sized freezer, so beware. Talk to friends and fellow self-supporters: would they like some joints? Have any of them experience of killing beef? By distributing some of your beef largesse you can add substantially to your "credit" with self-supporters.

Sadly, today, the bureaucrats have taken over and all but eliminated the traditional small local slaughterhouses; and this has greatly pushed up the price of having your animals slaughtered. You will only consider killing your own beast as a final resort or in an emergency. Nevertheless, we should all know how these essentials of life are managed.

In practice, you will ask around for information about the best and most convenient local slaughterhouse. The smaller the place, and the closer to you, the better. Use a place recommended by other self-supporters, somewhere where you can be on first name terms with the operators. Take some time to find out about their skills and problems – it will pay dividends in how well they hang and butcher your beast.

A beef animal will put on weight (and meat) very efficiently for its first 3 years of life. But most self-supporters will tend to have their animals killed rather younger than this. First, it is illegal in many countries to kill animals for beef when they are older than three years. Second, you will find a very large animal takes up a lot of space both in the field or cowshed and in the deep freeze. Kill an animal younger than 6 months and you get very tender meat – but very little of it.

You can, of course, rear and sell your beef animals for cash, but in the age of polluted and unreliable foodstuffs it seems mad to give away large quantities of meat you know to be wholesome. And you will find you pay at least twice as much for your meat as you get for an animal you sell, so you soon discover who makes the money out of farming.

Make your enquiries about slaughter in good time and make doubly sure you have all the paperwork that the slaughterman requires. This will vary from country to country. But in an age of mad cows and unknown prion pathogens we have been encumbered by a veritable confetti of forms and tags. If you do not have the right papers you will be stuck – and it can take weeks for the paperwork to be sorted. Do not be tempted to believe that logic has anything to do with all this – in the end it is simplest to jump through all the bureaucratic hoops.

About 10 days after the animal has gone to slaughter the butcher will contact you to ask how you want the carcass cut up. You will need to do a little homework here – is it to be T-Bone steaks, sirloin joints on the bone, chopped round steak, or mince? Everyone has their preferences but I would always get plenty of good quality mince, some nice large joints on the bone, and keep the round steak (on the legs) for braising.

So, paperwork and rules apart, here is the lowdown on how you can deal with your own meat if you really have to. I know of no other place where you will find simple instructions on butchery.

Slaughtering an ox
Before killing an ox you should really starve him for 12 hours, but it's not the end of the world if you don't. Bring the ox quietly into where you are going to kill it. Then either shoot it with a small bore rifle (which almost always means a .22) or with a humane killer. If you shoot it, it never knows anything is going to happen. Shoot just above the point at which imaginary lines from alternate eyes and horns cross.

The ox will fall immediately and lie on its side. Beware, all animals have violent death throes and those hooves can be dangerous. Once it has fallen, stand under the chin with one leg pushing the chin up so as to stretch the head upwards, and the other leg in front of and against the forelegs. If the animal starts to lash out with its forelegs now it cannot hurt you.

With the throat thus stretched, stick a pointed knife through the skin at the breastbone and make a cut 12 inches (30 cm) long to expose the windpipe. Then stick your knife right back to the breastbone, pointing it at an angle of about 45 degrees towards the back of the animal. Make a deep cut forwards as long as your other incision. Your knife will now be along one side of the windpipe. This cut will sever several main veins and arteries and the animal will bleed. If you have a block-tackle you should haul the animal up by the hind legs at this juncture to assist bleeding. Catch the blood and, if you don't like black pudding, add it to the pigs' mash or dump it on your compost heap.

Skinning an ox
Skinning is the most difficult part of the whole process, and you will wish you had seen an expert on the job. Lower the animal to the ground to skin it.

Skin the head first. You can do this with it lying flat on the ground, but it makes it easier if you cut through the nostrils, push a hook through the hole, and haul the head just off the floor with the block-tackle. Slit the head down the back, from the poll to one nostril by way of the eye and simply skin away until you have skinned it. I am not going to pretend this is an easy job, nor a very pleasant one, but there is a lot of very good meat on an ox's head and you can't just throw it away.

When you've skinned it, grab the lower jaw and stick your knife in the neck close to the head. Cut just behind the jaws first, then disjoint the atlas bone, the top bone of the vertebral column, and take the head off. Lay the carcass straight on its back with some wedges. Sever the tendons of the forelegs, cutting from the back, just below the dew claws. Slit the hide of the leg from this point to just above the knee. Skin the shank.

BEEF JOINTS

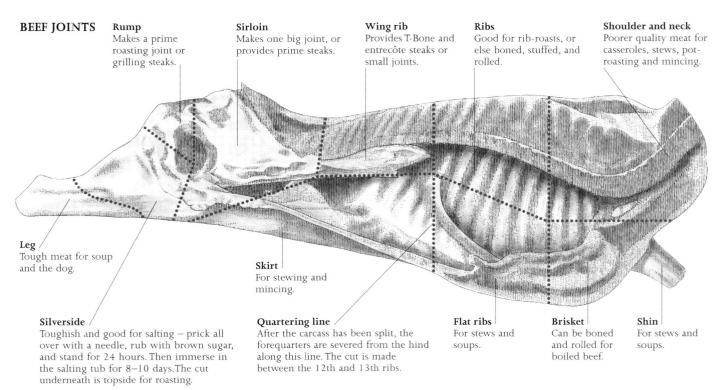

Rump
Makes a prime roasting joint or grilling steaks.

Sirloin
Makes one big joint, or provides prime steaks.

Wing rib
Provides T-Bone and entrecôte steaks or small joints.

Ribs
Good for rib-roasts, or else boned, stuffed, and rolled.

Shoulder and neck
Poorer quality meat for casseroles, stews, pot-roasting and mincing.

Leg
Tough meat for soup and the dog.

Skirt
For stewing and mincing.

Silverside
Toughish and good for salting – prick all over with a needle, rub with brown sugar, and stand for 24 hours. Then immerse in the salting tub for 8–10 days. The cut underneath is topside for roasting.

Quartering line
After the carcass has been split, the forequarters are severed from the hind along this line. The cut is made between the 12th and 13th ribs.

Flat ribs
For stews and soups.

Brisket
Can be boned and rolled for boiled beef.

Shin
For stews and soups.

Cut through the lower joint or, if you can't find it, saw through the bone. Skin all the legs out to the mid-line of the body. Slit the skin right down the belly, and haul as much of the skin off as you can. Use a sharp, round-pointed knife for this and incline the blade slightly so the edge is against the skin and not the meat. Hold the skin as taut as you can while you use the knife. Skin both sides as far as you can like this. Open the belly now, putting the knife in just behind the breastbone and protecting the knife with the hand so you don't pierce the paunch. Slit the abdominal wall right down the centre line to the cod or udder. Don't puncture the paunch! Then cut along the breastbone and saw along its centre. Saw through the pelvic bone at the other end. Cut the tendons of the hindlegs and the hock-bone, make loops of the tendons, and push the ends of your gambrel (which is a wooden or steel spreader) through them. Haul the hind quarters clear of the ground with the tackle. Slit the hide along the underside of the tail, sever the tail near the arse and haul the tail out of its skin for ox-tail soup.

Gutting an ox

With a pointed knife cut all the way round the rectum (bung) so as to separate it from the rest of the animal. After it is free, tie a piece of binder twine round it, tight, so that nothing can fall out. Then cut it free from the backbone. Now haul the beast up fairly high and pull the rectum and other guts forwards and downwards so they flop out. Cut the liver out carefully, and remove the gall bladder from it. Hang the liver on a hook, wash it, and hang it up in the meat safe. Pull out the paunch (guts and stomach) and all

the innards, and let it flop into a huge dish. You can clean the stomach thoroughly by putting it in brine, and then use it for tripe. The intestines make good sausage skins (don't eat the reticulum or "bible" stomach – it looks like the leaves of a book inside). Haul the carcass clear off the ground now. Cut out the diaphragm (the wall between chest cavity and abdomen). Heave out the heart and (lights) lungs, and hang them up: the lungs are for the dogs, but the heart is for you. Rip the hide right off the shoulders and fling a few buckets of cold water into the carcass.

Go to bed after a hefty meal of fried liver; only the offal is fit to eat at this stage. In the morning, split the carcass right down the backbone with a cleaver if you have faith in yourself, or saw if you haven't. Wash the two halves of the carcass well with tepid water and trim to make it look tidy. At this point I would strongly advise you enlist the aid of a butcher. Jointing is a complicated process, and there is no substitute for watching an expert do it. After all, you want to make the best use of the animal you have spent all that time fattening for this moment. And you can't really improve upon the recognized butcher's cuts.

Salting beef

The traditional salting tub has a loose round board with holes in it beneath the meat, and another on top of the meat. You can put a stone on the top board to keep the meat down, but never use a metal weight. To make your brine, boil salted water and allow to cool. Test the brine for strength; if a potato will not float in it, add salt until it does. You can pickle tongue by spicing the brine with herbs, cloves, lemon, onions, and soak the tongue in it for 6 days.

Goats

Goats are called "desert-makers" in some countries because they destroy what scrub there is and prevent more from growing. But where the goat is controlled it can fill a place in a mixed ecology, and if you want to discourage re-afforestation it has a very useful part to play. In cut-over woodland, for example, the goat can go ahead of the other animals as a pioneer, suppress the brambles and briars, prevent the trees from coming back, and prepare the way, perhaps, for the pig to come and complete the process of clearing the old forest for agriculture.

Goats will thrive in deciduous woodland (perhaps one to an acre), and give plenty of milk, but they will, have no doubt of it, prevent any regeneration of trees. In coniferous woodland, goats will find very little sustenance, but on heather or gorse-covered mountainsides they will thrive, and there is no doubt that a mixture of goats and sheep in such situations would make better use of the grazing than sheep alone. Goats, concentrated enough, will clear land of many weeds, and they will eat vegetation unsuitable for sheep and leave the grass for the sheep.

For the self-supporting smallholder the goat can quite easily be the perfect dairy animal. For the person with only a garden, the goat may be the only possible dairy animal. This is because goats are very efficient at converting roughage into milk and meat. Goat's milk is not only as good as cow's milk; in many respects it is better. For people who are allergic to cow's milk it is absolutely ideal. For babies it is very good. It makes magnificent cheese because its fat globules are much smaller than those of cow's milk and therefore do not rise so quickly and get lost in the whey. It is harder to make into butter, because the cream does not rise within a reasonable time, but with a separator, butter can be made and is excellent. On the other hand, milking goats takes more labour per gallon of milk than milking cows, however you do it, and so does herding, or fencing, goats.

Fencing and tethering

Restraining goats is the goat-keeper's chief problem, and we all know goat-keepers who also try to be gardeners and who, year after year, moan that the goats have – yet again – bust into the garden and in a few hours completely ruined it. Young fruit trees that have taken years to grow are killed and the vegetables absolutely ravaged. But this annual experience has no effect whatever on the beliefs of the true goat-keeper. Next year he will win the battle to keep goats out of his garden. What he forgets is that the goats have 24 hours a day to plan to get into his garden, and he doesn't.

Three strands of electric fence, with three wires at 15 inches (40 cm), 27 inches (70 cm) and 40 inches (1 m) above ground level, will restrain goats, and so will a 4-foot (1.2-m) high fence of chain link with a support wire at 4 foot 6 inches (1.4 m) and another support wire lower down. Wire netting will hardly deter goats at all.

Tethering is the other answer. Where you can picket tether goats along road verges, on commons, and so on, and thus use grazing where you could use no grazing before, you must be winning. But it is unfair to tether any animal unless you move it frequently; above all, don't keep putting it back on the place where it has been tethered before except after a long interval. Goats, like sheep, soon become infested with internal parasites if they are confined too long on the same ground. By tethering you are denying the animal the right to range and search for clean, parasite-free pasture. Picket tethering is very labour-intensive, but for the cottager with a couple of goats and plenty of time it is an obvious way to get free pasture.

Another form of tethering is the running tether. A wire is stretched between two posts, and the tether can run along the wire. This is an obvious way for strip-grazing a field, and also a good way of getting rid of weeds that other animals will not touch.

TOGGENBURG
A fairly small Swiss goat. Yields well and can live on grass.

ANGLO-NUBIAN
Gives very rich milk, but in relatively low quantities.

SAANEN
A large goat of Swiss origin. Capable of high yields, if given good grazing.

Feeding

A kid should have a quart (1 litre) of milk a day for at least 2 months, but as he gets older some of this may be skimmed milk. A reasonable doe should give from 3 to 6 pints (1.5–3.5 litres) of milk a day. In the winter a doe in milk should have about 2 pounds (0.9 kg) a day of very good hay (you should be able to raise this – about 750 pounds or 340 kg a year – on a quarter of an acre), 1 or 2 pounds (0.5–1 kg) of roots or other succulents, and from 1 to 2 pounds (0.5–1 kg) of grain, depending on milk yield. Goats should have salt licks available. It is a mistake to think that goats will give a lot of milk on grazing alone: milking goats need good feeding. They will, for example, thrive on the silage you can make by scaling grass clippings in fertilizer bags. As for the grain you feed to your goats; a good mixture, such as you might feed to dairy cows, is fine, or you can buy "cake" or "dairy nuts" from a merchant (and pay for it too). All debris from the market garden or vegetable garden can go to goats, but it is better to crush, or split, tough brassica stems first. Feed them their concentrates individually, or the goats will rob each other.

Housing

Goats are not as winter-hardy as cows and cannot be left out all the time in north-European or North American winters and expected to give any milk. They don't like cold and they hate rain. High-yielding goats (which personally I would avoid) need very high feeding and warm housing, but medium-yielding ones must have shelter from the rain, and an airy but fairly draught-proof shed to sleep in o'nights. Giving them a table to lie on, with low sides to exclude draughts, is a good idea and if they are lying in a fierce down-draught of cold air, adjust the ventilation until they are not. Maybe a board roof over the table to stop down-draughts would be a good idea.

Otherwise you can treat milking goats much the same as you treat cows. Dry them off 8 weeks before kidding. But a goat may milk for 2 or 3 years after kidding without kidding again.

Rearing orphans

One possible good use for goats is rearing orphans of many kinds. Goats are excellent for suckling other animals: calves thrive better on goat's milk than they do on their mother's, and it would be reasonable, if you kept cows, but had some wilderness or waste land on which cows could not thrive, to keep a flock of goats on the bad land and use them for suckling the calves so that you can milk their mothers. Goat milk is very digestible, and pretty good milk anyway, and orphan piglets, for example, which don't take very kindly to cow's milk, will thrive on goat's. Calves will suck straight from the nanny. Lambs will too, but don't let them – they may damage her teats and give her mastitis. Milk her yourself and bucket-feed the lambs. Milk the nanny and feed the milk to piglets through a bottle.

MILKING A GOAT
You can milk a goat that is standing on the ground, just as you would milk a cow (*see p.98*). But because goats are so much smaller, a stand helps. Coax her into position with some hay or grain.

You can rear foals on goats. The suggestion has been made that a suitable person could make a living – or half a living – by running a goat-orphanage, not for orphan goats but for other animals. Neighbours soon get to know such things, and orphan lambs are ten a penny in the lambing season in sheep districts and very often a sow has too many piglets.

Billy kids

Kids come whether you want them or not, and in the end you have to find something to do with them. What you can do is castrate them and then eat them. The most humane way to castrate a goat or anything else is to use rubber rings, put on with an "Elastrator," which you can buy from any agricultural supplier.

A billy becomes very male at 3 months, whereas a lamb does not develop specific male characteristics until about 6 months. Therefore many of us who produce fat lamb don't bother to castrate, but male goat kids should be castrated unless you intend to eat them before 3 months, or they will become strong-tasting. In my opinion, a goat wether (castrated billy goat) tastes as good as any mutton, particularly if you lard it well, or marinate it in oil and vinegar, or oil and wine, for it is short of fat compared to sheep. Up to 6 months old, it lends itself very well to roasting with various herbs or spicy sauces. If the billy goat has not been castrated, the meat has a strong gamey taste, and it is best to marinate it for 3 days in wine and vinegar and then curry it. Cooked in this way it is delicious. As a general rule, though, it is better to castrate if you are going to eat it.

Pigs

The pig fits so well into the self-supporter's economy that the animal almost seems designed with that in mind. It is probably the most omnivorous animal in the world and will thrive on practically anything. It is even more omnivorous than man, because a pig can eat and digest grass while we cannot. A pig will not thrive on grass alone, but it can make it a substantial part of its diet. And it will convert virtually anything that you grow or produce on the farm into good meat. Throw any vegetable matter, of whatever kind, to a pig, and he will either eat it – converting it within hours to good meat and the best compost in the world – or he will tread it into the ground, dung on it, and turn it into compost that way. Put a pig on rough grassland, or scrubland, that you wish to bring into cultivation, and he will plough it for you, and root it up for you, and manure it for you, and at the same time extract sustenance from it to live and grow on.

Feeding

Self-supporters should aim to produce all the food they need to feed the pigs from off their farm. Barley meat, maize meal, potatoes, Jerusalem artichokes, carrots, fodder beet, parsnips, turnips, or swedes – these are all crops that may be grown to feed pigs; and if supplemented with skimmed milk or whey you have a pretty good diet. I have fattened pigs very successfully on boiled potatoes and skimmed milk. I have fattened them on raw carrots and separated milk. Wheat "offals" such as bran and middlings, are good too, but there is no doubt that either barley meal or maize meal have no peers when it comes to fattening pigs. Even then they must have a "protein supplement" which may well be whey or separated milk, though any other high-protein food will do: meat meal or fish meal, cooked meat or fish, bean meal or any other high-protein grain. Soya is excellent for fattening pigs. If pigs are out of doors they don't want mineral supplements. If your pigs get some fresh greens, some milk by-products and scraps, they don't want or need vitamin supplements. Let sows run out over plenty of land, and in summer they will get nearly half their sustenance from grass, if they are not in milk. Keep sows or growing pigs on artichokes, or a field of potatoes, or where potatoes have been harvested, or other crop, and they will get half their food from that.

But for milking sows, or sows in late pregnancy, animal protein is absolutely essential, as it is the only source of vitamin B_{12}. A good rule for rations is that breeding sows, out of doors on grass and with access to scraps, surplus vegetables, and so on, will need 2 pounds (1 kg) a day of concentrate such as barley meal and some protein, but 6 pounds (3 kg) a day when they farrow. If they are indoors or having concentrates only, you can double these amounts. When the baby pigs are 3 weeks old or so you can start "creep feeding" them: that is, allowing them through fence holes too small for the sow, where they can eat concentrates unrestricted.

Fattening pigs can be given as much as they can eat until about 100 pounds (45 kg) in weight (half grown), after which their ration should be restricted or they will get too fat. Restrict them to what they will finish in a quarter of an hour. If they take longer than that, give them less next day; if they wolf it all in 5 minutes and squeal for more, increase their ration. Feed them like this twice a day. Do not restrict their intake of roots and vegetables, but only of concentrates. Watch your pigs constantly, and if they look too thin, or too hungry, feed them more. I have kept pigs successfully for over 30 years with nothing more elaborate than some rolled barley, skimmed milk, maybe at a push some fish meal, some bran, plenty of roots and vegetables, and whatever else there is to give away.

The pig bucket

Now I must touch on the high art of the pig bucket. For the man with a thousand sows it is irrelevant, but for the self-supporting family, fattening a pig or two in the garden to kill, or with one or two dear old breeding sows who become almost part of the family themselves, the pig bucket is very relevant indeed. As I have said before, nothing should be wasted on the self-sufficient holding.

DUROC
A breed of hogs with its origin in the eastern United States and in the Corn Belt. Compact, attractive colour and good for fleshing out.

GLOUCESTER OLD SPOT
A fine and beautiful English breed evolved originally for living in apple orchards and woodlands. It is hardy, prolific, and a fine baconer.

WESSEX SADDLEBACK
Also known as the British Saddleback, this is a hardy outdoor breed popular for crossbreeding.

WELSH
A popular commercial breed because it's white, long, and lean. It is a reasonable outdoor breed.

PIG HOUSING

A pig house should be very strong, either movable or else easily dismantled and re-erected, and, if possible, at little or no cost. Mine are made of scrap corrugated iron nailed to bush timber. Walls and roof can be double iron sheets with insulation (old paper bags or bracken, for example) stuffed between. When piglets are 3 weeks old, make a gap in the fencing that is too small for the sow but large enough for the piglets to get through and "creep feed" on concentrates.

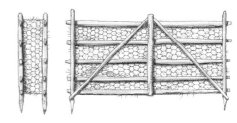

HURDLES

Two hurdles lashed together with straw between, and covered.

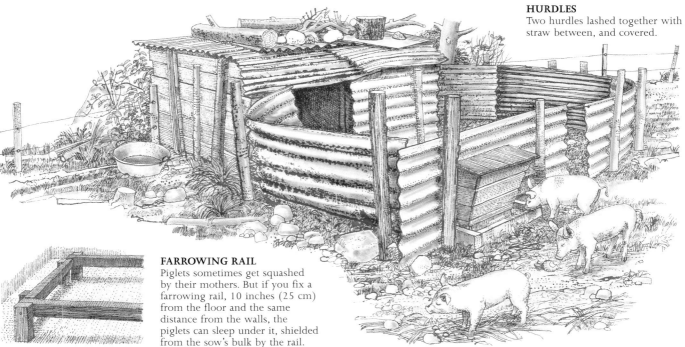

FARROWING RAIL

Piglets sometimes get squashed by their mothers. But if you fix a farrowing rail, 10 inches (25 cm) from the floor and the same distance from the walls, the piglets can sleep under it, shielded from the sow's bulk by the rail.

The dustman should never have to call. Under the kitchen sink there should be a bucket, and into it should go all the household scraps except such as are earmarked for the dog or the cats. When you wash up develop the "pig bucket technique". This means scraping all scraps first into the sacred bucket; then dribble – rather than run – hot water from the tap over each plate and dish so that the water carries all grease and other nutriments into a bowl (helped by a brush). Throw this rich and concentrated washing-up water into the pig bucket. Then finish washing the dishes any way you like and let that water run down the sink. The concentrated washing-up water from the first wash is most excellent food, and should not be wasted on any account.

Housing and farrowing

Except when sows are actually farrowing (giving birth to a litter) or have recently farrowed, they can live very rough. If they have plenty of straw or bracken and are kept dry, with no through draughts, and if possible with well-insulated walls and well-insulated roof in cold climates, they will do very well. Several sows together are, generally speaking, much happier than one sow alone. However, when a sow farrows she must have a hut to herself.

The hut needs to be big enough for the sow to turn round in comfort. If you like, you can have a farrowing rail, which prevents the sow from lying on the piglets (we kept six sows for 8 years and only lost 2 piglets during all that time from crushing). The books say the sow should have no litter when she farrows. All I can say is we give our sows access to litter, and they carry as much as they want of it indoors and make an elaborate nest. What they don't want they chuck out.

It is delightful to watch a sow making her nest, and there is no doubt that she is most unlikely to have any trouble if she is able to go through all her rituals before farrowing, and is then left alone in peace to farrow, with no other pigs jostling her, and no over-anxious owner fussing over her like an old hen. Sows eating their piglets, or lying on them, are generally products of an artificially organized system. If you break the chain of instinct here, you do so at your own risk.

The period when the sow wants a boar occurs at 21-day intervals, as in the cow. I do not like to put a gilt, or young female pig, to the boar until she is at least nearly a year old. As for the boar, he can be of a different breed, since pigs interbreed well.

Getting a sow "in pig"

If you keep six sows or more, you can afford the keep of a boar, so you can buy one and let him run with the sows. If you have fewer sows you can take a sow to a neighbour's boar when she needs it. Gestation will be about 115 days and litters can vary from six to 20 piglets: ten is about the average but we often had 12 (and reared them all) with monotonous regularity when we kept sows out of doors. You can, if you prefer, buy weaners (piglets from 8 to 12 weeks old) from neighbours to fatten. If you do buy weaners, you won't have to have a sow at all. We often bought and fattened three weaners, sold one and ate two, and the one we sold paid for the two we ate.

Slaughterhouse or do-it-yourself

First check the position with regard to local slaughterhouses (see also p.102). It is simplest to have the pig slaughtered, scraped, and split into two halves. You can then butcher the carcass yourself for joints, pork, hams, and bacon (see opposite). Remember that many modern slaughterhouses will only process relatively small pigs. If your pig is a large baconer, you may have to look further afield to find a suitable slaughterhouse.

Before all the regulations came in, the traditional way to kill a pig was to stick it in the throat, but I don't recommend this. Although I see nothing wrong with killing animals for meat, I see everything wrong with making them suffer in any way. If we kill an animal, we should do so instantaneously, and the animal should have no inkling that anything nasty is about to happen to it.

Were a pig to be killed in situ, I would use food to lure the pig into the place where he was going to be killed, and then, when he was interested in the food, shoot him in the head – either with a humane killer (captive bolt pistol) or a .22 rifle. Immediately the pig had dropped you would stick it, squatting squarely in front of the recumbent pig while someone else held it on its back, stick the knife just in front of the breastbone, and when you felt the bone let the knife slip forward to go under it. Then the knife would be pushed in a couple of inches and sliced forwards with the point of the knife towards the head. This severed the artery. Then you had to look out: a nervous reaction took place. The pig appeared to come to life and thrashed about, so care was needed not to be cut by its hooves.

Scraping

If you are going to scrape your pig yourself, you must scald him, and this is a ticklish operation. You can either dip your animal right into a hot water tub, or you can lay him on the floor or a scraping bench, and baste him with hot water. Not having a bath big enough to immerse a whole hog in, we lay the pig on its side and slowly and carefully pour hot water over a small part of him. The water should be 150°F (65°C) when it comes from the jug (so it should be slightly hotter when it goes into it). Keep pouring gently, and from time to time try some bristle with the thumb and forefinger. When the bristle begins to lift, scrape – and scrape furiously. It is better not to use a knife. A sharpened hoe-head or metal saucepan lid is sharp enough. Off comes the bristle, off also comes the outermost skin of the pig, and no matter what colour your pig started off as, he will become white as snow.

Keep working until the pig is absolutely clean. Put the legs right into the jug of hot water; then take out and pull the horny toes off with a hook. You really need two or three helping hands to scrape a big hog, with someone to bring on the hot water and another to fetch the home-brewed beer (vital on such an occasion). The head is difficult: you might set fire to some straw with methylated spirits. Hold the head over the flame to singe it, and scrub with a wire brush. When finished, douse the pig with cold water to get rid of all the loose skin, bristles, and any blood.

If you dip him, leave him in water at 145°F (63°C) for 5 minutes and then haul him out and scrape him. You can only do this if you know your water is going to stay at this temperature (pretty well exactly). If it is a few degrees too cold it won't loosen the bristles.

BUTCHERING YOUR PIG

It's a good idea to butcher your own pig (that is, cut up the carcass), even if local regulations forbid home slaughter, (that is, actually killing the animal). You need to gut and clean your carcass before jointing (see p.111). Home butchery is important because it's the only way to make sure the joints are exactly how you want them. To make good bacon and ham it is far better to cut up the meat yourself. You will need strong beams and an easily hoseable floor to do the job. Block-and-tackle (to lift the carcass) is available from any good chandlers.

1 Cut vertical slits on each side of the tendons of the hind legs in order to insert the gambrel.

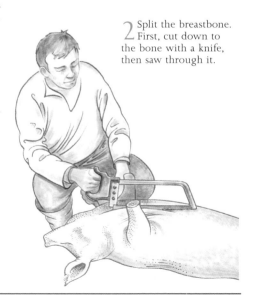

2 Split the breastbone. First, cut down to the bone with a knife, then saw through it.

3 Sling the pig up on the gambrel (hook). Your beams must be strong.

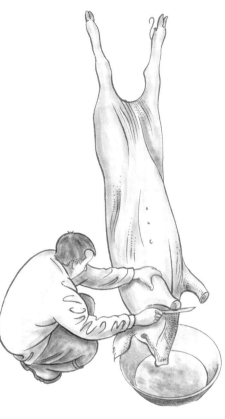

4 Cut the head off in order to sever the windpipe and gullet.

5 Cut round and tie off the rectum to prevent it from leaking.

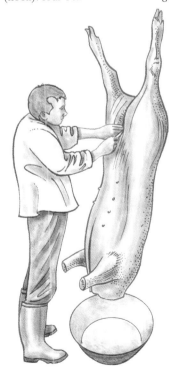

6 Slit along the line of the abdomen without cutting the guts. Have a large receptacle on the floor.

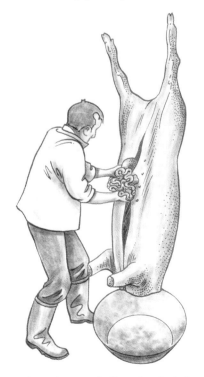

7 Haul out the innards. Keep the pluck (heart, pancreas, and lungs) separate. Throw a bucket of cold water inside the carcass.

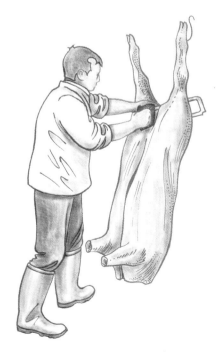

8 Saw down the backbone so as to cut the pig completely in half. Leave him to hang in an airy place overnight.

Hanging your pig

An inch or two above its foot on the back of the hind leg is a tendon. Cut down each side of this with a vertical slit through the skin and raise the tendon out with your fingers. Don't cut the leg above the hock as beginners do; this is barbarous and spoils good meat. Insert each cod of your gambrel through the two tendons in the legs.

Now, don't haul that pig off the ground until you have sawn through the breastbone or sternum! Cut down to the breastbone with a clean knife, then split it, right down the middle, with a saw. If you try to do this after you have hung up the pig all the guts will come flopping out and make the operation very difficult. Now grab hold of that tackle and heave away! Up goes your pig. Cut off the head just behind the ears, at the first spinal bone (the atlas) and you shouldn't have to use the saw. Put the head into brine.

EDIBLE PARTS OF A PIG

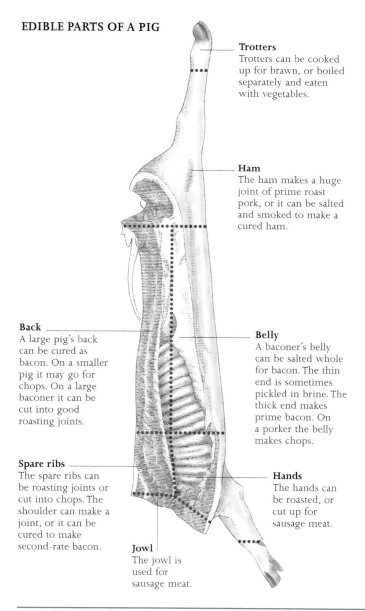

Trotters
Trotters can be cooked up for brawn, or boiled separately and eaten with vegetables.

Ham
The ham makes a huge joint of prime roast pork, or it can be salted and smoked to make a cured ham.

Back
A large pig's back can be cured as bacon. On a smaller pig it may go for chops. On a large baconer it can be cut into good roasting joints.

Belly
A baconer's belly can be salted whole for bacon. The thin end is sometimes pickled in brine. The thick end makes prime bacon. On a porker the belly makes chops.

Spare ribs
The spare ribs can be roasting joints or cut into chops. The shoulder can make a joint, or it can be cured to make second-rate bacon.

Hands
The hands can be roasted, or cut up for sausage meat.

Jowl
The jowl is used for sausage meat.

Before you have hauled the pig up too high, cut right round the bung (the anus) to sever it from the pig but not to pierce the rectum. Tie a string round it to prevent the excrement coming out. Now, haul the pig up further to a convenient height and score a tight cut right down from between the hams (haunches or hind legs) to the stick-cut in the throat. Don't cut through the abdominal wall! Cut right down it, keeping the guts back from the knife with your hand. You don't want to pierce the guts or stomach. Cut through the H-bone that joins the two hams, if necessary, with the saw, but don't cut the bladder. Then, gently haul out the rectum, the penis if it's male, the bladder, and all the guts, and flop the lot out into a huge basin. The penis and rectum can be chucked, or used for the dogs, but the rest is all edible or useful stuff.

Guts and brawn

Do not waste the guts. They need a good wash, after which you should turn the intestines inside out. You can do this by inverting them on a smooth piece of bamboo or other wood. Then scrape their mucus lining off with the back of a knife against a board. Get them quite clean and transparent. Lay them down in dry salt and they will come in handy, in due course, for sausage skins or "casings". The bladder can be filled through a funnel with melted lard, which will harden again and keep for months.

The stomach is edible, as are the intestines; together they are called "chitlings" or "chitterlings" and are quite nice. Turn the stomach inside out, wash it and put it down in dry salt until you want it. Now rescue the liver, which you can eat the same night, and peel out the gall bladder, which you can throw away. The heart should come away with the lights (lungs), and this is called the "pluck". Hang it on a hook. Carefully remove the caul — a beautiful filmy membrane that adheres to the stomach — and throw this over the pluck. The lungs are an extra treat for the dogs, while you can eat the heart. Throw several buckets of cold water inside and outside the carcass, prop the belly open with some sticks that are pointed at both ends, eat some fried liver, finish the home-brew, and go to bed.

In the morning the carcass should be "set" (stiff) if the weather is cold enough and you can split it right down the backbone. A butcher does this with a cleaver — if you are a beginner, you may like to do it with a saw. Take each half off, lay it on the table, and cut it up, as shown opposite.

Don't waste the head or the feet, for they make excellent brawn, or pork-cheese. Put the bones, skin, and all into a pan (and the bits of skin in a muslin bag so you can draw them out after all the goodness has been extracted) and boil and boil. Let it cool, break it into small pieces, boil again and add salt, pepper, and as many spices (marjoram, coriander, allspice, cloves, caraway) as your taste-buds can take. Boil it again, and pour into moulds such as pudding basins. The fat will rise to the top and form a protective covering and the brawn will keep a long time. Eat sliced cold. To deep freeze, pour straight into plastic bags.

JOINTING A PIG

This pig is a baconer, because it's too large for eating entirely as fresh pork. Therefore, most of it will be cured and preserved as bacon, ham, sausages, and so on. You will, of course, take a joint or two to eat as fresh pork, and if you have a deep freeze, this may be a larger proportion. Remember, a baconer is a fat animal and his meat is not as suitable for eating fresh as that of a little porker.

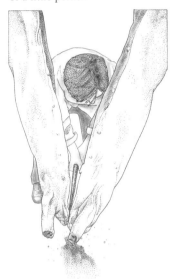

1 Split the carcass right down the backbone. A butcher might use a cleaver all the way, but amateurs will probably do better with a saw.

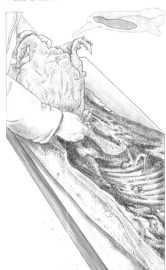

2 The leaf fat comes away very easily. It makes the best and purest lard. Next, take out the kidneys.

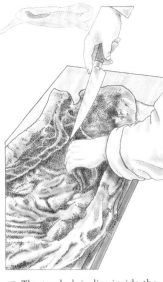

3 The tenderloin lies inside the backbone. It is delicious wrapped in the "caul", stuffed and roasted. The caul is a white, fatty membrane that supports the intestines.

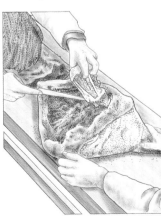

4 Cut the flesh of the ham using a saw.

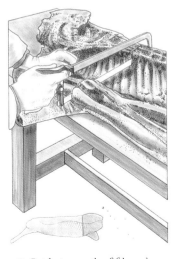

5 Remove the H-bone, which is one half of a ball-and-socket joint, to leave a clean, uncut joint.

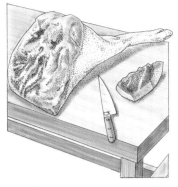

6 The ham must be trimmed so that a clean surface is left for the salt to penetrate. Work salt into every crevice.

7 This trotter is being cut off at the hock joint. You can saw the shank off below this joint to leave a cleaner surface for salting.

8 Cut between the fifth and sixth ribs to part the shoulder from the side, or flitch. The shoulders can be salted whole, or else boned and used for sausages.

9 Next saw off the chine, or backbone, and joint it.

10 Leave the ribs in the salted flitch. I like to cut them out and use them for soup. The rest of the side is salted for bacon.

11 Salt the whole shoulder, or you can halve it, roasting the top and salting the bottom.

SALTING PIG MEAT

Most of a big pig should be turned into bacon and ham, for there is no better way known to mankind of preserving large quantities of meat. The ham is the thick part of the pig's hind leg – the rump in fact. Bacon is the side of the pig. The shoulder can be cured too, or used for joints or sausage meat. But by far the most valuable parts of the pig are the two hams. There are two main methods of salting pig meat – dry curing and brine curing.

Dry curing

The first essential for dry curing to be effective is to guard against damage by blow flies. Even though your weather must already be cool, it is important to cover meat carefully with cloth or muslin. More important still, avoid loose pieces of meat, fat, or cavities which will provide perfect hiding places for maggots. Check all curing meat regularly. The easy way to make ham and bacon is by simply rubbing salt and sugar into the meat, then burying it in salt, then leaving bacon sides for 2 weeks in the salt, and hams for 3. It is better to do it more scientifically, because you save salt and the meat is not so salty when you eat it. I get over the problem of excess salt by slicing my bacon first, then soaking the thin slices for 10 minutes or so in warm water before frying them.

The proper way to do it, if you are going to be more scientific and efficient about it, is to use the following mixture for every 100 pounds (45 kg) of meat:

  8 lb (3.5 kg) salt
  2 lb (1 kg) sugar

Prepare this mixture with extreme care. Take half of it and put the rest aside. Use the half you have taken to rub the meat very thoroughly all over, rind-sides as well. Stuff salt hard into the holes where the bones come out in the hams and shoulders, and into any cavities. Success lies in getting salt into the meat quickly: it's a race between salt and bacteria. If the latter win you may lose an awful lot of very valuable meat. Unless the weather is too warm – 36°F (2°C) is an ideal temperature, but don't let the meat freeze – if you follow these instructions the bacteria won't win.

Cover all surfaces with the salt and leave on a salting tray, or a shelf. If you use a box there must be holes in it to allow the "pickle" – juice drawn out of the meat by the salt – to drain away. Make sure all the joints are carefully packed on top of each other. Be careful when you do this first salting to put roughly the right amounts of salt on each piece: not too much on the thinner bacon sides but plenty on the thick hams.

After 3 days, give another good rubbing with half the remainder of the salt (i.e. a quarter of the whole). Put the pig meat back in a different order to ensure even distribution of the salt all round. After another week haul out again and rub well with the last of the salt mixture. Put it back. Now leave it in the salt for 2 days per pound for big joints such as hams and 1½ days per pound for small joints and bacon. If you say that roughly a big side of bacon should cure for a fortnight and a large ham for 3 weeks you won't be far wrong.

Take the joints out at the allotted time, scrub them lightly with warm water to get the loose salt off, string them, and hang them up for a week or a fortnight in a cool, dry place. Then either smoke them or don't smoke them as the fancy takes you. Unsmoked or "green" ham and bacon taste very good, although personally I like them smoked. It's really all a matter of taste.

Brine curing

For every 100 pounds (45 kg) meat you should mix:

  8 lb (3.5 kg) salt
  2 lb (1 kg) sugar
  5 gallons (23 litres) boiled but cooled water

In theory thicker joints, like hams, should have a stronger brine – the above mixture with 4½ gallons (20 litres) of water – and the thinner joints, like bacon and bath-chaps which are the jowls of the pig, should have the mixture in 6 gallons (27 litres) of water. Put the meat in the brine, make sure there are no air pockets, put a scrubbed board on top and a big stone on top of this to keep the meat down (don't use an iron weight), and leave in the brine for 4 days per pound of each big joint. Thus you ought to weigh each joint before putting it in, and pull each one out at its appointed date. Bacon and small joints should only be left for 2 days per pound. After 4 or 5 days turn the joints round in the brine, and again every so often. If, in hot weather, the brine becomes "ropey" (viscous when you drip it off your hand), haul the meat out, scrub it in clean water, and put into fresh brine.

When you have hauled the meat out of the brine wash it in fresh water, hang it up for a week in a cool dry place to dry, then, if you want to, smoke it. You can eat it "green", i.e. not smoked at all. It should keep indefinitely, but use small joints and bacon sides before hams. Hams improve with maturing: I have kept them for 2 years and they have been delicious. Bacon is best eaten within a few months.

Cured hams and shoulders should be carefully wrapped in grease-proof paper and then sewn up in muslin bags and hung in a fairly cool, dry place, preferably at a constant temperature. If you paint the outside of the muslin bags with a thick paste of lime and water, so much the better. Like this they will keep for a year or two and improve all the time in flavour until they are delectable. Bacon can be hung up "naked" but should then be used within a few months. Light turns bacon rancid, so keep it in the dark. Keep flies and other creepies off all cured meat. Some country people wrap hams and bacon well, then bury them in bran, oats, or wood ashes which are said to keep them moist and improve the flavour.

Sheep

Sheep have a great advantage for the lone self-supporter without a deep freeze. In the winter time, a family can get through a fat lamb or a small sheep without it going bad. It is not that it keeps better than other meat, but just that the animal is smaller and you can eat it more quickly. However, I have kept mutton for a month, and in a climate where the day temperatures rose to over 100°F (39°C) in the shade. But the days were dry and the nights were cold. At night I hung mutton outside, in a tree – though out of reach of four-legged prowlers – and then, very early in the morning, brought it in and wrapped it in several thicknesses of newspaper to keep out the heat. The mutton was perfectly edible. You could do this in any climate where the nights are cold enough and the days are not muggy; and you could do it with any meat.

Sheep have two other advantages: they provide wool, and in various parts of the world they provide milk too. I have milked sheep, very often, and it's a very fiddly job. The milk tastes no better, and no worse, than cow's or goat's milk.

The chief disadvantage about keeping sheep on a very small scale is the problem of getting them tupped (mated). It doesn't pay to provide grazing for a ram if you have fewer than, say, half a dozen sheep. If you buy a ram to serve your sheep, you pay quite a high price for him, and when you sell him after he has served them, or next year say, you only get "scrap price", i.e. very little. And if you eat him you will find him very tough. I have eaten a 3-year-old ram and I know. So tupping is a problem with very small flocks of sheep.

Broadly, there are two quite sensible things you can do. You can keep some pet ewes and take them to a kindly farmer's ram to get them tupped in the autumn, or you might be able to borrow a ram. If you put a chest-pad on the ram with some marking fluid on it, or – in the old-fashioned way – rub some *reddle* (any coloured earth or dry colouring matter) on the ram's chest, you can see when all your ewes are served and then return the ram.

Alternatively, you can buy "store" lambs. Most hill farmers cannot fatten their lambs off in their first summer sufficiently to get them ready to sell and so they sell them as "stores". There are huge store marts in most hill-sheep countries. If you buy say 20 store lambs in the autumn, and keep them either on good winter grass or on rape, turnips, or other winter fodder crops, maybe allowing five lambs to the acre, you will probably find that you not only get your sheep meat for nothing, but you make a profit too. You can achieve this by killing one whenever you want some lamb to eat, and setting the fat lambs you haven't killed in the very early spring. If you also feed the store lambs a little concentrate (say, 1 pound or 0.5 kg a day of crushed barley and oats, or alternatively 2 pounds or 1 kg a day of maize and hay), they will fatten very readily.

Feeding

Sheep do well on pasture that has not had sheep on it for six months or so, because such pasture is free from sheep parasites. Five sheep will eat as much grass as a bullock, so an acre of good grass in the summer will easily support five sheep, but the stocking rate should be very considerably less in the winter for the grass doesn't grow. Sheep are very good for grass in the winter time – they "clean it up" after cows have been grazing on it all summer, because they graze much closer than cows. Sometimes you can fold sheep to great advantage on a "catch crop" of winter greens, turnip, rape, or hardy greens. This would be put in after a crop harvested fairly early in the summer. But you may decide you want this fodder for your cows.

During the winter the pregnant ewes need surprisingly little food, and if you have even a little grass they should make do on that. In very cold climates of course they need hay, and possibly corn, as well. People in very cold countries often winter ewes indoors and feed them entirely on hay, corn, and possibly roots.

DORSET HORN
A uniquely useful breed because it can lamb twice a year.

BORDER LEICESTER
A classic English sheep, a prolific breeder and good for mutton and wool.

SOUTHDOWN
A very small sheep, and therefore useful for the small self-supporting family.

LAMBING

The shepherd should not have to interfere at all, but if there is too long a delay in a birth the lamb or the ewe may die, so it becomes necessary to lend a hand.

1 Lay the ewe on her back, and preferably prop her up with a large bale of straw.

2 Wash your hands and the ewe's hind quarters very well. Lubricate your hand and her vulva with boiled linseed oil or carbolized oil.

3 If you see the forefeet, but the ewe cannot give birth after an hour, tie a soft cord round the feet and pull gently when she strains.

4 If you can't see anything, introduce the hand carefully while the ewe is not straining.

5 If the presentation is normal, grab the forelegs and pull the lamb out, but with care.

6 Pull more and more strongly while the ewe is straining, but don't pull when she isn't.

7 As the lamb's body appears, support it with your free arm. When he is half out, twist slightly to relieve the pressure.

8 Make sure the new lamb's nostrils are free of mucus, and leave him for the mother to lick.

ABNORMAL PRESENTATIONS

There can be many abnormal presentations. The shepherd must feel with his hand to find which one he is dealing with. In the case of twins both trying to get out together, he must gently push one back. Sometimes with a single lamb he must push the lamb back and adjust the limbs or the head. The whole thing is a matter of common sense and sympathy for lamb and mother. The lamb or lambs must be positioned so the head is not backward and the limbs are not doubled up.

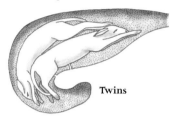

Twins

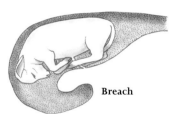

Breach

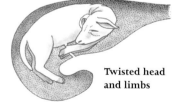

Twisted head and limbs

If you feed sheep, and they have no grass, a ewe will need about 4 pounds (2 kg) hay daily if there is nothing else, or say ½–1 pound (0.2–0.5 kg) a day, plus 15–20 pounds (6.8–9 kg) of roots. They will live on this without any corn or concentrates at all. They must not be too fat when they lamb, or else they will have lambing difficulties, but also they must not be half-starved.

In the spring, as soon as the ewes have lambed, put them on the best grass you have got, and preferably on clean grass: that is, grass that has not had sheep on it for some time. The grass is very nourishing at this time of the year and the lambs should thrive and come on apace. Within 4 months most of them will be fat enough either to eat or to sell to the butcher as "fat lamb".

Tupping

In cold climates sheep are generally tupped, or mated, in the autumn. If you have a flock of ewes it is best to cull them before mating: that is, pull any out that are so old that their teeth are gone. A full-mouth ewe, which has eight incisor teeth up, is 4 years old and should already have had three lambs. She may go for another year or two, or she may not, depending on the state of wear of her teeth.

Before putting the ram in with the ewes you should flush the flock. This means keeping them on very poor pasture for a few weeks, then putting them on very good pasture. Then put the ram in. All the ewes will take the ram then in fairly quick succession and you will not have a very drawn-out lambing season. A ram can serve up to 60 ewes in the tupping season. The gestation period is 147 days. Some people try to get ewes to lamb very early so as to catch the early lamb market, but unless you intend to lamb indoors, and with very high feeding, I would not recommend this. I like to see lambs coming in late February or March, and I find the later ones soon catch up with the poor little half-frozen winter lambs and pass them.

Lambing

Watch them carefully when they start to lamb. Leave them alone to get on with it: they generally can. But if a ewe is in labour for more than an hour and obviously cannot void her lamb, give her help. Get the ewe into a small pen where you can catch her. Lay her down (if she is not already down, *see opposite*). Wash your hands using carbolic oil or boiled linseed oil. If the lamb's forefeet are just showing, work them gently out, only pulling when the ewe strains. The feet are very slippery, so put a soft cord around them — a scarf or a necktie will do — and haul gently when the ewe strains: haul slightly downwards.

If you are not winning, insert your hand very gently into the vagina, feeling along the forelegs, and make sure that the head is not bent backwards. If it is, push the lamb back gently and try to pull the head forwards. The lamb should then come out.

You can get your hand right inside and feel about for abnormal presentations, but it is difficult because the uterus exerts great pressure on your hand. Still, if you only have a few ewes, and they are healthy, the chances are that you will have to do nothing with them at all.

Switching lambs

If a single lamb dies and you have another ewe with twins it is a good thing to fob one of the twins off on the bereaved ewe. Put the bereaved ewe in a small pen, rub the twin lamb with the dead body of her lamb, and try to see if she will accept the new lamb. If she won't, skin the dead lamb, keeping the skin rather like a jersey, and pull the skin over the live lamb. Almost invariably the foster mother ewe will accept him. The advantage of this is that the mother of the twins "does" (feeds) one lamb much better than she would do two, while the bereaved ewe does not get mastitis and have trouble drying up her milk. She is happy, the twins are happy, the twin-mother is happy, and so are you.

Orphan lambs

"Poddy lambs" are one answer for the self-supporter. Farmers will often let you have orphan lambs for nothing, or nearly nothing, and you can bring them up on the bottle, with a teat on it. You can give them warm cow's milk, diluted with water at first, later neat. Goat's milk is better than cow's milk, but don't let them suck the goat direct — milk her and feed from the bottle. Keep the lambs warm. They will grow up thinking they are humans.

Shearing

I start shearing at the beginning of July, but people further south start earlier. So see what your neighbours do. Most people don't shear the new lambs, but only the ram, the ewes, and any wethers (castrated males) left over from the previous year.

If you shear by hand (*see also illustrations on p.116*) you will find it much easier to sit the sheep up on a bench or a large box. Hold her with her back towards you and practise holding her firmly with your knees, thus leaving your hands free. Clip the wool off her tummy. Then clip up the throat and take blow after blow (a blow is the shearer's term for one row of wool clipped) down the left side of the neck, shoulder, flank, and right down as far as you can get it, rolling the sheep around as you clip. When you can get no further on that side, roll the sheep over and start down the right side, hoping to meet the shorn part from the other side as you roll the sheep that way. The last bit involves laying the sheep almost right over to clip the wool near the tail. The sheep should then leap away, leaving her fleece in one entire blanket.

Lay the fleece, body-side down, on a clean sheet or floor and cut away any bits of dunged wool. Fold the edges over towards the middle, roll up from the head end, twist the tail end into a rope, and wrap round the tight bundle and tuck the rope under itself. If the wool is for sale, pack the fleeces tight into a "pocket" (big sack). Put the dunged bits in a separate sack and mark it "daggings" (for preparation of wool for spinning *see p.268*).

Some hand shearing tips

If people tell you hand shearing is easy, you can tell them they lie. It is back-breaking, hand shears make your wrist ache like hell, and it is extremely difficult. Keep your shears sharp and cut as close to the sheep as you can without nicking her. If you do nick her, dab some stockholm tar on the wound. Beware of her teats. Of course mechanical clippers make the job much quicker, but it is still very hard work. But shearing is fun: if several of you are doing it, there is a great sense of camaraderie, and you have a sense of achievement when you get good at it. At first it seems quite impossibly difficult but just persevere and don't give up. You'll win in the end, if you don't bust a gut. Home-brew helps.

Sheep disease

Except on mountains, sheep suffer from a bright green blowfly which "strikes" (lays eggs on) them, particularly on any dunged parts. It is good practice to cut this wool off them before shearing. This is called "clatting" or "dagging". But if about a fortnight after shearing you either spray or dip your sheep in some proprietary sheep dip, you will protect them from fly strike for at least 2 or 3 months: in fact, probably until the cold weather comes and does away with the flies. If you don't protect them they will get struck and the maggots will eat right into the sheep and eventually kill them in the most unpleasant manner possible.

There are two other very common sheep diseases. Fluke is one of them. Sheep grazing in wet places can pick up large worms from a certain minute freshwater snail. These worms live in the gall ducts of the liver. Prevention is better than cure, so you have to drain land, or keep sheep from wet places, and use a regular worming programme (consult the vet in your area).

SHEEP SHEARING

Sheep should be shorn in the summer when the weather is warm enough.

1 Grab the sheep by the wool on her flanks, not by the wool on her back. Pick her up and sit her down on her rump for ease of handling.

2 Clip all the wool of her stomach down as far as the udder. Avoid cutting the udder on a ewe, or the penis on a ram or wether.

3 Open the wool up her throat and start shearing around the left side of her neck and head.

4 Keep on down the left side of the shoulder and flank as far as you can reach in that position. If you can hold the sheep with your knees, your hands will be free. Hold her skin tight with your left hand. Shear as close to it as you can get with the other.

5 Roll the sheep over and clip down her right side. The fleece should come right off her body, except at the hind quarter.

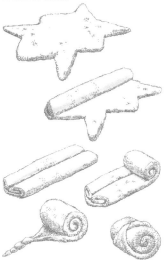

6 Lay the sheep flat on the ground and put your left foot over her to hold her between your legs. Finish taking the fleece off the hind quarters.

7 Trim her tail and hind legs separately for appearance. Keep the wool from these parts separate from the main fleece.

ROLLING A FLEECE

To roll the fleece lay it body-side down on a clean surface. Pick out any thorns, straw and so on. Turn the sides inwards and begin to roll tightly from the neck end. Twist the neck end into a rope. Tie it round the fleece and tuck it under itself.

DIPPING SHEEP

About a fortnight after shearing, the sheep should either be dipped or sprayed with a proprietary mixture. Dipping is better because it really soaks in. This is for various purposes. In areas where scab is present it is often compulsory, but in most regions it is necessary to guard against "strike" (see p.115) or blow-fly maggot infestation. It also kills keds and other parasites.

Another disease to which sheep are especially prone is foot-rot, likewise a scourge of sheep in damp lowlands: sheep on mountains seldom get it. To protect against foot-rot, trim the feet occasionally (better to use sharp pincers than rely on a knife) to remove excess horn. But if your sheep have got foot-rot, the best cure of all is to walk them through a foot-bath of formalin.

To kill a sheep for mutton and lamb
Depending on where you live, killing a sheep will be covered by the same regulations that affect cattle and pigs (*see p.102*). If you have to kill a sheep, stick it by shoving the knife into the side of the neck as close to the backbone and the head as you can get it. Keep the sharp edge away from the backbone and pull the knife out towards the throat. This cuts all the veins and arteries in the neck and the windpipe too. But I would never, never do this until I had stunned or killed the sheep with either a .22 bullet, a humane killer, or, in want of these, a blow on the head from the back of an axe.

Skinning
With the sheep on the ground or on a bench, slice a narrow strip of skin off the front of the forelegs and the back of the hind legs. Grasp the foreleg between your knees and raise this strip of skin off, right down into the

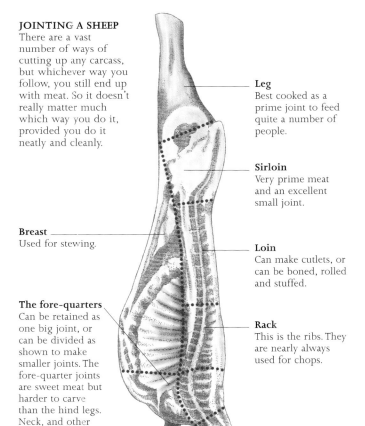

JOINTING A SHEEP
There are a vast number of ways of cutting up any carcass, but whichever way you follow, you still end up with meat. So it doesn't really matter much which way you do it, provided you do it neatly and cleanly.

Breast
Used for stewing.

The fore-quarters
Can be retained as one big joint, or can be divided as shown to make smaller joints. The fore-quarter joints are sweet meat but harder to carve than the hind legs. Neck, and other odd bits, go for stew or soup.

Leg
Best cooked as a prime joint to feed quite a number of people.

Sirloin
Very prime meat and an excellent small joint.

Loin
Can make cutlets, or can be boned, rolled and stuffed.

Rack
This is the ribs. They are nearly always used for chops.

brisket (chest). Keep the knife edge against the skin rather than the meat so as not to cut the latter. Hold the strip of skin tight away from the leg with the left hand. Do the same with the hind legs (along the back of them) as far as the anus. Then skin out the legs, still holding each leg in turn between your knees. Don't tear or cut the meat. Cut the feet off at the lowest joint, and raise the tendons of the hind legs for sticking the gambrel (hook) through. Pull the flap of skin you have raised between the hind legs off the carcass as far as it will go.

Fisting
Now "fist" the skin away from the belly. (Fisting is the forcing of your fist in between skin and the dead sheep.) Use the knife as little as possible. Keep your hand washed when you fist, and don't soil the meat. Cleanliness is essential in all butchery operations. When you are fisting a sheep skin off, make sure that you leave the "fell", or thin tough membrane, on the meat and don't take it off on the skin. On the skin it is a nuisance; on the meat it is a useful protection. Fist the skin off both the rear end and the brisket as far as you can. Then insert the gambrel and hoist the sheep up.

Slit the skin right down the belly, and then fist the skin off the sheep as high and low as you can. It will stick on the rump – fist here from below as well as above. Try to avoid using the knife as much as possible. If it's a nice fat lamb the skin will come off easily, but if it's a skinny old ewe, or a tough old ram, you've got problems. Free the skin with the knife from anus and tail. Then you can often pull the skin right off the sheep, down towards the shoulders, like taking a jersey off. Finally use the knife to take the skin off face and head. Then cut the head off at the atlas bone – this is right next to the skull.

Disembowelling
Cut around the anus. Pull several inches of the rectum out. Tie a string round it, and drop it back. Now rip the belly right down as you would do with an ox or pig (*see p.109*). Protect the knife point with your finger to stop it from piercing the guts or paunch. Pull the rectum down with the rest of the guts. Take the bladder out without spilling its contents, and carefully hoist out the paunch and guts and other innards.

Carefully haul the liver free from the back, and then haul the whole mass up and out of the sheep. You will have to cut the gullet to do this, and you should tie the gullet above the cut so the food doesn't drop out. Remove the liver carefully, and drop the guts into a big bowl. You can clean the guts and use all but the third stomach for tripe.

Cut down the breastbone. With a lamb you can do this with a knife, with an older sheep you may have to use a saw. Pull the pluck (heart and lungs) out and hang them up. Finally, wash the carcass down with cold water and go to bed. The next morning, early, split the carcass and cut it up. You can use the small intestine for sausage casings.

Poultry

CHICKENS

All hens should he allowed access to the great outdoors except in the winter in very cold climates. Not only is it inhumane to keep chickens indoors all the time, it leads to all the diseases that commercial flocks of poultry now suffer from. Some poultry-keepers go to such extremes of cruelty as keeping hens shut up in wire cages all their lives. Sunshine is the best source of vitamin D for poultry, as it is for us. Hens evolved to scratch the earth for their living, and to deny them the right to do this is cruel. They can get up to a quarter of their food and all their protein from freshly growing grass and derive great benefit from running in woods and wild places. They crave, and badly need, dust-baths to wallow in and fluff up their feathers to get rid of the mites. In well over 20 years of keeping hens running out of doors, I have yet to find what poultry disease is, with the exception of blackhead in turkeys. Our old hens go on laying year after year until I get fed up with seeing them, and put them in the pot.

Feeding

Hens running free out of doors on good grass will do very well in the summer time if you just throw some grain. In the winter they will need a protein supplement. You can buy this from a corn merchant, or feed them fish meal, meat meal, soya meal, other bean meal, or fish offal. I would recommend soya meal most of all, because soya is the best balanced of any vegetable protein.

CHICKEN BREEDS

Old-fashioned broody hens are hard to find because commercial breeders breed hybrids for egg production and nothing else. So you will have to search for those marvellous traditional breeds which can live outdoors, lay plenty of eggs, go broody and hatch their eggs, rear their chicks and make good table birds as well. You want breeds like the Rhode Island Red, a good dual-purpose hen, meaning it is a good layer and a good table bird; the Light Sussex, an Old English breed, again dual purpose; or the Cuckoo Maran, which is very hardy and lays large, deep-brown, very high quality eggs, but not in prolific quantities. US breeds like the various Plymouth Rock and Wyandotte fowl are popular and good for eggs and meat.

So if you live in a region where soya beans can be grown successfully, the problem of your protein supplement is easily solved. But soya must be cooked, as it contains a substance which, when raw, is slightly poisonous. Sunflowers are good too, particularly if you can husk the seed and grind it, but they're quite good just fed as they are. You can also feed the hens lupin seed (either ground or whole), rape seed (but not too much of it), linseed, groundnut or cotton seed (but this must be cooked first), crushed or ground peas or beans, lucerne or alfalfa, or alfalfa meal. These all contain protein.

From ten days old onwards, all chickens should have access to fresh vegetables, and after all, this is one thing we can grow. So give them plenty of vegetables whether they run out on grass or not. My method of feeding hens is to let them run outdoors, give them a handful each of whatever high-protein meat or grain I have in the morning, and scatter a handful each of whole grain in the evening. Wheat is best, or kibbled maize. Barley is very good, but it should be hummelled – banged about until the awns (spikes) are broken off. An equally good method is to let them feed both protein and grain from self-feed containers. These should be placed out of the reach of rats.

If you allow hens to run out of doors, or to have access to a good variety of foods, they will balance their own rations and not eat more than they need anyway. But if hens are confined indoors, you can balance their rations like this:

Laying mash

1 cwt (50 kg) *wheat meat*
1 cwt (50 kg) *maize meal (preferably yellow maize)*
1 cwt (50 kg) *other gram meal (oats, barley, or rye)*
1 cwt (50 kg) *fish meal*
30 lb (13.5 kg) *dried milk*
20 lb (9 kg) *ground seashells*
5 lb (2.3 kg) *salt*
Give them free access to this, and a handful each of whole grains to scratch out of their straw or litter.

Cuckoo Maran   Rhode Island Red   Light Sussex   Barred Plymouth Rock   Columbian Wyandotte

COOPS

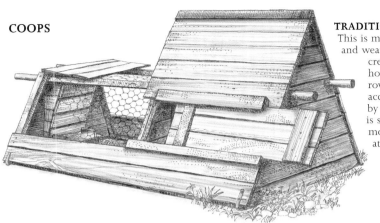

TRADITIONAL ARK
This is made of sawn timber and weather-boarding, is well creosoted, has a night-house with perches, a row of nesting boxes accessible from outside by a door, and a run. It is strong but easily moveable by handles at each end.

BROODY COOP
Individual sitting hens need a broody coop. It must have a rat-proof floor and slats in front to confine the hen, if needed, but admit her chicks.

HOME-MADE ARK
The homesteader should try to make his gear for nothing, and used fertilizer bags are free. We made this ark for nearly nothing. Thatch would work as well, and would look better.

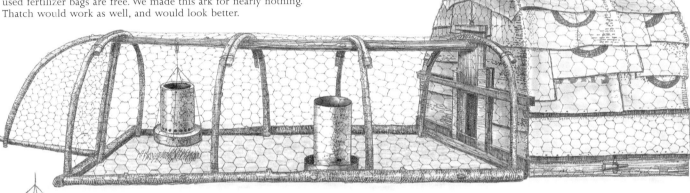

SELF-FEED HOPPER
You can buy galvanized hoppers, but you can make them yourself, for free, by hanging up an old oil-drum, bashing holes round its base to let the hens peck food out, and hanging the cut-off base of a larger oil-drum underneath to catch spilled food. Hang above rat-reach.

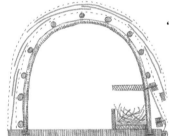

"FERTILIZER-BAG" ARK
A layer of wire-netting over the overlapping bags keeps them from flapping. The bags can be supported underneath by closely spaced horizontal rods of hazel or willow. An inspection door for nesting boxes can be made by hanging up fertilizer bags weighted with a heavy batten across their bottoms.

Fattening mash for cockerels or capons
Barley meal is the best fattener for any poultry, but can be replaced by boiled potatoes. Skimmed milk is also ideal. Feed the mash ad-lib:

3 cwt (150 kg) barley meal
1 cwt (50 kg) wheat meat
½ cwt (25 kg) fish or meat meal
30 lb (13 kg) dried milk; plus some time (ground sea shells) and salt

Little chick mash
30 lb (13 kg) meal (preferably a mixture of wheat, maize and oats)
12 lb (5 kg) fish meal or meat meal
12 lb (5 kg) alfalfa (lucerne) meal
2 lb (0.9 kg) ground seashells
1 lb (0.5 kg) cod liver oil
1 lb (0.5 kg) salt; plus a "scratch" of finely cracked cereals

If you give them plenty of wholemilk (skimmed is nearly as good) you can forget all but a little of the cod liver oil, the alfalfa meal, and half, if not all, the fish meal or meat meal. If you have food that is free to you, or a by-product of something else, it is better to use that (even if the books say it is not perfect) than something you have to pay for. I am a great believer in making the best of what is available.

Free range
If hens run completely free range, it is often better to keep them in their houses until midday. They generally lay their eggs before that time, so you get the eggs before you let them out, instead of losing them in the bushes where the rats get them. Hens will do a pasture good if they are not too concentrated on it, and if their houses or arks are moved from time to time over the pasture.

The chicken is a woodland bird, and hens always thrive in woodland (if the foxes don't get them). They can also be run very advantageously on stubble (land from which cereals have just been harvested). Keep them out of the garden or you will rue the day.

Limited grass
If you have fresh grassland it is a great advantage if you can divide it into two (a strip on each side of a line of hen houses, for example), and let the hens run on each strip in turn. As soon as they have really eaten the grass down on one strip let them run on the other. In the summer, when the grass grows so quickly they can't cope with it, let them run on one strip long enough to let you cut the other strip for hay. Or alternate hens with sheep, goats, cows, or geese. Poultry will eat any grass provided it is kept short, but ideally it should be of the tender varieties like timothy or the meadow grasses. There can be clover with it, although the hens will provide enough nitrogen in their droppings.

The Balfour method
This is suitable for the "backyard poultry-keeper" or the person who only has a small or limited garden. You have a pen around your hen house in which you dump plenty of straw, bracken, or whatever vegetation you can get. In addition, you have two (or three if you have the space) pens which are grassed down, and which can be approached by the hens from the straw yard. The hens will scratch in the straw yard and so satisfy their scratching instincts and spare the grass. Now let the hens run into one of the grass pens. Change them into another pen after, say, a fortnight or 3 weeks. They will get a bite of grass from this, and the grass in the first pen will be rested and have a chance to grow. The straw yard will provide half a ton of good manure a year from each hen. The old "backyard poultry yard" which is a wilderness of scratched bare earth, coarse clumps of nettles, rat holes, and old tin cans, is not a good place to keep hens or anything else.

Housing
The ordinary commercially produced hen house, provided it is mobile, works perfectly well. If used in the Balfour or limited grassland methods described above, it doesn't even have to be mobile, unless you intend to move it from time to time into another field. Utter simplicity is fine for a hen house. Hens need shelter from rain and wind, some insulation in very cold climates, and perches to sit on. Make sure the perches are not right up against the roof, and are placed so their droppings have a clear fall to the floor. The nesting boxes should be dark, designed to discourage hens from roosting in them and roofed so hens don't leave droppings in them. You should be able to reach in and get the eggs easily. There are patent nesting boxes which allow the eggs to fall down to another compartment. I think these are a very good idea; they prevent the eggs getting dirty.

We have moveable hen arks (*as illustrated*), with an enclosed sleeping area and a wire-netted open run. They hold 25 hens each and cost nothing but a handful of nails, some old torn wire netting and some free plastic fertilizer bags.

In countries which have heavy snow in the winter the birds will have to be kept in during the snow period. It is not a bad idea to confine some birds in a house in which there is electric light if you want eggs in the depth of winter. Give the birds, say, 12 hours of light and they will think it is summer time and lay a lot of eggs; otherwise they will go off laying as soon as the days get really short.

Rearing chickens
It is always a good idea to keep a rooster among your hens, and not just to wake you up in the morning. The hens will lay just as many eggs without a rooster but the eggs won't be fertile. Also if each batch of hens has a cockerel to marshal them and keep them together out of doors, they fare better and are less likely to come to harm.

If you leave a hen alone, and the fox doesn't get her, she will wander off into the hedgerows and wander back again in a few weeks with a dozen little chicks clucking at her heels. These chicks, being utterly naturally reared, will be the healthiest little chicks you will ever see. Alternatively you can watch your hens for broodiness. You can tell when a hen is broody by the way she squats tight on her eggs and makes a broody clucking noise when you try to lift her off. Help her by enclosing her in a broody coop (a little house with slats in front of it that baby chicks can get through and the mother can't). Give her nice, soft hay or something as a nest, and put a dozen fertile eggs under her. (They can be any kind of poultry egg you like). See that she always has water and food: she will eat very little.

Let her out once a day for a short walk, but get her in within half an hour or the eggs will get cold. Eggs should hatch out in 21 days from the start of brooding. As soon as the chicks are a few days old, you can let the hen out, and she will lead them around and teach them how to look for food. This is by far the best way to raise poultry, and beats any incubator.

If you are just starting with hens, you can either order "day-old chicks" or "point-of-lay pullets" which are young females just about to lay. When chickens are only a day old they don't need to eat and can be packed into parcels and sent about the country with impunity. A day or two older and they would die.

Keep such pullets as you need for your dock replacements and fatten the cocks for the table. Feed all little chicks on a fairly rich diet of high protein and finely ground meal. For the first few days add ground-up hard boiled eggs and milk by-products to their ration. Wheat meat mixed with milk is also a perfect food.

Always make sure that chickens have access to enough lime, in the form perhaps of ground seashells, and to insoluble grit like crushed flint. Hens running out of doors are not so dependent on artificial supplies.

KILLING AND PREPARING A CHICKEN

Grab the legs with your left hand, and the neck with your right hand so that it protrudes through the two middle fingers and the head is cupped in the palm. Push your right hand downwards and turn it so the chicken's head bends backwards. Stop as soon as you feel the backbone break, or you will pull the head off.

PLUCKING

Start plucking the chicken as soon as it is dead, while it is still warm; once it gets cold the feathers are much harder to get out. Be very careful not to tear the skin.

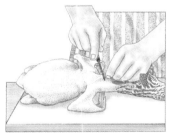

DRAWING

1 Slip the knife under the skin at the bottom of the neck and cut up to the head.

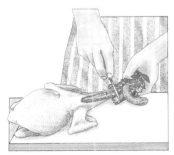

2 Sever the neckbone at the bottom end with secateurs or a sharp knife.

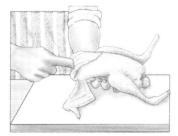

3 Remove the neckbone. Insert the index finger of your right hand, move it around inside, and sever all the innards.

4 Cut between the vent and the tail, being careful not to sever the rectum.

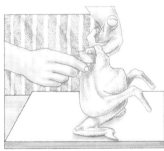

5 Cut all the way around the vent so that you can separate it from the body.

6 Carefully draw the vent, with the guts still attached, out from the tail end.

7 The gizzard, lungs, and heart will follow the guts.

8 Remove the crop from the neck end of the bird.

TRUSSING

If you put the bird in the oven and cooked it, it would taste just the same as if you trussed it. But to make a neat and professional job it is worth trussing properly.

1 Thread a large darning needle, force the chicken's legs forward, and shove the needle and thread through the body low down.

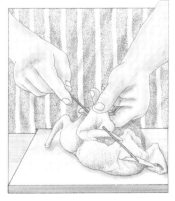

2 Push the needle through the wing and bring it across the skin at the neck.

3 Push through the other wing and tie the ends of the thread together.

4 Re-thread the needle and pass over the leg, below the end of the breastbone and round the other leg.

5 Cross the thread behind the hocks and tie around the parson's nose.

Cockerels should weigh 2–3 pounds (1–1.5 kg) at 8 to 12 weeks, and are called "broilers". Birds or table breeds should, at 12 or 14 weeks, weigh more – 3–4 pounds (1.4–1.8 kg) – and in America are called "fryers". After this you may call them "roasters". At 6 or 9 months old they are called "broilers"; so are old hens culled from your laying flock. An old hen can make good eating.

Good laying hens should have bright eyes, large red healthy-looking combs and wattles, wide apart pelvic bones (fairly loose so eggs can get out) and a white, large, moist vent. If they have the opposite, wring their necks. They won't lay you many eggs and you should certainty not breed from them. Don't stop hens from going broody. Finally, it goes without saying that eggs are much better eaten fresh, and it is quite possible to have new-laid eggs throughout the year. If you want to store eggs, clean them first and drop them into water-glass, which you can buy from a chemist.

GEESE
These are most excellent birds for the self-supporter. They are hardy, tough, self-reliant grazers and they make good mothers. The best way to start breeding geese is to buy eggs from somebody and put them under a broody hen. A hen will sit on five or six goose eggs and hatch them, but you want to make sure that she hasn't been sitting too long when you put the goose eggs under her, for goose eggs take longer to hatch than hens' eggs do (up to 30 days or even more). During the last week of the sitting, take the eggs out from under the hen every day and wet them with lukewarm water (goose mothers get wet but hen foster mothers don't). On the day when the eggs start to pip, wet them well. Some people remove the first goslings that hatch so that the foster mother doesn't think she's done her job, and then they replace them when the last egg has hatched. I've never bothered and always had good results.

Feed the goslings well for the first 2 or 3 weeks on bread soaked in whole milk (or skimmed milk). If they are fairly safe where they are, let the hen run around with them. If you fear they will get lost, confine the hen in a broody coop. I prefer the hen to run loose. When the young birds no longer need the foster mother, she will leave them and start laying again.

But geese, although such fierce and strong birds, are vulnerable to two enemies: rats and foxes. Rats will pull goose eggs out from under a sitting hen or goose, and they will kill baby geese whenever they get a chance. So poison them, deny them cover, gas their holes: do anything to get rid of them. These vermin are the enemy of everything wholesome on your farm.

As for foxes, they just love geese. They will snatch a sitting goose off her eggs whenever they can. So they cannot really co-exist with the self-supporter. Shoot them at night if you have a gun license. Use one of those powerful electric torches which illuminates the sights of the gun as well as the fox. If there are foxes in the area, you must confine sitting geese in a fox-proof place.

If you start reducing the goslings' food after 3 weeks, they will live on grass. As adults, they don't need any food except grass, but it is a good idea to chuck them some corn in January and February when you want to feed the goose up a bit to get her to lay well. Three weeks before you intend to kill them (generally Christmas), you should confine them, and feed them liberally with barley meal, maize meal, and milk if you have it to spare. They will fatten on this, and one of them will provide the best Christmas dinner in the world.

Geese pair for life, so I prefer to keep one goose and one gander together as a breeding pair, though many people keep a gander to two or three geese. They lay early: in February or March. If you leave them alone, they will sit on a dozen or more of their own eggs and hatch them with no trouble at all, but if you are greedy you can keep stealing their eggs and putting them under broody hens. But hens aren't always broody as early in the year as that.

When the time comes to kill a goose (or turkey for that matter), grab the bird by its legs with both hands. Keep the back of the bird away from you. Lower the head to the floor, and get someone else to lay a broomstick across the neck just behind the head. Tread on both ends of the broom stick, and pull the legs upwards until you feel the neck break. If you hold the tips of the wings as well as the two legs, the bird will not flap after it is dead. Then treat as with a chicken (*see* p.121).

DUCKS
To say that ducks don't need water is nonsense. Ducks do need water and cannot possibly be happy without it. It is inhumane to keep animals in conditions grossly different from the ones their species has evolved to live in. So give them access to water, but keep your ducklings away from it for their first week or two, until they have the natural protection of oil on their feathers. You must give them drinking water, however.

Swimming water for ducks is better if it is flowing, and renews itself. A stagnant pond is less healthy. Many eggs get laid in the water, or on the edge, and if the water is dirty, the eggs, which have porous shells, can be dangerous to eat. So don't eat eggs that have been lying in filthy water, no matter how much you superficially clean them. If there is no natural water on your holding, my advice is not to keep ducks. You can, of course, create an artificial pond, either out of concrete, puddled clay, or plastic sheeting buried in the ground, but if you do, make sure the water is renewable and does not become stale and stagnant.

One drake will look after half a dozen ducks and enjoy it, but ducks make rotten mothers. If you let them hatch their own eggs you must confine them in a broody coop, or they will kill the little ducklings by dragging them all over the place. Hens are much better duck-mothers than ducks are. Duck eggs hatch in 28 days and the baby ducks need careful feeding. Up to 10 weeks feed them as much barley or other meal as they will eat.

Add milk if you have it. Feed ducks about the same as you would hens when you are not fattening them. The duck is not a grazing bird to the extent that the goose is, but ducks will get quite a lot of food if they have access to water, or mud, or are allowed to roam around. They are partly carnivorous, and eat slugs, snails, frogs, worms, and insects. Don't let breeding ducks get too fat, or they will produce infertile eggs. They like a mash in the morning of such things as boiled vegetables, flaked maize, pea or bean meal, wheat meal and a little barley meat. Give them about half a handful each for breakfast, and half a handful of grain in the evening. If you find they get fat, give them less. If you find they get thin, give them more.

There are ducks (such as the Indian Runner) with very little meat on them that lay plenty of eggs, and table ducks (such as the Aylesbury) with plenty of meat but not many eggs. Then there's the Muscovy, a heavy, far too intelligent, very hardy bird, that's good to eat but has dark flesh.

Kill your young ducks at exactly 10 weeks. They are full of bristles before and after that time. They won't put on weight afterwards anyway. At 10 weeks they are easy to pluck and are at their prime. You can of course eat old ducks if you want to, but they are tougher and much fattier.

Housing for ducks can be extremely simple, but this does not mean that it should be reach-me-down. Ducks like a dry, draught-free but well-ventilated house. If it's mobile, so much the better, because otherwise the immediate surroundings get in a mess. Make it fox- and rat-proof.

TURKEYS
Compared to other poultry, these are very delicate birds. If they associate in any way with chickens, they get a fatal disease called blackhead, unless you medicate their water or food. If you want to have them without medicating them, you must keep them well away from all chickens.

You must even be careful about walking from the hen run to the turkey run without changing your boots and disinfecting yourself. It's hardly worth it. Turkeys do not seem to me to be a very suitable bird for the self-supporter, unless he wants to trade them. In this case he can rear them intensively in incubators and brooders, or buy them reared from another breeder.

PIGEONS
In America squabs are a delicious dish. They are young pigeons killed at about four and a half weeks. The parent birds are sometimes reared intensively, in a special house with a wire run or "fly" for the birds to flap about in. Kept like this the birds should be fed on grain, with some peas included, as they must have some protein. A pair of pigeons might eat up to a hundredweight of grain a year and raise up to 14 squabs a year, so if you had a dozen pairs you would be eating squabs until you were sick of them. If they are confined in this manner you must give them grit, as well as water for drinking and bathing in, and you must attend to hygiene by having some sort of litter on the floor that you can change.

Personally I would object strongly to keeping pigeons like that and would always let them fly free. Have a pigeon loft, get a few pairs of adult pigeons (already mated) from somebody else, and before letting them loose keep them for 3 weeks in some sort of cage in the loft, where they can see out (this is important). Then let them out, chuck them a little grain every day, and let them get on with it. Kept like this they are no work and little expense, and in fact do very little damage to crops, although you hope they eat your neighbour's and not yours. If he shoots a few, well, that won't break you. Harvest the squabs when the underside of the wing is fully feathered. Kill, pluck, draw, and truss, as if they were chickens (*see* p.133).

AYLESBURY DRAKE
The best British table breed. It is large, heavy and very hardy, and its ducklings grow exceptionally fast.

EMBDEN GANDER
The Embden is a good table breed. Its feathers and down are pure white and ideal for stuffing cushions and eiderdowns.

WHITE TURKEY
These can grow up to 38 pounds (17 kg). There is a small, quick-maturing form of this breed called the Beltsville White.

Rabbits

Rabbits are a very good stock for the self-supporting family to keep. They can be fed largely on weeds that would otherwise be wasted, and they make excellent meals. New Zealand Whites are a good breed to have because they are good meat and their skins are very beautiful when cured. Californians are also excellent. Such medium breeds tend to be more economical than very large rabbits (e.g. Flemish Giants) which eat an awful lot and don't produce much more meat. If you get two does and a buck, these should provide you with up to 200 pounds (90 kg) of meat a year.

Shelter
In the summer, rabbits will feed themselves perfectly well on grass alone, if you either move them about on grassland in arks, or else let them run loose in adequate fenced paddocks. The wire netting of the paddocks should be dug 6 inches (15 cm) into the ground to prevent them from burrowing; if you have foxes, you have problems. You can keep rabbits in hutches all the year round. They can stand cold but not wet, and they don't like too much heat but need a cosy nest box.

Breeding
You can leave young rabbits on the mother for 8 weeks, at which age they are ready to be killed. If you do this, you should remove the mother 6 weeks after she has kindled (given birth) and put her to the buck. After she has been served, return her to her young. Remove the latter when they are 8 weeks old and the doe will kindle again 17 days after the litter has been removed, gestation being about 30 days. If you keep some young for breeding replacements you must separate does and bucks at 3 months.

Sexing rabbits is easy enough. Lay the rabbit on its back, head towards you: press the fingers gently each side of where its equipment seems to be and this will force out and expose the relevant part. It will appear as an orifice in the female and a slight rounded protrusion in the male. When a rabbit the size of a New Zealand White is ready for mating she weighs 8 pounds (3.5 kg): don't keep her until she gets much heavier or she will fail to conceive. Always take the doe to the buck, never the buck to the doe, or there will be fighting, and always put a doe by herself when she is going to kindle. A doe should rear from seven to nine rabbits a litter, so if the litters are over 12 it is best to remove and kill a few, or else foster them on another doe that has just kindled with a smaller litter. If you do this, rub the young with the new doe's dung and urine before you give them to her, to confuse her sense of smell.

Feeding
Rabbits will eat any greens or edible roots. They like a supplement of meal: any kind of ground grain will do, but a pregnant doe should not have more than 4 ounces (115 g) of meal a day or she will get fat. Assuming that rabbits are not on grass and that you are not giving them a great quantity of greenstuff, feeding should then be in the order of 3 ounces (85 g) of concentrates a day for young rabbits over 8-weeks-old, plus hay ad-lib. Then, 18 days after mating, the doe should be given no more hay but fed on concentrates. She should have these until her eight week-old litter is removed from her, and at this time will eat up to 8 ounces (225 g) a day. Young rabbits can have a meal as soon as they are 2 weeks old.

Killing
To kill a rabbit, hold it by its hind legs, in your left hand; grab its head in your right hand and twist it backwards. At the same time, force your hand downwards to stretch its neck. The neckbone breaks and death is instantaneous. Before the carcass has cooled, nick the hind legs just above the foot joint and hang up on two hooks. Make a light cut just above the hock joint on the inside of each rear leg and cut up to the vent (anus). Peel the skin off the rear legs and then just rip it off the body. Gut the rabbit by cutting down the belly and removing everything except liver and kidneys (known as hulking). Remove the gall bladder from the liver.

FLEMISH GIANT
These are rather too large and unwieldy for meat production, but they are useful for crossbreeding.

CALIFORNIAN
A good meat rabbit weighing up to 10 pounds (4.5 kg). They are healthy and easy to rear.

NEW ZEALAND WHITE
Another good meat rabbit, and popular with breeders for its fur, which dyes well.

Bees & Honey

Bees will provide you with all the sugar you need, and as a self-supporter you shouldn't need much. A little sugar (or, preferably, honey) improves beer, and sugar is necessary if you want to make "country wines" (which I discuss on pp.226–230), but otherwise the part that sugar plays in the diet is wholly deleterious. It is such an accessible source of energy that we satisfy our energy requirements too easily, and are not induced to turn to coarser foods, whose valuable constituents are less concentrated and less refined. The ideal quantity of refined sugar in the diet is: nil.

Now honey will do anything that sugar can do and do it much better. Not only is it a healthier food, for beekeepers it is also free. It is sweeter than sugar, so if you use it for cooking or wine-making purposes, use about two thirds as much as you would sugar.

Before the sugar cane countries were opened up to the Western world, honey was the only source of sugar. For years I lived in central Africa consuming no sugar excepting that produced by bees. These bees were wild ones of course; all Africans know how to chop open a hollow tree with bees inside it and get out the honey. Many also hang hollowed-out logs in trees in the hope – generally realized – that bees will hive in them.

Beekeeping is really a way of getting something for nothing. It is a way of farming with no land, or at least with other people's land. You can keep bees in the suburbs of the city, or even in the centre of the city, and they will make plenty of excellent honey.

The medieval skep

The medieval method of keeping bees was in the straw skep. You plaited straw or other fibres into ropes, twisted the ropes into a spiral, lashing each turn to the next, until you formed a conical skep. You placed this in a cavity left in a wall, so as to protect it from being blown over or soaked from above by rain. In the autumn, if you wanted the honey, you either destroyed the bees you had in the skep, by burning a piece of sulphur under them, or you could save the bees by turning their skep upside down and standing another empty skep on top of it. The bees in the inverted skep could crawl up into the upper skep.

More sensibly, you could stand an empty skep on top of a full one, with a hole connecting them, and the bees would climb up. When they had done so you removed the old skep which was full of honey and comb. If you dug this comb out, you could wring the honey out of it by putting it in some sort of strainer (muslin would do), squeezing it, and letting it drip.

If the skep-inversion method worked without killing the bees, it would be quite a good way of keeping them. You need no equipment save some straw, a bee-veil, gloves, and a smoker. You get nothing like so much honey out of a skep as you do out of a modern hive, but then you could keep a dozen skeps with practically no expenditure, whereas a modern hive, even at its most basic, is a fairly costly item.

When hundreds of people kept bees in skeps, and probably every farm had half a dozen or more, there were a great many more bees about the countryside, and swarms were much more common than they are now in a new millennium. It was easy to find them and not so necessary to conserve the bees that one had.

Langstroth's method

In 1851, a Philadelphian named Langstroth discovered the key secret about bees, which was what he called the "bee space". This is the exact space between two vertical planes on which bees build their honeycomb without filling the space in between, yet still remaining able to creep through. This discovery made possible an entirely different method of keeping bees, and turned beekeeping from a hunting activity into a farming one.

The method Langstroth developed was to hang vertical sheets of wax down at the correct space apart. Instead of building their comb in a random fashion, the bees would build it on these sheets of wax. Then, with the invention of the queen excluder, which is a metal sheet with holes just big enough for the workers but not the queen to get through, the queen was kept down below in a special chamber (the brood chamber) so that she could not lay eggs in the cells above, which as a result were full of clean honey with no grubs in them.

You could then remove the frames, as the vertical sheets of wax were called, with their honey, extract the honey without killing any of the bees or bee larvae, afterwards replacing the emptied frames for the bees to build up and fill once more.

The modern hive

Langstroth's discovery has affected the construction of the modern hive. This has a base to raise it, and an alighting board, with a narrow slit for the bees to enter. On top of the base is the brood chamber, with its vertically slung "deep" or "brood" frames. These wooden frames have foundation inside them, like canvas inside a picture frame. The foundation is sheets of wax that have been embossed by a machine with exactly the pattern made by comb-making bees. Above the brood chamber is a super, which is shallower. The queen excluder divides the two chambers. You may have two or three supers, all complete with frames fitted with foundation one on top of the other. On the very top is a roof.

The roof has a bee-escape in it, through which bees can get out but not in. There should also be a clearer-board, which is a board with a bee valve in it. This will let bees through one way but not the other. Then you should have a bee veil, gloves, a smoker, and an extractor, which you may be able to borrow. The extractor is a centrifuge. You put your sections full of honey into it and spin them round at great speed, which flings the honey out of the comb onto the sides of the extractor. It then dribbles down and can be drawn off.

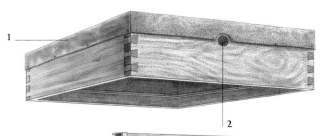

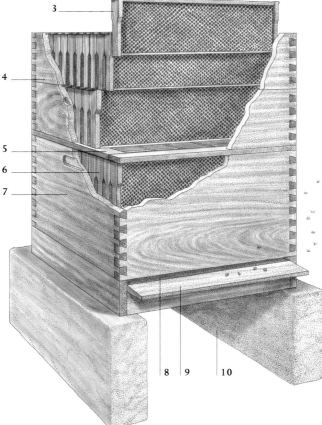

COLLECTING HONEY

Take the honey-loaded super out and bang, shake, and brush the bees out of it. Or else insert a clearing board the day before under the super, or supers, from which you wish to extract honey. The supers will then be free of bees when you want to take them out.

FEEDING

If you take all the honey from a beehive in the late autumn you will have to feed the bees sugar or syrup. The feeder allows the bees to lick the syrup without getting drowned.

ROBBING

Smoke, which is best applied with a special "smoker" quietens bees and makes them fill with honey and sting less readily. Use a screwdriver to break the top super off.

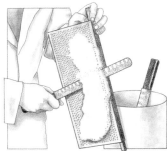

DECAPPING

To remove the honey, cut the wax capping from the comb with a hot knife. Use two knives – heat one while you use the other.

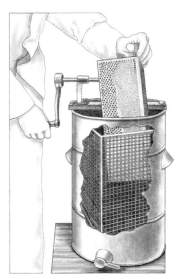

EXTRACTING

Put the decapped frames in the extractor. Spin very fast until the honey is all out of one side, turn the frames round and spin again.

THE HONEY TANK

This is useful if you have a large number of bees. Pour the extracted honey carefully through the strainer and let it settle before drawing it off into jars or containers.

THE HIVE

1 Waterproof roof
2 Ventilator and bee-escape
3 Shallow honey frame
4 Super
5 Queen excluder

6 Deep brood frame
7 Brood chamber
8 Entrance
9 Alighting board
10 Base blocks

THE SKEP

The original beehive, or skep, is made of twisted straw or rope sewn together into a conical shape with straw. If you use a skep your honey will be full of brood, or immature bees, because the queen can lay eggs in every cell. There is no queen excluder as there is in a modern hive. You can strain the brood out, but you kill a lot of bees. It is also impossible for the bee inspector to check a skep to see whether your bees have any diseases.

Capturing a swarm

If you are lucky enough to come across a swarm, what you will find will be a cluster of bees about the size of a football hanging on to a tree, or something similar. If it is a tree, you just hold a big, empty cardboard box under the swarm, give the branch a sharp jerk, and the swarm will fall *kerplomp* into the box. When this happens, turn the box upside down, put a stick under one side to keep it just off the ground, and leave it until evening. This is to let the scouts, out searching for a new home, come back to the swarm. Some ruthless people simply take the swarm away immediately. Swarming bees are unlikely to sting you, as they are loaded with honey and don't like stinging in this condition, but I'm not going to say they will never sting you.

To get a swarm into your hive, lay a white sheet in front of the hive, sloping up to the entrance, and dump the swarm out on to the sheet. They should all crawl up into the hive. Make sure the queen, who is bigger and longer than the others, crawls into the hive too: without her you won't have any bees.

Bees in a colony

The fine South African scientist Marais proved conclusively that a colony of bees, for all practical purposes, is one individual. Apart from the queen, the separate bees are more like cells of an organism than like individuals. One colony mates with another and produces a swarm, the bee-equivalent of a child. The queen lays the eggs, and exerts a strong hold over the rest of the colony; kill her and if the workers can't rear another queen quickly enough from an existing grub, the colony will just die. The drones are as expendable as spermatozoa. Each one tries to mate with a young queen of another colony; whether mating is successful or not, either way the drone is killed by the workers as he is of no further use. In a mature colony there are about 20,000 workers, and they spend their lives working: gathering nectar, building cells for storing the honey, feeding the queen, nursing the young bees, ventilating and cleaning the nest, guarding it, and generally doing everything that needs to be done. If a worker stings you, she dies. Her death is unimportant, for she is not an individual, but merely a cell. Her sacrifice means nothing.

The organism survives at the expense of the individual, so if you capture a swarm you can just leave the bees to get on with it, and they will establish themselves. Remember also the adage:

A swarm of bees in May is worth a load of hay,
A swarm of bees in June is worth a silver spoon,
A swarm of bees in July is not worth a fly.

which means that you won't reap much honey from a swarm of bees in July, but all the same do not despise one: hive it and it will establish itself and give you honey the next year.

Buying and feeding a nucleus

If you can't find swarms, you can buy nuclei of bees from other beekeepers or dealers, who are fairly common in most countries; then just follow the instructions on the box. If you do this you should feed the nucleus for a while. You can do this by giving them two parts sugar to one part water in a feeder, which you can buy and put in your hive on top of the brood chamber. In the case of a nucleus don't have a super: confine them in one brood chamber until that is full of honey and grubs before adding any supers on top.

Gathering the honey

As the frames get built up and filled with honey, and the brood chambers below with bee grubs, you may add a super, then a second super, and you may decide to take some honey. To do this take out one super, insert the clearer-board under it, and replace it. The next day, go and remove the super, which should be full of honey but empty of bees. Put the frames in the extractor and spin the honey out of them. You must first cut the capping off the combs with a hot knife. Each frame should be turned once to extract both sides. Then put the empty frames back in the super and return it to the bees so that they can start building on it again. Always work quietly and calmly when you work with bees. There is no substitute for joining a local beekeepers' group, or for making friends with a knowledgeable beekeeper and learning from him.

You should leave at least 35 pounds (16 kg) of honey in the hive for the winter. I rob my bees only once: in early August. After that I leave them alone, with one empty super, and they make enough honey to last themselves the winter. My one hive gives me 20–40 pounds (9–18 kg). The later honey in our case is heather honey which I could not extract anyway, because it will not come out in the extractor: it has to be pressed out. People who rob all the honey from their bees have to feed them heavily all winter on syrup or candy. In fact, some commercial honey nowadays is little more than sugar turned into honey by the bees. The honey you buy from small beekeepers is generally flower honey, though, and is much better and flavourful as a result.

Wax

The cappings which you cut off the combs are beeswax, which is a very valuable substance: it makes polish and candles (the best in the world), and is good for waxing leatherwork and other purposes. Gentle heat melts the wax and it will run down a slope for you to collect, minus most of its impurities, in a container. The heat can be supplied by the sun, shining through a glass pane into an inclined box. It has been said that the reason why the monks of the Middle Ages were such a jolly, drunken lot was that they had to keep lots of bees to provide the wax for their ecclesiastical candles; what could they do with the honey except make mead from it?

"French peasants do not 'garden', although they eat the best and most varied vegetables in the world. They grow their own vegetables on a field scale — a row of this down the field and half a row of that. And why not? A quarter of the labour. Up to now we have had a lot of pioneering work to do... but when only the routine, recurring work of stockmanship, cultivating, harvesting, and that is left to do, I believe that we could manage our little holding on somewhat the above lines with not very many hours of work a week.

And after all, if a family can grow all its food 'for free' off a piece of land which is no more than that family's fair share of the land surface of its country, and have some produce left over for other people, and still have time to do other work, it is in a very sound position and nobody can say that it is not pulling its weight."

JOHN SEYMOUR FAT OF THE LAND 1976

CHAPTER FOUR

FOOD
FROM THE
FIELDS

Clearing Land

Unless your holding is big and you plan to farm a proportion of it on the "dog and walking stick" principle, one of your first priorities will be to see if you can gain any extra usable land by clearing overgrown wood and bush land. Such land is worth clearing as long as it is not on a ridiculously steep slope, or irretrievably boggy or covered in boulders. Clearing land is hard but rewarding work, although it can be extremely expensive and time consuming.

Send in the pigs and goats

Your pig is your best pioneer. If you concentrate pigs in bush land they will clear it for you with no effort on your part at all. They won't, of course, remove trees, but all brambles, gorse, and undergrowth generally will yield to their snouts and they will manure the land at the same time. If there are any stubborn areas of thicket, try throwing some corn into them and the pigs will soon root them out.

Goats will kill small trees, and big ones too if they are concentrated, by barking them, and they will prevent trees from coming back. They will not, of course, get the trees out, any more than pigs will. You will have to do that.

Clearing woodland

It is essential to have a petrol-driven chainsaw for this work (of course you could hire a mechanical excavator, which is another consideration of cost against time, plus the back-breaking "slashing" of usable timber when the wood is green). A chainsaw is a lethal, noisy monster and you'll have to learn to use this with great care. Courses in chainsaw safety are available and you should ask your supplier about this. First cut off all light branches that are no good for firewood, then cut logs from timber that might serve as good firewood (this includes sections of large roots when the stumps are pulled out). Take all this away to your woodstore for cutting up and splitting later. You can then get a few helpers – kids often love this job – to take all the brashings (small branches) to make a bonfire. Make this in a corner somewhere where the flames will do no harm and you can then leave it for a couple of months until a cold dry winter day makes a good bonfire party.

If you leave the tree stumps reasonably high, say at about 2 or 3 feet (50 cm–1 m) this will make it much easier to lever them out. A good digger driver can clear a big area in a day at a reasonable cost. Have the stumps piled up in a corner for burning a few months later when they have dried out. It is cheaper to haul stumps out with a tree-jack or monkey-winch. You might hire or borrow one of these, or buy one if you had a lot of land to clear, but they cost in the hundreds. There are many varieties of them. Alternatively, you can dig stumps out with spade and mattock, but this is very laborious.

HAND TOOLS

If you haven't got pigs or machines to clear your land, you can do it by hand, but you need the right tools, especially a chainsaw (*opposite page*).

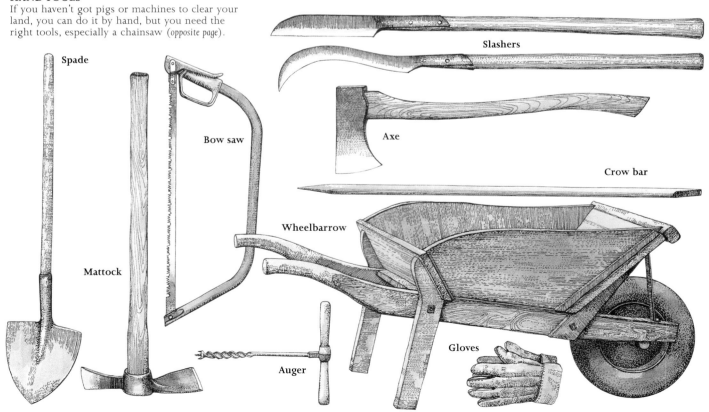

Spade

Bow saw

Mattock

Auger

Slashers

Axe

Crow bar

Wheelbarrow

Gloves

A more accessible method is sodium chlorate, which is a common weed killer, much used by terrorists for the manufacture of their infernal machines. If you drill holes in the stump and fill them with sodium chlorate, put some cover over the holes to stop the rain from getting in, and wait a month, you will find that the stump has become highly inflammable. Build a small fire on the stump and it will burn right away. A more organic option might be filling the holes with sugar, buttermilk, or dried milk powder to encourage decomposition.

Gorse, broom, and brambles

Areas covered with gorse, broom, and brambles can be cleared very effectively with a bow saw and a sharp scythe. Make sure you have a good pair of leather gloves and you can remove gorse fairly quickly with a small bow saw. A chainsaw is noisy and will quickly blunt when cutting gorse near to the ground (which is what you want to do). Your scythe will easily cut brambles if it is sharp and you always pull away from the roots. Cutting thorny brambles from a distance with a scythe is very effective: use a pitch fork to pile them up in a heap for burning. Once you have cleared an area, you may need to pass over it with your scythe once or twice every year for a few years after to keep down the prickly shoots that are re-emerging.

Removing rocks

Rocks can be very obstructive, particularly on boulder-clay or glacial till in which boulders have been left by the retreating ice in a completely random fashion. Again, a mechanical excavator can deal with these if they are not too big, hauling them out and dozing them to the side of the field. You can lift quite large rocks, of several tons or more, with levers. Dig down around the rock, establish a secure fulcrum at one side of it – a railway steeper will do, or another rock – insert a long beam of wood or a steel girder (a length of railway line is ideal) and raise that side of the rock a few inches. Now pack small rocks under the big rock, let the latter subside, and apply your lever to the other side. Do the same there. Continue to work your way round the rock, raising it again and again the few inches made possible by your lever and packing small stones under it each time you have gained a bit. You will eventually work your rock to a point above the surface of the surrounding ground. Once you have got a boulder out you may be able to roll it to the side of the field, again using levers. If it is too big for this, you can try lighting a big fire under it, heating it right through, and then throwing cold water on it. This should crack it.

Don't forget that clearance is not the only option. It is perfectly possible to renovate old woodland by judicious felling and replanting. You can then leave the old stumps to rot in situ. When the timber is more mature the wood will make an ideal holding ground for pigs or poultry. Or consider if it would not be better to replant old woodland as new woodland and farm it as forest.

CHAINSAW

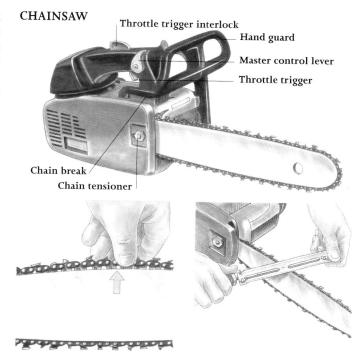

Throttle trigger interlock
Hand guard
Master control lever
Throttle trigger
Chain break
Chain tensioner

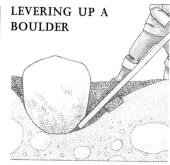

Tensioning Check the chain tension is right by pulling with your thumb, and make sure the chain lubrication is filled up and working properly.

Sharpening With a file holder leave the chain on the bar and lock it with the hand guard. Always file from inside to outside with the cutter in a horizontal fashion.

BURNING A STUMP

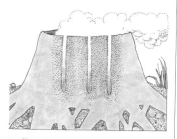

Drilling holes in the stump is a slow but effective way to remove it. For faster results build a fire on top.

LEVERING UP A BOULDER

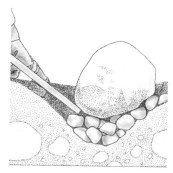

1 Use a rock or a chunk of wood as a fulcrum. Work a lever down beside the boulder.

2 Raise the boulder as far as possible. Prop up with stones. Take lever and fulcrum to the other side.

3 Repeat the process over and over, gaining a few inches each time. Once the boulder is out, roll or lever it off your field.

Draining Land

If you are lucky your land will not need draining at all. Much land has porous subsoil and possibly rock through which water can percolate, perhaps has a gentle slope, and is obviously dry. But land with an impervious subsoil, very heavy land, land that is so level that water cannot run away from it, or land with springs issuing out in it, may well need draining. Badly drained land is late land, meaning it will not produce plants early in the year. It is cold land and it is hard to work. You cannot cultivate it when it is wet – particularly if it has clay in it. In short, it will not grow good crops.

You can tell wet land even in a dry summer by the plants growing in it. Such things as flag irises, sedges, rushes, and reeds all give away the fact that, although dry in the summer, it will be wet and waterlogged in the winter time and should be drained.

Cut-off drains

Often, on sloping land, you can drain a field by digging a ditch along the contour above it (see illustration). The effect of this ditch is to cut off and take away the water that is percolating down from above. The rain that actually falls on the field is not enough to cause it to become waterlogged: it is the water that drains down from above that does the damage.

Springs

You can drain springs by connecting them by ditch or land drain (see illustration) to a stream that carries the water away.

You can tell where springs are by wet patches or by water-loving plants. If there is a large waterlogged area around the spring common, sense might tell you to make a larger hole around the mouth of your pipe and fill it with stones.

Land drains

Level land can be drained simply by lowering the water table. The water table is the level at which the surface of the underground water lies. It will be higher in the winter than it is in the summer, and in severe cases may be above the surface. You lower it by digging ditches, or putting in land drains, to take the water away. You can even do this with land below sea level, by pumping water from the deepest ditches up into the sea, or to raised-up rivers that carry it to the sea.

Obviously heavy soils (soils with a high clay content) need more draining than light soils, but even sand, the lightest of all soils, can be waterlogged and will then grow nothing until it is drained. The heavier the soil is, the closer together your drains will need to be, for the less is the distance water can percolate. A very few drains will suffice to drain light or sandy soil. If you have had no experience it will pay you to get the advice of somebody who has: in countries with government drainage officers these are the obvious choice. There can be heavy grants for draining, too.

There are three main types of land drain: open ditches, underground drains and mole drains. An open ditch is just what it says. You dig a ditch with battered (sloping) sides,

THREE SITUATIONS WHERE YOU NEED DRAINS
A Water runs downhill through porous soil or rock before hitting an impervious layer. This forces it generally sideways to the surface where it emerges as a spring.
B An impervious subsoil prevents rain from sinking in.
C Absolutely flat land has no slope to allow drainage.
The plants on the right are sure signs of wet land: (left to right) marsh orchid, marsh violet, flag iris, marsh marigold, jointed rush, wood sedge, common rush and bulrush.

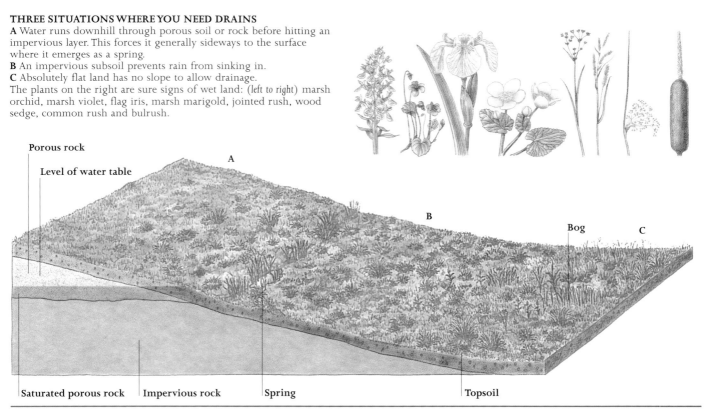

THE MOLE DRAINER
A torpedo-shaped steel object at the bottom of a narrow blade is dragged through the soil. The narrow slot made by the blade fills in but the drain remains. The drain lasts much longer in clay than in soft, sandy soil.

THE CHISEL PLOUGH
The chisel plough, or subsoiler, cuts a series of deep, evenly spaced furrows in the soil. This works very well with heavy clay, where the furrows last and ensure free drainage.

THE USES OF DRAINS AND DITCHES
A cut-off ditch will intercept water draining downhill, and lead it round your field to a receiving ditch at the bottom. An underground drain can be used to drain a spring, and a series of underground drains – herringbone pattern is ideal – can take sufficient water away to lower your water table. You want to get the water table at least 18 inches (45 cm) below the surface. Four feet (1.2 m) is ideal.

or have it dug by machine. On light land (sandy soil) the batter wants to be much less steep than on heavy land because the latter supports itself better. Common sense will tell you how much to batter. If the sides fall in, it's too steep.

Depth too is a matter of reasoned judgment. If the ditch is deep enough to lower the water table sufficiently for the crops to grow happily, it is deep enough. You certainly don't want standing water in the soil at less than 18 inches (45 cm) from the surface. It is better if you can lower the water table to 4 feet (1.2 m).

If you are having to dig the ditch by hand you won't want it too deep. And remember that open ditches need flashing-out (a thorough clearing of scrub and weeds) every year or two, and cleaning with a spade every 5 to 10 years. They also need fencing.

Underground drains are of many types. As long as they are deep enough not to be affected by deep ploughing or cultivating, and their slope is continuous to the outfall so that they don't silt up in the dips, they will require no maintenance and should last for centuries. Mole drains do not last for more than 5 to 10 years – less in sandy land.

But draining is simple common sense. Imagine what is going on down there. Dig try-holes to find how deep the water table is and where the springs are. Arrange to drain that water away to the nearest stream or river or whatever, or even let it debouch into wasteland below your land, and you will have well-drained productive land.

Plastic drain pipe

Semi-circular tile drain

Stone culvert drain

Bush drain

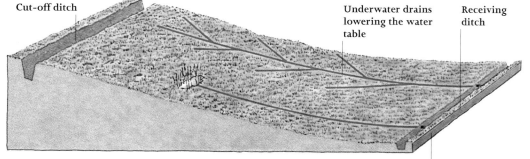

Cut-off ditch

Underwater drains lowering the water table

Receiving ditch

Underground drain capturing spring water

DRAINING A SPRING
Dig down to the spring. Lay a pipe or dig a ditch to carry the water away. If the spring covers a large area, fill in around your pipe with some stones.

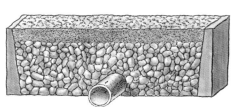

UNDERGROUND DRAINS
Stone culvert drains and tile drains are naturally porous. Plastic pipe drains have slits in them to let the water in. The Roman bush drain (simply bushes covered with earth) can be reinforced with a piece of perforated corrugated iron.

Irrigating Land

Wherever you live, your crops will benefit by irrigation, and in some countries they just won't grow without it. The luckiest cultivators are those who live in a hot, dry climate with plenty of water for irrigation. They have far better control over their husbandry than those who live in high rainfall areas. They have no serious weed problem: they simply kill their weeds by withholding water from them when the land is fallow. They can drill their seed in dry dust before they water it, and then immediately flood the land to make the seed grow. They can give the crop exactly enough water for its needs throughout its growing time, and then withhold water when it comes to harvest, and thereby harvest in perfect conditions. They have it made.

But the rest of us can also use irrigation to advantage. It takes 22,650 gallons (103,000 litres) of water to apply an inch-depth (2.5 cm) of water to an acre (0.405 hectare). If there is no rain during the rainy season it is nice to apply an inch (2.5 cm) a week during the period of hardest growth of the crop. In temperate climates with a fair rainfall, like most of Northern Europe and the eastern United States, the addition of from 2–6 inches (5–15 cm) during the growing season will probably be enough. In any case, the irrigator cuts his coat according to his cloth. Anything is better than nothing.

If you are lucky, you may be able to tap a stream above the land you wish to irrigate and lead the water down in a pipe, but unless your source is much higher than your land you won't get much pressure. On the other hand, contrary to popular Western belief, you don't really need a lot of pressure: you only need the water. By the simple means of laying a hose on the ground and moving it about from time to time, as patch after patch gets flooded, you can do a great deal of good. You can do more good by letting water run down furrows between your rows of crops, moving the hose each time the water reaches the bottom of another furrow.

Sprinkle irrigation

Broadly, there are two types of irrigation: sprinkle irrigation and flood irrigation. Western farmers tend to go for the former. They use pumps and either "rainers" rotary sprinklers, or oscillating spray lines, all of which need considerable pressure to make them operate. This is fine if you can afford the equipment, the fuel and have the water, which does not have to be above the field. But all this is expensive, and not really for the ordinary self-supporter. Personally, I could never see the point of squirting water up in the air at some expense just to have it fall down again, and have always practised some form or other of flood irrigation.

Flood irrigation

In countries where irrigation is really understood – and these are the countries where it is really needed – flood irrigation is what is used. If you have a stream running next to a field, it is not difficult to get a little petrol pump and a hose, and to move the pump along the bank of the stream as one stretch of the field after another is irrigated. Alternatively, you may be lucky and have a stream at a higher level than the field.

Ideally, the land should be either terraced in perfectly level beds or, if the field has a natural gentle slope, levelled into gently sloping beds with bunds separating each bed from the next. (A bund is a small earth bank not more than a foot high.) You can grass these bunds, in which case they are permanent, or level them down each year and build them up again. If you are working with tractors you will probably level them, because it gives you more room to manoeuvre. At the head of all the sloping beds is a water channel. To irrigate you build with a spade a little dam of earth about 1 foot (30 cm) high across this channel at the first bed, and break the bund which separates the channel from the bed with the spade. You sit there, in the sunshine watching the butterflies, until the water has meandered down the bed, covered all of it and has got to the bottom. If your bed is not properly levelled, and has no crop in it, you can use your spade to level it to spread the water evenly. On a hot day this is a delightful job.

Now you will have already built small dams level with each of the other beds. When the first bed is watered you close the gap in its bund, break its dam down, break a hole in the bund of the second bed, and let the water run into there. And so you go on.

Of course this pre-supposes that the water in your head-channel is higher than your beds. What if it is lower? Then you must do what many a Chinese or Egyptian does: just raise it that few inches. You can do this with a bucket, very laboriously, or a hundred other devices that ingenuity will lead you to. A small petrol pump might be one of them, a tiny windmill another.

If your field is very steep it is obvious that beds sloping down it will not do for flood irrigation. You will have to terrace it. This will involve stone- or at least turf-retaining walls and is a tremendous job. And if you have a very big field you may need two or more head channels on different contours, because the water won't be able to meander down from the top of each bed to the bottom.

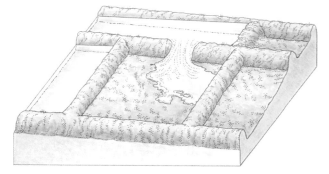

FLOOD IRRIGATION
Sloping beds with a water channel at the head are separated by earth. Make a dam across the channel at the first bed, break a hole in the bund separating the channel from the bed. Close the gap when flooded. Repeat when necessary.

Making Use of Woodland

The most useful trees for the self-supporter are, in order of importance: sweet chestnut (the best tree in the world for timber), oak, ash, and larch. In North America you would add hickory, sugar maple, and black cherry. If you have a saw bench capable of ripping down trees, then softwoods or any of the timber hardwoods are useful, too.

Hardwoods and softwoods

When considering timber for purposes other than fuel, you should look out for: a fairly quick rate of growth, hardness, and resistance to rot, and what I will call "cleavability" or "splittability".

For very many farm and estate uses it is better to cleave wood rather than rip-saw it (saw it along the grain). Cleaving is quicker, cheaper, the resulting wood is stronger, and lasts longer. Why? Because when you rip-saw you inevitably cut across some of the grain, or wood fibres. When you cleave, your cleavage always runs between the grain, which avoids "cross-graining" and leaves undamaged grain to resist the weather.

Sweet chestnut cleaves beautifully. It is fast growing, straight, hard, and strong. It also resists rot better than any other tree. Oak cleaves well, too, but not as well as chestnut. The heart of oak is as hard and lasts as long, but the white sapwood on the outside – most of a small tree – is useless. Oak is extremely slow growing and needs good soil to grow at all. Ash, however, is tough and resilient, but will rot if put in the ground. It is straight, grows fast, and splits well.

Above the ground, but exposed to the weather, ash will last a long time if you oil or creosote it every now and then. It makes good gates or hurdles. Larch is unusual in that it is a conifer but not an evergreen. It is very fast growing and is the best of the conifers for lasting in the ground, as long as it is creosoted. All the other conifers, or softwoods, are hopeless in the ground if not pressure-creosoted, and then they don't last many years.

Cherry and all other fruit woods are hard, and make fine firewood. They are good for making cog teeth in water mills, for example. Hickory is the best wood for tool handles. It doesn't grow in Europe (why, I don't know), and so is either imported or else ash is used, which is a pretty good substitute. Elm – alas now being killed off by Dutch elm disease – is good for any purpose where you want a non-splittable wood, such as for wheel hubs, chopping blocks, and butcher's blocks. It is great under water. Maple and sycamore are good for turning on a lathe, and making treen (carved objects). Walnut is a king among fine woods, and fit to harvest in a mere 150 years, though 350 is better if you have the patience to wait for it!

Firewood

Trees are your most likely source of fuel. If you have even an acre or two of woodland, you will find that, with proper management, the trees in it will grow faster than you can cut them down for your fire. A piece of woodland is the most efficient solar heat collector in the world.

THE FORESTER'S ESSENTIAL TOOLS
Fell your tree with axe and saw. Use hammer and wedge, or club and froe, for splitting. Adze and draw-knife are for stripping and shaping.

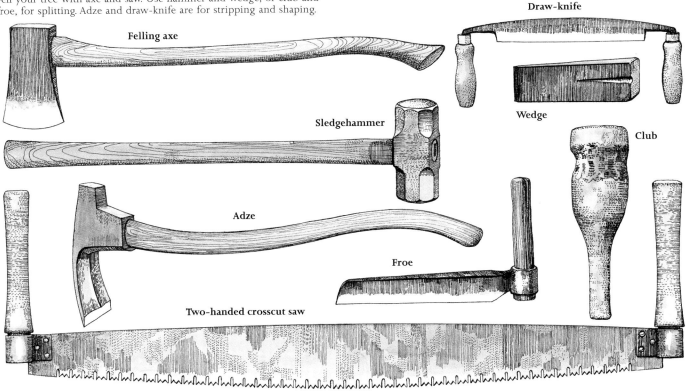

Draw-knife

Felling axe

Sledgehammer

Wedge

Club

Adze

Froe

Two-handed crosscut saw

Which tree to plant?

Ash is the best of all firewoods. "Seer or green, it's fit for a queen!" The loppings of felled ash are excellent. It burns as well when newly cut as when mature. Oak, when seasoned, is a fine and long-burning firewood, but it grows far too slowly to be planted for this purpose. Silver birch is good for firewood, though not for much else. It burns very hot when seasoned, and it grows fast.

Conifers aren't much good for firewood. They spit a lot and burn very quickly, but in the frozen north, where there's nothing else, that's what people have to use. Birch is better as firewood, and it will grow further north than any other tree. All the weed woods, like alder and goat-willow, are very sluggish when green, but can be burned when dry, though even then they don't burn well or give out prolonged heat. But what else is there to do with them? Any wood in the world will burn. But if you are planting trees especially for firewood, plant ash, and then coppice it.

Coppicing & planting

Coppicing means cutting down all your trees when they are about 9 inches (23 cm) in diameter, and then letting them grow again. They will "coppice" by putting up several shoots from each bole. Cut these down again in about 12 years and they will grow once more. This 12-yearly harvesting can go on for centuries, and help you harvest the greatest possible quantity of firewood from your wood-lot. Plant trees very close together and they will grow up straight and tall, reaching for the light: 5 feet (1.6 m) by 5 is fine. When they become crowded, you thin them and get a small preliminary harvest. In winter, plant trees at least 3 years old. You can buy them from a nursery, or the Forestry Commission, or you can grow them yourself from seed. Keep the grass and rubbish down every summer for 3 or 4 years, so the trees don't get smothered. Saw off low branches from the growing trees to achieve clean timber without knots. Feed with phosphate, potash, and tulle if needed. A scythe is ideal for clearing around young trees.

In existing woodlands uproot the weed trees (alder, goat-willow, thorn) to give the other trees a better chance. Wet land favours weed trees, so drain if you can. Keep out sheep, cattle, and goats to give seedlings a chance. Cut out undergrowth if you have time, or try running pigs in the wood for a limited period. They will clear and manure it and they won't hurt established trees. They will also live for months in the autumn on acorns or beech mast.

Seasoning wood

Stack the planks as they come out of the log, with billets of wood in between to let the air through. Kiln-drying is a quick way of seasoning, but time is better. Some wood, for example ash, can be laid in a stream for a few weeks to drive the sap out. This speeds seasoning, but some trees do take years to season. If you want woods for cabinet-making, for example, there must not be any subsequent movement. But for rough work, gates, or even timbers for rough buildings, seasoning is not so important. Always remember to treat trees as a crop. Don't hesitate to cut mature trees when they are ripe, but always plant more trees than you cut down.

TREES TO PLANT
These trees are among the most useful that you could grow on your land: **1** Ash
2 Larch **3** Silver Birch **4** Elm
5 Walnut **6** Sweet Chestnut
7 Shagbark Hickory **8** Oak

FELLING A TREE
Use an axe to trim off all roots to cut a face in the side of a tree.

A "face" is a forester's term for a deep, V-shaped notch and you cut it in the side towards which you want the tree to fall. Now begin sawing from the other side, making your cut a few inches above the deepest part of the face. When the tree "sits on" your blade, so you can't move it, use your sledgehammer to drive a wedge in behind your saw. Carry on sawing until you are close to the face.

The tree should feel like it is about to fall. Now pull out your saw, bang the wedge further in, and over she goes. A jagged piece of wood, called the "sloven", will be left sticking up from the stump. Trim it off with your axe.

RIVING WITH WEDGE AND SLEDGEHAMMER
Wedges and a sledgehammer are the best tools for "riving", or splitting, large logs.

Use the sledgehammer to drive a wedge into the end grain of the log. Then drive more wedges into the cleft thus made until the log splits right down its length.

Never use an axe as a wedge. The handle will break.

RIVING WITH FROE AND CLUB
For riving smaller wood the ideal tool is a froe. Whack the blade into the end grain with a club.

Or use a mallet. Now work the blade further into the wood by levering sideways with the handle of the froe.

You won't have got far before the wood splits down its length. This is much quicker than using wedges.

SAWING PLANKS
A pit-saw is a time-honoured tool for sawing logs into planks. One man stands on the log; the other is down a pit dodging the sawdust. Band saws and circular saws are easier but more expensive.

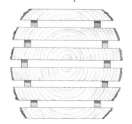

SEASONING PLANKS
Stack planks as they leave the log with spacers to let air through. Leave for at least 18 months.

5   6   7   8

Hedging & Fencing

Domestic animals can be herded: that means kept where they are supposed to be by human beings. But self-supporters will of necessity be a busy people. Fences will not only relieve them from the time-consuming task of herding, but will also give them a useful tool for the better husbanding of their land. Without the fence you cannot fold (enclose) sheep or cattle on fodder crops; you cannot concentrate pigs on rooting; you cannot even keep goats and chickens out of your garden.

Quickthorn hedge
The cheapest and most natural barrier you can build is a quickthorn hedge. Quick means alive, and such a hedge is established by planting thorn bushes, generally whitethorn (may), close enough together in a long line. Seedling thorns, about 6 inches (15 cm) high, can be planted in two lines, staggered, 9 inches (23 cm) between the two rows but 18 inches (45 cm) between the plants in the rows. You can buy the plants from nurseries or grow them yourself from haws, the seeds of the hawthorn. But the hedge must be protected from stock for at least 4 years, and this is what makes a quickthorn hedge so difficult to establish. Animals, particularly sheep and even more particularly goats, will eat a young quickthorn hedge. Therefore, some other sort of fence – probably barbed wire – must be established on both sides of a new quickthorn hedge: an expensive business.

Laying a hedge
But once the quickthorn hedge is established it is there, if you look after it, for centuries. You look after it by laying it. That is, every 5 years or so, cutting most of the bushes' trunks half way through and breaking them over. The trunks are all laid the same way – always uphill. They are pushed down on top of each other, or intertwined where possible, and often held by "dead" stakes driven in at right angles to them. Sometimes the tops of these are pleached (intertwined) with hazel or willow wands twisted through like basketry. In due course the pleaching and the dead stakes rot and disappear, but the hedge puts out new growth and can be very stock-proof.

The quickthorn hedge is a labour-intensive way of fencing, but labour is all it uses, and it lasts indefinitely. Also it looks nice, gives haven to birds and small animals, and serves as a windbreak (very important in windy regions). In days of old it supplied, with no extra work, faggot-wood, used for heating bread ovens and other purposes, to say nothing of blackberries. You can often restore old hedges to efficiency on a new holding by laying them, judiciously planting here and there an odd thorn bush to fill in a gap.

Dry-stone wall
If there is freestone (stone that cleaves out of the quarry easily in fairly even slabs) in your area, you probably already have got dry-stone walls. Dry means no mortar.

If you have dry-stone walls, you will need to maintain them. If you haven't, but you have the stone on your land, you can build some. It is backbreaking but costs nothing. You need tons of stone – much more than you think you are going to need – and a good hand and eye. Dig a level foundation trench first, then lay the stones carefully, breaking all joints, keeping sides vertical, and fitting the stones in as snugly as you can. Dry-stone walls can be quite stock-proof. They are enormously expensive in labour and need repairing from time to time.

Stone hedge
It is possible to build a cross between a wall and a hedge. You find these in areas where the natural stone is rounded or boulder-shaped, not the rectangular slabs which are found particularly in limestone country. Two stone walls are built with a pronounced batter – that is, they lean inwards towards each other. The gaps between the stones are filled in with turf, and the space between the two walls is filled with earth. A quickthorn hedge is then planted on top. After a year or two, grass, weeds, and scrub grow from the earth and the turf. The wall is quite green and not, to be quite frank, very stock-proof. If you look at a hundred such hedges I'll warrant you'll find a discreet length of barbed wire or two, or even sheep-netting, along ninety of them. These wall-hedges aren't really much good unless you fortify them with barbed wire.

Wattle hurdle
If you can get stakes from your own trees, a wattle-hurdle fence is free except for labour, and fairly quick to erect, but it doesn't last long. You drive sharpened stakes into the ground at intervals of about 9 inches (23 cm) and pleach, or weave, pliable withies (willow branches), hazel branches, holly, ivy, blackberry, or other creepers between the stakes so as to make a continuous fence. The weaving material soon dries out and cracks and gets rotten and you have to ram more in. The stakes themselves, unless of chestnut or heart-of-oak or other resistant wood, rot after a few years and break off. Where stakes or posts are expensive or hard to come by, it is an extravagant form of fencing.

Post and rail
A post-and-rail fence is stronger and, unless you are able to grow your own wood, more economical. It consists of strong stakes, either of resistant wood or else softwood impregnated with creosote, driven well into the ground, with rails of split timber nailed on to them. Abraham Lincoln, we are told, started his life as a "rail-splitter". The rails he split would have been for post-and-rail fences, for in his day that wonderful invention, wire, had not begun to encompass the world, and yet the new settlers heading west over North America had to have fences on a large scale. Today we have the benefit of simple but highly effective metal post rammers. Two people may be needed, but you can drive posts into what seems like extremely unlikely soil.

BUILDING A HEDGE

Cut stakes out of your hedge so as to leave strong bushes at intervals of about 12 inches (30 cm). Wearing a leather hedging glove,

bend each trunk over and half-cut through the trunk near the base with a bill-hook. Force the half-cut trunk down to nearly horizontal and try to push the end under its

neighbour, so keeping it in position. Be sure not to break it off. Take the stakes you have just cut and drive them in roughly at right angles to the trunks, and

interweave them with the trunks. Intertwine the tops of the stakes with pliable growth (e.g. hazel, willow). The stakes will rot but the living hedge will be secure.

OVERGROWN HEDGES

Tame a a runaway hedges with a slasher (left). Clear the under-growth with a bagging hook (above), but also hold a stick – or you might lose a finger or thumb.

USING STONE

A well-maintained dry-stone wall is even more stock-proof than an established hedge. You need stone that comes in even, flattish slabs. Dig down about 9 inches (23 cm) and make a level foundation trench. Lay the stones, neatly fitting them together. Make sure the sides are vertical and all joints are broken. If you can get hold of large, round stones, you can make a sort of stone hedge. Build two stone walls leaning towards each other about 12 inches (30 cm) apart. Plug the gaps between the stones with turf and the space between the walls with earth, and plant a hedge on top. To be really stock-proof, reinforce with barbed wire to ensure sheep don't walk straight over it, at least until the hedge is mature.

Steel wire

The invention of galvanized steel wire was the answer to the fencer's dream. It can be plain wire (often high-tensile), barbed wire, or netting. Plain wire is effective only if strained. Barbed wire is more effective if it is strained, but often a strand or two attached somewhat haphazardly to an old unlaid hedge is all there is between animals and somebody's valuable crop.

Netting is very effective but nowadays terribly expensive. Square-meshed netting is strongest for a permanent situation, but is awkward to move very often: diamond-meshed netting is much weaker, but stands being repeatedly rolled up and moved and is therefore ideal for folding sheep.

Straining wire

If you buy a wire strainer you can see easily enough how to use it, but there are several very effective ways of improvising one. A tool often used in Africa consists of a forked stick 2 feet (60 cm) long, with a 6-inch (15-cm) nail fastened with staples along its length just below the

fork. The wire to be strained is inserted under the nail and then wrapped twice round it for firmness. You then take up the slack by twisting the stick, using the fork like the handle on a tap.

Then you put the final stress only turning the stick round the corner post, using the stick as a lever. You can get short lengths of wire quite tight enough like this, although if you are straining extremely long lengths at a time, you will need a proper wire strainer, unless you pull the wire taut with a tractor.

Straining tips

If you strain wires on a post on a cold winter's day, you may well have to strain them again on a hot day next summer. Heat makes metal expand. Often, in practice, you can apply strain to wire by hauling it sideways – out of the line of the fence – to, say, a suitable tree with another snatch of wire. This is looked upon as very infra-dig by estate managers but is often useful just the same, especially when you are trying to make a fence stock-proof down in the depths of the woods on a pouring wet day.

Straining and anchoring fences

If you can't get a wire strainer, you can exert quite a little strain by using a post as a lever, or by using a block and tackle, or even by using a horse or a tractor. Many farmers use the tractor method. But do not strain wire too much. It breaks the galvanizing and takes the strength out of the wire. Always use common sense.

A strained fence is only as good as its anchor posts. A wire strainer, such as you can purchase, or easier still borrow from a neighbour, can exert a pull of two tons, and this multiplied by the number of wires you have in your fence will pull any corner post out of the ground unless it is securely anchored.

You can anchor a fence with a kicking post, a post placed diagonally against the corner post in such a way as to take the strain. The kicking post itself is secured in the ground against a rock or short post. Alternatively the strain can be taken by a wire stretched taut round a rock buried in the ground. A refinement of this, the box anchor, is the most efficient of all (*see illustration*).

Remember, if you anchor wire to any tree that is not fully mature, the tree will gradually lean over and the fence will slacken. It is bad practice to fasten wire to trees anyway: the staples and lengths of wire get swallowed up by the growing tree and ultimately break some poor devil's saw blade. Not that many of us are quite innocent in that respect. Using a post rammer is also easy and effective.

Electric fencing

The electric fence provides marvellous control over stock and land, making possible a new level of efficiency in farming.

You can get battery fencers, which work off 6-volt dry batteries or 12-volt accumulators, or mains fencers that work off the mains and will activate up to 20 miles (32 km) of fencing! One strand of hot wire will keep cattle in − it should be at hip height − and one wire 12 inches (30 cm) from the ground will keep pigs in if they are used to it. Until they are, use two wires. The wires needn't be strong, or strained, just whipped round insulators carried on light stakes, and the whole thing can be put up or moved in minutes.

Hurdles

Except for electrified wire-netting, which is expensive and hard to come by, sheep won't respect an electric fence. So when we wish to fold sheep on fodder fields we make hurdles (*see illustration*). It's cheaper than buying wire-netting. Some wood that rives (splits) is necessary: ash or chestnut is fine. If you use ash you should creosote it. To erect hurdles drive a stake in at the point where the ends of two hurdles, meet and tie the hurdles to the stake with a loop of binder twine. To carry hurdles, put as many as you can manage together, shove a stake through them, and get your shoulder under the stake. A fold-pritch is the traditional implement for erecting hurdles, and you can hardly do the job without it.

You can make wattle-hurdles out of woven withies or other flexible timber. These are light, not very strong, don't last long, but good for windbreaks at lambing time. To make them you place a piece of timber on the ground with holes drilled in it. Put the upright stakes of your hurdle in the holes and then weave the withies in. It is simply basket-work.

THE BOX ANCHOR

A fence is only really secure if its wires are strained, which means they can take a pull of two tons. Half a dozen strained wires will pull your corner posts straight out of the ground unless they have good anchors.

The box anchor is the best of all. Heavy soft wire (generally No. 8 gauge) goes from the buried rocks to the second post. A cross-piece morticed in this supports the two corner posts on which the wires are held.

TEN ANCHORS IN A FIELD

Every stretch of strained wire fence needs an anchor, and one anchor can only take a strain in one direction. Thus each corner of your field will need two anchors, and you will need one, each side of the gate.

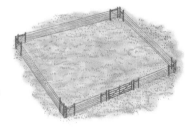

USING A POST RAMMER

Metal post rammers are the simplest and easiest way to drive posts into the ground. You can get them for most builders' or DIY hire shops, and they are cheap to hire. First you need to get the post embedded upright in the ground and you might need someone to help you. Use a mallet to drive the pointed end of the post into the ground. Your aim needs to be accurate here, or you may split the timber. Once the post feels secure, drive the posts in with the post rammer. You just hold the heavy rammer at head height above the post and let go so that the sheer force and weight of the hollow cylinder drives the post down into the earth. Quick and strong work!

HURDLES

Hurdles are movable fences which you can easily make yourself from any wood that splits. Use mortices to join the horizontals to the pointed uprights. Be sure that the ends of the horizontals are tapered in such a way that they apply pressure up and down and not sideways, otherwise the uprights will split. You can drive thick nails through the joints to hold them, or else use wooden dowels. Nail the cross-braces. Drill all your nail holes or you will split the timber. To erect your hurdles, drive stakes into the ground and fasten the hurdles to them with string.

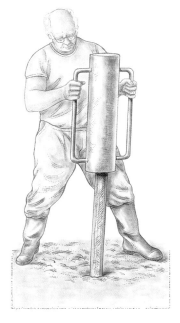

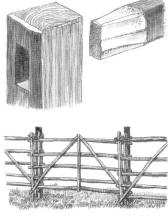

A FARM GATE

A cattle-proof gate for a field or farmyard is best built of split ash or chestnut. Use bolts to join the four main timbers to make up the frame, and also bolt the hinges on. Use clenched 6-inch (15-cm) nails for the other joints. Drill holes for the nails as well as the bolts and pour creosote through all holes. If you have a forked timber use the fork as the bottom hinge, but you must put a bolt through the throat to stop it splitting. The diagonal timbers hold the thing in shape; fit as shown.

WATTLE-HURDLES

Wattle-hurdles can be made of split hazel or willow withies woven on to uprights. Put a baulk of timber with appropriate holes drilled in it on the ground to hold the uprights while you are weaving.

POST-AND-RAIL FENCING

Strong uprights must be well tamped into the ground. Drive all nails right through and clench them.

WIRE-NETTING

Wire-netting is often convenient but always expensive. Square-meshed, or pig, netting (right) makes an excellent permanent fence, and coupled with a strand of barbed wire is completely stock-proof. Diamond-meshed, or sheep, netting (below) is weaker, but it can be rolled up and re-erected, which is what you need for folding sheep.

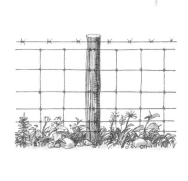

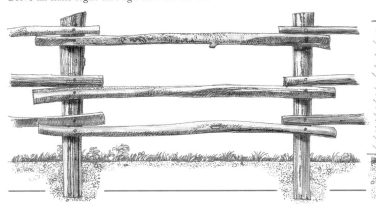

The Working Horse

There is great nobility about a working horse, and great beauty too. If you want to plough your acre a day for month after month with a pair of horses, you will need big horses such as the Shire, Suffolk Punch, Clydesdale, Percheron. If you have very little heavy farm work to do – just pulling a light pony-plough in already arable land, for example, or pulling a horse-hoe – then a light cob, or a Dale or Fell pony, will do. They are fun to ride and drive, and make a pet too.

Feeding

Horses, like other herbivorous animals, need to be fed often: they need a feed at least three times a day while they are working and must be given at least an hour to eat each meal. Also, they should have some hay to mumble at night if they are in the stable, or else be put out on grass. For working horses "good hay hath no fellow" as Bottom said in *A Midsummer Night's Dream*, but it must be good hay; dusty hay makes horses broken winded, mouldy hay upsets them, and too much clover hay, if too fresh, will cause them to scour (have diarrhoea). During the summer, when grass is good, working horses may be run on grass. They graze very close to the ground and should be put on pasture after cows have eaten off the long, lush grass, or else rationed severely as to how long they have on the pasture. Do not expect a horse to work hard, or get hard, on grass alone, for grass makes horses fat and soft. Take it that a horse on grass should have 6 pounds (2.7 kg) of oats a day for every half day that he works.

In the winter, or when grass is short, a horse is best kept in a stable and could eat: 16 pounds (7.2 kg) of hay, 12 pounds (5.4 kg) of oats, and maybe some swedes or carrots or even fodder beet as well, while on medium work. A large horse on very heavy work might well have more oats: perhaps up to 20 pounds (9 kg) and the same of hay. But whatever you do, do not overfeed a horse. If you do, you will kill him. A heavy horse doing only half a day's work should never get a whole day's ration. Hay will never hurt him but grain will. A horse on a high ration, working hard and continuously, may be killed by a disease called *haemoglobin urea* (he pisses blood) if you suddenly knock his work off but go on feeding him the same. The old practice was, on Friday night before the weekend, to give a horse a bran mash (bran is coarse husk taken off wheat to leave white flour, and bran mash is bran soaked in water) instead of the corn feed, and then to feed nothing but plenty of good hay over the weekend: to keep up the corn ration while the horse was idle could be fatal. Similarly, it is not fair to expect a soft horse, that is, a horse that has lived for weeks out on grass, to do immediate heavy work. If you try to do this he will sweat a lot, and puff and blow, and show signs of distress. Give him a little work each day, and feed a little more grain, and gradually harden him up.

Beans make a good feed for horses, but don't let them make up more than a sixth of the grain ration. It is very nice if you can get chaff (straw or hay cut up small by a chaff-cutter). Mix it with grain to make it last longer.

Foaling

Fillies can be put to the "horse" (horseman's word for stallion) at 2 years old (3 might be better), and a colt can serve a mare when he is 2 years old. Pregnancy is slightly over 11 months. She should not work "in shafts" when far gone in pregnancy – the pressure of the shafts might harm her. A mare will probably have less trouble in foaling if she is kept working than if she is not, because she will be in harder condition. If the mare is out on grass in the winter, and not working, she should be given a small ration of hay and maybe a little oats (not more than 4 pounds or 1.8 kg a day), as well as winter grass. In summer, grass should be enough.

After foaling, a mare should not be asked to work for at least 6 weeks; suckling the foal is enough for her. The mare might do very light work after 6 weeks. But wait until 4 months, preferably 6, before weaning the foal. Once the foal is weaned (taken right away from his mother and kept out of hearing) the mother should be put into work immediately in order to dry off her milk. If you don't want to work the mare, you can leave the foal on as long as you like. If you do wean the foal he should be given, say, 4 pounds (1.8 kg) of oats a day and the same of hay.

Breaking a foal

Colt foals should be castrated by a vet at about a year old, but this should never be done in the summer when flies are about, nor in frosty weather. A foal can be broken (trained) at about 2½ years old. But it is never too soon to get a halter on a foal (the first day is okay) to teach him to be led about. Foals have a great sense of humour and can be great fun. If you make a fuss of a foal and get him really tame, he is far less trouble to break. And the foal's feet should be lifted often, to accustom him to this necessary procedure.

To break him, get a bit into his mouth and drive him in front of you with a whip with long reins. After a few lessons get a collar on him. When he gets used to this, hitch him to some not too heavy object like a log or a set of harrows and get him to pull. Then put him next to an older horse and get them both to pull, say, a plough. Wait until he is pretty calm before trying him in shafts.

Kindness, firmness, and common sense are the qualities needed for breaking horses. It is absolutely essential you should not be frightened of the horse, for if you are, the horse will sense it immediately and you will never break him. If you have great trouble, try keeping him away from other horses in a loose box (a room in which he is kept loose) for a week or so, and spend some time daily with him, talking to him, feeding him, handling him, and getting to know him. He will then get used to you.

Shoeing

If a horse is to be worked at all hard he must be shod, perhaps every 6 weeks. The hooves grow under the shoes. If the latter have not worn too much they can be taken off, the feet trimmed, and the shoes nailed on again.

THE FULLY HARNESSED HORSE

The horse is now taken to the cart or implement and shut in. In other words, he is backed between the shafts and the tugs are hooked to the harness. Then the ridge chain which holds the shafts up is fastened, followed by the britchin chains. The tugs pull the implement or cart forward. The britchin chains hold it back if it tries to go too fast. A belly chain is passed from shaft to shaft under the horse's belly to prevent the shafts from skying if the cart or implement is back-loaded.

See that all is well adjusted before you move off. The shafts must not pinch the horse, the britchin must indeed take the weight of the cart going downhill, and the tugs, not the ridge chain, must exert the forward pull on the shafts.

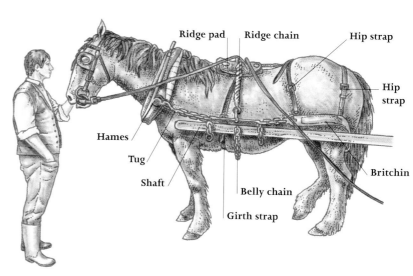

Ridge pad   Ridge chain   Hip strap

Hip strap

Hames

Tug

Shaft

Belly chain

Girth strap

Britchin

HARNESSING A HORSE

1 Use a halter to lead your horse for harnessing. Then take it off.

2 Have the collar in easy reach and the bridle hung on the crook of your left arm.

3 Hold the horse over his nose and fit the collar over his head upside down. Then strap the hames (the curved metal or wooden strips which take the strain) to the collar.

4 Tighten them at the top.

5 Slide on the bridle. To get the bit in the horse's mouth you may have to force his teeth apart by pushing your fingers in at the side. It takes practice.

6 Fit the saddle, or ridge pad, and tighten the girth, but not so much that the horse's wind is constricted. Sometimes the britchin will be kept fastened to the girth.

7 Here the britchin is being put on separately.

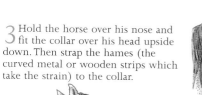

8 The crupper, a loop of leather that goes round the tail, must go on when the britchin is right back. It is then eased forward and strapped on to the saddle, if it is not already so strapped.

Horse & Tractor Power

There are three practical methods of powering instruments that have to be dragged over the land: farm tractors, garden tractors, and animals. Farm tractors are large, enormously expensive (unless they are very old), cost a lot to maintain, and are unsuitable for small plots of land or gardens, because when twisting and turning in a small space the wheels compact the soil to a damaging extent. I have sometimes been forced by circumstances to use a farm tractor in a garden and have always regretted it. A small, or garden, tractor is another matter altogether. It is light – lighter than a horse – and you can even pick it up if it is not too big. It does not compact the soil and is fairly cheap to buy and maintain. It will work in small corners and between row crops. It can often be adapted to many uses: one common type will mow grass, either by a reciprocating knife or a rotary knife; it will saw wood; drive all sorts of small barn machinery; rotavate; plough (rather inefficiently and very slowly); cultivate row crops; pull a small barrow; and scrape the snow off your paths. But many of these jobs you could do as quickly or quicker by hand, and compared with a large tractor or a horse, a garden tractor is extremely slow.

You can plough 3 acres a day with a small farm tractor or 5 with a large one, and it doesn't matter how rough the land is. With a horse you can plough perhaps half an acre: with two horses, an acre. But one horse will only pull a small plough and work on arable land. To plough grassland you need two horses. With a garden tractor it would take you days to plough an acre – and drive you mad with noise and boredom. It will rotavate old grassland but it will make pretty heavy weather of it.

Tractors have two major advantages. They are not eating or drinking when they are not being worked, and they do not use up your land in their search for food. But you do have to pay for their fuel, even though a garden tractor uses a tiny amount. On the other hand a horse – the best example of animal power – can be fuelled entirely from your land. Also, it can have another horse. It is unlikely that the tractor will ever be invented that can have another tractor. But a farm tractor is versatile. It will drive a really effective saw bench or any barn machinery (such as mills, chaff-cutters, and the like); it can be fitted with a powerful winch for hauling trees out, or down; it will pull a large trailer or cart; it will dig post-holes; and it will operate a digging arm that will dig ditches for you (but only just). Maybe we can sum up the whole complicated subject by saying: if you have only a garden, a garden cultivator will be valuable – at least if you don't want to do everything by hand, which you perfectly easily could do if you had the time. Don't think a garden cultivator is necessarily light work – some can be very hard to manage. If you have anything over 4 acres (1.6 hectares) – and like horses – a light horse might suit you very well. Once you have got your land broken up, either with pigs or with a borrowed farm tractor, a small horse will keep it that way, keep your land clean, and give you a lot of pleasure. On the other

hand, one of the larger garden tractors would do the work equally well, if not as pleasantly. A horse will eat the produce of an acre (0.4 hectare) of very good land in a year, or 2–3 acres (0.8–1.2 hectares) of poorer land. A horse wastes not an ounce of its food: what it doesn't convert into energy it puts back into the land as magnificent manure. On a very small holding you might consider keeping a horse and buying in its hay and oats or maize. You are thus buying-in fertility for your holding. If you have 10–15 acres (4–6 hectares) or over – and can pick up a cheap farm tractor in good condition (and know how to keep it going) – it might save you a lot of work, but two horses will do the job, but take three times as long.

Oxen are very good draught animals. They are much slower than horses but exert a strong, steady pull. Some horses are inclined to "snatch" with a heavy load and break things – I have seen oxen sink to their knees in hauling a heavy wagon out of the sand or mud and exert tremendous traction. Oxen are growing into meat during their working lives: horses are depreciating in value, but oxen require two people to work them – horses only one. So, two oxen will just about do the work of one horse – pull a very small single-furrow plough; a row crop cultivator; or a small cart. Four oxen will pull these things much better.

Mules are very hardy, particularly for hot and dry climates (they hate mud and constant wet). They walk fast, pull hard, and can live on worse food than a horse. But, they will not exert so much traction as a heavy horse and are inclined to snatch, kick, bite, and generally misbehave. Asses, or donkeys, can exert some traction, but walk very slowly. They can be used (as mules and ponies can) for carrying packs over ground too rough for a sled or a cart. One donkey won't pull very much at all – maybe a small row crop cultivator at best.

TRACTOR POWER
If you want mechanical power instead of horses, a rotavator is essential for any holding of more than 1 acre (0.4 hectares). There are two types, those that run on wheels and those that drag themselves forward by the action of their tines. Both do a good job, but the former, if big enough, is quicker and more powerful.

Preparing Land & Sowing

If you threw some seeds on grassland, or in a wood, all that would happen is that the birds would eat them. In order to sow seeds with any hope of success, you have got to do two things: eliminate the competition of existing plants, and disturb the soil so that the seed can get into it. If possible, bury the seed, although of course not too deep.

The most usual method of preparing grassland, or other land with an indigenous vegetation on it, is by ploughing or digging. But if you have pigs, use them. They will do the job even better than a plough. If I plough old pasture to sow a grain crop, I plough it to turn it over as completely as possible, although it still stands up in ridges.

After that, I drag disc harrows up and down the furrows for two passes, so I don't knock the ridges back again. At this stage I would add any fertilizer, such as lime or phosphate, that the land might require. Then I would disc again across the furrows. The reason for using a disc harrow at this point is that it cuts the hard turf furrows of the old grassland to pieces – instead of dragging it out as a spike harrow would do.

A pass or two with the spike harrow works the land down to a fine seed bed. It must not be too fine for winter-sown corn, nor even for spring-sown if it is wheat or oats. Barley needs a much finer tilth than either. I then broadcast the seed (sow it by hand), but if I had a seed drill I would drill it.

Remember: if you drill seed too deep, it will become exhausted before its shoots can get to the surface and it will die. So the smaller the seed, the shallower it should be. About three times the diameter of the seed is quite far enough. I would then harrow once again. A pass with a roller completes the sowing process. With row crops, hoe between the rows when the seedlings are about 6 inches (15 cm) high and then the gate can be shut on the bed until harvest time.

No-ploughing and no-digging

The no-ploughing or no-digging theory is now very popular. Adherents of this theory claim that the land should never be ploughed or dug because it is bad to invert the soil. Inverting the soil upsets the soil life, putting surface bacteria down so deep that they die, bringing deeper organisms to the surface where they die, too.

No-diggers and no-ploughers have great success, provided they have very large quantities of compost or farmyard manure with which to mulch their land. The seeds are virtually sown under a covering of compost. My own experience shows that for bringing grassland into cultivation, either the plough or the pig's snout is essential. The next year, if you still wish to keep that land arable, you can often get away with cultivating only, or even harrowing or other shallow cultivations.

The idea of very heavy mulches of compost is fine – providing you can get the compost. But the land itself will never produce enough vegetable material to make enough compost to cover itself sufficiently deeply and therefore you will have to bring vegetable matter in from outside.

Ploughing

The plough that has been used since Iron Age times has three main working elements: the coulter, the share, and the mould-board or breast. In Africa, Australia, and parts of North America, a similar plough, the disc plough, is much used. This is not to be confused with the disc harrow. It is a large, very dished steel disc, dragged through the soil at a certain angle. The angle is such that the leading edge of the disc acts as a coulter, the bottom edge acts as a share, and the belly of the disc acts as a mould-board. It is very good on trashy land.

Using the fixed-furrow plough

Now if you consider what happens when you actually take a fixed-furrow plough out into a field and begin to plough, you will realize that the operation is not as simple as you might think. Suppose you go into the middle of one side of the field and plough one furrow. Then what do you do? If you turn your horse or tractor round, face the other way, and go back, you will either simply plough the slice you have ploughed out back into its furrows so that you are back where you started from, or you will plough the other side of it and build up a ridge of two slices bundled up against each other, and under the ridge will be unploughed ground.

SOW THE SEED
Broadcasting is simply scattering the seed on the ground in the biblical manner. A seed drill drops the seed down pipes so that it is buried and safe from birds.

USE PIGS OR A PLOUGH
For bringing grassland into cultivation, the pig's snout is unbeatable – and a pig manures the land as it digs it. But, if you haven't a pig, you need to plough.

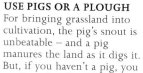

Making a ridge

The way to avoid the piece of unploughed ground is to plough your first furrow, then turn round and plough it back again with the piece of ground underneath it too. Then turn round again and plough the next furrow up against the first two. You have then made a ridge. You then simply go round and round this ridge, every time ploughing your furrow towards the ridge. It will be seen that, when you do this, you get further and further away from the ridge, and each time you have to go right round it. You soon find yourself having to travel a huge distance along the headland to get to where you are to start the furrows going back again. So what do you do then?

You go even further along the headland, and plough out another "setting out" furrow as the first furrow of a stetch of land is called, and make another ridge. (A stetch is the area of ploughed land around a single ridge.) You then plough round that one. As soon as you come to the last furrow of the first stetch you go back and travel on further beyond your second stetch and "set-out" a third stetch. And so on. The ridges are traditionally positioned 22 yards (20 m) apart.

It will be seen now that you will end up with a field with parallel furrows running down it, and in between each pair of furrows parallel ridges. In other words you have gathered the soil at the ridges and robbed it out of the furrows. If you go on doing this year after year you will end up with the typical "ridge-and-furrow" land of parts of the English Midlands. On wettish, heavy land this pattern of field has an advantage: the furrows, running up and down the slope, serve to carry the surface water away, and crops growing on the ridge part are well above the water table. In most parts of Southern Europe and North America, though, it would be criminal thus to plough up and down the slopes. It would lead to gully-erosion.

Turnwrest plough

Now if you want a simple life, and don't want your soil gathered into ridges, there is another kind of plough that will suit your needs and that is the turnwrest, or two-way

plough. This has two ploughing bodies, one in the ground and the other up in the air. One turns the furrow to the right, the other to the left. With this you simply plough one furrow, turn round, swing the two bodies over so they change places and plough back again. And when you do this both furrows are laid the same way. You avoid all the complications of "setting-out", "gathering-up", and all the rest of it, and when you have finished your field is quite level, with no "lands" or stetches at all. Many tractor ploughs are of this type nowadays, and the famous Brabant plough, widely used in Europe and drawn by horses or oxen, is also a turnwrest. The Brabant is a marvellous instrument. I have a tiny, one-horse one and it is worth its weight in gold.

Incidentally the ancient belief in the necessity of turning the soil deeply enough to bury all rubbish completely is becoming more and more discredited. Organic farmers, even in England and the Eastern States of the USA, prefer to leave their compost or manure on top of the soil rather than to plough it in. It is a fact that the enormous earthworm population in soil that has been organically farmed for some time drags all vegetable matter down into the soil without our help. Except for my potato field, I am tending more and more to leave manure on the surface and to disturb the soil deeply as little as possible. But, whatever their theories in this respect, practical growers find themselves forced, occasionally, to plough or to dig.

Sowing

There is no rule of thumb in farming, particularly when it comes to sowing seed. Examine the soil after every operation – remember the needs of the seed and the plant. If the seed is too shallow the birds will get it or it will dry out, that is if the soil is dry and dusty on top. If it is too deep it will use all its energy pushing its shoot up to the light and will die before the life-giving sun's rays can give it more energy. If the land is too wet the seed will drown. If it is too sticky the seed will not be able to push out its roots and shoots. The plant itself needs open soil, which permits the air to circulate and the water to rise.

DISC HARROW
Sharp steel discs break up lumpy ground more effectively than spikes. You can create a fine tilth without dragging turf or rubbish back up to the surface.

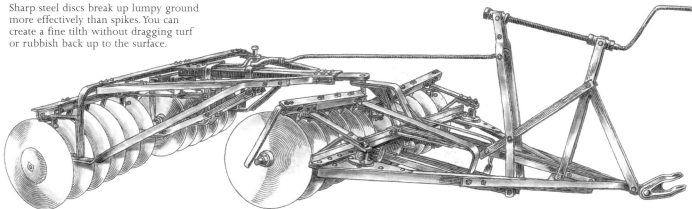

Don't forget that temperature is most important for seed germination. I knew an old farmer in Suffolk who used to drop his trousers in the spring and sit on his soil to see if it was warm enough to sow spring barley. He could sense the temperature, humidity, and so on with his bare backside more sensitively than he could with his hand. To drill his seed in too cold ground was only to have it rot, or the birds get it, or the hardier weeds grow away from it and smother it. To plant it too late was to have a late and not very heavy crop of barley. To get it just right he had to use his buttocks. He grew very good barley.

Hoeing

Having sown or planted your crop, you may have to "keep it clean", which is the farmer's way of saying suppress the weeds among it. Some crops don't need this because they grow quickly and densely, and smother the weeds by denying them light and soil space. You can often get away without hoeing cereals. But row crops (see pp.82–88) do need hoeing.

There are two kinds of hoeing: hand-hoeing and mechanical hoeing. The hand-hoe is simply a blade on a stick with which you cut through the surface of the soil. The cutting can be either a pushing action with a Dutch hoe, or a pushing action with an ordinary hoe (see p.42). In either case it is probably best to walk backwards to avoid treading on the weeds as you uproot them, thus leaving them to wither and die exposed. Walking forwards would simply do the weeds a favour by pushing them back into the soil and transplanting them.

Mechanical hoeing is the pulling of mounted hoe blades along between the rows of the crop by animal or tractor power. This only cleans the ground between the rows. It cannot do it in the rows but between the plants, because no machine has yet been invented that can tell the difference between a weed and a crop plant. It takes the eye of man, or woman, to do that. Therefore, even if you horse- or tractor-hoe, you will still have to hand-hoe as well, at least once. And on that point I have neglected to mention another mechanical hoe, the wheel hoe, which is most useful for the food-producing garden: it's the equivalent of the horse-hoe in the fields. It usually has a single wheel at the front and the cutting blade behind it. You push it up and down between the rows. This means, of course, that you still have to hand-hoe in the rows between the plants.

There is another consideration about hoeing. Not only does it kill the weeds, but it also creates a mulch of loose earth. This mulch conserves the moisture in the land, for it breaks the capillary crevices up which water can creep to the surface. It is very nice to see loose, broken-up soil on the surface rather than a hard pavement. It is true that some crops – onions are one, brassica, or cabbage-tribe, plants are another – like firm soil, but once they have got their roots in, it is best to shake up the surface around them. This lets the rain and air in and stops the moisture coming to the surface and evaporating too quickly. In Suffolk old countrymen say: "a hoeing is as good as a shower of rain". Experience shows that it is, too, and you cannot hoe too much.

Weeds

Selective weed sprays, or pre-emergence wood sprays, are the agri-businessman's answer to weed suppression in row crops. We organic farmers don't use them, because we cannot believe that it is good to douse our soils year after year and decade after decade in what are, after all, nothing more nor less than poisons. Also the hoe can do it, not only just as well but much better.

Weeds have been defined as "plants in the wrong place". Wrong, that is, from the husbandman's point of view. From their point of view they might be in the right place. But do not become paranoid about them. Weed competition, it is true, can ruin a crop. The weeds are so much more vigorous, and fitted for their environment, than our artificially induced crops can be. But also, under other circumstances, they do no harm and often do good. Bare, naked soil, with no crop on it, should be anathema to the husbandman. A good covering of weeds is as good as a crop of green manure.

Green manure is any crop that we plant merely in order to plough in again. A good crop of chick-weed, fat-hen, or many another annual weed, is just as good. And, in the summer when weeds are rampant no matter how many times we hoe, when you go along the rows and either hoe weeds out, or pull them up, and let them lie down between the crop rows to rot, you realize that they do a lot of good. They form a mulch, which covers the soil and stops the drying winds from getting at it, and eventually they rot and the earthworms drag them down and turn them into humus.

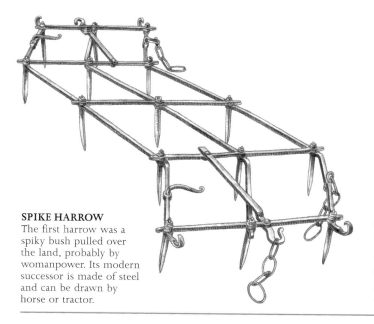

SPIKE HARROW
The first harrow was a spiky bush pulled over the land, probably by womanpower. Its modern successor is made of steel and can be drawn by horse or tractor.

Harvesting

The crown of the year is harvest time, and if you cannot enjoy that you are unlikely to enjoy anything. You sweat and toil, along with friends and neighbours, to gather in and make secure the fruit of the year's labours. The work is hard, hot, sometimes boisterous, always fun, and each day of it should be rewarded with several pints of home-brewed beer or chilled home-made wine or cider. All the cereals, with the exception of maize, are harvested in exactly the same way. When the crop is ripe, but not so ripe that it will shed its grain prematurely, the straw is cut. This may be done by scythe or combine harvester. Sickling a field full of corn is unbelievably tedious. A scythe used by a skilled man will cut 2 acres in a day, and if you can fit a cradle to it, it will dump the cut corn in sheaf-sized piles. Whet the blade at regular intervals with a rough stone, or hammer sharp as Alpine farmers do (see pp.278–279).

A horse- or tractor-drawn grass mower will cover a field fairly quickly, leaving cut corn all over the place. It must then be gathered into sheaves. A reaper-and-binder ties the sheaves for you, but it's a dirty great cumbersome machine and needs three horses to pull it for any length of time. If you are growing corn for home consumption, you should not need to grow much more than an acre. It is certainly not worth owning a reaper-and-binder to harvest such a small area, and it's hardly worth borrowing one either.

Sheaf and stook

Sheaves are bundles of a size you can conveniently grab, tied around the belly either with string or with a fistful of the corn itself. To tie a sheaf with corn, rub both ends of the fistful to make it pliable, put it round the bundle which is gripped between your legs, twist the two ends together tightly, then tuck the twisted bit under the part round the sheaf. (The reaper-and-binder, of course, ties its sheaves with string.)

Then walk along the field and stand the sheaves up into six or eight stooks or traives. Take two sheaves and bang the heads together so that they lean into each other at an angle and do not fall down. Lean two or three more pairs against the first two. Leave them like that for a week or two so that the corn can dry out in the sun and wind, and the grain can become dead ripe.

Mow and rick

In wet climates the practice is then to form a mow (which is intermediate between a stook and a stack) or rick. To make a mow stand about 20 sheaves up on their butts, leaning inwards in a solid circle. Then start building another solid circle on top of the first lot starting from the middle. You can build in a spiral if you like. Lay the sheaves of this second layer closer to the horizontal, with the ears inwards towards the middle. To stop these sheaves from slipping, snatch a handful of corn from each sheaf and tuck it under the band (string or corn tie) of the next sheaf.

Arrange all the time so that the centre of the mow shall be higher than the outside edges. Put layer after layer on like this, drawing each layer in a little so that the mow rises steeply to a point. It will then be crowned by, say, four sheaves with their ears upwards, waving in the wind like flags. Very pretty. Any rain that blows into it will run downhill along the sloping straws and on to the ground.

But before the stupendous gales of winter blow up, you should get the corn out of the mows and into ricks or stacks, in your stackyard. When working on a fairly small scale, stacks are better circular.

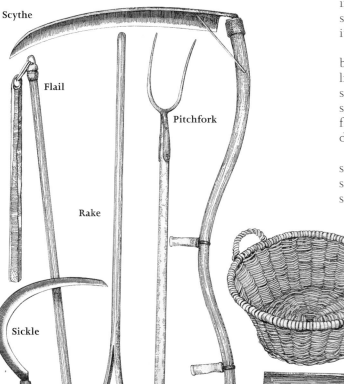

Scythe

Flail

Pitchfork

Rake

Sickle

Winnowing tray

Basket

Corn scoop

HAND TOOLS

Time-honoured equipment works well and is adequate for the homesteader. Cutting with a scythe is old-fashioned, honest labour, and time-consuming, but it produces better grain than that cut and threshed with a combine harvester.

Ricks

In ricks you lay the sheaves horizontally, ears inwards, layer after layer, always keeping your centre firm and high. Work from the centre out this time, and do not pull in as you did with the mow until the eaves are reached. Then pull inwards sharply until the apex is reached. You must then thatch this rick, or provide it with some waterproof cover, just on top. Corn will keep well in such a rick for years, as long as the rats don't get into it. To keep rats away you can build a raised platform on staddle stones.

Grain in the stook, mow, and rick is maturing naturally all the time, slowly ripening and drying out, and it is better grain than grain cut and threshed with a combine harvester.

Threshing and winnowing

Then you must thresh your grain. This is the business of knocking the grain out of the ears. You can do this by bashing the ears on the back of a chair, putting them through a threshing drum, beating them with a flail, or driving horses or oxen about on them.

A threshing drum is a revolving drum that knocks out the grain. A flail is two sticks linked together: the longer stick, which you hold, may be anything – ash or hickory is fine; the short stick, with which you wallop the grain, is often holly. The link which joins them can be leather, and eel-skin was traditionally used for this, being very tough. (Eel-skin is excellent for "leather" hinges by the way.) You try to lay the short stick flat on the ears of the corn.

After you have threshed (or knocked out) the grain, you must winnow it. Traditionally, this is done by throwing the grain, which is mixed up with chaff, broken bits of ear and straw, thistle seeds, and all the rest of it, up into the air in a strong breeze. The light rubbish blows away, and the grain falls in a heap on the ground. Common sense suggests you do this on a clean floor, or else have a sheet of canvas or suchlike on the floor to catch the grain. The chaff (the stuff that blows away) can be mixed with corn and fed to animals. Alternatively, a winnowing machine has a rotating fan that produces an artificial wind to winnow the grain. It also has a number of reciprocating sieves. These extract weed seeds, separate "tail" (small) corn from "head" (large) corn, and thoroughly clean the grain. They can be turned by hand or power.

Storing

When grain has been harvested naturally by the above means, it will keep indefinitely so long as it is kept dry and away from vermin. You can store naturally dried corn in bins, on a grain floor, in sacks: anything will do provided rats and mice and other vermin cannot get to it. The processes described above are just the same for wheat, barley, oats, rye, field beans, rice, buckwheat, sorghum, millet, linseed, oil-seed rape, and many other seed crops. Harvesting by combine is another kettle of fish. Here the machine cuts the corn and threshes and winnows the grain in one operation, while moving round the field. It saves an awful lot of labour.

STOOKS
Get your sheaves into stooks quickly or the grain will sprout and rot. Rub two sheaves' heads together so they stick firmly and don't blow down.

MOW
After their initial drying in the stook, you can put your sheaves into a mow. It's quick to build, safe and waterproof.

A THATCHED CORN STACK
If your stack is out of doors, a thatch is the most effective protection you can give it. A good thatch will easily last the winter, whereas even strong plastic may tear in a gale. Thatching is described on *pp.276–277.*

A CIRCULAR CORN STACK
Lay a base of sheaves with their butts outwards, and build on up, keeping the centre high. Pull inwards when you reach the eaves and let them overhang slightly so the rain runs off.

Use rat-proof staddle stones to keep your circular corn stack off the ground and you will keep out the damp as well as any rats.

The Cereals

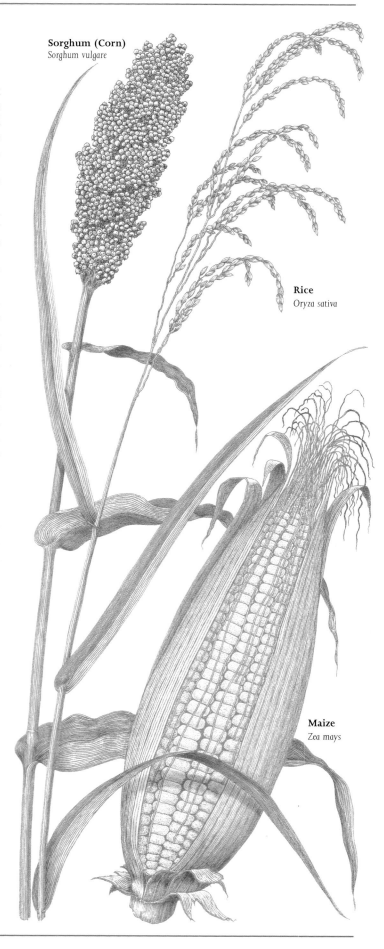

Sorghum (Corn)
Sorghum vulgare

Rice
Oryza sativa

Maize
Zea mays

Cereals are the staff of life for most of us. Even our milk and our meat derive largely from them. They are grasses nurtured and bred by people so that their grain is large and nourishing. Except in parts of the tropics where such roots as tapioca and yams are the staple carbohydrate, and in wet, cold places where the potato plugs the gap, wheat, barley, oats, rye, rice, maize, and sorghum are what keep us all alive and kicking.

The cereals have all been bred from wild grasses, and bred so far away from their parent stocks that they are now distinct species. In fact, it is sometimes difficult to guess which wild grass a particular cereal is derived from, and, in some cases, maize for example, the wild species is now probably extinct.

It was inevitable that the seeds of the grasses should become humankind's staple food. After all, grass is the most widespread of plants, its seeds copious and nourishing and easily stored. When the Bushmen of the Kalahari find a hoard of grass seed in an ants' nest, stored there by the industrious insects for their future supplies, the Bushmen steal this seed, roast it on a hot stone, and eat it themselves. Our Stone Age ancestors no doubt did the same. And it is but a short step to harvesting the grass oneself and threshing the seed out of it. Then it was found that if you put some of the seed into the ground, in the right conditions, it would grow where you wanted it to. Agriculture was born – and with it civilizations. This was made possible by people's ability to grow and store the food they had grown reliably.

Many smallholders feel that grain growing is not for them: it requires expensive machinery, is difficult, and cannot be done effectively on a small scale. This is just not so. Anyone can grow grain, on no matter how small a scale, provided they can keep the birds off it. Harvesting can be done quite simply with the sickle or even an ordinary carving knife. Threshing can be done over the back of a chair and winnowing outdoors in the wind. Grinding can be done with a coffee grinder or a small hand mill. Baking can be done in any household oven. It is very satisfying to eat your own bread baked from grain you have grown and milled yourself, from your own seed.

When the Roman armies wanted to conquer Britain they waited until harvest time, so that their soldiers could spread out over the country, reap the native wheat, take it back to camp, and make bread out of it. If the Roman legions could do it with such apparent nonchalance, there is no reason why we cannot do it too.

It is fortunate that grasses are widespread: they grow in practically any climate our Earth affords, and therefore different cultures have been able to find, and adapt, one particular grass to suit each area. So if we live in the wet tropics, we may choose rice; the dry tropics, sorghum; temperate heavy lands, wheat; temperate dry and sandy lands, rye; cold and rainy lands, oats; temperate light land, barley; and so on. There is an improved grass for nigh on every area and climate in which the human race can survive.

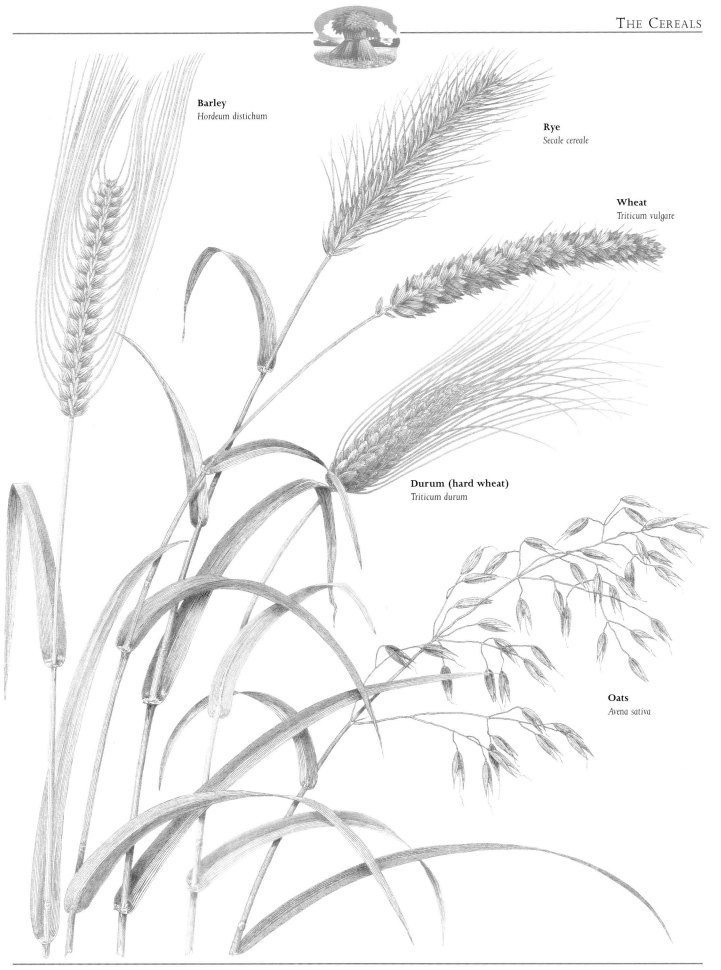

Barley
Hordeum distichum

Rye
Secale cereale

Wheat
Triticum vulgare

Durum (hard wheat)
Triticum durum

Oats
Avena sativa

Wheat

Ever since our Stone Age ancestors found that they could bang the grass seeds collected by seed-collecting ants between two stones and eat them, we have used cereals for food, and in all those parts of the world where wheat will grow, wheat is the favourite.

Hard and soft wheat

Hard wheat grows only in fairly hot and dry climates, although there are some varieties that are fairly hard even if grown in a colder climate. It is much beloved by commercial bakers because it makes spongy bread full of holes. It holds more water than soft wheat and a sack of it therefore makes more bread. In temperate climates soft wheat grows more readily and makes magnificent bread: a dense bread perhaps, not full of huge holes, not half water and wind but bread such as the battle of Agincourt was won on.

Sowing

Wheat grows best on heavy loam or even clay soil. You can grow it on light land, and you will get good quality grain but a poor yield. It will also grow on very rich land, but it must have land in very good heart.

In temperate climates wheat – and it will be one of the varieties called winter wheat – is often sown in the autumn. Winter wheat grows quite fast in the autumn, in the summer-warmed soil, then lies dormant throughout the winter, to shoot up quickly in the spring and make an early crop. In countries such as Canada and the northern United States, where the winter is too severe, spring wheat is grown, and this is planted in the spring. It needs a good hot summer to ripen it, and will come to harvest much later than winter wheat. If you can grow winter wheat, do so. You will get a heavier crop and an earlier harvest.

I prefer to get winter wheat in very early: even early in September in Britain, because it gets off to a quick start, beats the rooks more effectively (rooks love seed wheat and will eat the last seed if they get the chance), and makes plenty of growth before the frosts set in. Frost may destroy very young wheat by dislodging the soil about its roots. If the early-sown wheat is then "winter proud" as farmers say, meaning too long, graze it off with sheep. Graze it off either in November or in February or March. This will do the sheep good and will also cause the wheat to tiller (put out several shoots) and you will get a heavier crop. You can sow winter wheat in October and sometimes even in November. The later you sow winter wheat, the more seed you should use.

Spring wheat should be sown as early as you can get the land ready and you feel the soil is warm enough. I would say not before the beginning of March, although some people sow it in February. The earlier you sow it, the more you will lose from rooks, who have little other food at that time of the year, and the longer it will take to get established. But wheat needs a long growing season, and therefore the earlier you sow it, the better.

In other words, if you don't want a very late harvest, as always in farming, you have got to find a compromise between tricky alternatives.

Wheat needs a fairly coarse seed bed, that is, it is better to have the soil in small clods rather than fine powder. For autumn-sown wheat the seed bed should be even coarser than for spring-sown. This is so that the clods will deflect the winter rain and prevent the seed from being washed out and the land becoming like pudding.

So plough, if you have to plough, shallowly, and then do not work your land down too fine. In other words, do not cultivate or harrow it too much. Aim at a field of clods about as big as a small child's fist. If you are planting wheat after old grassland, plough carefully so as to invert the sods as completely as you can, and then do not bring them up again. Disc the surface, if you have discs, or harrow it with a spring-tined harrow, or ordinary harrow if you haven't got that. But do not harrow too much. Then drill or sow into that. The earlier you can plough the land before you put the wheat in, the better, so as to give the land a chance to settle.

You can either drill wheat, at a rate of about three bushels of seed to the acre, or else broadcast it at about four bushels to the acre. Whichever way you do it, it is a good thing to harrow it after seeding and also to roll it – that is, if you don't think the rolling will break down the clods too much. If it is wet, don't roll it. Discing is quite good after broadcasting seed but only do it once: if you do it twice you will bring the seed up again.

Care of growing crop

You can harrow wheat quite hard when it has started to come up but is not more than 6 inches (15 cm) tall. After you have harrowed it may look as if you have ruined it, but you haven't. You will have killed several weeds but not the wheat, and the harrowing does good by opening up the surface of the ground. If frosts look as if they have lifted the surface of the ground in the early spring you can roll, preferably with a ring roller, but only if the ground is pretty dry.

Jethro Tull invented a seed drill and developed "horse-hoe husbandry". His idea was to drill wheat and other cereals in rows 12 inches (30 cm) apart (there was much experimentation with distances) and then keep the horse-hoe going up and down between the rows. Very good results were achieved. The practice has been discontinued because developments in husbandry have enabled the farmer to clean his land, meaning free it from weeds, more thoroughly. It is therefore not so necessary to weed the wheat. In any case, a good crop of wheat that "gets away" quickly will smother most weeds on reasonably clean land. Agri-businesses, of course, use selective weed poisons to kill weeds in wheat. I used them only once and have seldom had a crop of any cereal that has suffered badly from weed competition. Selective weed poisons are only necessary to cover up the effects of bad husbandry.

Milling Grain

Modern industrial grain milling is enormously complicated, and aims to remove everything from the flour that is ultimately used for bread except just the pure starch. The milling of wholemeal, on the other hand, is very simple: all you do is grind the grain: nothing is taken out and nothing is put in. Wholemeal flour also has more of every beneficial thing, except pure starch (carbohydrate), than white flour has. And wholemeal bread is better for the digestion than white bread because it has roughage. Here is a percentage comparison:

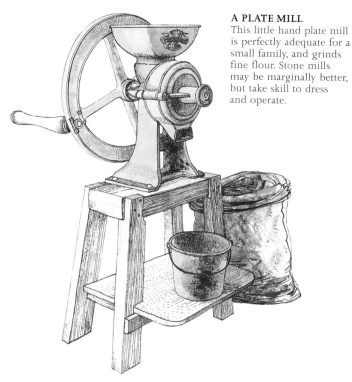

A PLATE MILL
This little hand plate mill is perfectly adequate for a small family, and grinds fine flour. Stone mills may be marginally better, but take skill to dress and operate.

	Protein	Fat	Carbohydrate	Calcium	Iron	Vitamin B1	Riboflavin	Nicotinic acid
WHITE FLOUR	2.3	0.2	15.6	4	0.2	0.01	0.01	0.2
WHOLEMEAL	3.1	0.6	11.2	7	0.7	0.09	0.05	0.6

There are four types of mill for grinding grain. Two of them are of little use to the self-supporter: the hammer mill, which will smash anything up, even feathers, but does not make very good flour; and the roller mill, as used in huge industrial mills, where steel rollers roll against each other and the grain passes between. The other two types – the stone mill and the plate mill – are both suitable for anyone who wants to make their own bread.

Stone milling

The stone mill is one of the oldest and most basic types of mill. It consists of two stones, one of which turns on the other, which remains stationary. The grain is passed between the two, generally being dropped down a hole in the top, or runner stone. The art of milling with stones, and particularly that of millstone dressing – crafting the patterns of "furrows" (channels) and "lands" (raised grinding edges) – has all but died out. The sooner it is revived, the better. However, in response to the new demand for such devices, several firms have put miniature stone mills on the market, both hand and electrically driven. These make very good flour, and they will grind very fine or coarsely as desired; the finer you grind, the longer it will take.

Plate milling

There are also some good hand-driven plate mills available. A steel plate with grooves cut in it revolves, generally vertically, against a stationary steel plate. Flour ground slowly with one of these seems just as good as stone-ground flour. If you have a tractor or a stationary engine, the ordinary barn plate mill found on nearly every conventional farm is quite satisfactory for grinding bread flour, provided you don't drive it too fast. If it goes too fast it heats the flour (you can feel it coming hot out of the spout). Heating the flour like this spoils the flavour.

There is one thing to remember that makes the milling of all grain much easier. That is: dry the grain first. In a warm, dry climate this may not be necessary, but in a damp climate it makes a great difference. When you are nearly ready to grind your wheat, keep it in a jute bag over your stove, or furnace, or dry the day's supply on a tray over the stove, or in a warm oven: anything to get the grain quite dry. Don't cook it of course. If you mill grain in larger quantities a brick kiln is not a bad idea, and it is also useful for kilning malt (which I describe later in detail on p.221).

There is no reason why anybody, even somebody living in a tenth-floor apartment, shouldn't buy a small stone or plate mill and a sack of wheat from a friendly farmer, and grind his own flour and make his own bread. Do not believe it when people tell you to do so does not pay. Whenever we have kept accounts about bread-making we have found it pays very well. You get your bread for considerably less than half of what you would pay in a shop and it is much better bread.

Bread made from freshly ground wheat and baked by home baking methods is superb bread. You are not interested in trying to sell as much holes and water as you can, as is the commercial baker. Your bread will be a lot denser than bought loaves, but well leavened nevertheless, and, if your oven was hot enough, well cooked. It will take far less of it to feed a hungry man than it takes of shop bread, and if you consistently eat your own good bread, you, and your family, will stay healthy and your visits to the dentist will be mere formalities. To see how wheat is harvested see *pp.148–149.*

Oats & Rye

OATS

Oats will grow in a damper climate than wheat or barley, and on wetter and more acid land. Thus it is a staple human food in Scotland, which led author and philosopher Samuel Johnson to rib his friend Boswell about how in Scotland men lived on what in England was only thought fit to feed to horses. Boswell replied: "*Yes – better men, better horses*". In North America and Europe oats tend to be grown in damper, colder places, and often on glacial drift, where the land may be heavy, acid, and not very well drained. Oats and potatoes have enabled people to live in areas where no other crop would have grown.

Sowing oats

In wetter areas it is most usual to sow spring oats: in drier and warmer areas winter oats are preferred and give a heavier yield and also are less attacked by frit fly, a common pest of oats. The only trouble with winter-sown oats is that it is likely to be eaten by birds. If it is possible to sow while other people are harvesting their spring-sown crops, it stands a better chance of survival because the birds are tempted by seeds dropped elsewhere. The cultivation of oats is exactly the same as that of wheat (*see* p.152).

Harvesting oats

But whereas barley should be allowed to get completely ripe and dry before harvesting, oats should not. There should still be a bit of green in the straw. Oats are far better cut and tied into sheaves with sickle, scythe, reaper, or reaper-and-binder, than they are combined, for combining knocks out and wastes a lot of the grain. When cut and bound, oats should be stooked and then "churched" three times. Churching is the old farmers' way of saying that it should be left standing in the stook for at least three Sundays. The purpose of this is to dry the straw thoroughly, and any grass growing among it, and to dry the grain itself, so that the grain will not go mouldy in the stack.

Many old-fashioned farmers, including me, feed oats to horses and cattle "in the sheaf". In other words we do not thresh the grain out but simply throw animals complete sheaves. One per beast per day during the winter, plus grass, will keep bullocks and dry cows looking fine. The animals eat straw and all. Oat straw, whether threshed or not, is the best of all the straws for feeding: good oat straw is better feed than poor hay. But, of course, working horses should be fed on the grain alone as well. You will find more details about feeding horses on p.142.

Milling oats

The Scots, and other sensible people, mill oats thus. They kiln it, that is, put it in a kiln until it is quite dry. It must be completely dried, so they kiln it at quite a high temperature: this is the most important part of the operation. They then pass it between two millstones set at quite a distance apart.

This gently cracks the skin of the oats off: They then winnow it. This blows the skins away and leaves the grain. Finally, they pass it through the stones again, but this time set them closer to grind it roughly, not too fine. This is real oatmeal, and it has satisfied some of the best appetites in the world over the centuries.

There are two ways of making porridge with it, very different but both equally efficacious. One is sprinkle the meal into boiling water, stirring the while, and the moment the porridge is thick enough for your taste take it off the fire and eat it. The other is: do this but then put the closed pot in a hay-box and leave it overnight. (A hay-box is a box with hay in it.) You bury the pot in the hay when the porridge is boiling and, because of the insulation, it cooks all night. Eat it in the morning. Eat porridge with milk or cream, and salt: never with sugar, which is a beastly habit, and not what porridge is about at all.

RYE

Rye is the grain crop for dry, cold countries with light, sandy soil. It will grow on much poorer, lighter land than the other cereals, and if you live on rough, heathy land rye might be your best bet. It will thrive better in colder winters than other cereals and stands acid conditions.

You might well grow rye to mix with wheat for bread: a mixture of rye and wheat makes very good bread. Rye alone makes a dense, dark, rather bitter bread; it's very nutritious and is eaten in pretty large quantities by the peoples of eastern Europe. It seems to do them good.

Sowing rye

You can treat rye in exactly the same way as the other cereals (*see* p.152). If you plant it in the autumn and it grows very quickly, which it often does, it is very advantageous to graze it off with sheep or cows in the winter when other green feed is scarce. It will grow very quickly again and still give you a good crop. It doesn't yield anything like as well as wheat, though, whatever you do to it.

Rye is often used for grazing off with sheep and cattle only. A "catch-crop" is sown, say, after potatoes have been lifted in the autumn. This is grazed off green in the spring during the "hungry gap" when a green bite is very welcome. Then the land is ploughed up and a spring crop put in. Its ability to grow well in the winter is thus utilized. One advantage of rye as a winter-sown crop is that it does not seem to be so palatable to birds as other grains. Wheat and oats both suffer badly from members of the crow family: rye seems to escape these thieving birds.

Harvesting rye

Rye ripens earlier than other grains. If you cut it when it is completely ripe, then it will not shed much. The straw is also good for bedding, and is a very good thatching straw. In the past years, I have grown rye specifically for thatching.

Barley

Barley has two principal purposes: one is feeding animals and the other is making beer. It doesn't make good bread because the protein in the grain is not in the form of gluten, as it is in wheat, but is soluble in water. It will not therefore hold the gases of yeast fermentation and so will not rise like wheat flour will.

You can mix barley flour with wheat flour though, say perhaps three to one wheat to barley – and make an interesting bread.

Barley will grow on much lighter and poorer soil than wheat, and it will also withstand a colder and wetter climate, although the best malting barley is generally grown in a fairly dry climate.

Sowing barley

An old saying goes: "Sow wheat in mud and barley in dust!" My neighbour says that the farmhands used to come to his old father and say: "Boss, we must get in the barley seed. The farmer over the valley is doing it."

"Can you see what horses he's using?" said the old man, who couldn't see very well.

"The roan and the grey gelding," they replied.

"Then don't sow the barley," said the farmer.

A few days later the same exchange took place, but when the farmer asked what horses his neighbour was using the men answered: "I can't see them for dust."

"Then sow the barley," said the farmer.

Don't take this too literally, but barley does need a much finer seed bed than wheat. There is such a thing as winter-sown barley, but most barley is spring-sown, for it has a much shorter growing season than wheat. It grows so quickly that barley will come to the sickle even if you plant it as late as May, but any time from the beginning of March onwards is fine, just so long as the soil is warm and sufficiently dry.

As I explained before, a certain Suffolk farmer used to drop his trousers and sit on the land before he drilled his barley, to see if the land felt warm and dry enough.

Where I used to live we have a festival, in the town of Cardigan, called Barley Saturday. This is the last Saturday in April. We were all supposed to have got our barley in by then, and to celebrate there was a splendid parade of stallions through the streets of Cardigan and the pubs stayed open all day.

Barley, particularly for malting for beer, should not have too much nitrogen, but needs plenty of phosphorus, potash and lime. I broadcast barley seed at the rate of four bushels, about 2 cwt (100 kg) to the acre. If I drilled it I would use less: about three bushels, or 1½ cwt (75 kg). Drilling it would probably be better, but we haven't got a drill, and in fact we get very good results by broadcasting. Of course after drilling or broadcasting you harrow and roll just as you do for wheat. Excepting that the seed bed should be finer, the treatment is in actual fact exactly the same as for wheat, but one tends to sow barley on poorer ground.

Harvesting barley

Harvesting is the same as for wheat. If you harvest it with a combine harvester it must be dead ripe. They say around here: wait until you think it is dead ripe and then forget all about it for a fortnight. An old method of harvesting barley is to treat it like hay, which isn't tied into sheaves but left loose. You turn it about until it is quite dry, then cart it and stack it loose just like hay. You can then just fork it into the threshing machine.

If you do bind it into sheaves, leave it in the stook for at least a week. But however you cut it, do not do so until the ears have all bent over, the grains are hard and pale yellow and shed easily in your hand, and the straw is dry. You can then put it into mows and the straw can be fed to livestock. It is better for feeding than wheat straw, which is useless, but not so good as oats. It is no good for thatching, and not as good as wheat straw for litter or animal bedding.

The grain is the beer-grain *par excellence*, but most of the grain gets fed to pigs and cattle. It can be ground (best for pigs) or rolled (best for cattle). If you haven't got a mill, just soak it for 24 hours. If you want to eat it, try:

Barley soup
This is one of the self-supporter's staple meals, not just a soup. It's warm and nourishing. You can vary the vegetables according to what you've got. Add more carrots if you haven't got a turnip and so on. You need:

2 oz (56 g) washed husked barley
1 lb (0.5 kg) stewing mutton
2 quarts (2.3 litres) water
1 teaspoon salt
3 or 4 carrots
2 or 3 leeks
3 or 4 onions
1 big turnip or 1 big swede

Put the whole lot in a stewing pot. Season slightly, cover, and simmer for 3 hours. Stir occasionally to make sure nothing is sticking to the bottom. Take out the meat at the end of the cooking time, remove the bones, and chop the meat into mouth-sized lumps. Put them back in the soup. Add chopped parsley if you've got it.

Northumberland barley cakes
If you haven't got a deep freeze, these will keep a lot longer than bread. They are like huge, thick biscuits and make a very good between-meal snack. Take:

1 lb (0.5 kg) barley flour
1 teaspoon salt
½ teaspoon bicarbonate of soda
¼ teaspoon cream of tartar
½ pint (0.3 litres) buttermilk or skimmed milk

Bung all the ingredients in a mixing bowl and stir into a soft dough. Make this into balls and press them out until they are about 10 inches (25 cm) across and ¼ of an inch (2 cm) thick. Bake on a griddle until the cakes are brown on one side. Turn over and brown the other side. Serve up cold, cut in pieces, and spread with butter.

Barley pastry
This is very light and crumbly and good for fruit pies and flans, and for the dentist's nightmare and child's delight, treacle tart. Take any shortcrust pastry recipe, substitute barley flour for wholewheat flour and slightly reduce the amount of fat. For example, with 8 oz (228 g) of barley flour use 3 ounces (84 g) of fat instead of 4 ounces (114 g). Roll out and cook in the same way as you would pastry made with wheat flour.

Maize

Besides the potato, and that horrible stuff tobacco, the most important contribution the New World has made to the Old is maize. The first white settlers in America called it Indian corn, and this was shortened to corn, and corn it now is. In England it is maize, although gardeners also refer to it as sweet corn.

Maize is grown for several purposes. First, for harvesting when the grain is quite ripe, and ready to be ground to make human or cattle food. Secondly, for harvesting before the ears are ripe to be eaten, boiled, and with lashings of butter, as "corn-on-the-cob". The grain in the unripe cobs is still soft and high in sugar, because the sugar has not yet turned into starch for storage, but is still in soluble form so that it can move about the plant. Thirdly, maize is grown for feeding off green to cattle in the summer time, long before the grain is ripe, just as if it were grass. Fourthly, it is grown for making silage. This is done when the grain is at what is called the cheesey or "soft dough" stage. To make silage the stems must be well cut up or crushed, in order that they can be sufficiently consolidated.

Maize will grow to the "corn-on-the-cob" stage in quite northerly latitudes, but it will not ripen into hard ripe grain (the grain is nearly as hard as flintstones) except in warmish climates. It is always planted in the spring and likes a warm, sunny summer but not one that is too dry. It will stand considerable drought, and the hotter the sun, the better, but in dry climates it needs some rain or irrigation.

Sowing maize

Maize likes good but light soil: heavy clays are not suitable. It must be sown after the last danger of frost as it is not at all frost-hardy. So plant 1 or 2 weeks after the last probable killing frost. You will need 35 pounds (15.8 kg) of seed for an acre, sown about 3 inches (13 cm) deep. Space between rows can be from 14 inches (36 cm) to 30 inches (76 cm): do what your neighbours do in this respect and you will not go far wrong. There should be about nine plants to a square yard.

Care of crop

Birds are a terrible nuisance, particularly rooks, and will, if allowed, dig up all your seed. Threads stretched about 4 feet (1.2 m) above the ground on posts hinder the rooks (as they do you too when you want to hoe), and shooting the odd rook and strewing its feathers and itself about the land puts them off for a short time. Rooks are a menace and there are far too many of them: the idea put about by rook-lovers that they are not after your seed but after leatherjackets is arrant nonsense, as the most cursory examination of a dead rook's crop will show.

Harvesting

Harvesting ripe cobs by hand, as the homesteader should be most likely to do, is delightful work. You walk along the rows, in a line if there are several of you.

You then just rip out the ears and drop them in a sack slung over your shoulder. It's best to tread the straw down with your foot so you can see where you have been (the straw is as high as you are). When you get hungry, why not light a fire of dried maize straw, or sticks, throw some cobs on it, without removing their sheaths, and when the sheaths have burnt off and the grain is slightly blackened, eat them? They are quite unlike the admittedly delicious "sweet corn" or "corn-on-the-cob" and are food fit for a king – or at least a hungry harvester (who has presumably got good teeth).

Maize in the garden

In cold latitudes you can grow maize for "sweet corn" in your garden. Plant it under cloches or else plant it in peat pots indoors and then carefully plant out pots and all after the last frost. Or you can plant direct into the ground after the last frost, say with two seeds in a station, stations 12 inches (30 cm) apart and 2 feet (60 cm) between rows. Plant in blocks rather than long thin lines because this helps pollination.

Maize likes well-dunged ground. Water it if the weather is really dry. Pick it when the silk tassels on the ears go from gold to brown.

Cooking maize

Boil in the sheaths (at least I do) for perhaps a quarter of an hour. Eat it off the cob with salt and oodles of butter. This is food that I challenge anybody to get tired of. For years we Seymours ate tons of it. It was our staple diet for the whole of the autumn. It is a crop that must be eaten as soon as possible after harvesting: if it is kept, the sugars begin to harden into starch and the fragrant elements of the succulent grains disappear.

Polenta or cornmeal mush

This can be made with ground maize or ground sorghum. It comes from north Italy and is rather stodgy unless well buttered and cheesed after cooking, but it is quite delicious. You need for six people:

8 oz (228 g) cornmeal
2 teaspoons salt
2½ pints (1.4 litres) water
3 teaspoons grated cheese and butter

Boil the water in a large pan with salt. Then sprinkle in the cornmeal, stirring all the time to prevent lumps. Keep on stirring. After 30 minutes it is so thick it is leaving the sides of the pan. Mind it doesn't brown on the bottom. Stop cooking and spread it out on a dish. Cover it with blobs of butter and grated cheese, and push it under the grill for a few minutes. It's very good by itself, and even better served up with fried spicy Italian sausages and plenty of tomato sauce.

Polenta gnocchi

Cook polenta as in the recipe above, but at the end of cooking stir in two beaten eggs and some grated cheese, and, if you want to make it more exotic, add 4 ounces (114 g) of chopped ham as well. Turn it all on to a flat wetted dish and spread it out so that it is about ¼ inch (1 cm) thick.

Next day cut it into squares, lozenges, or circles about 1½ inches (4 cm) across. Lay these overlapping in a thickly buttered ovenproof dish. Dot with more butter, heat in the oven or under the grill, and serve sprinkled with more cheese.

Rice

Rice, before it is milled, is called "paddy" by English-speaking people in Asia. There are for practical purposes two sorts: "wet" rice that grows in water, and "upland" rice growing on open hillsides, but only in places with a very high rainfall such as the Chin Hills of Burma. Wet (or ordinary) rice is grown on a large scale in the United States and in southern Europe, and there is no doubt that its cultivation could be extended to more northern latitudes. It will grow and ripen in summer temperatures of over 68°F (20°C), but these must cover much of the 4 to 5 months that the crop takes to grow and ripen.

It might well be that some of the upland rice varieties would grow in northern latitudes. The reason why we of the northern latitudes do not attempt to cultivate them may be because we are congenital wheat-eaters and do very well without rice. The wheat-eating peoples of India have a strong sense of superiority over the rice-eaters and look upon rice as food fit only for invalids!

Sowing rice

The best way to grow rice on a small scale is to sow the seed broadcast on a dry seed bed when the ground has warmed up in the spring, rake it well in, and then flood the seed bed but only just. As the shoots grow, always try to keep the water level below the tops of the plants. Rice survives in water by virtue of its hollow stem, which takes oxygen down to the rest of the plant.

When the plants are about 8 inches (20 cm) tall, pull them out in bunches and transplant them into shallow standing water in an irrigated field. Simply dab each plant into the soft mud 4 inches (10 cm) away from its neighbours. Billions of paddy plants are planted like this every year in India and China. Keep the paddy field flooded (never let it get dry) until about a fortnight before you judge the grain ripe enough to harvest. Then drain the field and let the grain ripen in the dry field.

Harvesting rice

Harvest with the sickle, thresh as you would other corn, "hull" (separate grain from husk) by passing through a plate mill or stone mill with the plates or stones open enough to hull the grain without cracking it, and you are left with "brown rice": that magical food of the yin-yan adherents. It is in fact a good grain, very rich in starch, but lower in protein and in several other qualities than wheat.

If you mill the brown rice more closely you get pearled rice, which is generally and wrongly called polished rice. This is almost pure starch and a very incomplete foodstuff, even less nutritious than white wheat flour (which is saying a lot). A further process, called polishing, produces true polished rice which is what most of the US buy in the shops. If you live on practically nothing else but pearled or polished rice you get beri-beri. So the sensible thing to do, if you live on rice, is to eat brown rice and not go to the trouble of removing the bran, which is the most nutritious part of it, and feeding it to the pigs.

Cooking rice

Unlike most other grains, rice does not need grinding before it is cooked. The Western way to cook your own home-milled rice is to wash the grain well in cold water and strain, then bring 1 pint (0.6 litres) of water to the boil, add a teaspoonful of salt, and throw in 6 ounces (170 g) of rice.

Bring this to the boil again and then allow it to simmer by reducing the heat. Cover the pan and simmer for 15 minutes. When the rice is tender, eat it. It will have absorbed all the water.

I personally use the Indian method which is to bring much more water than you really need to the boil, throw the rice in, bring to the boil again, allow to simmer until the grain is tender (but not reduced to that horrible rice pudding stuff) which will be in about a quarter of an hour, strain the water out, toss the rice up a few times in the strainer, and eat it. Each grain will be separate if you do this properly and the rice will be perfect.

You can colour and flavour rice very nicely by tossing a pinch of saffron into the rice while it is cooking. For brown rice, you need to allow at least 40–50 minutes cooking time.

Indian rice

North American wild rice (*Zizania aquatica*), or Indian rice, can be harvested when it is ripe, and dried in the hot sun or else "parched" by heating over a fire or kilning. This can then be boiled or steamed and eaten, preferably with meat. It is very nutritious, but very laborious to harvest.

Risotto

This is usually made with rice, as the name implies, but it is very good made with whole millet or with pearl barley. You need:

1 measure grain (1 lb or 0.5 kg should feed 8 to 10 people)
2 measures hot water or good clear stock
A little oil, salt and pepper
A variety of firm vegetables, such as onions, green peppers, peas, carrots, etc.

Use a solid pan with a lid (earthenware is good). Slice the vegetables and lightly fry them in a little oil. Put them aside in a separate dish when they are soft and slightly brown. Put some more oil in the pan and tip in the dry grains. Stir them until they are well-oiled and start to turn colour.

Put the cooked vegetables back in the pan with half the hot water or stock. Season well. Turn the heat low, or put the pan in a moderate oven, closely covered, for 15–30 minutes. Then add the rest of the broth, stir and cook another 15–30 minutes until all the liquid is absorbed, and the grains are soft and tender, but separate. Times vary with the hardness of the grain.

Rice griddle cakes

A good way of using up left-over boiled rice or rice pudding.

½ pint (0.3 litres) milk
4 oz (114 g) warm cooked rice
1 tablespoon melted butter or oil
2 eggs, separated
4 oz (114 g) wheat flour and a pinch of salt

Mix the milk, rice, and salt. Add the egg yolks, the butter, and the flour, and then stiffly beaten egg whites. Heat a griddle and drop the mixture in spoonfuls onto it. Cook both sides.

Growing Crops for Oil

If you have a small piece of land not taken up with growing food it is well worth planting some crop which will produce vegetable oil.

OIL SEED RAPE

Oil seed rape will grow in temperate climates. You plant it like kale (*see* p.162) and harvest it when it is still fairly green. Pull the plant out of the ground, dry it in a stack, thresh it and then crush the seeds to get the oil out. The residual "cake" can be fed to stock, but only in small quantities, as it can upset their stomachs.

FLAX

The seed of the flax plant is linseed which is very rich in oil, and a good feed for livestock in itself. It is high in protein as well as in fat. If you crush the seed in a mill or else scald it in hot water, you will have an excellent food for young calves, a good replacement for milk. This is good for most sick animals, and is fairly laxative. Linseed, mixed with wheat or mixed corn, makes a perfect ration for hens. As you can grow half a ton to the acre it is a crop well worth growing. It can be crushed for oil, but the oil is not very edible and is chiefly used in the manufacture of sundry products such as soap, paint and printer's ink.

I will deal with the production of flax for cloth and fibre on p.268.

SUNFLOWERS

Thirty-five percent of sunflower seed is edible oil, which is good for margarine, if you must have the stuff, and for cooking oil. Sow the seed 1 week before the last likely frost. Plant 12 inches (30 cm) apart in rows 3 feet (90 cm) apart. Harvest the crop when about half the yellow petals have fallen off the flowers. Cut so that you leave 12 inches (30 cm) of the stem on, and hang the flowers downwards in bunches under a roof.

To get the oil out you will have to crush the seed. Or you can feed the seed direct to poultry at a rate of 1–2 ounces (28–56 g) a day, and it is very good for them. You needn't even take the seed from the flowers. Just chuck them the complete flowers. You can sprout sunflower seed, then remove the husks and eat them.

POPPIES

Poppies can be grown for oil seed, as well as for more nefarious purposes, and up to 40 gallons (182 litres) of oil can be got off an acre. It is good cooking oil, burns in lamps with a clear smokeless flame, and the residual "cake," after the oil has been removed, makes excellent stock feed.

In a temperate climate sow the seed in a fine seed bed in April. Sow it fairly thinly, say 3 inches (8 cm) apart in rows 12 inches (30 cm) apart. Harvest by going along with a sheet, laying it on the ground and pouring the seed into the sheet from the heads. Go along about a week later and do it again.

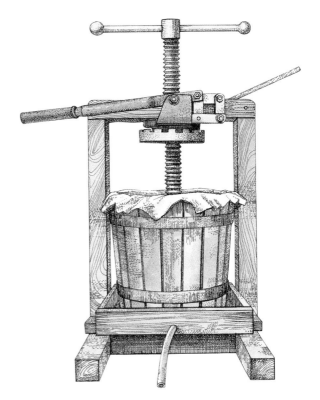

PRESSING OIL

To obtain oil from crops, the seeds need to be cracked first and then wrapped in cloth to make what are known as "cheeses", which are cheese-shaped bundles. These cheeses are then piled into the press and clamped down. A handle is turned to provide the required pressure.

Or you can thresh the seed out in the barn with a flail. I grew two long rows of poppies once and harvested about a bushel of seed from it, but the kids ate the lot. Whether they got high on it I never found out. My kids seemed to be high most of the time anyway.

Olives and walnuts both make fine oil. You will find details about growing olives on p.88.

Extracting the oil

One primitive method used in hot climates to get the oil out of olives, oil-palm and other oleaginous fruits is to pile the fruit in hot sunlight on absorbent cloth. The oil exudes and is caught by the cloths which are thereafter wrung out. The process sounds most unsanitary but is effective.

The other non-technological method is pressing. Before pressing, the seed must be cracked in a mill (plate or stone), or with a pestle and mortar. The cracked seeds are then put in what cider-makers call "cheeses" – packs of crushed seeds lapped round with cloth. The cheeses are piled one on top of the other in a press. The whole lot is then pressed and the oil is exuded. If you haven't got a press you can rig one up with a car jack. If the cracked seeds are pressed cold, the oil is of better quality than if they are heated first, but there is slightly less of it. The pressed residue is good for stock feed.

Growing Root Crops

In Europe in the Middle Ages there was an annual holocaust of animals. It was impossible to feed all these extra animals during the winter, so most of them were killed off in the autumn and either eaten then or salted down. Salt meat was about all that medieval man had in the winter or, indeed, until the first lambs could be killed in the early summer. And milk, too, was in very short supply.

The introduction of the turnip changed all this. If a proportion of your land was put down to turnips you could continue to feed and fatten animals all through the winter, and also keep up the milk yield of your cows. And the turnip was followed by all the other root crops.

With "roots" I am including all those crops known as fodder crops, such as kale, cow cabbages and kohl-rabi, as well as those crops of which the actual roots are the part of the plant that we grow to eat or to feed to our stock. This simplification is justified because all these crops can take the same place in our rotation, and serve the same purpose which is, broadly, to feed our animals in the winter time when there is very little grass. And we can eat some of the roots as well.

All these plants have this in common: they store up energy in the summer so they can lie dormant during the winter, and then release this energy early in the spring to flower and produce seed before other, annual, plants are able to do so. They are in fact biennials. We use them by making use of this stored nourishment for our winter feeding.

TURNIPS AND SWEDES
Swedes are more nutritious than turnips and I think are the better crop to grow. They are also sweeter and pleasanter for you to eat. Swedes have a neck, are more frost-hardy than turnips, and store better, being less prone to disease. Turnips yield slightly more.

Both these closely related plants are members of the brassica family, and thus are liable to club-root, a fungus disease sometimes called "finger and toe" in turnips and swedes. This is a killer and can reduce your yield very drastically, even to nothing. If your land is infected with it, don't grow turnips and swedes.

Sowing turnips and swedes
Turnips and swedes are sown quite late: swedes in May perhaps, and turnips about a fortnight later. In very dry and warm areas it is better to sow later still, for with early sowing in such areas there is a tendency for "bolting" to occur. Bolting is when the plant skips a year and goes to seed at once, when it becomes useless. But turnips and swedes are of most use in the wetter, colder areas.

The seeds are small and therefore want a fine seed bed. Get it by ploughing in the autumn and cross ploughing as early as you can in the spring. Or, if you can't plough in the autumn, then plough for the first time as early as you can in the spring and plough again, or rotavate, cultivate and "pull your land about" with whatever tines or discs you may have or can borrow.

Then drill in rows, preferably with a precision drill. This will drop the seed in one by one at a set interval. If you drill with such a drill at the rate of 1 pound (0.5 kg) an acre this will be about right, but to do this you must use graded seed. You can buy such seed at a seed merchant: it is much more expensive than ordinary seed, which you can grow yourself, but as it goes much further, it is cheaper in the end. In wet climates it is an advantage to sow on raised ridges, which can be made with a "double Tom" or ridging plough.

Singling
If you haven't got a precision drill, then drill the seed in rows as thinly as you can and rely on "singling" when the crop has declared itself. Singling means cutting out with a hoe all the little seedlings except one every 9 inches (23 cm). You can't go along with a ruler. so, because the plants are at all sorts of intervals, you will end up with some closer than nine inches and some further apart, but broadly speaking it doesn't matter that much. This singling of course cuts out the weeds as well as the surplus plants.

You will then have to hand-hoe at least once more, possibly twice, during the growth of the crop, to cut the weeds out in the rows, and you will save labour if you can horse-hoe, or tractor-hoe, several times as well.

Horse-hoeing is a very quick operation and does an enormous amount of good in a short time, but of course, in the end, you will always have to hand-hoe as well because the horse-hoe can't get the weeds between the plants in the rows.

Harvesting turnips and swedes
You can leave turnips and swedes out in the field until after Christmas if you like, except in countries which have deep snow or very heavy frost. You can feed the roots off to sheep by "folding", moving the sheep over the crop, every day giving them just enough to eat off in a day by confining them behind hurdles or wire netting. You will have to go along, however, either before the sheep have had a go at them, or afterwards, and lift the turnips out of the ground with a light mattock. Otherwise the sheep leave half the roots in the ground and they are wasted.

You lift the crop by pulling by hand and twisting the tops off. Then you can clamp (see p.199) or you can put them in a root cellar.

MANGOLDS
Mangolds are like huge beet and crops of 50 tons of them to the acre are perfectly possible. Scientists say they are "nearly all water". But experienced farmers answer to this: "Yes – but what water!" For they know that the moment you start feeding mangolds to cows the yields of milk will go up.

Mangolds are not suitable for humans to eat but make good wine. They are grown more in warmer, drier places than turnips are, but they are pretty hardy.

Sowing mangold

It is best, especially on heavy land, to plough for mangolds the previous autumn. You must then work the land down to a good seed bed in the spring, and then drill at the rate of 10 pounds (4.5 kg) of seed to the acre as near the first of April as you are able. If it is a cold, wet season you may not be able to get a seed bed till May, but it is not much good sowing the crop after the end of May: better sow turnips instead. Drill in rows about 22 inches (56 cm) apart and single out to about 10 inches (25 cm) apart in the rows. Then hoe and horse-hoe as for turnips.

Harvesting mangold

Lift in the autumn before heavy frosts set in, top, and put in little heaps covered with their own leaves until you can cart and clamp. In the old days farmers used to slice the mangolds in machines. Now we know that cattle can do it just as well with their teeth. Don't feed mangolds too soon: not until after New Year's Day. They are slightly poisonous until mature.

FODDER BEET

Fodder beets are very like mangolds, but smaller and far more nutritious. They are very high in protein, and are excellent for pigs, cows, or even horses. Personally I think they are a better grow.

Sowing fodder beet Sow and thin as for mangolds, but thin to 8 inches (20 cm).

Harvesting fodder beet Break the tap-roots with a beet-lifter or break out with a fork before you pull. Top with a knife, put in small piles, cover with leaves against frost, and cart to clamp when ready.

CARROTS

Carrots need a fine seed bed just like turnips, and not too much fresh muck: it makes them fork. Carrots will not thrive on an acid soil, so you may have to lime.

Sowing carrots

Sow them in rows 12 inches (30 cm) apart, or up to 18 inches (45 cm) if you intend to horse-hoe a lot, and sow the seed as thinly as you can. You may then avoid singling, but the crop takes a great deal of labour in hand-weeding, as it grows very slowly, far slower than the weeds in fact.

Harvesting carrots

In places with mild winters you can leave the crop in the ground until you want it, but if you fear hard frosts lift it by easing out with a fork and then pulling. Twist the tops off; don't cut them. Then clamp or store in sand.

It is really too laborious a crop to grow for stock feeding, but it does make very good stock feed, particularly for pigs. You can fatten pigs on raw carrots and a very little high protein supplement as well, and they will live on just raw carrots. And, of course, they are an excellent human food: very rich in vitamin A.

GOOD ROOT CROPS TO GROW

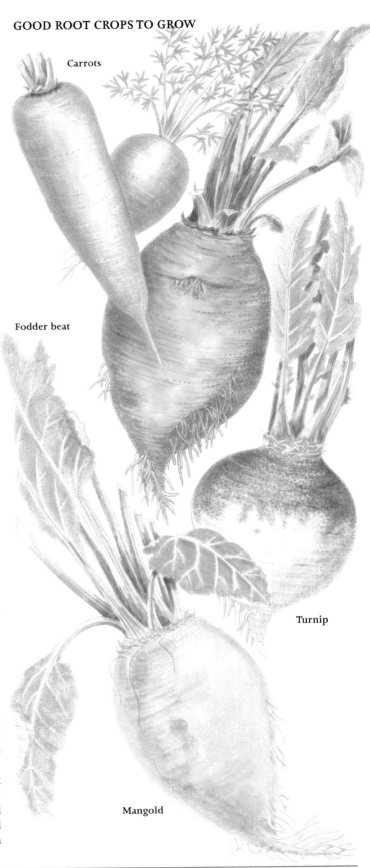

Carrots

Fodder beat

Turnip

Mangold

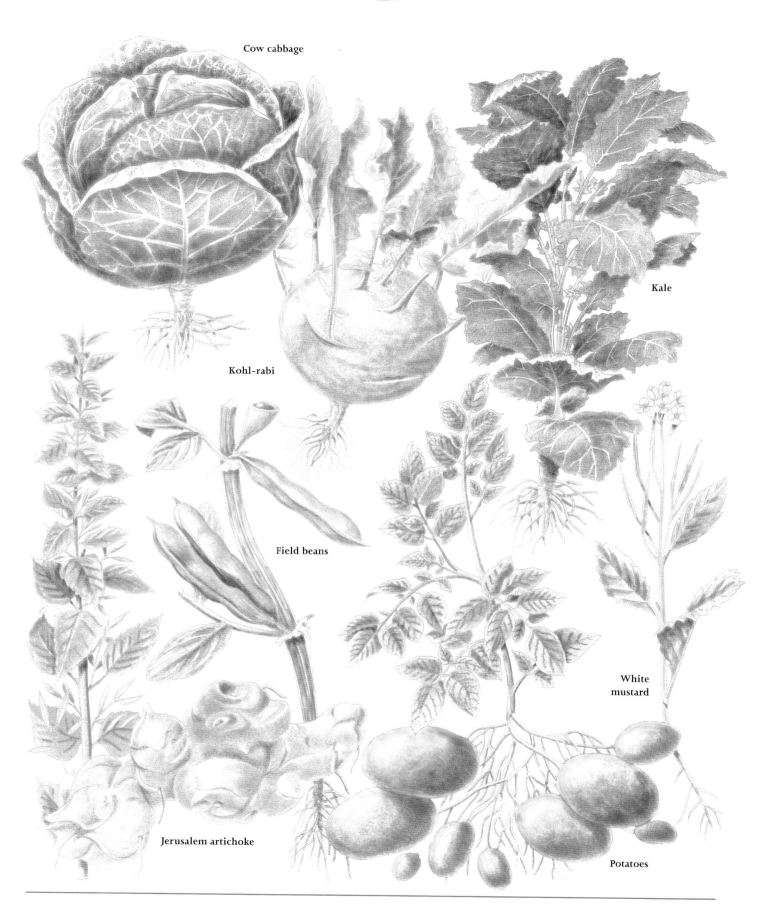

Cow cabbage

Kale

Kohl-rabi

Field beans

White mustard

Jerusalem artichoke

Potatoes

KALE

The most common of the myriad kales are: marrowstem, thousand-headed, rape kale, hungry-gap kale and kohl-rabi. There are lots of other curious kales around the world.

Sowing kale

Drill in rows about 20 inches (50 cm) apart, or broadcast, but drilling gives a better yield and if you have a precision drill you save seed. Sow from 2–4 pounds (0.9–1.8 kg) of seed per acre. Drill most kales early in April, although drill marrowstem in May so that it doesn't get too woody. Thin and hoe in rows and you will get a heavier crop, especially of marrowstem kale. Kale likes plenty of manure.

Harvesting kale

You can fold kale off in the winter. It is more used for folding to cows than to sheep – it is a marvellous winter feed for milking cows. Let the cows on to it a strip at a time behind an electric fence. Or you can cut it with a sickle and cart it to feed to cows indoors. Use marrowstem first and leave hardier kales like hungry-gap until after the New Year. After you have cut or grazed off a field of kale the pigs will enjoy digging up the roots. Or you can just plough in.

RAPE

Rape is like swede but has no bulb forms. It is good for folding to sheep or dairy cows. It can be sown (generally broadcast, not drilled) in April for grazing off in August, or can be sown as a "catch-crop" after an early cereal harvest to be grazed off in the winter but then it won't be a very heavy crop. Rape is too hot a taste for humans, so you can't eat it as a vegetable, although a little might flavour a stew.

COW CABBAGES

These are very similar to ordinary cabbages. Establish them like kale, or if you have plenty of labour you can grow them in a seed bed and transplant them out by hand in the summer. This has the advantage that you can put them in after, say, peas or beans or early potatoes, and thus get two crops in a year: a big consideration for the smallholder. You can get a very heavy crop, but remember that cow cabbages like good land and plenty of manure.

You can clamp cow cabbages and they make splendid human food, too, and are fine for sauerkraut (see p.209). All the above brassica crops are subject to club-root, and must not be grown too often on the same land.

MUSTARD

There are two species: *Brassica alba* and *Brassica juncea*, white and black mustard. They can be grown mixed with rape or grown alone for grazing off with sheep; or grown as a green manure to be ploughed into the land to do it good; or harvested for seed, which you can grind, mix with a little white wheat flour and moisten as required to produce the mustard that goes so well with sausages.

Remember that mustard is part of the cabbage-tribe though, and is therefore no good for resting the land from club-root. I would never grow it as green manure for that reason. It is not frost-hardy.

Cleaning crops

It must be understood that all the aforementioned crops (excepting rape and mustard when these are broadcast and not drilled in rows) are cleaning crops, and thus of great value for your husbandry. Being grown in rows it is possible to horse-hoe and hand-hoe, and this gives the good husbanding smallholder a real chance to get rid of weeds. So, although you may think that the growing of these row crops is very hard work, remember that it is hard work that benefits every other crop that you grow, and I would suggest that you should grow a crop in every 4 years of your arable rotation.

POTATOES

Where potatoes grow well, they can be, with wheat, one of the mainstays of your diet, and if you have enough of them you will never starve. They are our best source of storable vitamin C, but most of this is in the skins so don't peel them. You can even mash them without peeling them.

Seed potatoes

For practical purposes, and unless we are trying to produce a new variety of spud and therefore wish to propagate from true seed, potatoes are always grown from potatoes. In other words we simply plant the potatoes themselves. This is known as vegetative reproduction, and all the potatoes in the world from one variety are actually the same plant. They aren't just related to each other: they are each other.

We can keep our own "seed" therefore, from one year to the next, but there is a catch here. The potato is a plant from the High Andes, and grown at sea level in normal climates it is heir to various insect-borne virus diseases. After we have planted our potato "seed" (tubers) year after year for several "generations" there will be a build-up of virus infections and our potatoes will lose in vitality. Hence we must buy seed potatoes from people who grow it at high altitudes, or on wind-swept sea-islands, or in other places where the aphids do not live that spread these diseases. An altitude of over 800 feet (245 m) is enough in Britain for growing seed potatoes: in India most seed comes from Himachayal Pradesh, from altitudes of over 6,000 feet (1,830 m).

The cost of seed potatoes now is enormous and anybody who has land over 800 feet (245 m) would be well advised to use some of it for growing seed. In any case, many more of us ought to save the smallest of our tubers for "once-grown seed" or even "twice-grown seed". After we have carried on our own stock for 3 years, however, it will probably pay to import fresh seed from seed-growing areas rather than risk the spread of disease.

EARLY POTATOES

Potatoes that grow quickly and are eaten straight from the ground and not stored are called "early potatoes". To grow them you should chit, or sprout, them. They should be laid in shallow boxes, in weak light (at any rate not in total darkness, for that makes them put out weak gangly sprouts) at a temperature between 40°F (4°C) and 50°F (10°C). A cold greenhouse is generally all right.

It is an advantage to give them artificial light to prolong their "day" to 16 hours out of the 24. This keeps the "chits" green and strong and less likely to break off when you plant the spuds.

Planting early potatoes

Beware of planting too early, for potatoes are not frost-hardy, and if they appear above the ground before the last frost it will nip them off. On a small garden scale you can guard against this to some extent by covering them with straw, or muck, or compost, or cloches. If they do get frosted one night, go out early in the morning and water or hose the frost off with cold water. This will actually often save them.

Plant the earlies by digging trenches 9 inches (23 cm) deep and 2 feet (60 cm) apart, and putting any muck, compost or green manure that you might have in the trenches. Plant the spuds on top a foot (30 cm) apart, and bury them.

Actually, you will get the earliest early new potatoes if you just lay the seed on the ground and then ridge earth over them – about 5 inches (13 cm) deep is ideal. You couldn't do this with main crop because the spuds grow so large and numerous that they would burst out of the sides of the ridges and go green.

Spuds go green if the light falls on them for more than a day or two and then become poisonous: green potatoes should never be eaten nor fed to stock. The fruit of the potato plant, and the leaves, are highly poisonous, being rich in prussic acid.

Potatoes are usually planted by dropping them straight on muck, then earthing them up. But more people are now ploughing the muck in the previous autumn. Twenty tons of muck per acre is not too much. And always remember potatoes are potash-hungry.

Non-organic farmers use "artificial" potash: organic ones use compost, seaweed, or a thick layer of freshly cut comfrey leaves. Plant the spuds on top of the comfrey; as the leaves rot, the potato plants help themselves to the potash that they contain.

Harvesting early potatoes

You might get a modest 5 tons to the acre of early potatoes, increasing the longer you leave them in the ground: but if they make part of your income, the earlier you lift them the better.

You have the choice of either forking potatoes out, ploughing them out, or lifting them with a potato-lifter.

MAIN CROP POTATOES

The potato plant has a limited growing season, and when this stops it will stop growing. So it is best to let your main crop potatoes grow in the most favourable time of the year. This means the summer, and it is not advisable to plant them too early. Plant them in April and you won't go far wrong.

Planting main crop potatoes

If you use bought seed you can be a bit sparing with the amount: perhaps a ton an acre. Where you are using your own seed perhaps a ton and a half. Main crop potatoes should be planted 14 inches (35 cm) apart, in rows 28 inches (10 cm) apart if the seed averages 2½ ounces (70 g). If the seed is smaller, plant it closer: if it is bigger, plant it further apart. You will get the same yield either way. If you plant by hand you can control this very accurately. Ideally seed potatoes should go through a 1¼-inch (5-cm) riddle and be stopped by a 1½-inch (4-cm) one. Anything that goes through the latter is a "chat", and goes to the pigs.

If you have a ridging plough, horse-drawn or tractor drawn, put the land up into ridges. Put muck or compost between the ridges if you have it to put, although it would be better if you had spread it on the land the previous autumn. Plant your seed in the furrows by hand and, if you lack the skill or equipment to split the ridges – and by God it takes some skill – do not despair. Simply harrow or roll the ridges down flat. This will bury the potatoes. After about a fortnight take your ridging plough along again, and this time take it along what were, before you harrowed them flat, your ridges. So you make what were your ridges furrows, and what were your furrows ridges.

If you haven't got a ridging plough, use an ordinary plough. Plough a furrow, plough another one next to it, and drop your spuds in that. Plough another furrow, thus burying your spuds, and plough another furrow and drop spuds in that. In other words, plant every other furrow. Don't worry if your planted rows aren't exactly 28 inches (70 cm) apart. Potatoes know nothing of mathematics.

Care of main crop potatoes

Drag the ridger through once or twice more whenever the weeds are getting established. This not only kills the weeds, it earths up the spuds, gives them more earth to grow in, and stops them from getting exposed and going green. If you haven't got a ridging plough, and are skillful, you can ridge up with a single-furrow plough.

Go through the spuds with the hand-hoe at least once to kill the weeds in the rows, but after you have done this, and thus pulled down the ridges, ridge them up again. Up to 10 days after planting you can harrow your potato field to great advantage to kill weed seedlings, but after that take care because you might damage the delicate shoots of the spuds. It's all a matter of common sense. You want to suppress the weeds and spare the spuds. When the potatoes meet over the rows they will suppress the weeds for you and you can relax, but not completely, because of blight.

Blight

Blight is the disease that affected the Irish potato crop in 1846 and caused the deaths of two million people by killing their only source of food. You can be as organic as you like, but if it is a blight year you will still get blight. However, if you do get blight, don't despair; you will still get a crop. The virulent strain that killed the Irish has passed now. But your crop won't be such a big one.

You will know you have blight if you see dark green, water-soaked patches on the tips and margins of the leaves. If you see that, spray immediately, for although you cannot cure a blighted plant by spraying, you can at least prevent healthy plants being infected. These patches soon turn dark brown and get bigger, and then you can see white mould on them.

Within a fortnight, if you do nothing about it, your whole field will be blighted and the tops will simply die and turn slimey. Now the better you have earthed up your ridges the fewer tubers will be affected, for the blight does not travel down inside the plant to the potatoes but is washed down into them by rain.

Commercial farmers may spray the blighted tops with 14 gallons (64 litres) of BOV (brown oil of vitriol) to 85 gallons (386 litres) of water, or some such solution of concentrated sulphuric acid, or else some of the newer chemicals. This burns them off and the blight spores do not get down to your spuds. I cut my tops off with a sharp sickle (it has to be sharp, else you simply drag the spuds out of the ground when you cut) and burn them. Don't lift your spuds for at least a fortnight, preferably longer, after the tops have been removed. This is so that they don't come in contact with blighted soil.

But, of course, you will never get blight, for you will have sprayed your "haulms" (tops) with Bordeaux or Burgundy mixture, or one of the modern equivalents, before the first blight spore settled on your field, won't you? To make a Bordeaux mixture, dissolve 4 pounds (1.8 kg) copper sulphate in 35 gallons (160 litres) of water in a barrel, or in plastic dustbins. Then slowly stake 2 pounds (0.9 kg) freshly burned quicklime with water and make it into 5 gallons (23 litres) of "cream". Slowly pour the "cream" through a sieve into the copper sulphate solution. Make sure that all your copper has been precipitated by putting a polished knife blade in the liquid. If it comes out coated with a thin film of copper you must add more lime. Burgundy mixture is stronger and more drastic. It has 12½ pounds (5.7 kg) washing soda instead of lime. These mixtures must be used fresh, because they won't keep long. Spray with a fine spray very thoroughly. Soak the leaves above and below too. All the spray does is prevent spores getting in. Do it just before the haulms meet over the rows and again, perhaps a week later. An alternative is to dust with a copper lime dust when the dew is on. Blight attacks vary in date: it needs warm, moist, muggy weather, and in some countries agricultural departments give blight warnings over the radio.

If you do get muggy weather when your spuds are pretty well grown, spray for blight. You will get half as big a crop again as if you didn't. You can help prevent blight by not allowing "rogue" potatoes to grow: in other words, lift every single potato from your fields after harvest, for it is these rogues that are the repositories of blight. Pigs will do this job for you better than anything else in the world; they will enjoy doing it and they will fertilize the land too.

Harvesting main crop potatoes

Harvest as late as you can before the first frosts, but try to do it in fine weather. Fork them out, plough them out, spin them out or get them out with an elevator digger, but get them out. Leave them a day or half a day in the sun after lifting for the skins to set. Don't give them more than a day and a half or they will begin to green.

Storing main crop potatoes

Clamping (*see* p.199) is fine. Stacking in a dark shed, or a dark root cellar, is okay. The advantage of clamping is that if there is blight in your spuds, or any other spud disease, you don't get a build-up of the organisms as you would in a permanent building. Do what your neighbours do.

In intensely cold winters clamping may not be possible: no clamp will stop the frost, and potatoes cannot stand much frost or they will rot. But, if they are too warm they will sprout. If possible they want it just above freezing.

JERUSALEM ARTICHOKES

As a very few plants of these would give us more tubers than we would care to eat, they are seldom grown on a field scale except by a few wise people who grow them to be dug up and eaten by pigs. They are wonderful for this purpose, and provided the pigs have enough of them they will do very well on Jerusalem artichokes alone and a little skimmed milk or other concentrated food; ½ pound (0.2 kg) a day of "pig nuts", plus unrestricted rooting on artichokes should be enough for a dry sow.

Sowing Jerusalem artichokes

Drop them in plough furrows 12 inches (30 cm) from each other, and the furrows should be 3 feet (91 cm) apart. Put them in any time after Christmas, if the land is not frozen or too wet, and any time up to April. It doesn't much matter when, in fact. They won't grow until the warmer weather comes anyway and then they will grow like mad.

They will smother every imaginable weed and after the pigs have rooted them up, the land will be quite clean and well dunged too. But beware: pigs will never get the very last one, and next year you will have plenty of "rogues" or "volunteers". In fact, if you leave them, you will have as heavy a crop as you had the year before. I sometimes do, but never for more than 2 years running. They will grow in practically any soil and don't need special feeding, though they like potash. If you just put them into the ground and leave them, they will grow whatever you do.

BEANS

Beans are distinguished from peas and the other legumes because they have a square, hollow stem instead of a round solid one. There are hundreds of kinds and varieties of beans throughout the world. The various kinds of "runner" beans – scarlet runners, french beans, snap beans and so on – are seldom grown on a field scale, so I deal with them as garden vegetables (*see* pp.56–64).

FIELD BEANS

The crop which supplied most of the vegetable protein for the livestock of northern Europe and North America for centuries, and which should still do so today, is the field bean, tic bean, horse bean, or cattle bean: *Vicia vulgaris*. This is a most valuable crop, and it is only neglected nowadays because vast quantities of cheap protein have flooded into the temperate zones from the Third World. As the people of the latter decide to use their own protein, which they sadly need, temperate-zone farmers will have to discover the good old field bean again.

It produces very high yields of very valuable grain, and it enriches the soil in two ways: it is a legume, and therefore takes nitrogen from the air, and it forms deep tap roots which go well down and bring up nutrients, and themselves rot down afterwards to make magnificent humus. It is a most beneficial crop to grow, good for the land, and gives excellent yields of high-protein grain. As with other grains, there are two sorts of field bean: winter and spring; though farmers classify beans as "grain" or "corn".

Soil

Beans don't want nitrogen added, but they do benefit greatly by a good dressing of farmyard manure (muck) ploughed in as soon as possible after the previous crop has been harvested. In land already in good heart they will grow without this. They need lime though, as do all the legumes, and if your land needs it you must apply it. They need potash very badly and phosphates to a lesser degree. Non-organic farmers need to frequently apply 80 units of phosphorus and 60 units of potash, but we are more likely to rely on good farming and plenty of muck.

The seed bed need not be too fine, particularly for winter-sown beans. In fact, a coarse seed bed is better for the latter because the clods help to shelter the young plants from the wind throughout the winter. In very cold climates you cannot grow winter beans because very hard frost will kill them. You will get heavier crops from winter beans than from spring ones, less trouble from aphids, a common pest of the crop, but possibly more trouble from chocolate-spot, a nasty fungus disease.

Sowing beans

You can sow with a drill, if you have one, provided it is adapted to handle a seed as big as the bean, or you can drop the seed in by hand behind the plough. Plough shallowly – 4 inches (10 cm) – and drop the seed in every other furrow, allowing the next one you plough to cover them over. I find this method very satisfactory as the seed is deep enough to defy the birds. Jays will play havoc with newly sown beans, simply pulling the young plants out of the ground.

Care of crop

By all means hoe beans. Horse-hoe or tractor-hoe between the rows and try, if you can, to hand-hoe at least once in the rows. They are a crop that can easily suffer from weeds.

Harvesting beans

Wait until the leaves have fallen from the plants and the hilum, or point of attachment of the pods to the plant, has turned black. Cut and tie the crop with a binder if you have one. If not, cut with the sickle and tie into sheaves. You have to use string for this because bean straw is difficult to tie with. Stook and leave in the stook until the crop feels perfectly dry (maybe a week or two). Then stack and cover the stack immediately, either by thatching it or with a rick cover, or else stack it indoors. A bean stack without a cover will not keep the rain out, and if it gets wet inside the beans will be no good to you.

Threshing beans

Do not thresh until the beans have been in the stack at least 4 months. Many farmers like to leave the stack right through to the next winter before they thresh it, because beans are better for stock after they are a year old. Thresh just as you would for wheat, that is with a flail or in a threshing drum.

Feeding to animals

Grind or crack the beans before feeding. Mix them as the protein part of your ration. Horses, cattle, pigs, sheep, and poultry all benefit from beans. Cattle will pick over the straw and eat some of it. What they leave makes marvellous litter (bedding) and subsequently splendid manure.

SOYA BEANS

Soya beans, or soybeans, are grown in vast acreages in China and the United States in the warmer latitudes. They don't do much good in the climate of southern England and we must wait until a hardier variety is bred, if it can be, before we can grow them.

In climates where they can be grown, sow or drill soya beans well after the last possible frost about an inch (2.5 cm) deep, 10 inches (25 cm) apart in the rows and with the rows 3 feet (90 cm) apart. Keep them well hoed because they grow very slowly at first. Their maturing time can be anything from 3 to 5 months depending on the climate. If you need to, you can extend the season by covering them with glass or plastic. If you pick them young you can eat the pods whole. Otherwise shell them for the three or four beans inside. The same instructions apply to Lima beans.

Grass & Hay

By far the most important and the most widespread, crop grown in the world, is grass. Its ubiquitousness is amazing: it grows from the coldest tundra to the hottest tropic, from the wettest swampland to all but the driest desert. In areas where it only rains once every 5 or 10 years, grass will spring up within days of a rainstorm and an apparently barren and lifeless land will be green. This is why grass has been called "the forgiveness of nature". All the cereals are of course grasses: just grasses that have been bred for heavy seed yields. Sugar cane is grass and so is bamboo, but when a farmer speaks of grass he means the grass that grows on land and provides grazing for animals, and can be stored in the form of hay and silage.

Now the confusion here is that what the farmer calls "grass" is actually a mixture of all sorts of plants as well as grass. Clover is the most obvious and important one, and most "grassland" supports a mixture of grass and clover, and very often clover predominates over grass. Therefore whenever I write of "grass" I would ask the reader to know that I mean "grass and clover". Grass itself, too, is not just grass. There are many species of grass, and many varieties among the species, and it is of the utmost importance which species you grow.

Managing grassland

You can influence the make-up of the grass and clover species on your grassland by many means. For example, you can plough land up and re-seed with a chosen mixture of grass and clover seeds. But these will not permanently govern the pasture. According to how you manage that grassland, so some species will die out, others will flourish and what the farmer calls "volunteer" grasses — wild grasses from outside — will come in and colonize. But essentially the management of the grassland will decide what species will reign.

If you apply heavy dressings of nitrogen to grassland you will encourage the grass at the expense of the clover. If you go on doing this long enough you will eventually destroy the clover altogether. The reason for this is that normally the clover only survives because it has an unfair advantage over the grasses. This advantage is conferred by the fact that the clover has nodules containing nitrogen fixing bacteria on its roots and can thus fix its own nitrogen. The grasses cannot. So in a nitrogen-poor pasture, the clovers tend to predominate. Apply a lot of nitrogen and the grasses leap ahead and smother the clover. Alternatively, if you put a lot of phosphate on land you will encourage the clovers at the expense of the grass. Clover needs phosphate: grass nothing like so much. Clover-rich pasture is very good pasture and it also gives you free nitrogen.

If you constantly cut grassland for hay, year after year, and only graze the "aftermath" (what is left after you have cut the field for hay) you will encourage the coarse, large, vigorous grasses like perennial ryegrass and cocks foot, and you will ultimately suppress the finer grasses and clover, because these tall coarse grasses will shade them out. On the other hand, if you graze grassland fairly hard, you will encourage clover and the short tender grasses at the expense of the tall coarse ones. If your land is acid you will get grasses like bent, Yorkshire fog, mat grass, and wavy hair grass that are poor feeding value. Lime that land heavily and put phosphate on it and you will, with the help of mechanical methods too and perhaps a bit of re-seeding, get rid of these poor grasses and establish better ones. If land is wet and badly drained you will get tussock grasses, rushes, and sedges. Drain it and lime it, and you will get rid of these. Vigorous and drastic harrowing improves grass. It is good to do it every year.

Topping off

People frequently ask me how they can get rid of tiresome weeds such as docks and thistles that often ruin grassland. Well, as in so many things in life, the motto is "prevention is better than cure". There are two vital things you should remember about docks and thistle: first, don't let them seed, and second, they hate being regularly cut down. The practice known as "topping off" deals with both and is most definitely an important grassland maintenance routine. To top off a field you simply cut everything down to a height of about 9 inches (23 cm) frequently. I do this with a scythe and it makes a pleasant job on a breezy July day. Topping off is normally done in July when thistles and docks are fully grown but have not yet set viable seed. You can buy or borrow special mowers for topping off and over a period of time this will eliminate the large weeds and encourage tender shorter grasses.

Improving old pasture

You may inherit grass in the form of permanent pasture which has been pasture since time immemorial. Often this is extremely productive, and it would be a crime to plough it up. But you can often improve it by such means as liming, phosphating, adding other elements that happen to be short, by drastic harrowing (really ripping it to pieces with heavy spiked harrows), subsoiling, draining if necessary, heavy stocking and then complete resting, alternately grazing it and haying it for a season, and so on.

Now if you inherit a rough old piece of pasture, or pasture which because of bad management in the past is less than productive, the best thing to do may be to plough it up and re-seed it.

THE BALANCED PASTURE
Some of these plants are almost certain to be found in a good pasture.
Left to right, top row:
1 Meadow fescue (*Festuca pratensis*);
2 Perennial rye-grass (*Lolium perenne*);
3 Cocksfoot (*Dactylis glomerata*);
4 Timothy (*Phleum pratense*);
5 Italian rye-grass (*Lolium multiflorum*).
Bottom row:
6 Burnet (*Sanguisorba officinalis*); 7 Lucene/Alfalfa (*Medicago sativa*);
8 Red clover (*Trifolium pratense*); 9 Ribwort plantain (*Plantago lanceolata*).

One way to plough up is to "direct re-seed it", that is plough it and work it down to a fine seed bed, sprinkle grass and clover seed on it, harrow it, roll it, and let it get on with it. You can do this, according to the climate in your locality, in spring, summer, or autumn. What you need is cool, moist weather for the seed to germinate and the plants to get established. Or you can plough up, sow a "nurse crop" and sprinkle your grass seed in with it. The nurse crop can be any kind of corn or, in some cases, rape. When you harvest the corn you will be left with a good strong plant of grass and clover.

Seed mixture

As for what "seed mixture" to use when establishing either a temporary ley, which is grassland laid down for only a year or so, or permanent pasture, go to your neighbours and find what they use. Be sure to get as varied a mixture as you can, and also include, no matter what your neighbours say about this, some deep-rooting herbs. Ribgrass, plantain, chicory, yarrow, alfalfa, and burnet are ideal deep-rooting herbs for your seed mixture. You can rely on them to bring fertility up from down below, to feed your stock in droughts when the shallower-rooting grasses and clovers don't grow at all, and to provide stock with the minerals and vitality they need. On deep, lightish land, alfalfa by itself, or else mixed with grasses and clovers, is splendid for it sends its roots deep down below. What if it does die out after a few years? It has done its good by bringing nutriments up from the subsoil and by opening and aerating the soil with its deep-searching roots.

HAY

Grass grows enormously vigorously in the first months of the summer, goes to seed if you don't eat it or cut it, then dies down and becomes pretty much useless.

In the winter, in northern climates, grass hardly grows at all. In more temperate climates though, it may grow pretty well for 10 months of the year provided it is not allowed to go to seed.

Now there are two ways of dealing with this vigorous summer spurt: you can crowd stock on the grass to eat it right down, or you can cut the grass and conserve it. The way to conserve it is to turn it into hay or into silage. You can then feed it to stock in the winter.

Hay is a more practical proposition for the average self-supporter. You should get 2 tons of good hay off an acre of good grass. The younger you cut grass for hay, the less you will have of it, but the better it will be. Personally I generally cut hay before my neighbours, have less, feed less, but the cattle do better on it. In France and in places where a very labour-intensive but highly productive peasant agriculture prevails, grass is cut very young, made into hay very quickly, and then the grass is cut again, maybe three or four times, during the season. The resulting hay is superb: better than any silage, but the labour requirement is high.

Haymaking

To make hay: cut the grass before, or just after, it reaches the flowering stage. If it has begun to go to seed you will get inferior hay. Then pull it about. Fluff it up and keep turning it. Let the wind get through it and the sun get at it. If you are very lucky it may be dry enough to bale, or, if you are making it loose, to cock, in 3 days. Then bale it or cock it and thank God. The chances are, in any uncertain climate, that you will get rain on it, which is always bad for it, and then you have the job of turning it about again to get it dry again. In bad years you may have to go on doing this for weeks, and your hay will be practically useless for feed when you finally get it in.

HAYCOCKS AND TRIPODS
A haycock (*far right*) is a pile of hay, solid but loose enough to allow air to circulate. Another method of drying, which is particularly useful in wet climates, is the tripod. Take three light poles, say 6 feet (1.8 m) high each. Hold them together and tie a loop of string or cord very loosely round them near one end. Stand them up and make a tripod of them. Tie two or three strands of string or wire right round the tripod to hold the hay. Pile hay up round the sides, starting from a small circular base, keeping the outside walls as vertical as you can, and then bury the tripod completely with grass, rounding off the top nicely. Use bent tin to make air vents at ground level. There must be at least one on the windward side.

Haymaking hints

Even for small pieces of grass in the garden there will be opportunities to make useful hay. Each year we make a good hayshed full from our orchard and this is done totally by hand with scythe and fork. We learned a very useful trick from friends in Austria and that is how to carry a huge bundle of dry hay with using a doubled up length of rope. You simply lay the rope out on the ground and pile your hay on top with a fork. When you have sufficient for a strong man to carry, you take the two loose ends of the rope over the top and put them through the loop which is at the other side of the heap. When you pull on the free ends they tighten up the whole loop around your bundle and you sling it over your shoulder and march promptly to the hayshed. I will warn you now that haymaking like this is a scratchy and ticklish operation as all the bugs in the world seem to love to rest in the new hay.

One of the pleasures of orchard haymaking is watching the chickens follow the scythe. The birds very quickly realise that under the newly cut swathe they will find a wonderfully varied selection of insect life. And by scything the orchard regularly you will, of course, keep all nasty weeds like nettles and docks well under control. You may even get two cuts of hay in a good year.

Cocking

A cock is a pointed-topped dome of hay that you build with a fork. It sheds most of the rain and allows a certain amount of in-cock drying, but if the grass is too green, or wet from rain, you may have to pull the cocks open again and spread the hay about to dry. Then, if rain threatens, throw it up in the cock again. If you are worried about too much moisture in hay that you have cocked, thrust your hand deep inside. If the hay in there is hot, or feels wet and clammy, you must spread that hay about and dry it again. You can only stack it when it is dry enough – that is when it is no longer bright green and feels completely dry.

Baling

A bale is a compact block of hay which has been rammed tight and tied with string by a machine called a baler. You mustn't bale hay until you are sure it is dry enough. If you bale when it isn't dry it will heat in the bale and the hay will be spoiled. Once the hay is baled there is nothing you can do with it. Just get those bales inside as soon as you can: they will shed a certain amount of light rain. But once rain gets right into them you have had it – spoiled hay.

A word here about the modern trend towards huge round bales. These are fit for brutish giants and indeed can only be handled effectively by noisy diesel tractors with hydraulic lifting gear. We have found it is often difficult to buy in anything other than the big round bales these days because the old square balers are almost extinct. What you can do if this is the case in your area is to use your trailer sensibly to carry one bale at a time. Make sure of course it is well tied down in the trailer. When you get this home

you have to have a set-up where you (and a couple of strong friends) can roll it right into your hayshed. Once there you can pull out the miles of string used in its construction and feed it in the normal way.

Machines for hay

There is a great armoury of machines, both tractor and animal-drawn, for dealing with hay. There are machines for tedding (fluffing-up), windrowing (gathering together into long loose rows), turning (turning the windows over), and raking. But all you need if you don't have too much hay, or have enough labour, are some wooden handled rakes and some pitch forks. You can make the best hay in the world with just these. You can ted the hay with the pitch forks, rake it into windrows, then rake three or four windrows together, cock these with the forks, load the cocks on to a cart with the forks, and ultimately stack it. Haymaking in climates that have wet summers is always a gamble; a triumph if you win, and something to put up with if you lose.

Tripoding

In wet climates the tripod is a useful means of drying wet hay. Grass that has only had maybe a couple of days of air drying can be tripoded, even if it looks quite green, for the air continues to get through it in the tripod. I have seen hay left on the tripods in bad weather for a month; however, this does not necessarily make very good hay after that sort of treatment.

SILAGE

If you take grass, clover, lucerne, crushed green maize, kale, or many other green things, and press them down tight in a heap from which you exclude the air, it will not go bad, as you might think, but will ferment into a food very nourishing to animals. In fact good silage is as good as the very best of hay. And of course, because you can cut your green crop at any stage of growth, you can cut it young when its protein content is highest, so it makes good feed. Grass you can cut again and again during the season, instead of waiting until it is fully grown as most people do when they make hay.

Making silage

Silage making is not an easy option for any smallholder. You need quite a bit of fancy equipment to do the job well, and it is almost impossible to do it in small quantities. Farmers can use a special forage harvester that cuts and mashes up the grass, then blows it into another trailer pulled by another tractor. Alternatively, the farmers can use a machine to wrap the cut grass up in a great cocoon of plastic, which is then sealed to keep out the air. You can do this yourself with cut grass, using fertiliser bags and sealing them with sticky tape. It makes marvellous silage but is hardly an effective way to keep cows through the winter!

"Collecting wild food is part of our pleasure. If we go for a country walk we keep our eyes open for fungi. ...Wild mushrooms are becoming a very scarce bird in this country. This is due to the use of chemicals on the land, the lack of horses, and probably most of all to the periodic ploughing up of pastures and cropping and reseeding. The beneficent mycelium does not have time to establish itself. We are going to try this March to enspore our own piece of permanent pasture. The mushrooms grown on compost in sheds or dark cellars taste as much like real mushrooms as margarine tastes like butter.

Giant puff ball is the easiest food in the world to prepare, for it cuts cleanly into lovely firm white slices which can be fried in butter; and then it is delicious. Parasol mushroom is another favourite of ours. It tastes similar to field mushroom, but is stronger flavoured and I think better. Champignon, or 'fairy ring' toadstools are fine if you can find enough of them to make it worth cooking them for they go down as cooks say 'to nothing'."

JOHN SEYMOUR FAT OF THE LAND 1976

CHAPTER FIVE

FOOD

FROM THE

WILD

Game

Man should be a husbandman, not a bandit. We have no right to slaughter animals just for fun or to assuage our blood-lust. Nor have we the right to endanger the stock of any species of animal. Yet we have a part to play in maintaining the balance of nature (and if we fail to play it, nature will very rightly shrug its shoulders and shake us off). And we can also supplement our diet with good food (wild meat is a far better source of protein than the meat of domesticated animals) as well as protect our crops. True husbandmen will accept their responsibility in this matter. They will also accept responsibility in the way they hunt game. It is unforgivable to wound an animal instead of killing it outright, so don't go shooting until you are a good shot. And never take a shot unless you are absolutely certain of a kill. (Of course, it goes without saying that you must abide by the often complicated rules and regulations regarding firearms ownership and use, as well as having an appropriate game license.) You can't just shoot game willy-nilly; even if it runs on to your land, you need to know the laws protecting native birds – is it "in season" for instance – so get all the details first from your local authority.

Guns

A shotgun is a smooth-bore tube which fires a charge of shot. Lead shot should never be used: it is a pollutant. Always buy cartridges using shot of other metals.

The sizes of shot are numbered according to the number of individual pellets it takes to make up an ounce (28 g): thus no. 1 shot is very big (it is used, wrongfully in my opinion, for roe deer), no. 3 is about right for wild geese, no. 5 for duck, and no. 6 for pheasants, rabbits and small game, while nos. 8 and 9 are used for snipe or woodcock.

Shotguns are graded according to the size of their bore (size of their barrel). The bore depends on the number of lead balls in a pound (0.5 kg) that exactly fit a barrel. Thus the barrel of a 12-bore takes 12 balls to fit it, making up a pound. The 12-bore is by far the commonest size all over the world now and is a good all-purpose gun. Whatever the gun, it is all a matter of what people think is "sporting".

Cartridges are loaded with nitro-powder, which is smokeless and reliable, but some people load their own brass-cased cartridges (with an apparatus bought from a gun shop) and thus save a lot of money. Modern cartridges are fired by a percussion cap (small brass cap containing fulminite of mercury) pressed into the base of the cartridge.

A rifle has a series of spiral grooves cut down inside the barrel, and a single bullet of soft metal, or coated with soft metal. When the bullet is propelled out of the chamber of the gun into the barrel, the metal around it conforms to the shape of the spirals and this gives the bullet a spin. Without this spin the bullet will not travel accurately but will invariably veer off to one side or another. The "two-two" (.22-in bore) is common all over the world and its ammunition is cheap, light, and small. The rifle is very

effective up to several hundred yards. I have shot kudu, reed-buck, and various other big game with a .22, and have never once merely wounded one of them or failed to kill it; but then I would never use a .22 for such a purpose unless I was very close and quite sure of my target. For larger game, however, larger rifles are really much better. Seven millimetre is a very common size (the 7 mm Mauser has always been, in my opinion, the best small sporting rifle in the world); the 6.5 mm Manlicher is as good ballistically but has an inferior magazine. The 9 mm is fine for thick-skinned game. I used a .404 inch bolt-action rifle in Africa; it gave me a certain sense of security when being charged by a buffalo but, by God, it kicked!

Rabbits as game

When they multiply, rabbits become an all-pervading menace, so that no forester can plant a tree without enormous expense on extermination and rabbit-proof fencing, and 25 percent of the crops in many areas might go down the rabbits' throats. In any case they are also very good food. The most humane way of taking rabbits (along with hares, small deer or buck, or vermin such as foxes, crows and other marauders and birds) is to shoot them either with a shotgun or a .22 rifle. Very early in the morning is the best time to "walk up" rabbits with a shotgun, or to lie in wait for them with a .22.

Long-netting is perhaps the second best way of killing rabbits. It is humane, quiet, cheap (no expenditure on cartridges), and, properly done, can be very efficient. You set up the net between the rabbits' grazing ground and their burrows in the day time at your leisure. Keep it folded up out of the way where the rabbits can get underneath it, with a release string to let it drop when you pull it. You then go along at night, when the rabbits are out in the field grazing, and the net is between the grazing ground and the rabbit warren. You pull the string and down comes the net; your accomplice gets behind the rabbits, makes a noise; they all run off into the net and get tangled, whereupon you kill them. I have caught a dozen rabbits with one setting of a long net.

Ferreting

A good way of controlling rabbits is with ferrets. Furthermore, it is great fun (although my earlier comments about regarding the rules on capture and killing of game apply here, too). Keep ferrets in a hutch, keep them clean, feed them sparingly on fresh meat, and handle them often to keep them tame. Use deliberate steady movements when handling them, as they will sometimes bite your hand, thinking that you are giving them a piece of meat. Only a reliable ferret can be worked loose; an unreliable one may kill a rabbit down the hole and "lay up" with it. A line ferret has a collar round her neck with a long line on it. The disadvantage of this is that the line may get snagged around a tree root far down a burrow, in which case you will have a lot of digging to do. We used to work them loose but keep

SKINNING A RABBIT

Once you have killed your rabbit you will need to prepare it for the pot. Before you skin it you must "paunch" it (remove the guts). Skinning is not difficult; you will find the skin comes away from the flesh very easily. If the thought of preparing rabbit appals you, just brace yourself and think hard of rabbit pie.

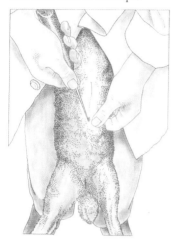

1 Hold the rabbit between your knees by its head, so that its tail hangs free and its belly faces upwards. Cut a hole in its belly.

2 Pull the skin apart at the cut, and insert two fingers into the hole.

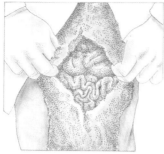

3 Prise open the belly to expose the guts and remove them. "Paunching" is now complete.

4 Cut off all four paws of the rabbit with a sharp knife.

5 Separate the skin and fur from the flesh of the rabbit, at the belly.

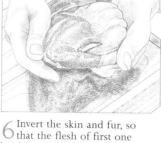

6 Invert the skin and fur, so that the flesh of first one hind leg, then the other, is exposed.

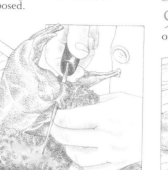

7 The hind quarters are now free from all skin and fur and now is the moment to cut off the tail.

8 Hold the hind legs in one hand and pull the skin down to the front legs.

9 Expose the front legs and cut away the last tendon joined on to them.

10 Pull the skin over the rabbit's neck and cut off its head.

11 Split the hind legs from the belly and cut out the anal passage. Then place your knife in the rabbit's breast, and remove the "lights" and heart. You must remove the gall from the liver.

one "liner" in reserve. If a ferret did lay up we would send the liner down and then dig along the line and thus find the errant ferret. Probably the best thing is just to use loose ferrets and trust to luck. The best way of recapturing a ferret if it lays up is probably a box trap, with a dead rabbit inside it, and a trap-door, so that when the ferret goes in the door shuts behind it. Rabbits "bolted" by the ferret are best caught in purse nets. These are simply small bags of nets staked around the entrances to the holes.

Assuming you can legally shoot them, game birds such as pheasants, partridge, pigeon, and wild duck (I stopped shooting wild geese when I discovered that they mate for life) make fine delicacies. Most game should be hung up – by the neck, not the feet – in a cool airy larder for some time before eating. The reason for this is so the guts do not press against the meat of the breast. Do not gut the birds then. In the winter in a northern climate, it is all right to hang a pheasant, or a wild duck, for as much as 10 days. After that, pluck them and gut them. Game birds hung up with their feathers on, look pretty, but, if you just want them to eat, it is quite a good plan to pluck them when you shoot them there and then (if you are on your own land), because the feathers come out easily when the birds are still warm.

Fish & Sea Foods

The self-supporter ought to make the most of any opportunity he gets, and fish should figure at the forefront of his healthy, varied, natural diet. The "sport" of angling, in my opinion, is a complete waste of time. Catching fish, weighing them and throwing them back does no one any good. Freshwater fish make wonderful food. People should be encouraged to take or farm freshwater fish for food. The methods I describe are not necessarily legal in every country, but all I can say is that they ought to be.

FRESHWATER FISH

Trout Plenty of people catch trout with their bare hands by "tickling". You lean over a bank and very gently introduce your hand into a cavity underneath it, waggling your fingers in a tickling movement as you do so. When you feel a fish with the tips of your fingers, you just gently tickle its belly for a minute; then you grab it and fling it on the bank. "Groping" is another method: wade along in a shallow stream, walking upstream, and grope with your hand under rocks, grabbing any fish you find there. You might get bitten by an eel while you are doing it, though.

Pike "Snaring" was a method much used in East Anglia when I was a boy. You have a wire snare hanging from a stick, and when you see a large pike hanging in the water, as pike do, you very carefully insert the snare in front of him and let it work slowly back to his point of balance. When you think it is there, you haul him out. If the wire does just touch him as you work it over him he thinks it is a stick because it is going downstream.

Salmon You can "gaff" salmon. To gaff a salmon first locate him; you will find him resting in a pool or under an overhanging tree. You then take the head of a gaff, which can be made from a big cod hook, out of your pocket. Cut a light stick from a bush, and lash the gaff on to the stick. You have a lanyard (light cord) running from the eye of the gaff to your wrist where it is tied round. You drag the gaff into the fish and then just let go of the stick. The line unwinds from the stick which falls away, and you haul the fish in with the line. If you try to haul him in with the stick he may well pull you in.

Eel Sensible people, among whom I include the Dutch and the Danes, account the eel the best fish there is, and indeed if you have ever eaten well-smoked *gerookte paling* in the Netherlands you must agree. You can take eels in "grigs" or "eel-hives": these are conical or square baskets made of osiers, wire-netting or small mesh fish netting on a frame, with an admission funnel like a very small lobster pot. Bait this with fresh fish or meat; whatever people say, eels don't like bad fish. Fresh meat or fresh chicken guts in a gunny sack, with the neck of the sack tied tight, and some stones inside it to sink it, will catch eels.

"Babbing" for eels is a good way of catching them. Get a bunch of worms as big as your fist, thread wool yarn through them, tie them in a bunch, and lower them into shallow water on the end of a string, which is tied to a stick. After a while, haul the "bab" gently out and eels may be found hanging to it, their teeth entangled in the yarn. Pull the bunch over your boat, or over the bank, and give it a shake. I have caught a hundredweight of eels this way in an afternoon. I have not discussed conventional angling with rod and line, for this is done more as a sport than for food production, although good anglers can sometimes get a lot of food this way too. But the fresh waters should be farmed for fish just as the land is farmed for crops and animals.

SEA FISH

From the viewpoint of the person who wishes to catch sea water fish, they fall into two groups; pelagic and demersal. The former swim freely about the seas, independent of the bottom. The latter are confined to the bottom of the seabed. Obviously the means of taking them are quite different.

Catching pelagic fish

Hooks and feathers Sometimes you can catch hundredweights in a few hours when hooking for pelagic fish. Mackerel in particular can be caught productively in this way, though traditionally they were caught by a last. This was a piece of skin, about 2 inches (5 cm) long, cut from near the tail of a mackerel you had already caught. The method was to move along at about 2 knots, dragging the last on a hook astern. Then somebody discovered the "feathers". With this new invention you have perhaps a dozen hooks, on snoods (short branch lines), tied to a line with a weight on it. Each hook has a white or coloured feather whipped to the shank (though almost anything will do: bits of white plastic or shiny tin). You lower the tackle from a stationary boat, find the depth at which the mackerel are biting, and plunge the feathers up and down with a motion of your arm.

Drift net Herring cannot be caught on a hook. Unlike mackerel, they don't hunt other fish, but live on small fauna in the plankton, so their mouths are too small for hooks. They are traditionally caught with a drift net. This fine net hangs down vertically in the water suspended from a float-line, which is a line with corks or plastic floats all along it. You can let it drop to any depth you like by hanging it on longer or shorter pendants. The whole net must have positive buoyancy. It will catch more at night, and a fine night is the best time for catching herring. "Shoot" (put into the water) the net from a boat and hang on to one end of it for an hour or two, letting both boat and net drift with the tide. Cast off occasionally and row along the net, and just lift a few yards of it to see if there are any fish. If a shoal hits the net, haul it in. Don't bother to try getting the fish out of the net into the boat, just pay the net down into the stern of the boat and go back to port. Then unload the net and shake the fish out into a piece of canvas, laid on the beach. Drift nets will take any pelagic fish if the mesh is the right size: mackerel, sprats, pilchards, salmon, sea-trout, and many other fish are all taken in drift nets.

Catching demersal fish

Trawl net Fish on the seabed can be taken by a trawl net. There are basically two sorts: the beam trawl and the otter trawl. The beam trawl is a net bag with its mouth held open by a beam which is supported on two "heads" which are like the runners of a sled. The otter trawl has two "otter boards" holding its mouth open: they swim through the water like kites, holding the trawl mouth open as they do so. Probably for the self-supporter with a small boat the beam is best, although many would dispute this.

You need considerable power to haul a trawl, particularly an otter trawl, which takes a certain minimum speed to keep the otter boards working. A small beam trawl can be hauled by sail alone, especially if you work downtide. Often the tide is enough to pull the net. Always trawl downtide anyway; the fish face uptide. A small-meshed beam trawl also takes shrimp.

Tangle net This has very strong, thin, man-made fibre. It is a very tight, large-meshed net which sinks to the bottom of the sea, where some of its width is supported by a submerged cork line and the rest just lies in a heap on the bottom. Anything that walks or swims near the bottom is taken by it, getting inextricably tangled, and then you have the lovely job of clearing the net! That is the disadvantage, for the net is hell to clear and always gets badly torn, so that you have to repair it. But it catches a lot of fish and will take crabs and lobster along with everything else.

Shore seine net This is another of the long-wall nets. You keep one end of it on the beach while the other end is taken out in a boat which goes round in a half-circle, coming back to the beach again. Both ends of the net are then pulled in and any fish that were caught are dragged up on to the beach.

Long-line You can shoot this from a boat. The line can have any number of hooks on it, each one in a snood, and each one baited. Coil the line down carefully in a basket, or tin bath, or plastic tub. As you coil the line down, lay each hook in order over the side of the receptacle, next to the following hook in the line. The snoods are long enough to allow this. Bait each hook. Then go uptide from where you wish to shoot the line, throw out one anchor, and let the line whip overboard as the tide drives it inexorably downtide. The baited hooks should fly over one by one. Have a short piece of stick in your hand to help them do this if they are reluctant.

If you get in a "fangle" let the whole line go over; don't try to unfangle it or you get a hook in your hand as sure as fate. If you work carefully and keep cool it should go over clear. When you get to the other end, throw over the other anchor, and the buoy and that is that. Come back the next day and haul against the tide, using oars or engine or the wind to carry the boat along at the right speed.

The size of hooks will depend entirely on the kind of fish you hope to catch.

HOOKS, LINES, AND SINKERS

1 Parlour pot, a type of lobster pot **2** Bab, a tied bunch of worms for eels **3** Feathers, weighted line with snoods **4** Feather (detail) **5** Barbless hook, for removing hook from fish **6** Last, a piece of shiny skin **7** Lug-worm bait **8** Treble hook for pike **9** Hook for cod **10** Hook for dabs.

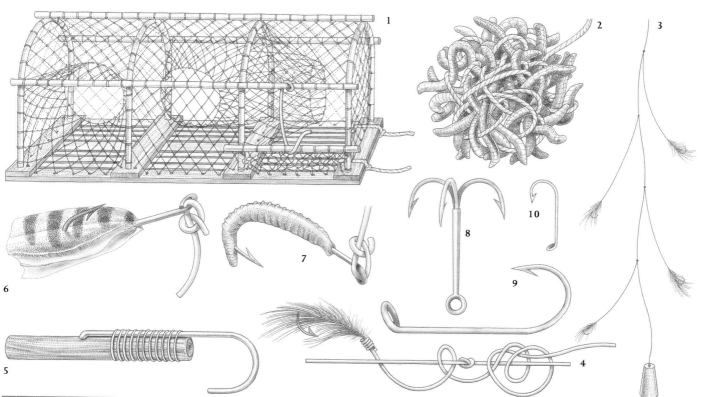

Hooks

Size 6 or 8 hooks are fine for dabs, plaice, and so on, while size 4/0 to 8/0 may be needed for conger-eel, or large cod. Conger are very apt to be caught on long lines: I once helped catch half a ton in a night. Mind you, there were 1,200 hooks. For such large fish it is good to have a swivel on each snood, so the hook can turn as the conger turns.

Getting a hook out of the throat of a large fish is very easy when you know how to do it. You need a small barbless hook securely attached to a handle. Get the hand-hook, as I shall call it, in the bite of the fish-hook, and yank the fish-hook out with the hand-hook, holding the snood firmly with your other hand so as to keep the two hooks engaged. You should carry a "priest" (small wooden club, traditionally of boxwood). It is so called because it administers the last rites, and is more humane than letting fish drown to death in air.

Hand-line Only once in a while is hand-lining for bottom fish productive. Those lines of hopefuls who lean endlessly on the rails of piers spend far more on bait and tackle than they take home in fish. Don't bother to hand-line unless you know there are fish there (ask a local). You can sometimes find a good mark for whiting, or codling, that makes hand-lining more than just a way of passing the time.

THE SEASHORE

You do not need a boat to benefit from the riches of the sea. A visit to the seashore provides ample opportunity to accumulate edible sea creatures of various sorts for a snack, if not a full meal. Indeed, someone without a boat may catch fish quite effectively with a beach long-line.

Go down to the bottom of a beach at low tide and lay a long-line (as described earlier) along the sand near the sea. As the tide comes back, demersal fish will follow the water, intent on helping themselves to such small beach animals as emerge from the sand to go about their business when it is covered. You will catch some fish, perhaps not many; you will be lucky to get a fish every 20 hooks, but after all one fish is a meal and better than no fish at all. If you really want to practise this fishery effectively, put down a lot of hooks; 100 is not too many.

The long-line must be anchored by a heavy weight at each end and should have a pennant (branch line) on it with a buoy on top so as to make it easy to recover. Remember here that the tide is not the same every day.

PLANTS AND CREATURES OF THE SEASHORE
1 Razor **2** Common whelk **3** Common limpet **4** Edible cockle
5 Common oyster **6** Common mussel **7** Common winkle **8** Edible crab
9 Lobster **10** Brown shrimp **11** Purple laver **12** Sea lettuce **13** Dulse

About the time of every full moon and about the time of every new moon there is a spring tide, when the water goes both higher and lower than in neap tides, which occur at half-moon periods. Even these spring tides vary: some come up much higher and go down much lower than others. So you may lay your line out at the bottom of the beach on one tide only to find that the tide does not go out far enough to uncover it the next day. If you have a buoy on a pennant, you will be able to wade out and recover your line.

As for bait, nearly everywhere on sandy beaches you will find the reliable lug-worm. This can be dug out at low water with a spade or fork. There is a trick to this: lug-worms throw up worm-shaped casts of sand. Do not dig under this. Instead, look for a small hole which should be a foot away from each cast. This is the worm's blow-hole. Dig there, dig fast, throw the sand out quickly, and you will get your lug-worm.

Other forms of bait are limpets, mussels, slices of herring or mackerel, whelks, and hermit crab tails. Limpets must be knocked off the rocks with a hammer by surprise; once you have warned them of your intentions they cling, well, like limpets, and you can only get them off by smashing them to pieces. Mussels are somewhat soft and some people tie them on to a hook with a piece of cotton.

Shellfish

Mussels Pick these as low on the rocks as you can get at low tide, preferably below the lowest tide mark, although this is not always possible. They must be alive: if they are firmly closed it is a sign that they are. They must not be taken from water in which there can possibly be any pollution from sewage, as they are natural filters and will filter any bacteria out of the water and keep it in themselves. Advice commonly given is cook them only long enough to make them open their shells is extremely dangerous. All mussels should be boiled or steamed for at least 20 minutes; otherwise, food poisoning can take place.
Cockles Rake these out of the sand with a steel rake. You soon get good at spotting where they are under the sand, which somehow looks different: it is often greyer than the surrounding area. Then rake them into small hand-nets and wash the sand out of them in shallow water. It's much easier to harvest cockles on the sand flats when shallow water still lingers over the sand. Boil or steam them for 20 minutes.
Razor fish These betray themselves by squirting water out of the holes in the sand. They do this when you tread on the sand nearby. They live very low down on the beach, right down where the sand is only uncovered at low spring tides. If you walk backwards over the sand you will see the spurts of water after you have passed. The best way to get them is with a razor spear. This is a pointed iron rod with small barbs near the point. You push it gently down the hole and the razor fish closes on it and is pulled out.
Oysters You should only eat oysters raw if you are sure they are unpolluted. But they are delicious cooked and

much safer to eat then. To open an oyster, hold it in a cloth in your left hand and plunge a short stiff blade into the hinged end. You can cheat by popping them in a hot oven (400°F or 200°C) for not more than 4 minutes, but if you intend to eat them raw this is desecration.
Winkles These can be picked up in small rock pools at low tide. Boil them for a quarter of an hour in water. Pick them out with a pin, sprinkle them with vinegar, and eat with bread and butter. However, I think they are pretty dull.
Whelks Whelks are a deep water shellfish and are caught in pots like lobster pots but smaller. Salt herring or mackerel make a good bait. Boil them for half an hour, or steam them. They taste rather like wet leather.
Lobsters and crabs These are normally caught in pots, which are cages with funnels into them so that the shellfish can get in but not out. The pots can be made of willow, steel mesh, or wire-netting. A more sophisticated pot is the "parlour pot" which is longer, with an entrance hall at each end; and the net funnels into the "parlour" which is in the middle. If you have to leave pots out for long because of bad weather the parlour pot is good, because the lobsters, on finding themselves confined in the entrance halls, try to get out, get into the parlour, and wait there: the bait is not eaten and attracts more lobsters.

When trawling you may catch hermit crabs. Fishermen, if they don't want the tails for bait, normally throw them overboard. This is nonsense – the tails are delicious boiled. Spider crabs are also tasty.

Seaweed

Many seaweeds are edible, but there are two plants that are excellent to eat: laver weed (*Porphyra umbilicalis*) and samphire (*Salicornia europaea*). Laver weed has thin, translucent purple fronds and grows on rocks on the beach. To cook it you soak it for a few hours in fresh water, dry it in a slow oven, and powder it in a mortar. Then boil it for 4 hours, changing the water. Drain it and dry it and you have made laver bread, the stuff that the South Wales coal miners used to think was good for their chests. Eat it with bacon for breakfast. You can just wash laver weed well and boil it for several hours in water in a double saucepan. Beat this up with lemon or orange and a little butter or oil, and it makes a good sauce for mutton.

Other more delicate seaweeds, such as sea lettuce (*Ulva lactuca*) and dulse (*Rhodymenia palmata*), can be treated in the same way as laver weed. The other really valuable seaweed, samphire, is not really a seaweed. It looks like a miniature cactus below high-tide mark, and can be eaten on the spot, raw, as is (provided the estuary is not polluted). Boil and serve like asparagus with butter; but if you eat it like this you must draw the flesh off between your teeth, leaving the rough fibres behind. Samphire also makes a most magnificent pickle: fill a jar with it, add peppercorns and grated horseradish; then pour in a boiling mixture of dry cider and vinegar in equal quantities, or else just vinegar.

Plants, Nuts, & Berries

There are innumerable wild foods that you can find growing in the woods and fields, and hedgerows, but my advice would always be: find out what the local people consider good to eat in your locality and eat that.

With fungi you really must know which ones are safe, and for this you need either a knowledgeable friend or the advice of local people. Besides the common field mushroom, a few fungi that are delicious to eat and easy to identify are: shaggy ink cap (try it boiled lightly in milk), giant puff ball, parasol, shaggy parasol, horse mushroom, cep, boletus (several species), morel, and chanterelle.

An enormous number of "weeds" can be eaten, so can all kinds of seeds, and of course a great many wild fruits, berries, nuts, and fungi (*see pp.180–181*). More "weeds" can be eaten than are a positive pleasure to eat, but a few that are excellent are: nettles, fat hen, and good king Henry. Treat all three exactly as you would spinach: pick them in the spring when they are young and tender, cover with a lid, and boil.

Some other wild substitutes for green vegetables are: shepherd's purse, yarrow, ground elder and lungwort. common mallow can be pureed and turned into a good soup; chickweed can be cooked and eaten like spinach or used in salads; Jack-by-the-hedge is a mild substitute for garlic. You will find many other varieties in your locality, and don't forget the dandelion: it is delicious raw in a salad. But use sparingly.

Of the edible nuts, walnut is the king in temperate climates. After picking, leave the nuts for some weeks until the husks come off easily, then dry them well. You can pick hazelnuts green when they are nice to eat but won't keep, or you can pick them ripe and bury them, shells and all, in dry salt. Sweet chestnuts are magnificent food. Pick them when they ripen in autumn. Shell the prickly covers off them, and store in a dry place. Of course the finest way to eat them is to roast them in the embers of a fire, but prick them first to stop them exploding. Raw, they are bitter. Puréed, they taste marvellous, and turkey is unthinkable without chestnut stuffing. Beechnuts are tasty but fiddly to eat; better to crush them in a mill, put the pulp in cloth bags and press it. It yields a fine oil. Ash keys make quite good pickle; boil them well and pickle in vinegar.

Of the wild fruits, the elderberry is perhaps the most versatile. The berries can be used for cooking in a number of ways. Mixed with any other fruit they improve the flavour; boiled in spiced vinegar they make an excellent relish or sauce which will keep well, if properly bottled when hot. The berries also make an excellent wine, as do the flowers, which add flavour to cooked gooseberries and also gooseberry jam. If you find blueberries or bilberries in the wild, do not ignore them: they make a wonderful pie. And if you find cranberries, you can preserve them but their flavour is nowhere better captured than in a fresh cranberry sauce. Mulberries and rowanberries make very good jam. And do not forget juniper berries, which can impart an agreeably tart flavour to all savoury dishes.

BIRCH SAP
Betula pendula

There are few people who realise that all around them are birch trees which can be the source of the most perfect country wine. Birch wine is made from the sap gathered in spring. Simply choose a good strong tree where the trunk is at least 12 inches (30 cm) in diameter. Using a battery drill or brace and bit, drill a hole about 1 inch (2.5 cm) in diameter into the tree at a point where you can conveniently hang a bucket. The sap will flow into your bucket.

BLACKBERRY
Rubus fruticosus

For many households the annual routine of blackberry gathering is an enjoyable signal that the autumn is near and the summer is almost gone. A walking stick with a good crook is useful for pulling down the higher berries and even the smallest child can join in the fun. There are a thousand and one uses for the blackberry, from excellent wine, a superb accompaniment to apples in pie, and straightforward jam.

CRAB APPLE
Malus sylvestris

Crab apples vie with sloes for the title of the bitterest of fruits. Some of this bitterness is caused by tannin, in which they are rich. One of their great uses is as an additive to tannin-poor wines. Mead, for example, ferments much better with a little crab apple juice added. But the best use is, of course, crab apple jelly made by boiling and sieving through muslin.

ELDERBERRY
Sambucus nigra

The elderberry is one of the most versatile of wild fruits. Mixed with almost any other fruit they greatly enhance the flavour and boiled in spiced vinegar they make an excellent relish. Elderberry wine is one of the kings of country wines – it matures well and can almost pass for a claret after 3 or 4 years in the bottle.

ELDERFLOWER
Sambucus nigra

The secret of using elderflowers to make a delicious fruit cordial drink is never to put too many into your brew and always harvest on a fine, sunny day when the fragrance and nectar are at their height. I find I can reach even high blossoms (usually the best) by taking a garden hoe and pulling down the branches.

GORSE
Ulex europaeus

Few things can beat the spectacular show of colour provided by gorse in full flower. On a hot spring day the sight and smell are simply intoxicating. And it is with the flowers that we want to make what is probably the most enticing and delicate of country wines. It is a long and prickly job to collect sufficient flowers, but children love it, and what better things could be found to pass the time on a fine sunny day?

HAZEL
Corylus avellana

Hazel is a moody sort of tree which seems to grow like a weed in some places and can be completely impossible to grow in others. Hazelnuts are a wonderful addition to a winter diet. You can use them in all sorts of ways. Nuts must be dry before you store them or they will rot.

OAK
Quercus robur

In their first flush of spring growth the leaves of the mighty oak tree provide the raw material for a very fine country wine. The acorns that come in autumn can be used to make a flour, but better by far, to feed them to the pigs, for whom they will provide an excellent source of natural protein.

WILD PLUM
Prunus domestica

Wild plums are more common than you might think. Look out for their distinctive small white blossom in the hedgerows very early in the spring. Mark your sport and get back there in late summer to see how the crop is doing. Keep the location of your trees secret, for news travels fast and others will be eager to share your bounty.

SLOE
Prunus spinosa

Sloes make a marvellous fruit wine and are a magical addition to strong spirits. The invigorating colour and flavour of sloe gin is something everyone should savour. Gather your sloes after the first frost. Take ½ pound (228 g) of sloes, prick them all over with a fork or pin. Each one must be pricked several times before being put into the gin. Half fill a bottle with them; now add an equal quantity of sugar, fill the bottle with gin, and in a few months you have a fine liqueur.

SWEET CHESTNUT
Castanea sativa

In most years the sweet chestnuts in northern countries are too small to make good eating. But with the right varieties and a good season, the chestnuts will stand you in good stead. Far more people should grow sweet chestnuts, for the wood is also one of the best for furniture that is stable and not liable to twist and shrink. Sweet chestnuts make the best stuffing you can get for that Christmas turkey.

WALNUT
Juglans regia

For sheer volume of crop you cannot beat a good walnut tree. The trees grow much better and faster than you might think. Plant from seed and grow on in pots before planting. Young trees are very easily damaged by frost, so avoid frost pockets and cover in cold weather: 15 years later you will begin to harvest. The nuts need to be separated from their green skins and well dried before you store them. Walnut oil is super for salads.

Mushrooms

There are few more remarkable sights in this world than to pass by a field of pasture that has been miraculously covered with perfect white field mushrooms. Sometimes, if the weather and soil make the right combination, you can find literally thousands of beautiful mushrooms.

The edible parts of fungi are the fruiting bodies that are produced very dramatically by huge spreading masses of mycelia, which draw their nutrients as parasites from roots and decaying vegetation. If you are a beginner at mushroom picking, it really is a good idea to learn from someone with more experience, as there are a huge range of excellent fungi to be found.

Where to look

You will always harvest your best specimens early in the morning. Fungi grow in a wide variety of places, but they will not tolerate chemical fertilizers or sprays. They say it will take 20 years for the horse mushroom to appear in grassland after the use of chemicals has been stopped, and I have found this to be true. In fact, the majority of edible fungi grow in the proximity of woodland and many have close symbiotic relationships with particular tree roots. But wild grassland does always produce an excellent crop of fungi every autumn and you will find each season's crop in similar places to the previous year's.

What to avoid

There is probably no need to warn you of the fly agaric, as this bright red and white spotted fungus is so well known. The most dangerous of all fungi is the death cap (*Amanita phalloides*). A single death cap contains enough toxins to kill several people. Usually it grows in woodland, particularly with oak trees. It can vary in colour, being similar in size to a field mushroom, but its characteristic features are white gills on the underside and a "volval" bag at the base. Any fungi growing from a "volval" bag are best left well alone for many are poisonous. A death cap has white spores, not brown like most edible mushrooms.

Another mushroom to avoid is the "yellow stainer", easily confused with field or horse mushrooms. It has the distinctive feature of turning bright yellow when bruised or cut. It also smells rather like disinfectant.

BAY BOLETUS
Boletus badius

Usually found in woodlands, this fungus is pale to brown in colour. It has light yellow pores on the underside and these stain blue if damaged, making the fungus easy to recognize. The flesh also stains a bluish colour when cut and smells very mushroomy. The stalk has no frills but is smooth from base to cap. The flavour is very good and the season is a long one, from early summer to autumn. You can store by slicing and drying or flash freezing. They taste fine raw when sliced and make great soup.

SHAGGY INK CAP
Coprinus comatus

This is a very common but distinctive mushroom, and easy to identify with its egg-shaped shaggy cap. It often grows on newly disturbed ground in large clusters. The cap is covered with beautiful white scales and there is no veil on the stem when the cap opens to a bell shape with a dark black underside. These mushrooms need to be young and fresh to make good eating — and very good food they are too. Shaggy ink caps make a wonderful mushroom soup. These mushrooms do not store well, so again they are best used fresh.

GIANT PUFFBALL
Langermannia gigantea

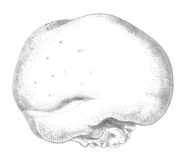

As its name suggests, this mushroom can grow to a truly amazing size. I have seen them well over 18 inches (45 cm) across, which makes a major feast, but they can grow over twice this size! The huge white ball of a giant puffball is not hard to identify. They must be used young before the spores have time to develop and the insects have time to take their share. You have to slice them up like rump steak to cook them. By themselves they have little flavour, but fried quickly with a little bacon they are delicious.

HORSE MUSHROOM
Agaricus arvensis

This mushroom is a beauty and my personal favourite. You will only find it on old pasture that has been grazed by horses or cattle. Some of my earliest memories are of the excitement of finding these superb large fungi in early morning forays with my parents and then frying them up for breakfast. The horse mushroom has a slight aniseed smell and, unlike a field mushroom, does not shrivel up when cooked. Just beware you don't over-indulge if you are lucky enough in these times of chemical farming to find a crop of these. The cap of the horse mushroom may be yellowy in colour, but be careful not to confuse it with the "yellow stainer" fungus, which will make you ill.

CHANTERELLE
Cantharellus cibarius

These wonderful fungi are much loved by the French may be found in woodland clearings. Seasoned mushroom hunters will keep their locations a close secret as they tend to grow in the same places each year. Chanterelles are fairly small – up to 4 inches (10 cm) across but usually smaller – with a distinctive yellow colour and a slight smell of apricots. The caps become like small, fluted trumpets as they age and the gills are heavy, irregular and run down the stems. They are best stored in good olive oil or in spiced alcohol. They taste delicious and are especially good for vegetarians because of the protein. Also delicious, but less common, is the winter chanterelle. This is a smaller and greyer version that grows in woodland much later in the year.

ORANGE PEEL
Aleuria aurantia

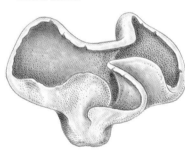

An extraordinary, brightly coloured and very striking fungus, you will find the orange peel quite commonly in large clumps in grassland and on bare earth from autumn to early winter. The caps soon become wavy and are of fairly robust texture. Quite small – up to 2 inches (5 cm) across – the fungus is bright orange on top and a lighter shade on the velvety underside. These store well if dried.

PARASOL
Macrolepiota procera

Where the horse mushroom gives a blast of texture and flavour as befits the king of the fungi, the exquisite parasol is the most delicate and desirable queen. The mushroom is usually found in open fields and has large brown scales in a symmetrical pattern around a pronounced central bump. The cap can grow up to 10 inches (25 cm) across and the gills are white. The stem is long and tough with a large ring around it. The parasol will dry well for storage. You can make a delicious dish by dipping pieces of the parasol in batter and deep frying.

WOOD MUSHROOM
Agaricus silvicola

Only found in woodland, the wood mushroom is a more delicate version of its close relative, the horse mushroom. It does *not* grow out of a volval bag like the death cap and its gills are pink to brown in colour, not white. The flesh does not discolour when cut and the smell is of a slight aniseed. The cap is a creamy-yellow colour that darkens as it ages and is smaller than the horse mushroom, growing to only 4 inches (10 cm).

PENNY BUN
Boletus edulis

The penny bun is also known as the "cep" mushroom and is a great prize for the mushroom hunter, as it has an unusual nutty flavour. Found in woodland or sometimes in heather with dwarf willows, the "cep" can grow quite large – over 2 pounds (1 kg) in weight. When picking, cut the cap in half to check for maggots. These work their way up through the stems. You will recognize the penny bun because the cap looks just like freshly baked bread. The colour darkens as the mushroom ages. The underside will have yellow pores, not gills. The stem is bulbous and solid white with brown stripey flecks. It stores well if dried in thin slices.

THE PRINCE
Agaricus augustus

Resembling a stocky version of the parasol, the prince is a fine mushroom. It grows up to 10 inches (25 cm) wide and is found in woodland. The top is flecked with brownish scales. The gills are off-white when young, turning dark brown with age. The flesh is strong white and smells of mushroom. The stem is very strong and often scaly with a large floppy ring under the cap – it is too tough to make good eating unless cooked in stews. It has a strong flavour and can be frozen or dried for excellent winter meals.

HONEY FUNGUS
Armillaria mellea

This yellowy-brown fungus is a tree-killer – but highly edible for humans! The active part of the fungus is a black cord-like rhizomorph that covers huge areas under the soil and seeks out trees, which it destroys. It normally grows straight out from trees and stumps, usually in large clumps. The flesh is white and smells strong and sweet. The gills vary from off-white to brown and the stalks are tough, often fused together at the base and with a white, cotton-like ring below the cap. The caps become tough if you dry them so it's best to freeze.

FIELD MUSHROOM
Agaricus campestris

Undoubtedly the best known of all mushrooms, before the days of chemical farming whole fields would be covered by the prolific field mushroom. Get up early after a hot summer spell has been followed by rain to pick. The silky white caps grow up to 4–5 inches (10–12 cm), the gills are pink, and the smell mushroomy. The ring around the stems is very fragile and often missing. Maggots can be a problem – check older specimens by cutting through the stems. I like to store by flash freezing or drying.

"We kept blundering about trying to buy a cow. It is difficult buying a cow if you know nothing about it, and we don't want to be robbed. We went and looked at a herd of pedigree Jerseys and were offered one — a cull — at just the hundred and twenty guineas. And you can buy an awful lot of milk for a hundred and twenty guineas. And you can pay an awful lot of veterinary bills on a pedigree Jersey or a pedigree any other breed. We did have enough sense — or instinct — to steer away from over-bred stock.

... When you learn to milk comfortably, which you do in about a week, it becomes a pleasant job. I look forward now to the morning and evening milking. There seems to me to be a friendliness between the cow and me. I put my head in her old flank and squirt away, and there is a nice smell, and a nice sound as the jets hiss into the frothing bucket, and I can think, and sum things up, and wonder what I am going to have for supper. In the winter it is dark and cold outside, but warm in the cowhouse and the hurricane lantern throws fine shadows about the building. The whole job takes perhaps ten minutes — night and morning."

JOHN SEYMOUR FAT OF THE LAND 1976

CHAPTER SIX

IN THE
DAIRY

The Hayshed & Cowshed

THE HAYSHED

For all sorts of reasons the hayshed has always been a rather romantic place. Of course the hay itself embodies the bounty of the earth in spring. Good hay smells wonderful and a full hayshed gives the self-supporter a marvellous gut feeling of satisfaction – looking forward to those long winter nights when the stock depends on hay for daily meals. Hay also feels good, and how many of us can remember an afternoon spent relaxing on this natural mattress? My own memories of the hayshed come in two parts. The first is the hot and sweaty time I spent as a teenager lugging heavy bales into those high corners under a hot roof. The ends of the compressed grass were rough and the sweat would sting as it ran into the scratches. The second is of many gentle times relaxing with a friendly farm cat or, if luck was going my way, a nice young lady.

Hayshed hints

The hayshed must be functional and it must be in the right place. Choose your location so as minimize the distances the hay must be carried. I've tried to have the back of the shed opening on to a roadway for easy loading whilst the front is only a few yards from the cowshed door. This is ideal. With a bit of extra cunning you can also use one wall of the shed to make the back of your compost heap. It is important to make sure the hayshed is well-ventilated. Still, musty air is not going to improve your hay as it goes through the winter. Many farmers leave two or three walls open to the air. I have found that vertical wooden slats, with a couple of inches space between each, make excellent sides for a hayshed. The slats keep out most of the rain, which simply runs down them to the ground (if you had horizontal slats this would not work).

Keep your construction cheap, strong, and simple. If you can find a source of whole pine trees (perhaps a local sawmill) then you can use these to make strong corner posts. Slap plenty of creosote on to the bottoms before you bury them 3 feet (1 m) deep in the earth. Get them accurately vertical before you ram back the earth. You can then saw off the tops to the exact heights you need. The roof is usually made of corrugated iron in any one of its modern manifestations. Now, if you can, find a source of offcuts from a sawmill to make your sides. Keep a big opening in one side, and if possible have this opening closest to your access road for easy loading. Your hayshed door must be large enough to allow easy access with a pitchfork full of hay. And it must be strong enough to withstand a good shaking by the wind.

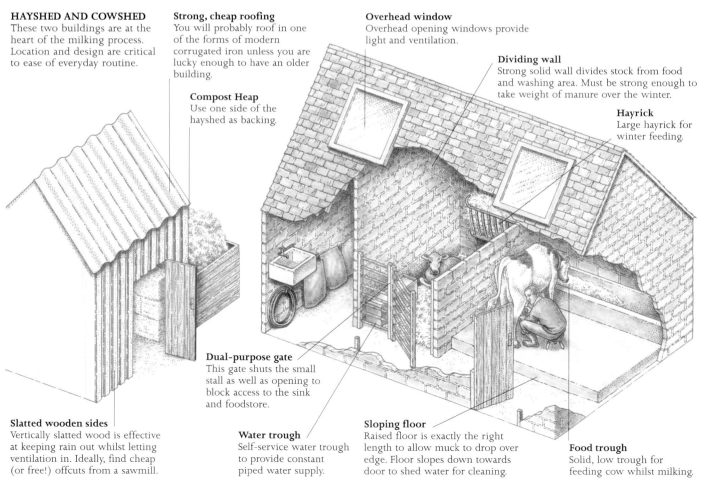

HAYSHED AND COWSHED
These two buildings are at the heart of the milking process. Location and design are critical to ease of everyday routine.

Strong, cheap roofing
You will probably roof in one of the forms of modern corrugated iron unless you are lucky enough to have an older building.

Compost Heap
Use one side of the hayshed as backing.

Overhead window
Overhead opening windows provide light and ventilation.

Dividing wall
Strong solid wall divides stock from food and washing area. Must be strong enough to take weight of manure over the winter.

Hayrick
Large hayrick for winter feeding.

Dual-purpose gate
This gate shuts the small stall as well as opening to block access to the sink and foodstore.

Slatted wooden sides
Vertically slatted wood is effective at keeping rain out whilst letting ventilation in. Ideally, find cheap (or free!) offcuts from a sawmill.

Water trough
Self-service water trough to provide constant piped water supply.

Sloping floor
Raised floor is exactly the right length to allow muck to drop over edge. Floor slopes down towards door to shed water for cleaning.

Food trough
Solid, low trough for feeding cow whilst milking.

THE COWSHED

The modern cowshed is an extremely functional place where each part is specifically designed for a particular purpose. Design and layout has evolved over many years of trial and error, so pay careful attention to the key points. In order to milk a cow efficiently you must ensure various things: the cow must be contented, she must be fastened in a way that is secure without being too constricting, she must not be fretting for her calf and, last but not least, her back end must be in a place where manure will not splash on to you or into the fresh milk. Finally, you must have a milking area that can be easily cleaned.

These objectives are all achieved by careful layout. The cow is content because she has a big, strong manger of food in front of her nose. So the top end of your milking area must have a good feeding trough just about floor level. We use a built-in concrete trough. Remember that anything which is not really robust will almost certainly be broken at some stage when the animal gets into a muddle over something. We use a built-in concrete trough which which is 30 inches (76 cm) wide measured on the outside. Any food-bins, especially in the cowshed, should be securely latched in case the cow slips her tie.

The cow is fastened to a wall with a chain and ring which can slide up and down on a vertical steel rod. This allows the cow to move her head up and down easily even though she is firmly and closely attached to the wall. This way she cannot romp about, but she does not feel so restricted that she will panic. Her peace of mind is also much improved if she senses that her calf is close beside her – hence the need for a loose box adjacent to the milking area. This is where you can keep the calf whilst milking. The sensible width for a loosebox is about 5½ feet (1.7 m), the length 9 feet (2.8 m) – just big enough for a full-sized animal and plenty big enough for a calf.

To ensure that cow manure does not contaminate the milk (or the person milking) you have the milking area raised about 9 inches (22 cm) above the dunging passage at the rear of the animal. You must also make sure the length of the raised milking area is exactly correct (it is 5½ feet or 1.7 m measured from the outside edge of the feeding trough in our cowshed and this works fine). Naturally the floor must be sloping away from the head of the cow so that dirt and loose straw can be hosed away easily into the dunging passage. Make sure there is a convenient drain to take the waste water. Obviously you must have a tap and powerful hose close at hand to keep the cowshed clean. A modern electrical powerhose would be ideal, so make sure you have power points within convenient reach. You will need a tap and sink too for cleaning implements and equipment. And if you want to avoid feeding all the rats in the neighbourhood you will take steps to keep all foodstuffs (rolled barley, chicken food, and so on) in rat-proof, closed containers. We have found that old chest deep freezes are excellent for this purpose – and you can pick these up for next to nothing.

I have found it convenient to keep all our animal food in the storage area of the cowshed where there is water for mixing and a sink for cleaning. Don't forget that your loose box must have water (a small self-filling trough will do) and a good, large hayrick.

Towards a working dairy

The dairy is not where you milk cows, but where you process milk. Most homesteaders have to make do with the kitchen. We made butter, cheese, yoghurt, and so on in our kitchen for 20 years, and pretty successfully, too. But it has to be said it is a messy and difficult business doing dairy work in the kitchen, and a special room is a great convenience and luxury if you can create one.

The dairy should be as cool as possible and very airy. Working surfaces should be marble, slate, or dowels (round wooden rods). Ideally the dairy should have a concrete or tiled floor with a drain for taking the water away. You must be able to swill it down with plenty of cold water and sweep the water out. Concrete rendering requires a very fine finish. Use four parts of sand to one of cement on the floor, and five to one on the walls. Smooth it off very carefully with a steel trowel. It is better not to whitewash or paint it. The ceiling should not have any cracks in it, or it will let down the dust.

Traditional dairies of the old farmhouse days were always on the north or east corner of the house – this is the coldest place as it receives virtually no direct sunlight. Windows could open to let in plenty of air but all ventilation would be protected by fly-screens made of fine mesh wire small enough to keep out flies. Commonly found on caravans, similar screens are now increasingly available at any DIY store. Adjacent to the dairy would be a pantry; again with fly-screened, ventilated windows facing north. This was where cheese and vegetables could be stored before the advent of fridges. Solid stone floors or heavy tiles also helped to keep the space cool. And there was always at least one door shutting off the whole area from the heated parts of the main house.

All your shelving in the dairy should be easy to wipe clean. Use either well varnished wood or a modern synthetic surface as found on kitchen work surfaces. Remember that any chipboard products are useless in the dairy: they absorb water and soon deteriorate.

There should be hot and cold water, with the hot preferably boiling, as sterilization is the most important thing. There should be a big sink and I prefer a draining board made with dowels. The water then drains straight down on to the floor through the gaps between the dowels, and the air comes up into the utensils.

The ideal dairy has the minimum of cupboards, fridges, and the like resting on the floor. This is so the whole floor can be hosed down and swept clean. Don't go to a lot of trouble to get milk, and then let it go bad because of lack of hygiene. Always avoid keeping anything in a dairy that is not absolutely needed, because everything catches or retains dust.

The Dairy

DAIRY EQUIPMENT

A cream separator or settling pans are a great advantage. Mount the separator on a strong bench. You will need a butter churn (any device for swishing cream about), and a butter worker (see opposite) or a flat clean table, ideally marble or slate, for working butter. A cheese vat is a great labour-saving device. It is an oblong box, ideally lined with a stainless steel jacket. The milk inside the vat can be heated or cooled by running hot or cold water through the jacket. There should be a tap at one end and it should be easy to lift the vat so that the liquid whey which accumulates during cheese-making can be drained off.

You will also need a cheese press, a chessit and a follower, unless you are going to make nothing but Stilton, which doesn't need pressing. The chessit is a cylinder open at the top and full of holes to let out the whey. The follower is a piston which goes down into it and presses the cheese. And the press itself is a complex combination of weights, levers, and gears which can put pressure on the follower, and therefore the cheese, inside the chessit.

Cheese presses are now hard to come by, but you can improvise. Drill holes in the bottom of an empty food can or an old saucepan. This is your chessit. Cut out a metal disc to fit inside it. This is your follower. Your pressing will have to be done with weights, which can be bricks, rocks, encyclopedias – anything that is heavy. Some of the small cheese presses you see for sale and that operate using a spring system are very difficult to use in practice. As the cheese is compressed the pressure drops as the spring expands. The ideal press operates with weights and levers

so that a constant pressure is maintained. We had a local carpenter make ours to a design we saw in the Netherlands (*see illustration*). It works perfectly.

Other requirements will be a dust-proof cupboard for keeping thermometers, an acidimeter, and a fly-proof safe for your butter and cream cheese. Hard cheese should not be stored in the dairy, because it is generally stored over long periods and it would get in the way.

Hygiene for dairy containers and utensils

1 Physically remove any cream, milk, dirt, and so on which may be adhering to the vessel, either inside or out, with hot or cold water and some sort of brush.
2 Scald the vessel inside with boiling water.
3 Rinse the vessel well with cold water to cool it down.
4 Turn the vessel upside down somewhere where it can drain and where the air can get into it.
5 Leave it upside down until you want it. Never wipe any dairy utensil with a cloth or rag, no matter how clean you think the cloth is.

Always clean dairy utensils the moment they are empty, but if you do have to leave them for some time before washing them, fill them with clean, cold water. Never leave utensils wet with milk. Remember, milk is the perfect food for calves, babies, and bacteria! The moment milk leaves the teat of the animal, bacteria attack it and it begins to go sour. If they are the wrong sort of bacteria it will turn not sour, but bad. Sour tastes nice, but sour. Bad tastes horrible. Enough bacteria of the sort that you need (chiefly *Bacillus lacticus*) will occur naturally in any dairy.

PLAN OF THE DAIRY
Butter and cheese can be made in the kitchen, but if you have a room or outhouse you don't know what to do with, it is well worth equipping it as a dairy. If possible the floor should be made of tiles, or concrete, with a drain. Install plenty of taps for hot and cold water, a large sink, and draining boards made of spaced dowels.

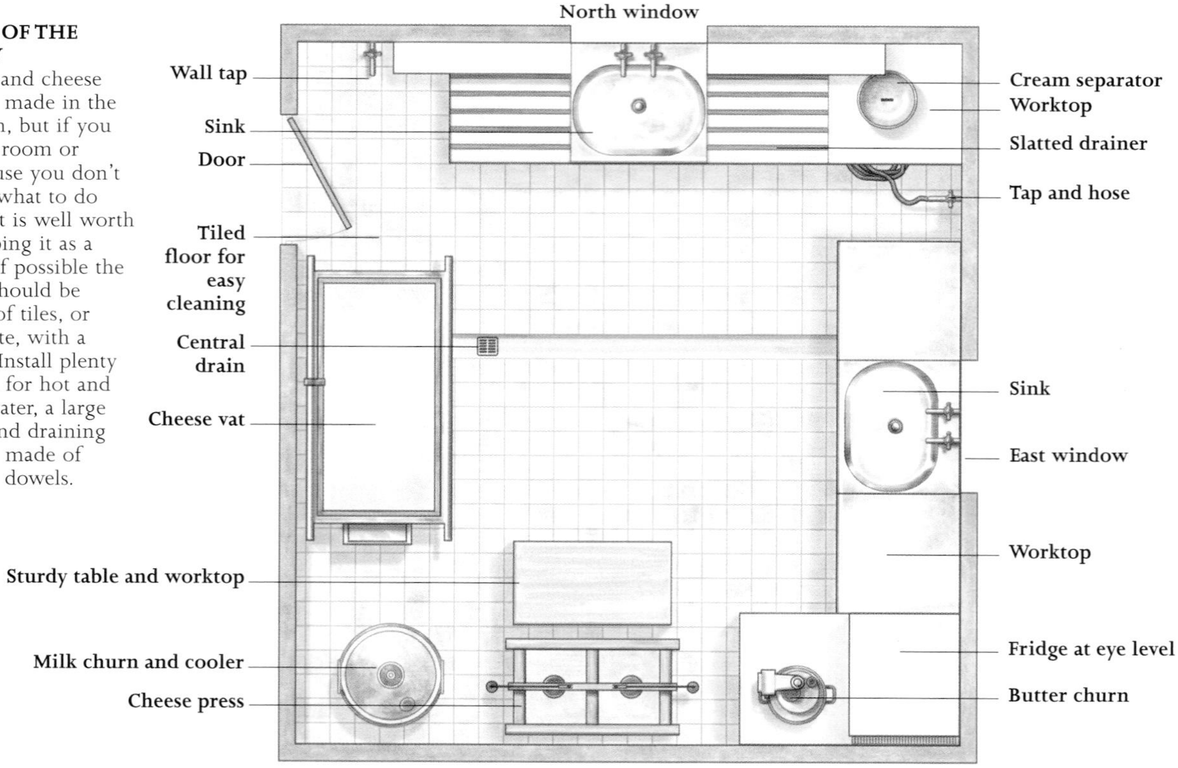

North window

Wall tap
Sink
Door
Tiled floor for easy cleaning
Central drain
Cheese vat
Sturdy table and worktop
Milk churn and cooler
Cheese press

Cream separator
Worktop
Slatted drainer
Tap and hose
Sink
East window
Worktop
Fridge at eye level
Butter churn

CHEESE VAT
Large double-skinned tank that can be cooled or heated slowly and effectively by circulating water around the milk. Often mounted on wheels to make it easier to manage the heavy mass.

IN-CHURN COOLER
This device has long loops of piping that can be inserted into the milk churn and, when circulated with cold water, rapidly bring down the temperature of the fresh milk.

MILK CHURN
This familiar-shaped large container is used for fresh milk and was once a common sight on country roadsides as it awaited the daily milk collection. Now comes in polypropylene and stainless steel.

CHEESE PRESS
A vital piece of equipment that must work perfectly in order to make good-quality cheese. Various designs are available, ours is cribbed from an excellent Dutch design.

BUTTER CHURN
Large, strong container that can be turned about in order to shake cream into butter. Modern electric versions now make the task much easier, as well as managing smaller quantities.

CREAM SEPARATOR
This is basically a clever series of scoops designed to remove the rising cream from milk. Modern electric versions are smaller and neater than the traditional separator.

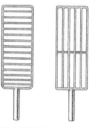

CURD CUTTERS
Curd knives are simply an array of knives or long blades which are used to cut curd into the correctly sized pieces for making cheese. They come in stainless steel or plastic.

RENNET
Enzymes removed from a calf's stomach and used to begin the process of making cheese from milk. Vegetarian rennet is available from some specialist suppliers.

SKIMMER
This is a simple tool for removing solids from the body or surface of milk. It is made of stainless steel, and is about 4 inches (10 cm) in width.

THERMOMETER
The standard device for measuring temperature of liquids. It floats in a protective plastic cage. It needs to have a fairly large and sensitive scale up to 100°C (212°F).

CHESSIT
Stainless steel tube into which curd is placed to be pressed into cheese. Each different size has a correctly fitting "follower" which takes the pressure. Drainage holes allow liquid to escape.

MUSLIN
Fine cloth which is free of fluff. Used to strain soft cheese. Cheesecloth is sold in roughly 9-foot square (1-sq m) units.

MOULD
These are shaped containers used to form cheeses. Modern rectangular or round plastic moulds are designed specifically for different soft cheeses (such as brie, camembert, ricotta).

COOKING PAN
Large robust stainless vessel which can be put on to a hot ring to warm liquids. Other milking sundries include squat buckets and strainers made of aluminium or stainless steel.

STRAW MAT
Small mats are very useful for putting under freshly made cheese because they allow the cheese to "breathe". The straw mat must be kept clean and aired at all times.

SETTLING PAN
This is a stainless steel, shallow, wide topped pan suitable for leaving cream to rise. A useful tip is to make a plywood top to keep out the flies and other aerial bugs.

Making Butter & Cream

CREAM

If you leave milk alone the cream will rise to the surface and you can skim it off. You do this with a skimmer, which is a slightly dished metal disc with holes in it to let the milk run out but retain the cream. Or if you leave the milk in a shallow dish with a plug hole in the bottom, you can release the milk and leave the cream sticking to the dish. Then you simply scrape the cream off.

Alternatively, you can use a separator: a centrifuge which spins out the heavier milk, leaves the lighter cream and releases them through separate spouts. (Cream put through a separator is 35 percent butter fat (15 percent more than skimmed cream). Milk should be warm to separate.

The colder the milk is, the quicker the cream will rise to the surface. It is a very good thing to cool milk anyway, as soon as it comes from the cow. Cooling it slows down the action of the souring organisms. And, of course, the wider and shallower the pan that you settle the milk in, the faster the cream will rise to the surface.

Devonshire or clotted cream

Leave fresh milk for 12 hours, then heat it to 187°F (92°C), and immediately allow it to cool. Leave for 24 hours, then skim. What you skim is Devonshire cream.

BUTTER

Butter is made by bashing cream about. But it will not work until the cream has "ripened": in other words, until lactic acid bacteria have converted some of the lactose, or milk sugar, into lactic acid.

Commercially, cream is pasteurized to kill all bacteria, lactic acid bacteria included, and then inoculated with a pure culture of bacteria. We can't, and don't want to be, so scientific, but we make equally good butter by keeping our cream until the oldest of it is at least 24 hours old. It can be kept twice as long as this if everything is clean enough. We add more cream to it at every milking, at a temperature of more or less 68°F (20°C). Then we make sure the last batch of cream has gone in at least 12 hours before we start churning it.

Churning

The best-known churn is simply a barrel in which the cream is turned over and over so it flops from one end to the other and bashes itself. But there are churns, like the blow churn, which have paddles that whirl round to beat the cream. You can make butter on a tiny scale by beating it with a wooden spoon or paddle, plunging a plunger up and down in a cylinder or using an egg-beater. Anything, in fact, which gives the cream a good bashing. If the cream is more or less at the right stage of acidity, and at the right temperature, it will "come", meaning suddenly turn into little butter globules, in as little as 2 or 3 minutes. If it hasn't come in 10 minutes, take its temperature, and if it is wrong, bring it to 68°F (20°C). Then try again.

It doesn't matter how sour the cream is when you churn it, provided it's not bad. Taste it. If it's bad, it's useless. When the butter has come, drain the buttermilk out. (If your cream has been kept right, it will be the most delicious drink in the world.) Then you must wash the butter. Washing should be continued until every trace of cream or buttermilk or water has been removed.

Butter worker

There is a fine thing called a butter worker, which is a serrated wooden roller in a wooden trough. This squodges the water out. Keep putting more cool, clean water on and squodging until the water that you squodge out is absolutely clean and clear, without a trace of milk in it. Traditionally a wooden scoop was needed to dig out the butter from the butter worker, after which it was formed into blocks, rolls, or rounds or pressed into storage dishes. Your butter is made when the last drop of water has been pressed or squeezed out of it. From now on, don't expose it to the light or the air too much, and if you keep it wrapped it will last much longer.

If you haven't got a butter worker don't despair. Do your washing and squodging on a clean board with a "Scotch hand" or a wooden paddle. Very few beginners at butter-making ever wash the butter enough, and so their butter generally has a rancid taste, particularly after a week or so. Squodge, squodge, and squodge again.

Salting

To make salty butter, use brine for the last washing, or sprinkle dry salt on the butter and work it in thoroughly. If you find the butter is too salty, wash some of the salt out. If you wash too much out, put some more back. It's as simple as that.

To keep butter, incorporate 2½ percent of its own weight of salt in it, and follow this method: scald out an earthenware crock, tub, or barrel and dry outside in the wind and sun. Throw a handful of butter into the vessel as hard as you can to drive the air out of the butter. Repeat, sprinkling more salt in after each layer and pummelling the butter down with your fist to drive out the air.

When your crock is full, or you have no more butter, cover the butter with a sprinkling of salt and some grease-proof paper or other covering. It will keep for months. If it is too salty, simply wash some salt out before you eat it. It will be just as good as fresh butter. But remember that it must always be well washed in the first place.

Ghee

Ghee is a great Indian standby. Put butter in a pot and let it simmer gently over a slow stove for an hour. Skim off the scum. Pour the molten butter into a sterilized container, cover from the air, and it will keep for months. It won't taste like butter. It will taste like ghee. It's very good for cooking, and helps to give real curries that particular taste.

MAKING BUTTER

Butter is made by bashing ripened cream. So a churn of some kind is essential. To shape the finished butter, use two wooden paddles, called "Scotch hands", or, better still, an old-fashioned wooden butter mould.

1 I use a blow churn. Fill the churn with cream and turn the handle.

2 When the butter "comes" or coagulates, drain off the buttermilk.

3 Dump butter out on a clean draining-board or on a butter worker.

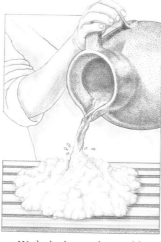

4 Wash the butter thoroughly by repeatedly mixing with cold water and squeezing.

5 Squodge, or press, to remove all water and traces of buttermilk.

6 Add salt to taste, or if you want to keep the butter for a long time, add a lot.

7 Work the salt well in. You can always wash salt out again.

8 Shape the butter with wetted "Scotch hands", otherwise known as paddles. Take care to squeeze any remaining water out. The secret of making good butter is to wash and squeeze all milk and water out of it.

9 There are various churning moulds and wooden blocks for the final shaping and imprinting of butter. With this mould you pack the butter in tightly, so as to fill in all the airholes.

10 Then it may even take two of you to force the butter out on to grease-proof paper.

11 The butter can be patted into oblongs or put in a mould that impresses a design into the butter, such as thistles, cows, or wheat sheaves.

YOGHURT

Yoghurt is milk that has been soured by *Bacillus bulgaricum* instead of the more usual *Bacillus lacticus*.

To make 2 pints (1 litre) of yoghurt, put 2 pints of milk in a bowl. The bacteria need a warm climate. So if it's too cold for them, you should warm your milk to about blood heat. Stir in 2 tablespoons of a good live yoghurt that you have bought in a shop, and at this point mix in any fruit and nuts that you want for flavouring, though to my mind nothing beats a bowl of natural yoghurt served up with a spoonful of honey. Cover the mixture, and keep it at blood heat for 2 or 3 days. A good method of keeping it warm is to bury it in straw. When it has gone thick, it is yoghurt.

Take some out every day to use, and put the same amount of fresh milk back and so keep it going. But the milk should be very clean and fresh, the vessel should be sterile, and you must keep it covered. It may go bad after a while. If it does, start again.

Thick milk or curds and whey

If you leave clean milk alone in the summer it will curdle, and become "curds and whey" such as Miss Muffett was eating when she had that unfortunate experience with the spider. Curds and whey are slightly sour and delicious to eat. Sprinkle cinnamon over the top, or a little salt.

ICE-CREAM

I remember a strange experience I had in Oregon on the west coast of the USA. While visiting friends, I was amazed by the terrific noise made during the night by tree frogs so I set off into the woods to investigate. These frogs are tiny creatures but they certainly make a loud racket.

After walking through the woods for 20 minutes or so I saw lights up ahead. Houses are not common in these parts, where you can drive for 3 or 4 hours through woods without seeing a single home, so I went closer. The household was having a big party and I soon found myself in the thick of it.

The hosts handed me an ice-cream-making barrel that was duly loaded up with a cream mixture that was surrounded by salt and ice. It was then my job to look out for all the children in the place and ask each one to give the barrel a good shaking. In this way the children were occupied for at least an hour shaking away happily, especially as there was the imminent prospect of fresh ice-cream as a reward. Sure enough, as if by a miracle, the ice-cream mixture set firm and we feasted on the results. It was the local way to keep the kids entertained while the adults had a knees up – and very effective it was, too!

Ice-cream – a history

Ice-cream is said to have been invented by Catherine de Medici. It was made by placing a tin or pewter container inside another container, which was then filled with a mixture of ice and salt. Cream was then poured into the inner container and sugar and flavouring, such as fruit juice, liqueur, or even jam, added. The mixture was then stirred continuously with a spaddle or revolving paddle, generally made of copper, and at the same time, the inner pot was turned by a handle. By keeping the mixture constantly on the move, the ingredients would not separate before congealing and no lumps would form. As the cream was churned, it also froze slowly to make the ice cream.

Ice-cream made with cream is very different from the stuff you buy from itinerant vendors and is really worth eating. Pure frozen cream is pointless. Ice-cream should be sweet and fruity; white of egg, gelatine, and even egg yolks can be used to enrich the texture. Ice-cream is not difficult to make and the kitchen makes an adequate dairy so long as you keep all surfaces and utensils scrupulously clean. Here is a typical recipe for delicious ice-cream:

Strawberry ice-cream

4 oz (115 g) sugar
1 pint (0.5 litres) water
1 lb (0.5 kg) strawberries
¾ pint (0.4 litres) single cream or ½ pint (0.25 litres) double cream

Use the following method:

Make a syrup of the sugar and water.
Mash up the strawberries and strain the seeds out, and when the syrup has cooled pour it into the strawberry mash.
Add the single cream as it is, or if you use double cream, beat it first and fold it in.
Then you have to freeze it. This can be done with a deep freeze, a refrigerator, or even with just ice. If you do it with ice you must mix the ice with salt, because salt makes the ice colder. Two pounds (0.9 kg) of ice to 1 pound (0.5 kg) of salt is about right

William Cobbett, in his marvellous *Cottage Economy*, described how to preserve ice from the winter right through the summer in a semi-subterranean ice house, insulated with a great thickness of straw, and with a provision for draining off the water from melting ice. Most large country homes and mansions had ice houses in their grounds. The ice would be collected from nearby frozen ponds, lakes and rivers by cart and then buried in the ice house. This meant that ice-creams and ice for cooling drinks was available all year round, particularly in the summer months.

To freeze ice-cream with ice you need a metal container, with a good cover on it for the ice-cream and some means of stirring it when it is inside the container. This container must be buried in a larger container filled with the ice-and-salt mixture. The latter must be well insulated from the outside air. To freeze ice-cream in your fridge, set the fridge at its coldest and put the ice-cream mixture in the ice compartment. Open it up and stir it from time to time to prevent large ice crystals from forming. A deep freeze can be treated just like the ice compartment of a fridge. Today, the widespread availability of fridges means that we all have the ability to freeze food – and make ice-cream – very easily.

Making Cheese

A pound (0.5 kg) of cheese has 2,000 calories of energy in it. Meat from the forequarters of an ox has a mere 1,100 calories. And cheese, hard cheese at any rate, is easy to store and, within certain limitations, improves with age. Cheese is made from milk whose acidity has been increased either by an additive, or by simply being left in the warm so that it does it itself. The extra acid causes the formation of curds and whey. Cheese is made from the curds. The whey is drained off and can be given to the pigs.

SOFT CHEESE

Soft cheese is made by allowing milk to curdle, either just naturally which it does anyway in the summertime, or by adding rennet. Rennet is a chemical which occurs in the stomachs of calves and has the property of curdling milk. Milk curdled with rennet is called junket. Milk which curdles naturally forms curds and whey. If you simply hang up some curds and whey in a muslin bag, the whey will drip out and the curds will turn to soft cheese. This is tasteless without flavouring, but seasoned with salt and herbs, or garlic, or chives, it is delicious. Eat it quickly because it won't keep long and is therefore no good for preserving the summer glut of milk for the winter.

Cream cheese and poor man's cheese

Cream cheese is simply soft cheese made with curdled cream instead of curdled milk. The result is smoother, richer and more buttery.

Poor man's cheese got its name because it could be made from the milk of one cow. A lot of it was eaten in England in the Middle Ages. Warm some milk slowly in a pot and let it curdle. Leave the curds in the whey overnight and drain the whey off in the morning. Then cut up the curd, salt it, tie it up tightly in a linen cloth, and leave it all day to drip. Re-tie it more tightly in the evening and leave it to hang for a month. You can eat it at the end of the month and poor man's cheese will taste even better if you work some butter into the curd and leave it for 3 or 4 months to mature.

HARD CHEESE

Hard cheese is important as a method of preserving the summer flush of milk for the winter and also as a very valuable source of protein and a marvellous food. Everyone needs cheese and lacto vegetarians can scarcely do without it. Hard cheese is difficult to make and better cheese is made from the milk of many cows than from the milk of but one cow.

The reason for this is that, for bacteriological reasons, the best cheese is made from the milk from two milkings only: the evening's milking and the following morning's. If you have to save up more than those two milkings to get enough milk to make a cheese you will almost certainly run into trouble with such things as over acidity and "off" flavours – the cheese will have a nasty taste.

CHEDDAR

If you make cheese on a fairly large scale, say, from the milk of six or seven cows, you should have certain equipment and do the job scientifically. Over the page I describe how to make hard cheese if you have approximately 5 gallons (23 litres) of milk from two milkings and don't feel the need to be too scientific. A marvellous cheddar cheese can be made in this way, but luck, as well as skill, and common sense come into it. If you don't make good cheese using the method overleaf you will have to try using starters (see p.192).

Caerphilly

Once you have made cheddar successfully, you might enjoy trying to make other cheeses. Caerphilly originated as a semi-hard cheese made by the wives of South Wales coal miners, for their husbands to take down the pit. It is an easy cheese to make, but will not keep as long as cheddar.

To make Caerphilly, strain the evening's milk into a vat and cool it if the weather is hot. The next morning skim off the cream and add a starter to the milk at a rate of about ½ percent of the milk. Warm the cream of last night's milk and pour it in together with that morning's milk. The purpose of skimming last night's cream off and warming it up and putting it back again is that only in this way will you get it to mix back with the milk and enrich the cheese.

Heat the vat to 68°F (20°C). Measure the acidity with an acidimeter. When it reaches 0.18 percent add a teaspoon of rennet extract to every 5 gallons (23 litres) of milk. About 45 minutes after renneting the curd will be ready for cutting. Cut both ways with the vertical knife but only one way with the horizontal knife. After cutting leave for about 10 minutes, gradually raising the temperature of the vat to 88°F (31°C): the "scalding". When this is complete, and there is 0.16 percent acidity, draw the whey off, scoop the curd into coarse cloths and leave it in a dry vat to drain. After half an hour cut the curd into 3-inch (8-cm) cubes, tie again in the cloths, and allow to drain another half hour.

Mill the curd (break it up into small pieces) and add 1 ounce salt (28 g) to every 3 pounds (1.4 kg) curd, mix well, and put it in cloth-lined chessits or moulds. About 10 pounds (4.5 kg) of curd go into each chessit and the chessits are traditionally fairly small and of flattish section, not the huge half-hundredweight cheddar ones. Two hours later apply pressure of 4 hundredweight (203 kg). The next morning take the cheeses out, turn and put on fresh cloths, put back in the press, and give them 5 hundredweight (254 kg). That afternoon turn again, put on more clean cloths, and apply 15 hundredweight (762 kg) for the night. The next day take out and store it for a month at as near 65°F (19°C) as you can get, turning it two or three times a week and wiping it with a cloth dipped in salt water. After that your home-made Caerphilly is ready to eat.

Stilton

Properly made, this blue-moulded cheese is one of the finest in existence. Factory made, however, it's pretty boring. You need about 15 gallons (68 litres) of milk using the one- or two-curd system. Let the evening's milk curdle, and let the morning's milk do the same. Mix the two together and the rest of the process is the same. With the one curd system, take the milk straight from the cow and put it in a tub or vat. Heat to about 85°F (30°C). Add a teaspoon of rennet extract for every 5 gallons (23 litres) of milk. Dilute the rennet with ½ pint (0.3 litre) of cold water before adding it to the milk.

After half an hour try dragging the dairy thermometer upwards through the curd. If it leaves a clean cut and no curd sticks to it, then it's ready for cutting. But don't cut it. Ladle it out into vessels lined with coarse cheesecloths so the whey can drain through the cloths but not away completely (you've left the plug in the vessel or sink). About 3½ gallons (16 litres) of curd should be ladled out in fairly thin slabs. After ladling, the curd should be left there soaking in its own whey for 30 minutes. Pull out the plugs and let the whey run away. Draw the corners of the cloths tighter around the curds. Replace the plugs and let the curds have a second draining for half an hour. If the curd feels soft, leave it longer in the whey; if it is firm, draw the plugs and drain the whey off. Now keep tightening the cloth round the curd – hauling one corner of the cloth around the other three corners and pulling tight. Each time you do this you gently expel some whey. Do it five or six times. When the curd contains 0.18 percent acidity, turn it out of the cloths. Pile the bundles of curd up on top of each other, then cut into about 3-inch (8-cm) cubes.

Keep turning the pile every half hour until the acidity reaches 0.14 or 0.15 percent. This may take from 2 to 4 hours. If you haven't got an acidimeter just go on until the curd is fairly solid but still moist and has a nice flaky look about it when it's cut. Now break the curd up into small pieces as for other cheeses, add an ounce (28 g) of salt to every 3 pounds (1.4 kg) of curd and mix well. Place the curd in hoops or moulds: 15 gallons (68 litres) of milk should produce about 26 pounds (11.8 kg) of curd. Place it on a wooden board. By now the curd should be cool, not more than 65°F (19°C). Don't press, just let it sink down.

Take the cheese out and turn it twice during the first 2 hours, then once a day for 7 days. When the cheese has shrunk away from the sides of the mould, take it out and scrape the surface of the cheese with a knife to smooth it. Then bandage it tightly with calico. Put it back in the hoop and mould. Take it out of the mould and re-bandage it every day for 3 days. Then take it to the drying room which should have a good draught and be about 60°F (16°C). Take the bandage off once or twice to help drying, and leave it off for a day. Then put it on again.

After 14 days remove the cheese to a cellar, again about 60° F (16° C), but with not too much draught, and plenty of humidity. Leave it there for 4 months before eating it.

SEMI-SOFT CHEESE

For me *Pont-l'Eveque* is one of the best of these Continental cheeses. To make it take 6 pints (3.5 litres) of 12-hour-old milk. Heat to 90°F (32°C) and add a teaspoonful of cheese-maker's rennet diluted with three times its own volume of water. Leave for half an hour to curdle. When the curd is firm enough (when it comes cleanly away from the side of the vat) cut it both ways with a curd knife.

Spread a cheesecloth over a wooden draining rack and ladle the curd on to the cloth. Fold the corners of the cloth over the curd and gently squeeze. Progressively increase the pressure until you get a lot of the whey out. Place a mould, which is just a collar 1½ inches (4 cm) deep and about 6 inches (15 cm) square on a straw mat on a draining board. After the curd has been draining for about an hour in the cloth, break the curd up and put it in the mould in three layers, with a layer of salt between each. Use 2 ounces (56 g) of salt for the 6 pints (3.5 litres). Pack the curd well down and into the corners.

When the mould is full, turn it upside down on to another straw mat on another board. Both mats and boards should have been washed in boiling water. Repeat this turning process every 10 minutes for an hour. Turn the mould upside down again once a day for 3 days. Then take the cheese out of the mould and scrape the surface gently with a knife.

You can eat it there and then, but it is far better if you can keep it at a temperature of more or less 58°F (15°C) for 2 weeks, turning it on to a clean mat daily. The outside of the cheese will be covered with mould. Wrap the cheese in waxed paper and keep for another month, turning it over each day. Before eating or selling, scrape the mould from the surface. The outside should then be quite firm but the centre should be soft and buttery and utterly delicious.

Starters

Starters are batches of milk rich in lactic acid bacteria; you can buy them or make them. Take a quart (1 litre) of milk from a healthy cow and allow it to get sour in a clean well ventilated dairy. Don't include the very first milk to come from the cow. See that the udder, and you, are washed thoroughly before milking. Strain the milk straight from the milking bucket into a sterilized container. Leave this fresh milk in the dairy for 24 hours. It is best if the temperature of the dairy is about 70°F (21°C). This quart of milk becomes almost a pure culture of *Bacillus lacticus*.

Put some fresh milk through your separator (don't worry if you haven't got one). Heat this milk to exactly 185°F (85°C) and cool it quickly to 70°F (21°C). This pasteurizes it. Skim the top off the first quart of milk (give it to the cat). Pour the rest of the first quart into the new, now pasteurized, milk. This inoculated milk must be covered with a cloth and kept for 24 hours at about 70°F (21°C). This is then your starter. Add a pint (0.5 litre) of this everyday to a new batch of pasteurized milk to keep the culture going for months.

MAKING A HARD CHEESE

1 Put the evening's milk in a settling pan and leave it overnight. In the morning skim the cream off with a skimmer, put it in a separate pan and heat it to 85°F (30°C). Pour it back into the milk and stir it in.

2 Now add the morning's milk and any *starter* if you have it – it "starts" the lactic acid bacteria working much more quickly and enables them to defy competition from unworthier organisms. Gently heat the milk to a temperature of 90°F (32°C).

3 Put 1 teaspoonful of rennet in a cupful of cold water and pour it in. Stir with your hand for about 5 minutes. As soon as

the milk begins to cling to your fingers, stop stirring.

4 Immediately start stroking the top of the milk with fleeter. This stops the cream from rising to the surface. Stroke gently for about 5 minutes. After this the curd should set enough to trap the cream.

5 When the curd feels firm to your hand (about 50 minutes after you have stopped stirring), cut it with curd knives or a long-bladed kitchen knife, into cubes about ¾ inch square (5 sq cm).

6 Warm the milk – now "curds and whey" – *extremely slowly* –

to 100°F (38°C). If you haven't got a cheese vat the best way is to scoop a saucepanful of whey out, heat it, and pour it back again. As you do this, stir very gently with your hand.

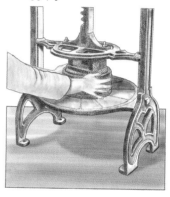

7 Pitching raises the acidity of the curd. You simply leave the curds to soak in the whey, and test every now and then for acidity. If you haven't got an acidimeter, use the hot iron test. Take a bit of curd, touch it on hot iron so that it sticks, and draw it away. If the thread is less than half an inch long when it breaks, leave the stuff to go on pitching. When the thread breaks at just about half an inch, the acidity is right (0.17 to 0.18), and you can drain off the whey.

8 A cheese-mill is two spiked rollers working against each other to break the curd up into small pieces. if you haven't got a mill you will have to do this by hand, which is a rather laborious task. To do this, break the curd up into small pieces (about the size of walnuts) with your fingers. Once separated, mix an ounce (28 g) of fine salt to every 4 pounds (1.8 kg) of curd and blend well.

9 Line your chessit with cheesecloth, put your bits of curd into it, cover with the cloth, put on your follower, and apply pressure.

10 If you've got a cheese press, apply pressure with this, or improvise. For the first 6 hours 20–30 pounds (9.0–13.6 kg) of pressure is enough, then pull the cheese out, wash the cheesecloth in warm water and wring it out, wrap again and replace the cheese upside down. Put on half a hundredweight (25 kg) pressure. After a day turn the cheese and replace it. After another day turn it and give it ½ ton pressure for 2 days, turning it once.

11 Paste the surface of the cheeses with flour and water, and then wrap with calico or clean cloth. Store at about 55–60°F (13–16°C). Turn every day for a week and thereafter about twice a week.

"*After all these romantic yearnings for the fine and the primitive we ended up with an Aga. It sits silent and brooding as though it has some atomic pile in its belly. It almost never goes out, it keeps the water beautifully and constantly hot, it will boil a kettle in a few moments at any time at all, night or day, it is marvellous for making bread and — an enormous economy — it is fine for boiling up potatoes for the pigs. Now, every night, we simply put a big iron pot full of spuds and water in the slow oven, and take it out in the morning when we make the early morning tea.*

The great objection to fattening pigs on potatoes — labour — is out.

And any time of the winter or summer, our living kitchen is warm to go into, there is a fine hot cupboard for drying clothes, and the stove fairly economical with fuel, which of course we have to buy.*"

JOHN SEYMOUR FAT OF THE LAND 1976

CHAPTER SEVEN

IN THE
KITCHEN

The North-Facing Storeroom

Sound storage is an absolute must in my book for any would-be self-supporter. Whilst rearing, growing, and brewing are the nitty-gritty of self-sufficiency, good and effective storage has its own rewards, too. Sadly, this is often only learned by bitter experience, as wonderful produce from the garden becomes useless fodder for the compost heap long before it can be eaten. You simply cannot eat all the food you grow as you harvest it – and you wouldn't want to, would you, since the whole point of being self-sufficient is having home-grown produce all year round!

Pantries, larders, dairies, and cellars

For thousands of years humans managed to live without powered fridges, deep freezes, tin cans, plastic wrapping, and all the other modern-day conveniences of consumerism. And it wasn't so long ago when houses were always built to include pantries, larders, dairies, and cellars – the storerooms. If you look at traditionally designed old farmhouses you will see how carefully the builders took advantage of the lie of the land. All the old Northumbrian farmhouses were aligned east-west: the desire was simply to ensure that the front of the house faced due south and had the benefit of what sun's warmth there was throughout the day. All of the rooms (apart from the pantry and the dairy) faced south. The corridor and stairs were on the north side of the house, forming an insulating barrier against cold north winds.

If you are fortunate enough to have a house with a cellar then many of your storage problems will be eased. If you are building your own house, or an extension, then think carefully how you could incorporate a cellar (at least part of which should be below ground level). A cellar keeps its cool and constant temperature by being under the earth. An old cellar will tend to be very damp and under-ventilated. You need to watch out for this and take steps to keep damp out of your stored produce. You may even want to increase ventilation by knocking a couple of holes through to the outside and putting in mesh to avoid flies and mice.

Unfortunately, cellars are not that commonplace in some countries, so most self-supporters will have to improvise: and the north-facing storeroom is my suggestion to solve the problem. If you cannot muster the resources to build such a space then you can at least make do with the smaller and simpler device of a "meat safe". My preference is to construct a storage area against the north wall of the house, or the east wall if there is no convenient north wall, to provide the following six ingredients:

1 Temperature Good storage temperature should be cool, but without frost, and should not change quickly. This will work for wines, beers, jams, pickles, and all vegetables. The underground cellar is the perfect answer as far as temperature is concerned, but it will suffer from lack of ventilation and may be difficult to protect from flies and rats.

A north-facing storeroom or traditional pantry can avoid these problems even though it may be warmer than we would like if the weather is hot.

2 Humidity If a place is too wet, then moulds and fungus will thrive, string or muslin will rot, and paper will be worse than useless. On the other hand, if your storage is too dry your vegetables will quickly shrivel up. For most purposes (in a northern climate) the ambient air humidity is reasonable enough. If there are sufficient vegetables in the store, then they will also create their own humidity as water slowly evaporates from them.

3 Security against pests Rats and mice get in through the tiniest of tiny holes but they will also remorselessly gnaw away at wood, cement, or plastic when they scent the prospect of food. Plastic and wood will always be suspect materials at the base of doors, whereas aluminium sheeting is a good resistant material if you really want to keep the pests out. Hanging food from a rafter is useless protection against mice, but if you hang from rings screwed into a flat, smooth ceiling then you will keep the little beasts at bay. Having cats, terriers, or hawks about will also disrupt rat or mice sorties.

4 Ventilation Fly-proof mesh (and plenty of it) is essential. Make sure you have ventilation at the top and bottom of the store or in line with the prevailing wind in order to keep a flow going through your store. Ventilation prevents mould and fungus even though it will tend to dry things out; so if the weather is very warm and windy you may want to cover some of your vents.

5 Flies Unless you plan to store uncooked meats, it doesn't have to be perfectly fly-proof. My preference would be to have a special "meat safe" kept specifically for this purpose and that should be as fly-proof as you would want. Obviously, the more you discourage flies, the better, and not just blow flies but also the moths that could lay eggs of vegetable-eating grubs on your produce. Cover individual items with muslin if you are worried about them.

6 Sunlight Direct sunlight not only creates unwanted warmth but it can also discolour and dry out foodstuffs. Wines and beers should be kept out of the sun whilst they mature - hence the traditional use of brown glass bottles.

Other traditional cooling methods

Don't forget that you can keep things cool by simply standing them in water, in the shade, and covering with a cloth. The water will soak up into the cloth, which is then dried by the wind. As those of you may have managed to get through a physics course in secondary school will know, water has a high latent heat of evaporation. Water cannot simply turn to water vapour without sucking in heat from its surroundings. So when the wind evaporates the water, it creates a significant cooling effect. The same principle applies when large trees are grown around city squares – the thousands of gallons of water that are evaporated every hour by the sun and wind cool down the air temperature.

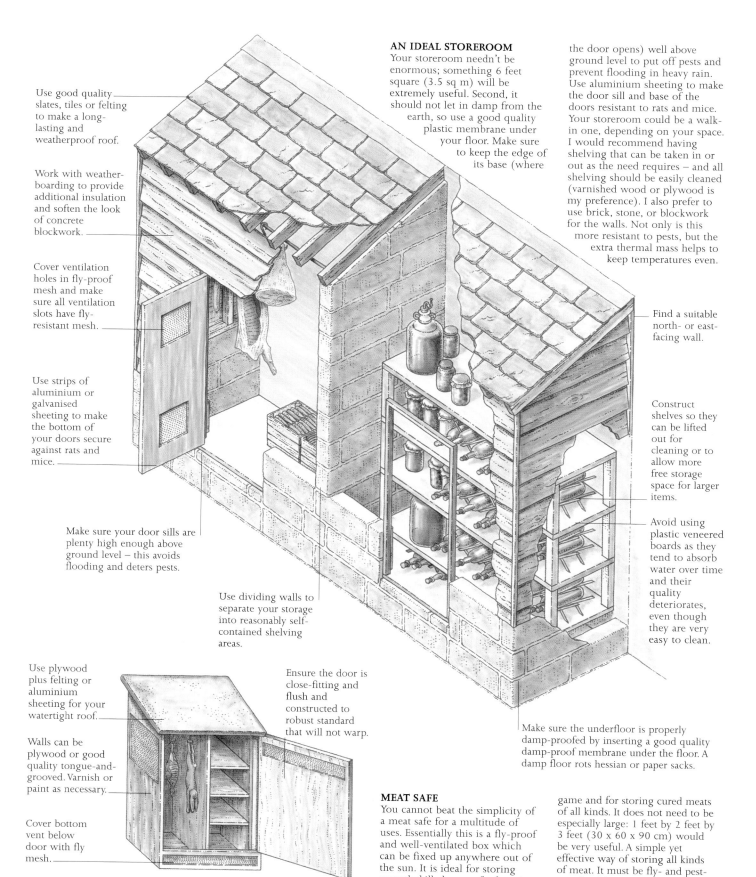

Use good quality slates, tiles or felting to make a long-lasting and weatherproof roof.

Work with weather-boarding to provide additional insulation and soften the look of concrete blockwork.

Cover ventilation holes in fly-proof mesh and make sure all ventilation slots have fly-resistant mesh.

Use strips of aluminium or galvanised sheeting to make the bottom of your doors secure against rats and mice.

Make sure your door sills are plenty high enough above ground level – this avoids flooding and deters pests.

Use dividing walls to separate your storage into reasonably self-contained shelving areas.

AN IDEAL STOREROOM

Your storeroom needn't be enormous; something 6 feet square (3.5 sq m) will be extremely useful. Second, it should not let in damp from the earth, so use a good quality plastic membrane under your floor. Make sure to keep the edge of its base (where the door opens) well above ground level to put off pests and prevent flooding in heavy rain. Use aluminium sheeting to make the door sill and base of the doors resistant to rats and mice. Your storeroom could be a walk-in one, depending on your space. I would recommend having shelving that can be taken in or out as the need requires – and all shelving should be easily cleaned (varnished wood or plywood is my preference). I also prefer to use brick, stone, or blockwork for the walls. Not only is this more resistant to pests, but the extra thermal mass helps to keep temperatures even.

Find a suitable north- or east-facing wall.

Construct shelves so they can be lifted out for cleaning or to allow more free storage space for larger items.

Avoid using plastic veneered boards as they tend to absorb water over time and their quality deteriorates, even though they are very easy to clean.

Use plywood plus felting or aluminium sheeting for your watertight roof.

Walls can be plywood or good quality tongue-and-grooved. Varnish or paint as necessary.

Cover bottom vent below door with fly mesh.

Ensure the door is close-fitting and flush and constructed to robust standard that will not warp.

Make sure the underfloor is properly damp-proofed by inserting a good quality damp-proof membrane under the floor. A damp floor rots hessian or paper sacks.

MEAT SAFE

You cannot beat the simplicity of a meat safe for a multitude of uses. Essentially this is a fly-proof and well-ventilated box which can be fixed up anywhere out of the sun. It is ideal for storing recently killed game, for hanging game and for storing cured meats of all kinds. It does not need to be especially large: 1 feet by 2 feet by 3 feet (30 x 60 x 90 cm) would be very useful. A simple yet effective way of storing all kinds of meat. It must be fly- and pest-proof and kept in a cool place.

Harvesting & Storing

Here are a few golden rules that I suggest you follow when it comes to harvesting your hard-won crops. Don't harvest more than you can process or store, or it will simply be wasted unless, of course, you can give it away and improve your social credit ratings. Be absolutely careful not to damage the crop when you pick it or dig it. You must never try to store substandard or damaged produce. Harvesting, processing, and storing are part of a seamless process, each dovetailing into the other and all useless without the others. In many ways the excitement of the harvest is often tempered by the realities of processing and storage.

VEGETABLES

Clamping is the traditional way to store a variety of root crops. The problem for the smallholder is that this is not a very effective method for small quantities. Hence my suggestion that you try using an old deep freeze or two as a vegetable store. It can work surprisingly well, and it keeps out the rats. For beetroot and other roots, I have found hessian or stout paper sacks very good if kept in the confines of a good storeroom or cellar.

Beans and peas should be dried and stored away in great quantities every autumn. When they are thoroughly dried, threshed, and winnowed, store them in crocks, barrels, bins, or other mouse-proof places. Mushrooms and most fungi dry out at an ideal temperature of 120°F (50°C) so crumble them afterwards into a powder and store them in closed jars. The powder is marvellous for flavouring soups and stews. Sweet corn is excellent dried: it really is a thing worth having. Boil it well on the cob, dry the cobs in a slow oven overnight, cut the kernels off the cobs, and store them in closed jars. When you want to eat them, just boil them.

FRUIT

As a rule the early-maturing varieties of apples and pears will not store well. So eat them as you pick them, and store only late-ripening varieties. Leave these on the trees as long as possible and only pick them when they are so ripe that they come off if you lift them gently. Pick them and lay them carefully in a basket. Then spread them out gently in an airy place to let them dry overnight. The next day store them in a dark, well ventilated place at a temperature of 35°–40°F (2°–4°C). Pears like it very slightly warmer.

Ideally each fruit should be wrapped individually in paper to isolate any moulds or bacteria. Only perfect fruit can qualify for storing: so disqualify any with bruises, cuts, or missing stalks. If the floor is earth, stone, or concrete, you can throw water on it occasionally to keep the air moist. Storing fruit in a hot, dry attic is simply giving pigs a treat.

Apples may well keep until spring. If you fear they won't keep long enough, you can happily dry them. Core them, slice thinly, string up the slices, and hang over a stove, or in a solar-heated drier (see p.244), at 150°F (65°C) for 5 hours. When crisp and dry put them into an airtight container and store in a cool place.

LIFTING POTATOES

Any damage to your potatoes or other root crops, especially carrots, will be multiplied twenty-fold by storage – roots are very easily bruised. Always sort the crop carefully into three piles: the first for storage, these are perfect and large; the second for more immediate consumption, these may be damaged or too small to store; and the third for the compost heap or burning, these are diseased, too small or too damaged to use. Remember that your crop must have time to dry out, preferably on a breezy day, before being put into storage. Also remember that the places you are going to store your crop must be scrupulously cleaned and dry – as must any straw.

DIGGING CARROTS
When digging carrots for storage make sure your fork is well away from the roots. Put the fork right under and pull gently up with your hand grasping the foliage.

SELECTING CARROTS
Shake or brush the soil free, never wash in water, before you start selecting. Now sort your crop carefully into three piles: one to store, one to use, and one to dispose of.

STORING POTATOES

1 Use an old deep freeze for storing potatoes. First place a layer of clean, dry straw across the base.

2 Tip in several baskets of dry, undamaged tubers until you have a layer perhaps 1 foot (30 cm) deep. Then add a further layer of dry, clean straw.

3 Tip in the next layer of spuds. Then put a final layer over the top before closing down the lid on to a small chock to allow in air but keep out rats.

CLAMPING

Clamping is a method of protecting root crops in the open, where diseases do not build up as they can in a cellar. But no clamp keeps out hard frost, so in very cold winters you must store indoors.

1 When you pick potatoes for clamping you should let them dry for 2 or 3 hours first. Prepare the clamp by putting a layer of straw on the ground.

2 Heap the potatoes (or other root crop) up on top of the straw in the shape of a pyramid so that when it is finished, rain will drain off.

3 Cover with a layer of straw or bracken. Allow a period for sweating before covering with earth.

4 Cover with a layer of earth 5 or 6 inches (13–15 cm) thick. Beat the earth flat with the back of a spade.

5 Make sure that bits of straw protrude from the clamp to admit some air to the crop inside.

STRINGING ONIONS

You can store onions on trays with slats, on polythene netting, or on a wooden stand. But the ideal way of keeping them is to string them up in a cool place with access to plenty of air. Before you store your onions, always remember to dry them thoroughly first, either by leaving them on the ground in the sun, or covered but in the wind if it is wet.

1 Make sure that all the onions you want to string have long stalks. Start by knotting four of them firmly together.

2 Add onions one by one to the original four. Twist their stalks and knot them tightly round string, baler, or binding twine.

3 Continue adding individual onions to the growing bunch, ensuring that each one is securely tied on, and that the bunch does not become too heavy.

4 Plait the knotted stalks round the end of a long piece of string so that the onions hang evenly when you hold them up.

5 Hang the string up when you decide the your bunch is complete. The onions should keep indefinitely.

OTHER STORING METHODS

HANGING IN NETS

Squashes (marrows or pumpkins) will keep best if hung in nets, although they can be stored on shelves, if turned occasionally.

STORING BEETROOT

Use dry sand so that the roots don't touch and keep safe from frost.

FLATBED STORING

Late-ripening apples last all winter if you keep them in a cool, dark place, but be sure that they aren't touching each other. A storing bed (see below) is ideal for this. You could even wrap each one in paper also. Store onions on wire netting. If it rains, they should be in the wind, but under cover.

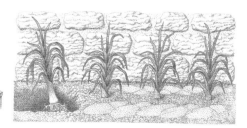

Apple storing

Wire-mesh onion drying bed

HEELING IN LEEKS

If you fear that your leeks, celery, and Jerusalem artichokes might be exposed to frost in the open, "heel" them into dry, sheltered ground near the house, where they will derive some protection from harsh weather. Otherwise they are generally best left in the ground until required.

Making Bread

There is white bread and wholemeal bread, and many gradations between the two. There is leavened bread and unleavened bread, and again many gradations. There is sourdough and soda bread, pitta bread and flat bread, but the great thing for the self-supporter to remember is that whatever kind of bread you choose to make, and whatever kind of grain you make it from, the process is simple. It is also fun, and even the most ham-fisted cooks can take pleasure and pride in their efforts.

Undoubtedly the first breads were unleavened, and undoubtedly the first person who discovered yeast discovered it by accident. If you make a dough with flour and water without yeast or baking powder and then bake it, you will be left with something very like a brick. People got over this by rolling the dough out very thinly and cooking it that way. (In Baghdad to this day you will see bakers putting great sheets of thin dough – as big as small blankets – into enormous cylindrical ovens.) But no doubt one day someone mixed up some dough, didn't cook it immediately, and found the stuff began to ferment.

What had happened was that wild yeasts had got into it and were converting the sugar (in the flour) into alcohol and carbonic acid gas. The alcohol evaporated, but the carbon dioxide blew the glutinous dough up into bubbles. This unknown ancient took up the bubbling doughy mass and placed it upon his hot stone or maybe into a little hollowed-out stone oven and made what was the first leavened bread.

It was then found that bread could be made not in thin sheets but in thick loaves, and was still good to eat. Furthermore, it was discovered that leavened bread stays palatable longer than unleavened bread: good home-baked wholemeal bread can taste fine for 5 days or more, while unleavened bread tastes very dreary unless you eat it when it is still absolutely fresh.

Yeast

How long it took mankind (or womankind) to figure out the true nature of that lovely stuff yeast, we will never know. But certainly they must have found that if they were lucky enough to get a good strain of wild yeast in their dough, they could go on breeding it – simply by keeping a little raw dough back from each baking to mix in with the next batch of bread. The old pioneers in the Wild West were called "sourdoughs" because they made their bread thus. And even today, people out of touch with bakeries and yeast suppliers commonly make bread with sourdough.

If you live near a bakery, always buy your yeast fresh. It should be a creamy putty colour, cool to the touch and easy to break, with a nice yeasty smell. Don't buy any that is crumbly or has dark patches. It will keep for 1 week to 10 days in a screwtop jar in the fridge. Or cut it into 1-inch cubes and freeze it. Both yeast and bread freeze well. If you cannot obtain fresh yeast, you can still make a perfectly good bread with dried yeast.

This is widely available in packets, and it will keep for up to 3 months. But it is a good idea to test dried yeast if you have had it around for some time. Drop a few grains into a little warm liquid dough mix; if it is still "live" it will froth in under 10 minutes.

If you are using fresh yeast for any recipe specifying dried yeast, always double the quantity. Or halve it if the recipe asks for fresh yeast and you are using it dried.

Yeast flourishes in a warm atmosphere in temperatures between 48–95°F (9–35°C), but strong heat – over 140°F (60°C) – will kill it. Set your dough to rise in a warm place: on top of the stove, in the airing cupboard, even under the eiderdown on the bed.

If you are brewing beer (see pp.222–225), you can use your beer yeast for your bread-making. Conversely, you can use your bread yeast in your beer-making. Neither is ideal because they are two different sorts of yeast, but we have done it often and we get surprisingly good beer and good bread.

Kneading

A word about kneading. Kneading is important because it releases the gluten and distributes the yeast right through the dough. Don't be afraid to treat your dough fiercely when you knead it. Push and pull it about until it seems to take on a life of its own, becoming silky and springy in your hands. Then leave it alone to prove, that is, to rest for a few hours until it has risen. When it has risen enough it should jump back at the touch of a finger.

Keeping

If you don't have a freezer keep bread in a dry, cool, well-ventilated bin. Don't put it in an airtight container or it will go mouldy. Make sure the bread is quite cool before you stow it away or the steam in a warm loaf will make it turn soggy. Keep your flour in a dry, dark, cool cupboard.

There's much more to bread than white sliced or wholemeal. And we should all be thankful for that. Bread can be made from soya, rye, wheat, corn, sorghum, or oats. If you vary your grain, you vary your bread. It's as simple as that. Have it leavened or unleavened, plain or fancy, or try a mixture of flours.

Bread at its most basic, as we have just seen, is simply yeast, flour, salt, and water. Add milk, butter, eggs, sugar, honey, bananas, carrots, nuts, and currants and you will enrich your bread, change its taste and texture. Roll it in wholewheat grain, or poppy seed, sesame, dill, celery, caraway, sunflower, or aniseed as you please. Brush it with milk, paint it with egg yolk. Shine a currant loaf with sugar syrup. Knot it, twist it, plait it. Experiment, and you will find that being your own baker is one of the great joys of the self-sufficient way of life.

On the following two pages I describe some of the different flours and a variety of breads you can make in your own home – and how to make bread the traditional and tasty way.

Bread made with different flours

For people who grow rye, barley, oats, maize, rice, sorghum, and the rest, it is useful and interesting to try some breads made with these grains, or with them mixed with wheat flour (*see also p.214*). It must be remembered that of all the grains, only wheat has enough gluten to sustain the gas generated by the living yeast sufficiently to make fairly light, or risen, bread.

You can try a combination of two or three different flours, but it is usually worth adding some wheat flour. And always add salt. Oil, butter, lard, or margarine help to keep bread moist. Water absorption varies with the sort of flour. Here is a rundown of the different flours:

Wheat flour Wheat flour is rich in gluten, which makes the dough stretch and, as it cooks, fixes it firmly round the air bubbles caused by the leavening.

Rye flour Rye flour gives bread a nice sour taste, and can be used on its own, although a lighter bread will result if half or a third of the flour is wheat flour. Maslin, flour made from rye and wheat grown together and ground together, was the staple English flour of the Middle Ages. Only the rich ate pure wheaten bread.

Barley flour Barley flour alone makes very sweet-tasting bread. A proportion of a third barley flour to two-thirds wheat produces good bread. If you toast the barley flour first your bread will be extra delicious.

Oatmeal Oatmeal is also sweet and makes a very chewy, damp bread, which fills you up very nicely. Use half oat and half wheat flour for a good balance.

Cornmeal Bread made from cornmeal has a crumbly texture. Try half cornmeal and half wheat flour.

Ground rice Ground rice bread is a lot better if the rice is mixed half and half with wheat flour.

Cooked brown rice Like the whole cooked grains of any other cereal, cooked brown rice can be mixed with wheat flour to make an unusual bread.

Sorghum By itself sorghum (or millet) flour makes a dry bread. Add wheat flour and you will get nice crunchy bread.

Soya flour Soya flour too is better mixed with wheat. The soya flour adds a lot of nourishment.

Bread made without yeast

Unyeasted bread is really solid stuff, quite unlike yeasted bread which is, after all, half full of nothing but air. To my mind it can only be eaten cut very thin. Warm or even boiling water helps to start softening the starch in the flour. Kneading helps to release the gluten. If unyeasted dough is allowed to rest overnight the bread you make will be lighter, as the starch will soften more and a little fermentation will begin. The carbon dioxide released will provide a few air holes.

I suggest the same proportions of whole wheat to other flours as with yeasted bread. Other ingredients need be nothing but salt and water, and perhaps oil to brush the tops of loaves. Knead well, and leave to prove overnight.

Unyeasted bread may need longer and slower cooking than yeasted bread. It will also need good teeth.

Standard wholemeal bread

I never measure my flour because what matters is getting the dough to the right consistency, and flour absorbs more or less water according to its fineness, quality, etc. But for people who must have exact quantities of everything this is what Sam Mayall, an experienced English baker, who grows and mills his own wheat, uses:

2½ lb (1 kg) of wholemeal flour
1 oz (28 g) salt
½ oz (14 g) dried yeast
2 teaspoons soft brown sugar
1¼ pints (0.7 litres) water

Put the flour and salt in a large bowl. Put the yeast in another bowl, add the sugar and some warm water. Leave in a warm place to rise.

When the yeast is fermenting well, add it to the flour and the rest of the water, and knead it till it is soft and silky in texture. Return it to the basin and leave it to stand in a warm place until it has about doubled its size. Knead it again for a few minutes and mould into loaves. Place in warmed greased and floured tins, and, if it is soft wheat flour, leave it to rise for 5 minutes. If it is hard wheat flour, allow longer, up to 20 minutes. Put in an oven of 425°F (220°C) for 45 minutes.

Maize bread

Maize bread tastes good. It is crunchy and rather gritty and should have a nice brown crust. You will need:

1½ pints (0.8 litre) boiling water
2 lb (0.9 kg) maize flour
2 teaspoons baking powder
3 eggs (optional)
½ pint (0.3 litre) buttermilk (optional)

Mix the maize meal with the baking powder and pour on the boiling water. Adding eggs and/or buttermilk improves the bread. Bake in a greased tin at 400° F (200° C) for about 40 minutes.

Sorghum bread

This is a rather dry bread, and only really worth making if sorghum is all you've got. Sorghum is much better mixed with wheat flour. You need:

12 oz (340 g) sorghum flour
1 teaspoon baking powder
1 teaspoon salt

Mix the ingredients and wet with warm water to make a stiffish dough. Bake for about 50 minutes in a moderately hot oven at 350°F (177°C).

Oat bread

In those damp parts of the world where nothing else will grow, oat bread is common. It is heavy and sweet-tasting. To make it you will need the following ingredients:

1 lb (0.5 kg) rolled oats or oat flour
3 oz (84 g) sugar or honey
1 tablespoon salt
4 oz (114 g) butter
1 pint (0.5 litre) boiling water
    (use a little less if you use honey)
1 oz (28 g) yeast or ½ oz (14 g) dried yeast

Mix the dry ingredients well, rub in the butter, and add the boiling water. Dissolve the yeast in a little tepid water. When it begins to froth, mix it well in with the other mixture. Leave to rise for a few hours. Then dump dough on a floured board and knead it for about 10 minutes. Cut and shape into rounded loaf-sized lumps, and allow for some expansion.

Put on a baking tray in a warm place and allow to expand for about an hour. Then bung in a very hot oven of 450°F (232°C) for 45 minutes. Test as usual by tapping the bottoms of the loaves to see if they are hollow. Stand to cool on a wire tray so that the air can circulate all round them.

MAKING BREAD

If you can boil an egg, you can bake bread. There is absolutely nothing difficult about it. To make six medium loaves, take 4½ pints (2.3 litres) of water, warmed to blood heat, 2 ounces (56 g) of salt, and the same amount of brown sugar, 1 tablespoon of fresh yeast (or half this amount of dried yeast). You can even use yeast from the bottom of your beer kive.

1 Put all the ingredients into a large mixing bowl. When the yeast has dissolved, pour in enough flour to make a fine, sticky mash. Stir this well with a wooden spoon.

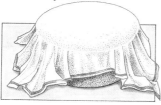

2 The spoon should stand just about upright. Cover dish with a cloth and leave it overnight in a warm place free from draughts.

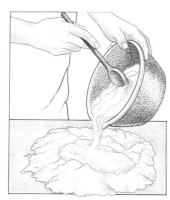

3 Come morning the yeast will have the dough spilling over with enthusiasm. Heap some dry flour on to a table and dump the dough into the middle of this.

4 Sprinkle dry flour on top of the dough and it is ready for kneading. Start by mixing the dry flour with the wet dough.

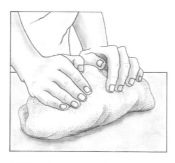

5 The aim is to make a fairly stiff dough, dry on the outside. You do this by pushing the dough away from you with the palms of your hands (*above*) and then pulling it towards you again (*below*). This is kneading and it is a very sticky process. When the dough sticks to your hands (and it will), fling on some flour. Whenever it feels sticky, sprinkle flour.

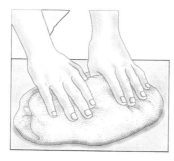

6 Kneading must be done thoroughly – you need to push and pull and fling on the flour until you have a dry, satisfying little ball. Roll it about to your heart's content. But after 10 minutes the fun has to stop. It is nearly ready for baking.

7 Divide the dough into six equal portions. Grease the baking tins and shape your dough. Fill the tins just three-quarters full. Score patterns on top with a knife and leave covered for about an hour in a warm place.

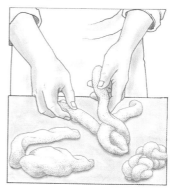

8 If you want to be more decorative, make a plaited loaf. Divide the dough into three, make each sausage-shaped, and plait. Just like that. If you want to, brush the top with milk to make it shiny and sprinkle with poppy seeds.

9 Shape little rolls with left-over dough. Put them on a baking tray and leave on top of the stove to rise. After 30 minutes put them in a very hot oven (450°F/ 232°C). In 10 minutes you will have magnificent breakfast rolls.

Now pick up your bread tins very carefully. If you jog them they will collapse and you will have solid bread, so gently ease them into a hot oven (425°F /220°C). Half an hour later take a look to see if they are cooking evenly. Turn them round if necessary. Wait another 15 minutes and they should be done.

10 To test them, tap the bottoms. If they sound hollow they are done. Or push in a skewer – it should come out clean. If it doesn't it is not a disaster; put them back for a few more minutes.

11 When you are sure your loaves are good and ready, take them out and stand on top of their tins to air.

TEMPERATURE AND TIME

Bread rises (and yeast ferments) best at 80°F (27°C). Yeast will die at any temperature much over 95°F (35°C) and it won't multiply under 48°F (9°C). So the place where you set the bread to rise must fall within these temperatures. Usually the top of the stove is ideal. The oven should be – well, hot. Apart from the time you spend waiting for things to happen, you probably don't spend more than half an hour working the dough, and the result is six beautiful wholemeal loaves.

Preserving

It is autumn and you have a surfeit of all the crops you have been gathering through the summer. What more fulfilling than to bottle, pickle and preserve in all possible ways for the dark days of winter ahead? Now read on… The harvest season is short for most things, although in a temperate climate it is possible to pick fresh green things every day of the year. Few things can equal the pleasure of coming fresh to new green peas at the beginning of their season after 6 months of pea-abstinence. The palate, jaded and corrupted by months of frozen peas, or quick-dried peas masquerading as fresh garden peas, gets no real pleasure from these specimens. True dried peas, cooked as pease pudding, or put in soups and stews, are quite another thing. They are a traditional time-honoured way of preserving plant protein for the winter months, and eating them all winter does not jade the palate for the fresh garden pea experience every June.

At the same time there is, potentially, a vitamin shortage in the dark winters, and those dark, cold days should be enlivened by nice tastes and odours besides that of salt bacon. So the self-supporter will wish to preserve certain things, preferably by a process which improves their natural flavour, such as bottling, pickling, chutneying, or wine-making. There is nothing more encouraging in autumn than the sight of shelves heavily laden with full jars and crocks. More than anything, they give you the feeling that you are likely to survive the winter. This may sound like a contradiction but it isn't.

There is a place – and certain suitable foods, especially berries – for freezing (see pp.206–207). Although it is an effective storage system, you may not improve any food by deep freezing, but you actually improve fruit and vegetables by making them into chutney, jam, and the like. Freezing meat is another matter: unless you are very hungry you cannot eat a bullock before it goes bad. In more sensible times people killed meat and shared it. Now the whole principle of sharing with neighbours is forgotten and the cold of the deep freeze often replaces the warmth of neighbourly relations.

Wine

Wine-making (see pp.226–229), much like beer-making (see pp.220–225), turns sugar into alcohol. Some fruits, such as grapes grown in a warm climate, have so much natural sugar in them that you don't have to add any. But many of the things you can make wine from are low on sugar. So you will have to add sugar if you want alcohol of a decent strength. And remember that weak wine won't keep, it just goes bad. Some "wine" described in books of wine recipes is simply sugar-water fermented and flavoured with some substance. Most flower wines (see p.228) are made like this, and people even make "wine" of tea leaves – that sugarless substance!

Fruit wines have their own sugar, but usually not enough, so you must add some. The same goes for root wines. Parsnip, which is by far the best, has quite a lot of sugar.

What country wines do is to preserve and even enhance the flavour and bouquet of the things they are made of. They cheer us up in the dark days of winter and are very good for us too.

Chutneys and pickles

You make both chutneys and pickles by flavouring fruit or vegetables, or a combination of both, with spices and preserving them in vinegar (see pp.210–211). The methods of preserving, however, do not resemble each other.

Chutneys are fruits or vegetables which have been cooked in vinegar, often heavily spiced and sweetened. They are cooked until all excess liquid has evaporated, leaving a thick pulp, the consistency of jam. The flavour is mellow. Pickles are put down whole or in large chunks in vinegar, but not heated in it. Anything which is to be pickled must not have too much moisture in it. So sometimes moisture must be drawn out first with salt. The resulting flavour is full and sharp.

Both chutneys and pickles are an excellent way of preserving things for the winter and of enhancing their taste as well. They are delicious with cold meats and meat pies, and also offset the taste of curries or cheeses.

Ketchups and similar sauces are strained juices of fruits or vegetables spiced and cooked in vinegar. These, too, if well made, can give a lift to plain food.

Bottling

The principle of bottling is very simple. Food is put in jars, and both jars and their contents are heated to a temperature which is maintained long enough to ensure that all bacteria, moulds, and viruses are destroyed; at this point the jars are completely sealed to prevent any further pathogens from getting in, and then allowed to cool. Thus the contents of the jars are sterilized by heat, and safe from attack by putrefactive organisms.

The same principle applies to tinning, or canning, except that the product is preserved in an unattractive steel box. It is also a process that the self-supporter will find considerably less easy than bottling.

Fruit bottles very well. Vegetables are far more difficult, because they are low on acid, and acid makes food preservation easier.

My own feeling about the bottling of vegetables is: don't do it. What with salted runner beans, sauerkraut, clamped or cellared roots or cabbages, and, in all but arctic climates, quite a selection of the things that will grow and can be picked fresh out of doors all winter, there is no need for the rather tasteless, soggy matter that vegetables become when they have been bottled. On the other hand, tomatoes, which aren't strictly speaking a vegetable, are a very good thing to bottle indeed. They give a lift to otherwise dull winter dishes like nothing else can. They are easy to bottle, you can grow a big surplus during their short growing season, they are rich in vitamins, and they taste delicious.

Freezing

Virtually all fruits, vegetables, and meat can be stored by freezing. Some freeze better than others, and you will discover just what you like and what you do not. Freezing preserves by making it too cold for bacteria and moulds to function – but the downside is that ice takes up more space than water so the cells within foodstuffs are broken and the texture, and even the flavour, of foods can change.

You may think that by using freezing as a major method of storage we are courting disaster if your electricity supplier cuts off your power, or if the weather or other natural elements disrupt it. There are two remedies for this: the paper remedy is to take out insurance; the practical remedy is to organize yourself with a standby generator. You can buy reliable standby generators at a fairly reasonable price and the power requirements of a deep freeze are fairly minimal. We are so dependent on electricity, which is, after all, a very convenient power source for many things, that a standby generator is well worth the effort.

As with all food storage, we should only process good quality raw materials. Be fussy about this. Damaged fruit and vegetables will not freeze well and you risk wasting time and space that could be better used for other things.

The deep freeze is like a bizarre frozen filing cabinet and, like all filing systems, it needs to be kept in order. Not only must you have your foods clearly labelled – with labels that do not come off when frozen – but you must also rotate your usage so the oldest items are used first. To give you some idea of how we manage, our smallholding has three fairly large chest freezers: one is kept for meat; the other two for fruit, vegetables, ice-creams, and other delicacies we have a mind for. At least once each year we arrange to empty one of the freezers to give it a good defrosting – and almost every year we panic to use up or give away fruit we have been carefully hoarding through the winter because June has arrived and the next crop is due. Possibly we have links to prehistoric squirrel species.

TIPS FOR FREEZING
• Do not skimp on the quality of raw foodstuffs.
• Use foods in rotation.
• Defrost freezers regularly.
• Use good-quality freezer bags.

The longer food is stored, the more moisture it loses; so, after much trial and error, we now use the expensive zipper bags because they are tough, they lie flat, and you can use them several times.

Large boiling pan

Bowl

Knife

Scissors

Jar of sugar

Chopping board

Colander

Zip top freezer bags

FLASH FREEZING

1 For best results many foods are best "flash" frozen, especially mushrooms and soft fruits. Spread the fruit on a tray or on newspaper and put straight into the deep freeze uncovered for 24 hours.

2 Put the frozen foods into plastic bags. As they are already frozen they will not stick together, making them more attractive when cooked.

FREEZING BLACKCURRANTS

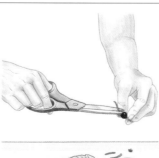

1 A freshly picked basket of blackcurrants is full of stalks, leaves, and bugs. Take out as much of this unwanted debris when you are picking and you will save time later.

2 Cut off stalks, take out leaves and debris, and chuck bugs into the compost bin. This is actually a very pleasant job to do socially after supper on a fine summer evening.

3 Fortunately, fruit does not need any special treatment before being frozen. However, it is a good idea to spread out the fruit to be frozen so they don't touch before flash freezing.

4 Now put the fruit into good-quality plastic bags ready for storage in the freezer. Write the contents and date on the bag. Self-sealing or "tuck-and-lock" freezer bags are best.

PLUMS

Everyone looks forward to the plum harvesting season. After all, what can be better than a fragrantly fresh plum straight from the branch? I pick my plums by making sure the grass under the tree is well cut and cleared, then I simply give the tree or the branch a gentle shake. Ripe plums will rain down on your head – possibly with a few angry wasps to boot. Picking them up from the ground is a pleasant job – and if your turf is soft there will be little bruising. Watch out for dozy wasps and get the plums inside for sorting and processing as soon as you can. Plums keep reasonably well for a few days but the flavour definitely suffers over time.

BEANS, MANGETOUT, AND SUGAR SNAPS

All these vegetables freeze very well and the basic approach to preparation and freezing is the same. First of all, harvest your crop when it is perfectly ripe. Check the crop regularly, for you do not want food that is stringy. This may mean picking every day. Get into a good routine of picking and processing a little every day as the crop progresses. Young and tender food is the essential aim and the more often you pick, the better the plant will respond by producing more. Different varieties of pea crop in different ways – there are dwarf varieties, which can crop for weeks on end, but others may produce almost the whole crop in just two or three pickings.

1 Split each plum with a sharp knife and take out the stones. It's a great way to sit down with friends and enjoy a chat while working.

2 Put the halved plums straight into your freezer bags, making sure that no leaves or stalks manage to creep in, too.

3 Add 3 or 4 tablespoons of sugar to each bag. Give it a shake to spread out the sugar, squeeze out the air, and seal for the freezer.

1 Sort and trim the pods, getting rid of unwanted leaves, damaged pods, and stalks. Top and tail by cutting off each end of the pods. You can slice three or four beans at a time but watch your fingers!

2 Beans must be sliced diagonally into thin slices with a sharp knife or by using a cunning bean slicer. Mangetout and sugar snaps are frozen whole.

3 Put a large pan of water on the stove to boil and tip your pods into the boiling water. Leave for a minute or two until it comes back to the boil.

4 Drain off the pods after they have been "blanched". Plunge them quickly in cold water to stop them cooking, give them a good shake to get rid of any water and let them drain. You may need to plunge into cold water a few times to cool them enough.

5 Once cooled, you are ready to bag up the pods for freezing. Be careful to squeeze out all the air in the bag before you seal it. Purpose-made freezer bags with resealable tops and strong plastic that resists tearing are a great help for this.

MANGETOUT

Simply top and tail mangetout, or sugar snaps with a sharp knife before you blanche them in boiling water. Try to remove any "strings" from the peas at the same time as you top and tail.

PEAS

Ordinary peas freeze very well. Pod the peas, take out any leaves or pod and blanche them before freezing. Have your pig bucket handy to put the "waste" in straightaway.

Bottling

Glass jars for bottling must have airtight tops, capable of supporting a vacuum, and arranged so that no metal comes into contact with the contents of the jar. If you examine the common "Kilner" jar, or any of its rivals, you will find quite a cunning arrangement ensuring that the above requirements are met. A rubber ring compressed by a metal screw-cap forms an airtight seal, and only the glass disc inside the screw cap comes into contact with the jar's contents. Kilner and other proprietary jars need the metal parts smearing with vaseline to prevent them from rusting, both when in use and when stored away. Keep the rubber rings in the dark, for light perishes rubber.

To bottle you also need a container in which jars can be boiled. If you buy one it should have a false bottom, so that the jars are not too close to the source of heat. Alternatively, put a piece of board in the bottom, or else just a folded towel. When bottling fruit pack the jars as tightly as you can; tapping the base of the jar on the table helps to settle the fruit, and drives air bubbles out.

BOTTLING FRUIT
Cold water bath method
Put the fruit into jars of cold brine or syrup and put the jars in cold water. Take an hour to bring water to 130°F (54°C), then another half hour to raise it to the temperature given on the chart below.

Oven method
Fill the jars, not putting any syrup or brine in them yet, and covering them with loose saucers only. Put them in a low oven at 250°F (121°C). Leave them for the time given in the chart, take out and top up with fruit from a spare jar that has undergone the same process, then fill up with boiling brine or syrup, screw on the tops, and leave to cool.

Hot water bath method
If you have no thermometer, and no oven, use the hot water method. Fill packed jars with hot syrup or brine, put the lids on loosely, lower into warm water, bring to the boil, then simmer for the length of time shown on the chart below.

For fruit other than tomatoes, use a syrup of sugar and water if you wish. Water alone will do, and if you pack the fruit tightly you won't need much. But if the fruit is sour, a weak syrup does help.

BOTTLING VEGETABLES
I strongly advise against the bottling of vegetables, but if you insist upon doing it you must heat in a pressure-cooker, as boiling at atmospheric pressure is not enough to make it safe.

Sweet corn can be bottled (although I prefer the oven-drying method I described on p.198). Husk your corn, remove the silk, wash well, and cut the corn off the cob with a knife. If you force the cob on to a nail sticking up from a board at an angle, you will have it steady for slicing. This will leave a little of each grain on the cob, but that's all the better for the pigs. Pack the corn in the jar to within an inch of the top, add half a teaspoonful of salt to each pint of corn, fill up to half an inch from the top with boiling water, put the lid on loosely and heat in a pressure-cooker at 240°F (115°C), at 10 pounds pressure, for an hour. Remove the jars from the cooker and seal.

Salting runner beans
Use a pound (0.5 kg) of salt to 3 pounds (1.4 kg) of beans. Try to get "dairy" salt or block salt; though vacuum salt will do. Put a layer of salt in the bottom of a crock, a layer of stringed and sliced beans (tender young French beans do not need much slicing, whereas runners always do) on top, another layer of salt, and so on. Press down tightly. Add more layers daily. When you have enough, or there are no more, cover the crock with an airtight cover and leave in a cool place. The beans will be drowned in their own brine, so do not remove it. To use, wash some beans in water and then soak them for no more than 2 hours.

Basic method	Cold water bath		Hot water bath		Slow oven	
	Take 90 minutes to bring water from cold to required temperature, then follow instructions given below.		Starting at 100°F (39°C), take 25–30 minutes to reach required temperature of 190°F (88°C). Follow instructions.		Preheat to 250°F (121°C). Leave bottles according to times given below.	
Liquid in bottles	Put cold syrup or water in before processing.		Put hot liquid at 140°F (60°C) in before processing. For tomatoes, liquid is optional.		Add boiling liquid at end of processing.	
Soft fruit Blackberries, raspberries, currants, etc. and apple slices	Temperature 165°F (74°C)	Time 10 min	Temperature 190°F (88°C)	Time 2 min	Temperature 250°F (121°C)	Time 45–55 min
Stone fruit Cherries, plums, etc. Citrus fruit	180°F (83°C)	15 min	190°F (88°C)	15 min	Heat oven to 300°F (149°C) and put hot syrup in before processing them.	40–50 min
Tomatoes	190°F (88°C)	30 min	190°F (88°C)	40 min	250°F (121°C)	80–100 min
Purées and tight packs	Allow 5–10 minutes longer than times shown above and raise temperature a little.					

BOTTLING TOMATOES

Jars of bottled tomatoes on your shelves in winter are a cheering sight. They are easy to bottle and it even improves their flavour.

1 Remove the green tomato stalks, and nick the skins with a knife.

2 Put the tomatoes in a bowl and pour over boiling water. Leave until the skins have loosened.

3 Drain and cover with cold water. Don't leave them very long or they will soon go soggy.

4 Peel off the skins carefully so that the tomatoes retain their shape and do not lose any juice. Make up a brine by mixing half an ounce (14 g) of salt to a quart (1 litre) of water.

5 Pack tomatoes in jars very tightly. Push large fruit into place with the handle of a wooden spoon.

6 If sterilizing in water, fill the jars with brine, cover with sealing discs, and screw lids on loosely; if in the oven, add brine afterwards.

7 Put jars in a pan of water, or stand on newspaper in the oven. Now cook.

8 When cool, try lifting the bottle by the disc only. The vacuum should hold.

MAKING SAUERKRAUT

Winter cabbages (see p.58) can be clamped but, if greens are scarce, make a sauerkraut standby.

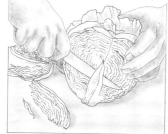

1 Shred hard white cabbage hearts finely, and estimate ½ ounce (14 g) salt for each 1 pound (0.5 kg) of cabbage.

2 Pack layers of shredded cabbage into a stone crock or wooden tub; sprinkle salt between the layers.

3 Spread one big cabbage leaf across the top, put a cloth over it, and cover that with a plate.

4 Weigh the plate down and leave in a warm place. In 3 weeks put the cabbage in jars and sterilize.

Making Pickles & Chutneys

Pickles and chutney are another way of preserving produce. They add flavour to cold meats, meat pies, cheeses and curries. The principle of both involves flavouring fruit and vegetables with spices, and then storing them in vinegar. Ideally you would make your own vinegar (*see pp.230–231*), but if you cannot do this, and have to buy it, you should note that there are vinegars of different strength, cost, and flavour.

Distilled or fortified vinegar is much the strongest (it is also the most expensive). Wine vinegar is the strongest natural vinegar, and more expensive than cider or malt vinegar. Remember that vinegar leaves its flavour in chutney, and even more so in pickles, so if you want to have the best-tasting accompaniments to your cold pies, you may find yourself paying for your vinegar. And the best-flavoured vinegar is wine. However, when you make chutney, much of the liquid is evaporated during the cooking, so malt vinegar could be a more economic proposition.

PICKLES

The vinegar is first steeped with spices and sometimes cooked with sugar, to improve and mellow its sharpness. To make a spiced vinegar suitable for a variety of pickles, you can add any spices you like. Ground spices make vinegar go cloudy, so if you want the pickle to be attractively presented, and clearly recognizable, use whole spices.

The ideal way of making spiced vinegar is to steep all the spices in cold vinegar for a couple of months, after which time the liquid is ready to be strained and used. Since this is not always practicable, here is a speeded-up version. For 2 pints of vinegar take 2–3 ounces (56–84 g) of spices and tie them in a little muslin bag. Include:

a piece of cinnamon bark
slivers of mace
some allspice
6–7 cloves
6–7 peppercorns
½ teaspoon mustard seed
Add garlic, or any herb, if you like the flavour, and for a hot taste add chilli, ginger, or more mustard.

Now put the vinegar and spices into a jug or heatproof jar which can be covered with a lid or a plate. Stand it in a panful of water. Bring the water to the boil, then take it off the heat. Leave the whole thing to cool down for 2 hours, by which time the spices should have thoroughly flavoured the vinegar. Remove the little bag and the vinegar is ready to use. You can pickle fish, eggs, fruit, and vegetables, and pickle them whole or in pieces. Moist vegetables and fish are usually salted first to draw out some of their water. Crisp pickles like cucumbers, beetroot, cabbage, and onion are put straight into cold vinegar. Others, like plums, tomatoes, and pears, are cooked till soft in spiced vinegar; this is then reduced to a syrupy consistency before finishing. When adding sugar to sweet pickle, use white sugar: it keeps the pickle clear and light. Pickle jars need close sealing to prevent evaporation, and the vinegar must not come in direct contact with metal lids. Eat all pickles within 6 months; after this they are likely to soften.

Pickled eggs

Hard-boil as many new-laid eggs as you like; you need about a quart (1 litre) of vinegar for every dozen. Shell them. Pack them in jars and cover them with spiced vinegar. Add a few pieces of chilli if you like. Close tightly and begin to eat after 1 month.

Pickled onions

Choose small button onions, and don't skin them at once. Instead, soak them in a brine of salt and water using 4 ounces (114 g) of salt to each quart (1 litre) of water. After 12 hours skin them. Put them in a fresh brine for 2 to 3 days, with a plate on top so that they stay submerged. Then drain and pack in jars or bottles with spiced vinegar. A little sugar added to the vinegar helps the flavour. They are good to eat after 2 or 3 months.

Pickled apples

This is a sweet pickle. Use small apples (crab apples are good). For 2 pounds (1 kg) of apples use 2 pounds (1 kg) of sugar and 1 pint (0.5 litre) of spiced vinegar. Cook the sugar and vinegar until the sugar is just dissolved. Prick the apples all over, using the prongs of a carving fork. If they are too big for the jar, cut them in half. Simmer in the vinegar-sugar mixture until they are soft but not falling apart. Put them in jars gently. Reduce the syrup to ½ pint (0.3 litre) by boiling. Pour it hot over the apples, but not so hot that it cracks the glass.

CHUTNEYS

Chutney is a concoction of almost any fruit or vegetable you like, flavoured with spices and cooked with vinegar to a thick, jam-like consistency. Soft, over-ripe fruit and vegetables are suitable, as they turn into pulp quickly. Ingredients for chutney can be marrows, pumpkins, swedes, turnips, peppers, onions, beetroot, carrots, celery, aubergines, mangoes, tomatoes, apples, rhubarb, blackberries, pears, bananas, lemons, damsons, gooseberries, plums, dried fruit, peaches, elderberries, cranberries, oranges, and grapefruit. The herbs and spices can be any of these: bay leaves, chilli, cumin, coriander, cardamom, cinnamon, cloves, ginger, allspice, peppercorns, mustard seed, horseradish, paprika, cayenne, juniper, and garlic. It is best to mince vegetables or fruit for chutney finely and then cook them slowly for a long time to evaporate the liquid. Sugar plays a large part in chutney. Most chutneys go dark as they are cooked, so if you want an even darker one, use brown sugar, or even black treacle.

MAKING TOMATO CHUTNEY

The secret of good chutney is to use contrasting ingredients. In this example, the spice and garlic offset the tomatoes and apple.

You need: 2 lb (1 kg) tomatoes, 2 onions, 1 cooking apple, raisins, 2 cloves garlic, ½ oz (14 g) fresh ginger, 2 oz (56 g) brown sugar, ½ pint (0.3 litre) vinegar, salt, and some spices.

1 Skin the onions, peel, and core the apple. Then chop them up finely.

2 Simmer the onion in a small pan with a little water. Add the apple and the raisins, and cook gently until they soften.

3 Skin the tomatoes, then chop them up roughly into chunks.

4 Crush the garlic and fresh ginger in a pestle and mortar with salt. If you are using dried ginger instead, add ½ oz (7 g) to the bag of spices.

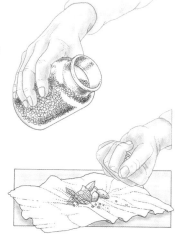

5 Tie up in a little muslin bag: 1 crushed bay leaf, 2–3 crushed dried chillies, ½ teaspoon mustard seed, 4–5 cloves; add cardamoms, cinnamon, coriander, peppercorns as you wish.

6 Tie the muslin bag to the handle of a large saucepan, so as not to lose it in the chutney.

7 Pour the softened ingredients into the pan, then everything else.

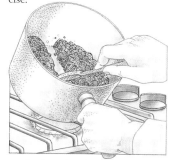

8 Cook on low heat for an hour or so, until mixture thickens, so when you draw a spoon through it you can see the pan.

9 Pot at once in hot, clean jars. Seal and label.

Cooking chutney

Use stainless steel or enamelled pans. Vinegar eats into copper, brass, or iron pans, so don't use them. Simmer hard ingredients such as apple and onion in a little water before mixing with softer ingredients such as marrow or tomato, and before adding salt, sugar, and vinegar, which tend to harden fruit or vegetables.

Put whole herbs and spices in a muslin bag, which you can tie to the handle of the pan so that you don't lose it in the chutney. If you prefer to use powdered spices they can be added loose to the other ingredients. Crush garlic and fresh ginger in a pestle and mortar before adding to chutney. Soak dried fruit in water before cooking it. Use sufficient vinegar just to cover the ingredients. Cook until the consistency is of thick jam, and there is no free liquid.

Be careful it doesn't burn towards the end. Stir well while it cooks. Pot while still hot in clean hot jars, cover and label.

Storing chutney

Chutney improves with keeping, so store it in a cool, dark place in glass jars. Make sure they are tightly sealed or the vinegar will evaporate, leaving an unappetizing, dry shrunken mess. Cellophane papers such as are sold for jam covers are not suitable. I use twist-on metal caps from old jam or pickle jars. Check that the metal from the lid is well lacquered or protected with a waxed cardboard disc, otherwise the vinegar will corrode the metal. You can also use synthetic skins, or waxed paper circles underneath a greaseproof paper tie-on cover. Cover the jars with a cloth that has been dipped in melted candle-wax.

Making Jams & Syrups

Jams and conserves of all kinds are a very useful way of preserving fruit. Usually the fruit is cooked first without any sugar, to soften it and to release the pectin, which is what makes it set. Sugar is added next, and the whole thing boiled rapidly until setting point is reached. As long as jams are properly made, well covered, and kept in a cool, dry place, they keep for ages.

Fruit should be under- rather than over-ripe, and clean. Bruises on damaged fruit don't matter as long as they are cut out. It is important to weigh the fruit before you begin cooking, otherwise you don't know how much sugar to add. Don't add more water than necessary to cook the fruit. The sugar should be preserving sugar, as this dissolves fastest. Brown sugars are okay, but bear in mind that they add a flavour of their own and in some cases are damp; therefore, adjust the weight.

Some fruit has more acid and pectin in it than others. Fruit which is low in acid or pectin usually needs extra acid or pectin added to it (see below).

Basic jam-making

In general, jam-making goes like this: clean, sort, and prepare fruit. Weigh it. Cook it with sufficient water to make it tender. Put it in a large, wide pan, and when it is boiling add the required amount of sugar. Stir until all the sugar is dissolved. Bring to a rapid boil. Don't stir. Test from time to time to see if setting point (see below) is reached. Stop cooking when it is, and allow to cool a little so that pieces of fruit will not float to the top of the jam in the jars. Fill hot clean jars to the brim with jam: cover, seal, label.

Testing for pectin

Put into a little glass a teaspoonful of strained, cooled fruit juice from the cooked fruit, before you add the sugar. Add 3 teaspoons methylated spirits, and shake together. Wait a minute. Pour the mixture out into another glass. If the fruit juice has formed one solid blob, the pectin is good. If it is several blobs, it is not so good, so add less sugar. If it is all fluid, it is useless, in which case boil the fruit again. Even add commercial pectin at a pinch.

Testing for set

Put a little jam from the pan on a saucer to cool. If the surface wrinkles when you push it with your finger, it is done. Examine the drips from the spoon: if a constant stream flows, it is no good; if large thick blobs form, it is okay. The temperature of the boiling jam should reach 222°F (105°C). It is best to use all or at least two of these methods to be absolutely sure your jam is ready.

Potted fruits or conserves do not keep as long as jam, but because they are only cooked briefly the flavours are very fresh. It is not so necessary to worry about pectin with conserves, so you can make them with low-pectin fruit like raspberries, strawberries, blackberries, and rhubarb. Also, note that there is more sugar per pound in conserves.

Damson or plum jam

Much of the pectin in plums is found in the stones, so if you can, extract the stones first, crack some of them, and tie the kernels in a little bag. If this is difficult, never mind; they will float to the top when the jam cooks and you can skim them off with a slotted spoon at the end. You will need

6 lb (2.7 kg) damsons or plums
6½ lb (3 kg) sugar
½ pint (0.3 litre) water

Wash the plums, and cut them in half. Simmer with the water until tender. Add the sugar, stirring until dissolved, then boil hard until setting point. Remove floating stones, or if you put kernels in a bag, remove the bag. Leave the jam to cool a little before potting so that the fruit will not rise to the top of the jars. Pot, seal, and label.

Raspberry conserve

4 lb (1.8 kg) raspberries
5 lb (2–3 kg) sugar

You can use damaged but not mouldy fruit. Warm the sugar in a bowl in a low oven. Butter a large pan, put in the fruit, and cook over a very low heat. As the fruit begins to give up its juice and bubble, slowly add the warm sugar. Beat hard until the sugar is quite dissolved. It should remain a lovely bright colour and taste of fresh raspberries. It should be quite thick. Pot and cover in the usual way, but examine for mould after a few months. Another way is to put sugar and raspberries in layers in a large bowl. Leave overnight, and bring just to the boil the next day, before potting.

Lemon curd

4 oz (114 g) butter
1 lb (0.9 kg) sugar
4 eggs
3 to 4 lemons depending on size and juiciness

This is not a jam, but a good way of using up eggs. Grate the rind from the lemons and squeeze out their juice. Put rind, juice, butter, and sugar into a small pan and heat until the butter melts and the sugar just dissolves. Let it cool. Beat up the eggs. Put them in a bowl which will just fit over a saucepan of simmering water, and stir in the juice. Beat over the saucepan of water, or use a double boiler, until the mixture thickens to curd consistency. Pot and cover. Lemon curd doesn't keep long, so use it up quickly. Don't make too much at a time. Richer curds can be made using eight egg yolks instead of four eggs. Variations include using oranges or tangerines instead of lemons. With sweeter fruits, use less sugar.

Lemon and carrot marmalade

8 oz (228 g) thinly sliced lemon
8 oz (228 g) shredded carrot
2 pints (1 litre) water
1 lb (0.9) kg sugar

Mix the lemon, the carrot, and the water. Cover and allow to stand overnight. Cook in a covered saucepan, bring to the boil, then simmer for about half an hour or until tender. Then add sugar, and simmer until it completely dissolves; boil rapidly until setting point. Try a little on a cold plate to see if it jells; it may take 15 to 30 minutes. Pour into clean, warm jars, cover with waxed paper and seal.

The flavour of carrot and lemon is very fresh and fairly sweet. Eat within 3 months.

Making fruit butters and cheeses

Fruit butters and cheeses are jams made from puréed or sieved fruit. Butters are softer than cheeses. Cheeses, if they are firm enough, can be turned out of their moulds as little "shapes". Fruit butters and cheeses are delicious when

THREE FRUIT MARMALADE
Make this from oranges, lemons, and grapefruit as a tasty change from Seville orange marmalade.

1 Squeeze out the juice from eight oranges, two lemons, and two grapefruit. Strain it and save the pips.

2 Shred the peel coarsely or finely, depending on how thick you like your marmalade.

3 Tie the pips in a bag, and soak with peel and juice for a day in 10 pints (5.7 litres) water. Boil for 2 hours.

4 Test for pectin by adding 3 teaspoons meths to 1 of juice. Shake. The juice should solidify.

5 Remove the bag of pips from the pan. Boil the mixture, add 7 lb (3 kg) sugar and stir until dissolved. Cook until it sets.

6 Let some marmalade drip from a spoon. If it falls in thick flakes it is properly set.

7 Or cool a little on a saucer. It is done if the surface creases when touched.

8 Put into hot, clean jars, cover with grease-proof paper and cellophane seal, and label. Start eating it as soon as you like.

they are eaten as puddings with cream or even spread on bread.

Blackberry and apple cheese
You will need equal amounts of blackberries and apples. Wash the apples but don't bother to peel or core them. Cut them up roughly. Pick over the blackberries and wash them if they are dusty. Put both fruits into a pan, just cover with water, and stew, stirring occasionally, until the apples have gone mushy. Sieve the cooked fruit. You should have a fairly thick pulp. Weigh it. Add 1 lb (0.9 kg) sugar to each pound of pulp. Boil together. Stir all the time, as this burns easily. When it thickens enough for you to see the bottom of the pan as you draw the spoon across it, it is done. Pot and cover like jam. It sets quite firmly, like cheese, and will last for ages.

Making jellies
Jellies are simply jams which have had all the solids strained from the cooked fruit. When the juice is boiled up with sugar it forms jelly, which can be used in the same way as jam.

Blackberry and apple jelly
This recipe will suit any high-pectin fruit, such as crab apples, redcurrants, citrus fruits, quinces, gooseberries, sloes, damsons and rowanberries. You can also experiment with mixtures of fruits. Cook them separately if one needs more cooking than the other.
   Proceed as for blackberry and apple cheese to the point where the fruit is cooked and soft. Then strain the juice through a cloth. Don't succumb to the temptation of squeezing it to speed it up or the finished jelly will be cloudy. Measure the juice and add 1 lb (0.9 kg) sugar to each pint (0.5 litres) of juice. Cook until setting point is reached and pot and label in the usual way.
   If you are very economically minded you can stew up the residue of fruit in the jelly bag with more water, then either extract more juice or make a fruit cheese by sieving it. Follow the instructions given above if you want to do this.

Fruit syrups
Fruit syrups are made in the same way as fruit jellies, though you don't need to add so much sugar to syrups. To prevent spoiling by fermentation (when you would be on the way to making wine) you have to sterilize syrups and keep them well sealed. They make very refreshing drinks and milk shakes in summer, or you can use them as sauces for puddings and cereals.
   Extract the juice from any unsweetened cooked fruit you fancy, as for jelly, or, if you wish, extract it by pressing then straining. Measure the amount and then add about 1 pound (0.9 kg) sugar per pint (0.5 litre) of juice. Heat it until the sugar is just dissolved – no more or it will start to set like jelly. Let it cool. Sterilize the bottles and their lids, preferably the screw-cap sort, by immersing in boiling water for 15 minutes. Drain, then fill with syrup. Screw up tightly then unscrew by half a turn, so that the heating syrup will be able to expand (leave about a 1-inch/2.5-cm –gap at the top of each bottle).
   Stand the bottles in a pan deep enough for the water to come up to their tops. If possible use a pan like a pressure-cooker that has a false bottom. Bring slowly to the boil and keep boiling for 20–30 minutes. Take out the bottles and screw the lids on tightly as soon as they are cool enough. If you are doubtful about the tightness of the seal, coat with melted candle wax.

Cakes, biscuits, & puddings

Cakes, biscuits, and puddings – now this is where the real magic of cooking lies for me. Somehow, over thousands of years, dedicated humans have experimented with all sorts of mixtures and ovens, and today we can enjoy the results. I've no doubt the chemistry of flour, eggs, yeast, and baking powder is complex indeed – and I certainly never studied it in school – but the end products are straighforward enough. With modern mixing machines there is no excuse for missing out on the excitement of a weekly session of making your own desserts and treats. So what is the key to all this magic? All equipment must be clean and cool. Try to keep one specific board, preferably made of marble or slate, for rolling pastry. The rolling pin should not be too heavy – in fact, a round glass bottle filled with cold water is ideal. You have to make pastry in as cool a place as possible so that the cooking process creates lightness by expanding the cool air trapped in the dough. People with hot hands simply cannot make good pastry! Keep plenty of flour to hand for dusting over the pastry to prevent it from sticking. Roll it lightly and try to press evenly with both hands; always roll away from you, taking short, quick strokes and lifting the rolling-pin between each stroke.

There are endless recipes for cakes, pastry, and puddings, so have a go. With a little practice you will soon find recipes your family will enjoy.

Lemon Drizzle Cake

You can make this cake using oranges instead of lemons but, either way, it makes a zesty treat for all the family – and it is another wonderful way to use up all those fresh eggs from your ever-productive chickens. You will need:

finely-ground rind of 2 lemons (preferably organic), reserving some shredded rind for decoration later
6 oz (170 g) caster sugar
8 oz (225 g) unsalted butter, softened
4 fresh eggs
8 oz (225 g) of self-raising flour
¼ teaspoon of salt
1 teaspoon of baking powder

Preheat the oven to 160°C (320°F). Grease a 2 lb (900 gm) bread tin or a round 7–8 inch (18–20 cm) cake tin and line with greaseproof paper. Mix the lemon and the caster sugar together. Cream the butter with the lemon and sugar mixture and add the eggs. Blend the ingredients until the mixture is a smooth, pale yellow colour. Sift the flour, together with the baking powder and salt, into a bowl and fold into the mixture one third in at a time.

Turn the mixture into the baking tin, smooth it out on the top so that it is level, and bake for approximately 90 minutes, or until golden brown and springy to the touch.

Farmhouse Carrot Cake

This is another great way to use up the eggs and carrots from your garden as it makes a marvellously moist and flavoursome cake. You will need:

1 lb (450 g) caster sugar
4 eggs
1 cup olive oil
8 oz (225 g) carrots, finely grated
8 oz (225 g) plain flour
1 heaped teaspoon of baking powder
1 teaspoon of ground allspice
1 teaspoon of ground cinnamon

For the icing you need:
8 oz (225 g) icing sugar
1 cup of softened cream cheese
2 oz (56 g) of butter
1 teaspoon of vanilla essence
6 oz (170 g) of chopped nuts (try walnuts)
A little milk, if required

Mix together the sugar, eggs, oil, and grated carrots in a large mixing bowl. Sift all the dry ingredients together in a separate bowl. Add a small amount of the dry ingredients to the mixture a bit at a time and and mix together until you have blended all the ingredients.

Grease and flour two 9-inch (22-cm) cake tins. Divide the mixture between the two tins. Cook for about 40 minutes in a preheated oven at about 190°C (375°F). Check if it is cooked by inserting a skewer into the centre of the cake. If it comes out clean, it is ready; if the mixture is attached to the skewer you need to let the cake cook for a little longer.

Once cooked, let the cakes cool in their tins for a few minutes and then tip them out onto wire racks to let them cool thoroughly.

To make the icing, combine all the ingredients, except the walnuts, together until you have a smooth, spreadable consistency. You may need to add milk if the mix is too dry. Add the walnuts and blend together. Spread over the top of one cake layer, top with the other cake, and then spread the remaining icing over the surface.

Doughnuts

You won't eat a shop doughnut again after you've made your own. These doughnuts are made from a normal bread dough (see page 203) but you also include the following ingredients during mixing:

2 oz (56 g) melted butter
1 egg
2 oz (56 g) sugar
milk (substitute this for the water used in making a normal bread dough)
pinch of salt
1 lb (0.5 kg) flour
1 oz (28 g) yeast or ½ oz (14 g) dried yeast

Mix the dough until smooth and soft. Cover with a cloth and leave it to rise in a warm place for about one hour. After it has risen, knead it vigorously for about 4–5 minutes. Roll out the dough on a floured board. Use a large glass or cutter to press out a circular shape. To make the rings, use a bottle top and press down in the centre of the circle. Don't worry if the holes look small as they will more than double in size during cooking. Cover them with a cloth and leave in a warm place to prove for 20 minutes.

Once proved, fry them on both sides, a few at a time, in deep, boiling fat. The spherical ones will turn themselves over at half time, leaving a pale ring round the equator. Drain the doughnuts on kitchen paper and then roll in caster sugar. Eat at once.

Flap Jacks

These make a wonderfully healthy and quick snack.

4 oz (112 g) butter
2 ½ oz (70 g) sugar
1 tablespoon syrup
6 oz (170 g) of porridge oats

Melt the butter, sugar, and syrup into a pan. Stir in the porridge oats and mix well. Spread onto a greased baking tin, press down firmly, and , with a sharp knife, divide into squares. Bake in a moderate oven at 190°C (375°F) for 15–20 minutes. Allow to cool and remove by breaking each square off the block.

Cheese Scones

Quick to cook and delicious to eat warm and crumbly from the oven, these savoury scones make an appetizing addition to any teatime feast. You will need:

1 oz (28 g) butter
6 oz (170 g) self raising flour
3 oz (85 g) strong cheddar cheese, grated finely (reserve a little for sprinkling over the top)
¼ teaspoon of mustard
½ teaspoon of salt
1 egg
2 good tablespoons of water

Rub the butter into the flour and then add the grated cheese and seasonings. Add the egg and the water and mix together. Once you have a firm dough, roll it out so it is about half an inch thick. Sprinkle a little more grated cheese over the top and cut out circular shapes with a cutter or a glass. Place onto a greased tray and bake in the oven at 220°C (425°F) for 15 minutes.

Blackberry Charlotte

This is a classic pudding for autumn when blackberries, walnuts, and apples are in good supply. You need:

2 ¼ oz (65 g) unsalted butter (plus some for greasing the dish)
65 g (175 g) fresh white breadcrumbs
2 oz (56 g) of soft brown sugar
2 fl oz (60 ml) of golden syrup
finely-grated rind and juice of 2 lemons
2 oz (56 g) of walnut halves
15 oz (450 g) of blackberries
15 oz (450 g) of sliced cooking apples

Preheat the oven to 180°C (355°F). Grease a 15-oz (450-ml) dish with butter. Melt the butter and add the breadcrumbs. Sauté gently until the crumbs turn golden brown and slightly crisp. Leave to cool.

Put the sugar, syrup, lemon rind, and lemon juice into a small saucepan an gently warm through. Add the crumbs. Chop the walnuts very finely and add to the mixture. Put a layer of blackberries in the dish, followed by a layer of the crumbs and another layer of sliced apples. Continue layering, finishing with a layer of the crumb mixture on the top. Cook in the oven for 30 minutes.

Meats

To the self-supporter, meat is not merely "flesh". Each animal has its own life saga with escapes, injuries, and always a shared sympathy between living beings. Seeing an animal happily content with its accommodation, its food, and its "carer" is a pleasure indeed. But there will have been frustrations and furies, too – the day the beast escaped into the strawberry bed or the time it broke its water trough and water flooded the food store. Finally comes the day when animal becomes meat to be eaten or preserved.

Smoking One very useful way of helping to preserve the meat is to smoke it. This also helps it dry out, and probably helps it mature quicker. It is also much easier than people seem to think. If you have a big open chimney, simply hang the meat high up in it, well out of reach of the fire, and leave it there for about a week, keeping a wood fire going the while. There is a lot of mysticism about which wood to use for smoking: the Americans swear by hickory, the British will hear of nothing but oak. In my experience it matters very little provided you use hard woods and not pine.

Whatever you use don't let the temperature go above 120°F (50°C): 100–110°F (39–43°C) is fine. Building a smoke-house is a matter of common sense and a little ingenuity. For years we used a brick outdoor lavatory, tybach, jades, or loo. (We didn't use it for its original purpose of course.) We had a slow-combustion wood-burning stove outside, with the chimney pipe poking through the wall of the jakes, and we hung up the meat from lengths of angle-iron under the roof. It does seem a pity though not to make use of the heat generated by the smoking fuel, so surely it is better to have your burning unit inside a building, even if the smoke chamber itself is outside. Often a slow-combustion wood burning stove can heat a house and, with no increase in fuel consumption, it will automatically smoke whatever you like. There are two kinds of smoking: cold-smoking and cooked-smoking. The latter is common in America and Germany but almost unknown in Britain. It consists of smoking at a higher temperature, from 150–200°F (65–93°C), so that the meat is cooked as well as smoked. Meat thus smoked must be eaten within a few days because it will not keep as cold-smoked meat will.

Thawing meat Always take your meat from the freezer to the fridge 24 hours before you intend to cook it. Meat should be thawed slowly, overnight at room temperature. Once thawed the meat should be cooked immediately, as it will not keep.

Bones Many modern butchers like to "bone" joints of meat. This is not something I recommend. I know housewives hate the idea of paying for the bone but if they would just pause for a moment, they would realize that boned meat actually costs more per pound and they are really paying for the butcher's time in doing the boning. The plain fact is that meat on the bone cooks better and the marrow and presence of the bone add an extra flavour.

Larding and barding I don't hear these terms used much today because real bacon is such a rarity. But if you make your own bacon from your own pigs then you are likely to have a fair bit of "fatty" bacon which can be used in roasting other meats. Larding is simply threading pieces of bacon through a joint using a "larding needle". "Barding" means covering lean or exposed parts of a roasting joint with a few slices of fatty bacon. This provides fat to prevent burning and imparts an extra boost to the flavour.

Roasting I try to have an open air roast over a fire once or twice a year, and very delicious and exciting it is, too. If you want to roast a pig or lamb like this you need at least 12 hours burning time: the longer the better. So start your fire early in the morning; you will need two strong metal supports to hold your rotating spit rod and you must fasten the carcass to this, using wire. Keep the meat high up to start with and progressively lower it during the day. Always have a good supply of well dried timber for the fire and appoint a responsible person to be "in charge" at all times. With an oven, start at a moderately hot temperature, say 190°C (375°F), to "seal" the joint. Keep an eye on the meat to ensure it does not dry out and baste with fat as often as necessary. If you want to make fat or skin crisp then rub in some salt before cooking. If you suspect your oven temperature may be uneven then use cooking foil to cover parts of the joint exposed to "hot spots". Do not think that there are fixed times for making a perfect roast: each piece of meat will have its own characteristics and a long thin joint will cook much quicker than a fat, round one. Something like 20 minutes to the pound will not be far wrong plus 20 minutes extra. Pork, lamb, and veal should cook longer than beef.

Braising This is an excellent way of cooking the tougher less attractive joints of meat by effectively combining roasting and boiling. If possible use a heavy cast-iron cooking pot with a close fitting lid. Tie your meat up with string if necessary. Partially fill the pot with water suitably spiced up with whatever takes your fancy: I find dried tomatoes and black pepper corns brilliant. Then cook for however long it takes for the meat to become tender. Your oven temperature will not be as high as for a roast, perhaps 300°F (150°C). Make sure the liquid does not boil away. If the meat is fatty you can allow the whole lot to cool overnight before reheating the next day. All the fat will solidify on the surface and can be removed.

Stewing For me this is long, slow, cool cooking to tenderize the less attractive and tougher parts of the animal. The first job is to cut up the meat into lumps less than 1 square inch, removing all unnecessary fat and gristle. These must now be "sealed" with a quick burst of strong heat using good fat or olive oil and perhaps a few chopped onions. Now make up your chosen mixture of vegetables and seasoning before allowing everything to simmer gently in a slow oven for 3 or 4 hours. You can keep a good stew going for several days simply by adding fresh vegetables and stock – but remember, it must be boiled up strongly once each day, or it will go bad.

Fish

I vividly remember catching mackerel from a yacht as I entered a small anchorage on the Scottish island of Gomera near Mull. Within a couple of minutes I had the fish filleted; the sails came down and I made anchor. So quick was the journey from ocean to the grill that the fillets actually jumped off the grill. Freshly grilled with lemon juice, mackerel is delicious. What did this tell me? That well-cooked fresh fish is one of the great delicacies of the world. But how often have we been disappointed in restaurants to find the meal either pappy and overcooked or almost raw?

So, aside from finding a good fishmonger or local fisherman and timing to perfection, bear in mind these tips:

Gutting fish It is essential to clean and gut fish properly. For most fish gutting is done by slitting the fish on the underside from the head halfway to the tail and carefully removing the insides. Always handle fish gently to avoid bruising. If there is a roe in the fish you can either replace it if it is small or cook it separately.

Make sure you take out all of the guts, including any black skin lining the body cavity. Rub this out with salt if you have to because if you leave it, there will be a bitter taste. Wash the fish out well with cold water; use a running tap rather than soaking the fish or you will find they become watery.

Scaling If there are scales on the fish, then these must be removed by scraping from tail to head with a sharp knife. Angle the knife slanting against the fish and scrape slowly so the scales do not fly everywhere. Sometimes the scales are tough to remove: if so, dip the fish quickly into boiling water. If you are going to serve the fish whole then you should cut off the fins with a sharp pair of scissors. Either cut off the head or take out the eyes.

Skinning Round fish are skinned from head to tail. You make a cut along the sides of the fish close to the fins and then make a cut in the skin just below the head. Now start pulling the skin away downwards, using a knife to hold down the flesh. Put your hands in salt if the fish is too slippery – or use a cloth to hold it. Flat fish are skinned in similar fashion but starting at the tail. The skin of flat fish can be pulled off quickly once the sides have been cut and loosened – just hold down the tail to keep it steady.

Filleting You will need a very sharp knife and a fair bit of practice to make a good job of separating the flesh from the bones. The basic technique is to make a cut down the whole of the back of the fish down to the backbone. Then carefully scrape away the fillets of flesh from either side; you make two fillets from each side. It is usually easiest to work the first fillet from the left-hand side of the fish working from head to tail. Then turn the fish around and work the second fillet from tail to head. Now turn the fish over and repeat for the other side.

Fish stock Making a good fish stock is the essence of cooking a fine fish sauce. The best stock is made from fresh fish trimmings after filleting. Discard any black-looking skin and break the bones up into small pieces.

Put the trimmings into a saucepan with water and/or milk, a small piece of onion, white peppercorns, and some parsley. Simmer the mixture on a low heat for about half an hour and then strain off your stock. Add white wine instead of milk if you prefer.

Boiling fish This is a good way to cook large fish. They should be left unskinned with the heads on and eyes removed. Fish should be put into the water when it is piping hot but not boiling, and the water should be salted, with a little vinegar or lemon juice added. If you do not make the water slightly acid the fish will not be white and firm when cooked.

A fish kettle is the best way to boil fish. This will have a drainer to take off the water and avoid breaking the fish. If you do not have a fish kettle then you can use an ordinary large saucepan. Put a plate in the bottom and rest the fish on this tied up in muslin. By hanging the ends of the muslin over the sides of the cooking vessel you can remove the fish whole without breaking it.

Weigh the fish before you boil it: 8–10 minutes per pound and 10 minutes over, is a good guide. Do not put too much liquid over the fish or the boiling will shake it about and damage the skin. About 2 inches above the fish should be sufficient.

Grilling fish This is a good way to cook smaller fish like herring, mackerel, and trout. Prepare the fish carefully and score the skin on both sides to prevent it from cracking when cooking. Season the fish with pepper and salt and brush it all over with oil or melted butter. Alternatively you may want to split the fish open, removing the bone and coating it lightly in flour or fine oatmeal. Make sure the grill is hot before you start cooking. Allow 7–10 minutes for cooking and turn the fish at least once. It should look nicely browned when ready and should be served immediately.

Frying Probably the most popular way to cook fish, as well as one of the trickiest to do well. There are two important things to remember: first, make sure your fish is as dry as possible before you cook it, and second, coat the fish in some mixture which will prevent it absorbing the cooking fat. There are several different ways of coating fish for frying. The easiest way is simply to dip the fish into sieved flour. Alternatively, you can use the familiar batter of the fish and chip shop: this is a weak pancake mix of flour, milk, and egg. I prefer to use egg and breadcrumbs and this certainly looks the best when the fish is cooked. Make sure the fat is always kept very hot: do not put too many pieces of fish in the pan at any one time, and let the fat heat up again before putting in a fresh piece of fish. Drain the fish on kitchen roll before serving so as to remove excess grease.

Baking fish Fish can be baked very simply in an open dish in a moderate oven. Add the seasoning of your choice. Baking is a dry way to cook fish, so you may need to add some butter, fat, or milk to provide moisture.

Poaching or steaming With the correct equipment these are two other excellent ways to cook fish.

Vegetables

We talk easily about "vegetables" but we are really talking about a huge range of completely different foodstuffs. At one extreme we have simple salads, fresh and uncooked. At the other end there are things like globe artichokes that need a lot of cooking and a lot of eating. What we do know is that vegetables are essential for our supply of vitamins, trace elements, and roughage. It is, alas, a sad fact that all vegetables begin to deteriorate the second they are picked, some much more than others: the vibrant sugars produced by the growing plant are quickly converted into starches as soon as the vegetable is taken from the soil. So vegetables must be stored properly in humid, cool, fly-free spaces and they must have all damaged material carefully removed before cooking. Exactly the same priority attaches to the preparation of vegetables prior to storage in the deep freeze. So if you are a home producer you should be able to take your veg straight from the deep freeze all ready for a quick burst of boiling water (usually) to become deliciously ready for the table. Generally, vegetables should not be washed until just before they are to be prepared for cooking. Close-leaved plants like Brussels sprouts and cauliflower can benefit from a short soaking in vinegared water, which will draw out slugs and bugs (if you use salt it tends to toughen the leaves). Always carefully remove all damaged parts before cooking and wash thoroughly in cold water to remove all traces of the garden.

If you do not have a good set of purpose-built stainless steel pans for steaming vegetables, then "steamers" are something you should consider, for in my opinion, most vegetables are best cooked by steaming rather than immersion in boiling water. Vegetables can, of course, be roasted, baked, fried, or grilled. For oven temperatures I take moderate to be 170°C/325°F, moderately hot 190°C/375°F and hot 220°C/425°F. Almost certainly the major sin when cooking vegetables is to overcook and boil into oblivion. Better a little crunchy than soggy bland mush!

Artichokes

Brilliantly simple to cook and prepare, these make a most tasty and sociable starter. Choose the larger flower heads but make sure they are still tender and there is no sign of the actual flower appearing. Soak these in water (probably 1 or 2 maximum per person) with a little vinegar added to bring out bugs. After half an hour take out, wash, and drop into a large pan of rapidly boiling water. After 5 or 6 minutes lower the heat and simmer for another 40 minutes to one hour, depending on the size of the artichokes. You can tell when they are cooked when the leaves peel off easily. Remove from the water, drain, and serve with a choice of dressings. Butter is good or alternatively you can try our favourite, which is a mixture of good-quality soy sauce, lemon juice, and olive oil in equal parts.

Broad beans

Do not pod these until just before you are going to cook them. In all cases other than when beans are very fresh and young, you will have to remove the skins of each bean. Do this by dropping the podded beans into boiling water for a few minutes, then remove and the skins will be removed easily. Now you can boil in salted water. Remove any scum as it appears and test regularly to see when they become tender. Drain and serve with melted butter and salt and pepper as you like.

French beans

Pick these before they are fully grown so you avoid the risk of the dreaded "string" bean. Small and tender is better than big an stringy! If the beans are very small and young you can simply cut off the heads and tails and serve them whole. Older beans must be thinly sliced diagonally into lozenge-shaped pieces. Drop the sliced beans into a pan of salted boiling water. When the beans are ready they will sink to the bottom of the pan. Serve with hot butter and plenty of salt and pepper.

Beetroot

Small, well-shaped beetroot make an excellent hot meal. Choose your beetroot carefully, scrub clean, and then boil in salted water for at least one hour, possibly more. Test if they are cooked and tender by using a finger – if you spear them with a fork all the colour will tend to leak out into your cooking water. Once cooked remove from water and take off the skins. You can then quarter them and serve with a dressing made by frying a few well-chopped onions in butter and adding a little tarragon vinegar.

Carrot croquettes

Choose some nice red carrots and do not cook them too soft. Grate about 1 cupful per person. Now melt together butter and flour in a saucepan (about equal parts, 1 ounce/28 g per person) together with about one cupful of milk per person. Cook steadily until the mixture begins to draw away from the sides of the pan, stirring constantly. Now add your grated carrot plus the yolk of one egg for each person to be served. Add seasoning to taste (salt and pepper plus some sugar) and turn out on to a plate to cool. When cool you can form into shapes as you choose – use a little flour if you need to stop them sticking together.

Baked eggplant (aubergine)

This is a very quick and delicious way to cook aubergines, especially if you have an aga or large stove that is always hot. Wipe or wash the aubergines and cut off the stalk ends. Cut the aubergines down the centres to split into two halves. Put the halves into a baking tray, cut side uppermost, and add a little olive oil to stop them from sticking. Now grate some cheese on to the upturned flat sides of aubergine. Put into a moderate oven at about 170°C (325°F) for around 40 minutes.

Creamed leeks

Wash 5 or 6 leeks and clean them carefully. Cut off the roots and most of the green leaves. Split them open lengthwise and cut into pieces about 2 inches (5 cm) long. Throw these pieces into boiling water, slightly salted, and cook for 10 minutes, then drain. Now put them into a pan with half a pint of milk seasoned with pepper and salt. Simmer them slowly until they are tender. Now strain off the milk and arrange the leeks carefully in a warmed vegetable dish. Melt 1 ounce/28 g of butter in a pan and mix with one dessert spoon of flour. Then pour in the milk you have strained off the leeks. Stir the mixture until it boils and cook for a couple of minutes. Now add the cream at the end and pour the entire sauce over the leeks. Sprinkle with parsley and serve hot.

Baked mushrooms

Wash and peel your mushrooms, then remove the stems. Now put them hollow side uppermost in a baking tray greased with a little olive oil. Sprinkle them with salt and pepper and a few drops lemon juice, then add a small piece of butter to each one. Give them about 15 minutes in a moderate oven. Of course, you can add crushed garlic to the butter and produce the dish which is so popular in restaurants.

Potatoes au gratin

Make mashed potatoes in the normal way but add plenty of milk and butter together with grated cheese – add as much cheese as you fancy depending on the strength of the cheese. Season with salt and pepper and add a little mustard. Now pour the mixture into an ovenproof dish. Grate cheese to cover the top and add a few handfuls of breadcrumbs. Spread a little melted butter over the surface and then brown in a hot oven at 220°C (425°F). This is delicious.

Baked tomatoes

Choose moderately sized tomatoes and cut out the stalk and the hard part at the root of the stalk. Place the tomatoes on a buttered baking tray and put a small piece of butter into the hole left after removing the stalks. Sprinkle with pepper and salt and bake in a moderate over for about 15 minutes. Serve in a hot dish.

Baked cauliflower or aubergine

This is a brilliant quick and tasty way to cook large vegetables. Get your oven reasonably hot. Grease a baking tray (I like to use olive oil). Place your prepared vegetable in the centre (large lumps for cauliflower, cut in half for aubergine) and sprinkle grated cheese (cheddar is fine) over the top with a little pepper and salt. Bake it for about 30 minutes.

"Since we have been here we have made wine of cowslips, parsnips, elderberries, crab apples, wheat, marrow, broom flowers, grapes, and sloes. The cowslip was superb, but never since the first year have we had enough cowslips. The crab apple was unfortunate. Firstly, the apples probably have quite a lot of native sugar in them and so we should not have added so much. Secondly, I added what we decided to add without telling Sally [John's wife at the time], and she then came along and added the same amount again! The yeast cannot turn all that amount of sugar into alcohol — because when the alcohol gets to a certain strength it kills the yeast.
So the stuff tasted like very sickly alcohol grenadine. It was revolting and we gave it to the pigs which made them as drunk as people. Knowing what we now know we would not have given the pigs a treat. We would have tried refermenting it, mixing it first with another load of vegetable matter — adding some lemons perhaps — anything the flavour of which would not have clashed with the apples and which would have given more non-sugar matter for the yeast to work."

JOHN SEYMOUR FAT OF THE LAND 1976

CHAPTER EIGHT

BREWING
&
WINE-MAKING

Brewing Basics

Brewing, wine-making, and distilling have been major influences on the development of entire civilizations. Today they remain the cornerstone of success for several of the largest companies on Planet Earth, not to mention vital sources of revenue for the national governments. How fortunate, then, that alcohol is so conveniently produced in the natural world by yeasts that are ever present on the food we eat and the air we breathe. We may think wine, beer, and spirits are important today, but their value was even greater for our forebears, who faced a real challenge in storing foodstuffs through the winter. By converting sugars to alcohol the value of food could be preserved and its flavour even enhanced.

Some plants and fruits seem to lend themselves particularly effectively for wine-making and brewing. Malt and hops have formed a staple part of Western diets as "beer" for many years. Meanwhile, in warmer climes, the grape is pre-eminent with its strong flavours, natural yeasts, and excellent health-giving qualities. In some very special way the smell, taste, and presence of beautiful wine or beer seems to symbolize the very essence of bounty from the earth.

Yeast works in mysterious ways

Different strains of yeast behave in different ways and we need to choose one that suits our purposes best – or at least encourage that type above all others (see also p.200). Some yeasts will float on the top of the brew, others will sink to form a mud on the bottom. Still others will remain suspended as a colloidal mess – definitely not the brewer's favourite. Some yeasts work quickly but cannot stand high temperatures or too much alcohol. Other yeasts work slowly but go on to produce much more alcohol over many months. The first types of yeast are excellent, aggressive plants ideal for beer, whilst the second are what we want for wine but very easily displaced and upset by interlopers. This is why we can ferment beer quickly in relatively unsealed containers while we have to treat wine much more carefully. And the whole brewing process involves managing the complex microfauna of a sugary fruit solution in the best way we can to achieve the results we want. Sometimes the power of an aggressive early fermentation gives way slowly to the gradual pace of tough wine yeasts. And all the time the millions of fungal and bacterial spores that float through the air are awaiting their chance to contaminate our drinks. The vinegar flies lose no chance to splash their little feet in our future dream of wine, spreading harmful bacteria. And the oxygen of the air itself rapidly oxidizes many of the volatile flavours in our fruit soup.

Racking off

When we make beer (or wine for that matter) we must carefully separate the delicious-tasting part from the build-up of sludge and yeast. A fine, clear end result can only be achieved by managing a process called "racking off".

To do this well the first point I would make is not to rack off beer or cider until the major fermentation process has stopped. Whilst fermentation continues, the minute bubbles of carbon dioxide gas carry sediment with them into the body of the brew. As fermentation calms down, there are less bubbles and less sediment. With wines we may want to rack off several times during the much slower fermentation – and we do this once a solid body of sludge has built up.

The second point I would pass on is that a cool liquid holds less sediment than a warm one. Put your brew in a cool place at least 24 hours before you finally rack off the finished product. And if possible put the demi-john or fermenting bin on a high table in this position so it's ready for racking off without having to move it again. The more height you have, the more quickly you can syphon off the brew! Finally (and perhaps an obvious point), make sure you have some device for preventing the end of your syphon from going too far down into the sludge (or near it) for suction will pull it up. There are all sorts of special devices sold for this purpose in brewing shops, but I prefer to use a good solid copper nail fastened on to the pipe with elastic band. A copper or galvanized nail will not taint the brew and its weight helps keep the syphon pipe down.

Another important tip I have learned (from bitter experience) is not to put newly corked bottles straight into storage. Remember you will be storing them horizontally with labels uppermost so that the cork breathes properly and the sediment settles on the opposite side from the label. In this position you will lose everything – and make a big, smelly mess – if the corks blow out. So leave freshly corked bottles upright for at least 48 hours, and examine them carefully to make sure the corks are secure before laying them down.

When pouring out your beer to consume and enjoy, it is best to decant the whole bottle into a suitable container for the dining room – this way you have something pleasant to look at on the table and you avoid stirring up sediment and wasting a good half pint of each bottle. I have long used a lovely old German 2-litre flask with a porcelain and rubber clip top.

Distilling

If you get a big copper with a fire under it, half fill it with beer, float a basin on the beer, and place a shallow dish wider than the copper on the top of the copper, you will get whisky. Alcohol will evaporate from the beer, condense on the undersurface of the big dish, run down to the lowest point of it, and drip down into the basin. It is an advantage if you can run cold water into, and out of, the top dish to cool it. This speeds up condensation. And if distilling is illegal in your part of the world and some inquisitive fellow comes down the drive, it doesn't take a second to be boiling clothes in the copper, making porridge in the floating basin, and bathing the baby in the big flat dish. And what could be more innocent than that?

Malting Barley

Something that has contributed over the millennia to keeping humans human, even if it sometimes gives them headaches, is the invention of malt. One imagines that very soon after men discovered grain, they also discovered that if you left it lying about in water, the water would ferment, and if you drank enough of it, it would make you drunk. In fact, you can make beer out of any farinaceous grain whatever. During the war we had a company brewer in every company of the King's African Rifles. He brewed once a week, and would brew beer out of absolutely any kind of grain or grain meal that he could lay his hands on. Most of it was pretty horrible stuff but it kept us sane.

Later on in history some genius discovered that if you sprouted the grain first, it made better beer and made you even more drunk. He didn't know the reason for this, of course, but we do. It is because alcohol is made from sugar.

Yeast, which is a microscopic mould or fungus, eats sugar and turns it into alcohol. It can also do the same, in a much more limited way, with starch. Now grain is mostly starch, or carbohydrate, and you can make an inferior sort of beer out of it before it sprouts by fermenting it with yeast. But if you cause the grain to sprout, that is start to grow, the starch gets turned, by certain enzymes, into sugar. It then makes much better, stronger beer much more quickly.

So to make beer, we civilized people sprout our barley before we ferment it. This process is known as malting, and the sprouted grain is called malt. You can malt any grain, but barley, being highest in starch, makes the best malt.

Malting barley

Put your barley, inside a porous sack if you like, into some slightly warm water and leave it for 4 days. Pull it out and heap it on a floor and take its temperature every day. If the latter goes below 63°F (17°C) pile it up in a much thicker heap. In the trade this is called "couching" it. If the temperature goes above 68°F (20°C) spread it out more thinly and turn it often. Turning cools it. Keep it moist but not sodden: sprinkle warmish water on it occasionally. Remember you want to make it grow. After about 10 days of this, the acrospire, or shoot of the grain (not the root, which will also be growing), should have grown about two-thirds the length of the grain. The acrospire is to be seen growing below the skin of the grain. Couch it for 12 hours when you think it has grown enough.

Kilning the malt

After this you must kiln the grain. This means bringing it to a temperature of 120°F (50°C) either over a fire or stove, or in an oven with the door open to keep the hot air moving through the grain. Keep it moving in the kiln, which is simply a perforated steel plate over a fire, until it is dry.

Kilning for different beers

Now the colour and nature of beer can be altered one way or the other by the extent of the kilning of the malt after it has been sprouted. This kilning is necessary to kill the grain. If you didn't kill it, it would go on growing into long gangly shoots. Kilning also makes it keep, and you almost always have to keep it before you use it. If you just put it, wet and growing, in a bag, it would rot. Not only would it be useless, it would also smell nasty.

A light kilning makes a light-coloured malt and consequently a light-coloured beer, while heavier kilning makes darker malt and darker beer. If you want to make lager, keep the temperature down to slightly under 120°F (50°C). If you want dark ale take it up as high as 140°F (60°C), but not over. Why not over? Because over would kill the enzymes which are to go on turning even more starch into sugar when you mash the malt.

The maltster watches his malt in the kiln, constantly turning it and looking at it, and he stops the kilning at the right stage for the kind of beer he intends to make. You can stop kilning when you can bite a grain and it cracks between the teeth, but if you want a darker beer you simply go on kilning until the grains turn browner. If you want stout, you actually go on kilning until the grain turns nearly black – but do not allow the grain in the kiln to go over 140°F (60°C). Just give it longer, that's all.

When you have kilned enough, just crush the grain in a mill: don't grind it fine. Now you have malt and you are ready to start brewing your beer.

MALTING BARLEY
Soak the grain for 4 days. Pile it in a heap on the floor and by alternately spreading it out and piling it up again, or "couching" it, keep it at a temperature between 63°F (17°C) and 68°F (20°C). You will need to do this for about 10 days, until a shoot about two-thirds the length of the grain can be seen growing beneath the skin of the grain. Dry the grain completely in a kiln (*below right*), crush it in a mill, and you have malt.

Making Beer

Before Tudor times there were no hops in Britain and the stuff people drank – fermented malt – was called ale. At about that time hops were introduced from Continental Europe and used for flavouring and preserving ale, and the resultant drink was called beer. Beer is bitterer than ale was and, when you get used to it, much nicer. Nowadays the nomenclature has got confused and the words beer and ale are used indiscriminately. But make no mistake in our day and age, beer and brewing are big business – the lifeblood of huge transnational corporations and the source of billions in government taxation.

There are few areas where the self-supporter can make such real cash savings by creating a more delicious product to his or her own taste – at just a fraction of the commercial cost. And what self-supporter would be without a constant and ample supply of this great lubricant for sociability and wise reflection? Making your own beer is the blue riband activity for many a self-supporter, especially city dwellers, for whom it can be a first small act of "freedom" from mass production. Of course you can go further and prepare all the ingredients from your own land. Here's how.

Soil for hops

Hops like a deep, heavy, well-drained loam and liberal manuring, preferably with farmyard manure. But they will produce some sort of a crop on most land, provided they are well fertilized and the land is not waterlogged, and if you grow your own hops for your own beer, some sort of a crop is all you need: you need pounds, not tons.

Planting hops

Clean your piece of land thoroughly first. Make sure you get out all perennial weed roots and grass. Beg, borrow, or steal a dozen bits of hop root. Bits of root about a foot in length are fine. Hops produce an enormous mass of roots every year and an established hop plant just won't miss a foot or two of root.

Plant these bits of root at intervals of 2 or 3 feet (about 60–90 cm), with plenty of farmyard manure or compost. Arrange horizontal wires, some high and others down near the ground. Put vertical strings between the wires for the hops to climb up, three or four strings for each bit of root. When the hops begin to grow they will race each other up the strings and you can place bets with your family on the winner – they grow so fast you can almost see them move. Watch for aphids. If you get them, spray with derris, nicotine, pyrethrum, or other non-persistent insecticide.

Harvesting hops

Pick the flowers when they are in full bloom, and full of the bitterly fragrant yellow powder that is the virtue of the hop. Dry the flowers gently. If you put them on a wire, hessian, or some other perforated surface over a stove, that will do. When they are thoroughly dry, store them, preferably in woven sacks.

Malt and malt extract

You can brew beer from malt extract, which you can buy from the chemist, or in "brewing kits" from various enterprises. The beer you brew will be strong (or can be), will taste quite good (or can do), but it will not be the same as real beer brewed from real malt. The best beer will be the stuff you brew from the malt you have made yourself (*see p. 221*). But you can also buy malt in sacks, and this is preferable to malt extract. The difference between beer brewed from malt and beer brewed from malt extract is great and unmistakable, and if you once get used to beer brewed from malt you will not be content to go back to extract beer – nor to the liquid you buy from the pub.

Brewing beer

In the evening, before you go to bed, boil 10 gallons (45 litres) of water. While it is boiling, make a strainer for your mash-tub or kive (otherwise known as a brewing vat). This is a tub holding 20 gallons (90 litres), but with the top cut off. You can make the strainer by tying a bundle of straw, or hay, or gorse leaves with a piece of string, poking the string through the taphole of the kive, pulling it tight so as to haul the bundle hard up against the hole inside, and banging the tap in. The tap then holds the piece of string. Or, if you like, you can have a hole in the bottom of your kive with an ash stick pushed down into it to close it. When you pull the ash stick out, of course, it opens the hole. If you lay a layer of gorse in the bottom of your kive, some straw on top of this, then a flat stone with a hole in the middle of it, and then poke your ash stick through this hole, you have a magnificent strainer.

GROW YOUR OWN HOPS
Hops must be given strings to grow up, otherwise they get into a hopeless mess and the harvest is drastically reduced. Fix horizontal wires to sturdy posts and then arrange vertical strings between the wires, three or four to each root planted. The hops will do the rest. You just watch and keep them clear of aphids if necessary. Harvest when the flowers are in full bloom (*above*). Inside they will be full of bitter yellow powder – sweet nectar to a serious home brewer.

BEER-MAKING

To make consistently good home-brew you must start off with scrupulously clean kives and barrels. They should be scrubbed, scalded, and then disinfected by exposure to wind and sunlight. Choose a quiet evening and boil up 10 gallons (45 litres) of water in your copper.

1 While you are waiting for the water to boil make a strainer for the kive (brewing vat). Tie a small bundle of gorse, hay, or straw with a piece of string and drop it in the kive, poke the loose end through the bung-hole and pull hard. Then bang in the tap (wooden cock).

2 When the water boils let it cool to 150°F (66°C) and pour half into the kive.

3 Dump in 1 bushel (½ cwt or 25 kg) malt, the rest of the hot water, and stir thoroughly.

Then tuck the kive up for the night. Cover with a clean sheet and a blanket. The enzymes in the malt plus the water will then go to work extracting the malt sugar.

4 Next morning, open the cock and drain the "wort" (liquid) into a bucket, or even better into an "underbuck", the traditional wooden vessel.

5 Now "sparge" (sprinkle) the spent malt with kettle after kettle of boiling water to remove all sugar, until 10 gallons (45 litres) of wort have drained out into the bucket. And thence into the boiler.

6 Pack 1 lb (0.5 kg) of hops into a pillowcase and plunge it into the wort. If you want to cheat by stirring in sugar, honey, or malt extract (6 lb or 2.7 kg to 10 gallons or 45 litres of wort) now is the time to do it. Boil for at least 1 hour. Meanwhile, get on with cleaning out the kive. The mash makes splendid food for pigs or cows.

7 Draw a jugful of boiling wort and cool quickly by immersing in icy water. When it has cooled to 60°F (16°C), plop in your yeast – either packet beer yeast (about 1 oz or 28 g will do) or a couple of tablespoons of "barm" which you have strained off the top of your last brew and kept covered in a cool place. Then transfer the rest of the boiling wort back into the kive.

8 Cool the bulk of the wort as fast as you can by lowering buckets of cold water into it, but don't spill a drop if you want your beer to be worth drinking.

9 As soon as the bulk of the wort has cooled to hand-hot, 60°F (16°C), pour in the "starter", a jugful of foaming yeasty wort and stir. Cover with blankets to keep out vinegar flies.

10 Leave for 3 days. Then skin off the "barm". After fermenting stops (5–8 days) "rack" (pour off without stirring up the sediment).

Brewing beer – "mashing" the malt

When the water has boiled let it cool to 150°F (66°C). Then dump one bushel (about half a hundredweight or 25 kg) of cracked malt into it and stir until the malt is wet through. This is called "mashing", and the malt is now the "mash". It is most important that the water should not be hotter than 150°F (66°C) because if it is, it will kill the enzymes. Cover up the kive with a blanket and go to bed.

Early in the morning get up and open the cock, or draw the ash stick, to allow the wort, as the liquid is now called, to run out into buckets. Pour it from the buckets into the boiler, together with a pound (0.5 kg) of dried hops tied in a pillowcase, and boil it. While the wort is dribbling out, "sparge" (brewer's word for sprinkle) the mash with boiling water. (You don't care about the enzymes now – they have done their work and converted the rest of the starch into sugar.) Go on sparging until 10 gallons (45 litres) of wort have drained out. Much of the original 10 gallons has been absorbed by the mash.

Boil the 19 gallons of wort, and the hops in the pillow case, for an hour. If you want the beer to be very strong, add, say, 6 pounds (2.7 kg) of sugar now, or honey if you can spare it. Or, another way of cheating is to add 6 pounds (2.7 kg) of malt extract. But you needn't add anything at all. You will still get very strong beer. Clean the mash out of the kive and set it aside for the pigs or cows.

Transfer the boiling wort back into the clean kive. Take a jugful of wort out and cool it by standing it in cold water. When it is hand-hot, or about 60°F (16°C), dump some yeast into it. This can be yeast from a previous fermentation, or yeast you have bought especially for beer. Bread yeast will do, but beer yeast is better. Bread yeast is a "bottom fermenting" yeast; it sinks to the bottom in beer. Beer yeast is "top-fermenting" and is marginally better. The faster your bulk of wort cools now, the better. An "in-churn" milk cooler put into the wort with cold water running through it is very helpful. If you haven't got this you can lower in buckets of cold water, but be sure no water spills out and that the outside of the bucket is clean. Quick cooling allows less time for disease organisms to get into the wort before it is cool enough to take the yeast.

When the main body of the wort has cooled to 60°F (16°C), dump your jugful of yeasty wort into it and stir. This is the time when you should pray. Cover up very carefully to keep out all vinegar flies and dust. Try to keep away from the stuff for at least 3 days. Then skim the floating yeast off. Otherwise it will sink, which is bad. When it has stopped fermenting, after 5 to 8 days, "rack" it. That means pour it gently, without stirring up the sediment in the bottom, into the vessels in which you intend to keep it and cover these securely. From now on no air must get in. You have made beer.

You can use plastic dustbins instead of wooden or earthenware vessels; I don't like them, but they have their advantages for hygiene and accessibility. If you use wooden vessels, though, you must keep them scrupulously clean.

Beer from kits

Most good chemists or health food shops sell beer kits and usually all the equipment to go with them. The modern equipment is mostly plastic but this is at least one beneficial by-product of the petro-chemical industry! You will want to try quite a few different recipes – and do not think the most expensive is always the most delicious. For years I bought the cheapest tins of malt/beer extract made by Irish monks – they were delicious, but now alas, the monks have died out. Always keep a note of what you have brewed and how you brewed it.

Your key items of equipment are a large fermenting vessel, normally 40 pints (23 litres), with a well-fitting airtight top, a decanting syphon tube with rigid pipes at each end and a tap at the end you will put into bottles, and a heating mat for keeping the fermentation at the right temperature. Your kit will come with its own yeast and its own instructions. All you need to provide is "love", plenty of hot water, and a couple of pounds of sugar – then plenty of thirst to drink the stuff.

I normally make one brew each week and somehow it all seems to disappear. First swish out your fermenting vessel (which should be kept nice and clean – replace it if it gets old and worn) with a kettle full of boiling water to kill any beasts and bugs. Pour in your malt extract (after pre-warming it on the stove to make it pour easily), add 2 pounds (1 kg) of sugar and then two kettles full of boiling water. Stir the lot together and when well mixed, fill up the bin with cold water before adding the yeast. Try to avoid leaving the top of the bin open and you will avoid contamination and vinegar flies. Pop on the top and place on your heating mat in a calm place for about 1 week.

Modern beer kits have yeast which sinks to the bottom when fermentation is complete. Test your brew after 7 days: it should be a clear brown, with all foam gone, and have a bitter taste (all the sugar gone to alcohol) and a lovely, beery smell.

If it is ready, turn off the heating mat and put in a cool place, preferably high up where you can decant into bottles the next morning. Get all your decanting equipment ready before you start. You will find clear plastic fizzy-drink bottles ideal for storing your beer; make sure you give them a good wash and that they are stored full of clean water. Smell each one as you pour out the water and chuck any doubtfuls. You will find many a friend glad to give you these items of "waste".

Syphon off your beer carefully to avoid disturbing the sediment. Put extra sugar into each bottle; the exact amount is a matter of taste and judgement. I like beer without too much fizz, so half a teaspoon does for me in a 2-litre bottle before screwing down the lid. Keep the filled bottles in a warm place for 24 hours to get the secondary fermentation going before you put them out into storage. They will take another 10 days to fizz up. You can tell how much fizz there is by simply squeezing the bottle; a hard bottle means lots of fizz.

Making Wine

Strict cleanliness is essential in wine-making, for wine is made by a living organism (yeast) and if other living organisms (wild yeasts or other moulds or bacteria) get into the act, either the tame yeasts that you want to use for your wine cannot do their job, or you get putrefaction, bad tastes, and odours. Aside from that important point, you can ignore the plethora of books about home-made wine-making, each one blinding us with science more effectively than the last. You really only need to remember the following few essentials:

1 *As already stressed, keep all wine-making equipment scrupulously clean. Use boiling water whenever possible.*
2 *You are unlikely to get more than 3 pounds (1.4 kg) of sugar to ferment in 1 gallon (4.5 litres) of water, so keep to approximately this ratio if you want strong wine.*
3 *You must ferment at the temperature most favourable to vital yeasts.*
4 *You must give your special cultivated yeast every help and an unfair advantage over the wild yeasts and other organisms that might ruin your brew.*
5 *You must keep all contaminants out of your wine, especially vinegar flies, those little midges that hang round rotting fruit, carrying the bugs that turn wine to vinegar.*
6 *You must "rack" or pour off the wine from the lees and sediments before the latter spoil its flavour.*
7 *You must allow the wine to settle and clear in the cool after the yeast has done its work.*
8 *Finally, having safely bottled your wine, you must try to keep your mitts off it for a year with red wine, if you can, and at least 3 months with white.*

Equipment
You need jars, barrels, or bottles for fermentation. You also need fermentation locks (if you can get them). The purpose of these is to allow the gases produced by fermentation to escape while keeping out air, which is always germ-laden, and vinegar flies. Many a vat of fine wine has been made without a fermentation lock and with just a plug of cotton wool stuffed in the neck of the vessel. Many a gallon of wine has been ruined this way, too. A fermentation lock is a very useful thing. A thermometer is not to be despised, either. You also need a flexible tube – rubber or plastic – for "racking" or syphoning, a funnel or two, and containers for the final bottling of the wine. A corking gun is very good for driving in corks, which have to be driven in dead tight or air gets in and the wine goes bad. Polythene sealers are quite a good substitute for corks if you do not want to invest in a corking gun.

Materials
You will need yeast. Old-fashioned country wine-makers, including myself, have used all kinds of yeasts – bread yeasts and beer yeasts and so on – but undoubtedly it is best to buy wine yeasts from a shop. For very good and strong results some people use yeast nutrients, also bought from a shop. Acid is another thing you may have to add.

Lemons will provide this, as will citric acid, which you can buy. Tannin, too, can be bought, but tea or apples – particularly crab apples – will provide it. The reader may say that it is not being self-sufficient to buy all this stuff from a shop. True, but I would say that a trivial expenditure on this sort of thing is necessary if you are going to make a great deal of fine, drinkable wine.

GRAPE WINE
There is no wine like grape wine. Red grape wine is made by fermenting the grape skins in with the wine. White wine is made by taking the skins out. White wine is often made with red or black grapes, for all grapes are white inside. It is easier to make red or rosé wine than white because the tannin in the skins helps the "must" (wine-to-be) to ferment better, and the quicker it ferments, the less chance there is of bad organisms getting to work.

Crushing
Crush your grapes any way you like. Personally, I could not drink wine if I had seen somebody treading it with his bare feet, so I would use some sort of pestle and mortar for this job. If you want to make white wine, press the broken grapes in a press (a car jack will do), having first wrapped them in strong calico "cheeses" (*see* p.193). In the case of red or rosé, press in the same way, but then add a proportion of the skins to the wine. The more you add, the deeper the red colour of the wine, but, in cold climates at any rate, the deeper red ones may contain too much tannin and will be a little bitter as a result. Now in real wine-growing climates (where you will not be reading instructions like this anyway, since your neighbours will initiate you), you don't need to add any sugar. In less sunny climates add between 4 and 6 pounds (1.8–2.7 kg) of sugar to every 10 gallons (45 litres) of wine-to-be. If there has been a hot season and the grapes are sweet, you need less; if a bad season, more.

Fermenting
Let the juices and skins ferment in a vat. Grapes have their own yeasts in the "bloom" on their skins, but you had better add a wine yeast culture bought from a shop, if you can get one. Warm a bottle of the must (juice) to 75°F (24°C), dump the yeast culture into it, and stand it in a warm place with some cotton wool in the neck. Meanwhile, try to get your main body of must to 75°F (24°C). When the "starter" or culture in the bottle has started to fizz, pour it into the main body. If you keep the temperature at about 75°F (24°C) fermentation will be so active that there is no danger of air getting to the must, for the carbon dioxide given off will prevent this. Don't let the temperature rise above 80°F (27°C) or some of your yeast will be killed. Don't let it fall below 70°F (21°C), or your yeast will get sleepy and foreign yeasts will gain advantage. Always keep the skins stirred into the must. They will float on top, so don't let them form a dry, floating crust.

WINE-MAKING EQUIPMENT
Don't attempt to make your own wine without arming yourself beforehand with plenty of containers, to hold the must at each of its many stages.

Bottles and bottling are only the end stages of a long fermentation process, during which you will need at least several containers, such as jugs and jars – and quite possibly vats and barrels, too.

Key
1 Fermentation barrel
2 Corking gun
3 Jug
4 Bottle
5 Sieve
6 Bottle brush
7 Funnel
8 Hydrometer
9 Measuring cylinder
10 Screwtop bottle
11 Plastic or rubber syphon
12 Earthenware vessel
13 Barrel and tapped vat
14 Fermentation jar and lock
15 Cork and plastic sealer

Racking
When the first violent fermentation has ceased, rack off the must, squeeze the juice out of the skins so as not to waste it, and pour the juice into a barrel or carboy, so that the must fills it completely. Be careful not to leave an air space above it.

Let the temperature fall now to a temperature of about 60°F (16°C). Check, and when you think most of the sediment has sunk to the bottom, rack the wine into another container. At this stage people in continental climates often put wine out of doors in winter so that it almost freezes, because this hastens the settling down of sediment. Now rack it again. After another month or two of it sitting quietly, you can then bottle it in the way I will now describe.

Bottling
Bottles must be completely cleaned and then sterilized. It is no good "sterilizing" anything with dirt in it; the dirt must first be removed. Sterilize by heating slowly so you do not crack the bottles. You can do this in an oven if you like. Then pour in boiling water, or put in cold water and slowly bring to the boil and boil for 5 minutes. Hang the bottles upside down immediately to let them drain and stop dust from floating down into them. Either use as soon as they are cool or cork until you want them. Boil the corks before you use them and wack them in with a special corker. Store bottled wine on its side to keep the corks wet. If they dry out they will shrink, and air and vinegar bacillus will get in. Store wine in the dark, at a cool, even temperature. A cellar or basement is ideal.

COUNTRY WINES

I am going to give you some recipes for "country wines" that work, as I know from long experience. I would not put anybody off "scientific wine-making", which is reliable and produces good wines, but country people all over Europe and North America have used the sort of recipes I give for centuries, and very seldom have failures; indeed, their wines are very good. One point worth noting is that the larger the bulk of wine you make, the less likely you are to have a failure. My old friends in a Worcestershire village who all brew rhubarb wine in the summer and parsnip in the winter, in batches of 60 gallons (273 litres) stored in huge cider barrels, have never known what a failure is. Their spouses cry in vain for them to grow something else in their gardens, but their wine is superb.

Flower wines

Pour a gallon (4 litres) of boiling water over an equivalent quantity of whatever flowers you wish to use; allow to cool, and press the water from the flowers. Add 4 pounds (1.8 kg) of sugar, ½ pound (228g) of raisins (optional), and the juice of three lemons. As the flowers don't give much nutriment for the yeast, and sugar alone is not enough for it, add some yeast nutriment if you have some. A tablespoon of nutriment to a gallon of wine is about right. Then, when the temperature has fallen to 75°F (24°C), add yeast. A bought wine yeast is best. Put the wine in a vessel with a fermentation lock, and leave it to ferment. Rack off and bottle when ready. I have made wines from broom flowers, gorse flowers, elderflower (superb), cowslip, and dandelion, and I have drunk good rose wine.

MEAD

To supply what in your estimation is about 3 pounds (1.4 kg) of honey to 1 gallon (4.5 litres) of water you want comb cappings, odd bits of "wild comb" that you can't put through the extractor, and perhaps some pure honey stolen from the main storage pot when your wife isn't looking. Melt the honey in the water and ferment.

Honey is deficient in acid, so put the juice of two or three lemons in a gallon, or some citric acid. Mead also likes some tannin to feed the yeast, so some crushed crab apples are a good idea. I have heard of people putting tea in mead. I once dumped some rose hip syrup that the children decided they didn't like into my mead, which wasn't fermenting very well, and it started to ferment like blazes. Mead goes on fermenting for a long time, so do not try to hurry it, and if you can leave it in a bottle for a few years, so much the better. But can you?
Here are some wine recipes to try for yourself:

Rhubarb wine
15 lb (6.8 kg) rhubarb
2 ½ lb (6 kg) sugar
1 gallon (4 litres) water
and add yeast

Chop up the rhubarb, pour boiling water over it, and mash. Don't boil the mixture any further. Leave it to soak until the next day, then strain off your liquor and press the "fruit" to get as much out as you can. Stir in the sugar and bung in the yeast. Leave the whole thing to ferment, then rack it and bottle it.

Nettle wine
4 lb (1.8 kg) nettle tips
4 lemons
2 lb (1 kg) sugar (preferably brown)
1 oz (28 g) cream of tartar
2 gallons (9 litres) water
1 tablespoon dried yeast or brewer's yeast

Put nettles and cut-up lemons in the water and boil for 20 minutes. Strain liquor out and add cream of tartar and sugar. When cool enough add yeast and ferment for 3 days in a warm place. Then let it settle for a couple of days in a cooler place before bottling in screwtop bottles. You can drink it in a week, and it doesn't keep long. It is extremely pleasant and refreshing. If you add some ginger to it, it is even better.

Parsnip wine
4 lb (1.8 kg) parsnips
3 lb (1.4 kg) sugar
1 gallon (4.5 litres) water
some lemons or citric acid
and add yeast

Cut the parsnips up and boil them without letting them get too soft. They should just be easily prickable with a fork. Boil a couple of lemons up with them if you have them. Strain off the liquor, and while it is still hot, stir in the sugar, so that it dissolves. Put in some lemon juice or citric acid, and some raisins if you like. The purpose of the lemon juice or citric acid is to give the yeast enough acidity to feed on, as parsnips are low in acid. Put everything in a vessel, wait until the temperature drops to blood heat, then add your yeast and allow to ferment. Like all other wine, ferment under a fermentation lock, or put a wedge of cotton wool in the neck of the vessel, to keep the vinegar flies out and let out the carbon dioxide. Rack it well a couple of times and then keep it as long as you can lay your hands off it.

Elderberry wine
6 lb (2.7 kg) elderberries
3 lb (1.4 kg) sugar
1 gallon (4 litres) water
2 oz (56 g) citric acid or lemon juice
and add yeast

You are meant to get all the berries off the stalks, but I've shoved in stalks and all and it's made no difference. After all, if you can save a lot of work by departing from slavish convention, why not do so? Pour the boiling water on, mash hard with a potato masher, cover and leave to soak for 24 hours. Put the sugar and yeast in and leave it alone. The longer you leave it, the better. When it has finished fermenting, rack it into bottles or other containers, so as to leave the sediment behind. You do this with all wines. The above recipe can be applied to any wine that is made from berries or currants.

Elderflower "champagne"
12 heads of elderflowers (in full bloom and scent, picked on a hot day)
1½ lb (0.7 kg) sugar (white sugar is less obtrusive than brown in such a delicate drink)
1 lemon
2 tablespoons wine vinegar

This is nothing like champagne, of course, but it's a very refreshing summer drink, and it does not have to be kept long before you can drink it. Put blooms in a bowl with the juice of the lemon. Cut up the rind of the lemon and put that in (minus the white pith). Add the sugar, vinegar, 1 gallon (4 litres) of water and leave for 24 hours. Strain liquor in screwtop bottles, cork up, and leave for a fortnight. Don't add yeast – weak yeasts on the flowers are enough. Drink before three 3 weeks old.

MAKING ROSE HIP WINE

The principle of wine-making does not vary much according to the main ingredient. The addition of a wine yeast to your brew starts off the fermentation process, which can take as long as 3 months.

1 Take 3 quarts (3.4 litres) of rose hips, clean them, and chop them up finely. Crush with a wooden spoon or mallet.

2 Put the crushed hips into a deep bowl and pour 1½ gallons (6.8 litres) of boiling water over them. You can add the rind and juice of an orange.

3 Add 2 lb (0.9 kg) of sugar, and heat it to 75°F (24°C).

4 Stir in a teaspoon of fresh yeast. You can put this first into a bottle of "starter", which you add to the brew when it starts fermenting. Add 1 teaspoon of citric acid and ½ teaspoon of tannin.

5 Cover the must overnight to keep out vinegar flies and all other contaminants.

6 Strain the must from the hips through a sieve or muslin cloth. For even clearer must, use both these methods.

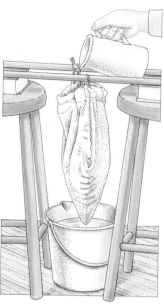

7 Or you can strain through a jelly bag, suspended from two stools. Don't press it, or it will go cloudy.

8 Strain the must into fermentation jars. Use a funnel. Keep at 75°F (24°C).

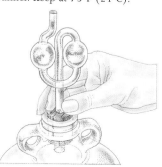

9 A fermentation lock keeps air out but allows gases to escape.

10 When fermentation stops, rack the wine off the lees into bottles. Use a rubber or a plastic tube for this.

11 If you have no tube, use a hand jug and a funnel. Leave an inch (2.5 cm) at the top for corks when filling the bottles.

12 A corking gun is excellent for driving corks in tight, but a wooden mallet will do. Date, label, and leave for a year.

Making Cider & Vinegar

CIDER

Cider should be made from a mixture of apples. The ideal mixture is a selection of apples rich in acid, tannin, and sugar, so a good combination mixes very sweet apples with very sour ones, perhaps with some crab apples thrown in to provide the tannin.

Cider can be made with unripe apples but it is never very successful. Ideally, the apples should be picked ripe and then allowed to lie in heaps for 2 or 3 days until they begin to soften a little. A few bad or bruised apples in the press don't seem to affect the quality of the cider at all. Apples vary greatly in juice content, so it is not possible to tell exactly how much cider you will get from a given number of apples. As a rough estimate, 10–14 pounds (4½–6 kg) of apples make one gallon (4.5 litres).

Over the centuries cider-making was one of the best ways of preserving the "food" value of apples over the winter. Alcohol and the fermentation process are like a magic preserving process. By the miracle of nature, cloudy and dirty apple juice become a delicious (if slightly intoxicating) golden liquor. Apples are produced in great quantities during the autumn. Some varieties store fairly well, Cox's for example, but most will only last a couple of months, and then only if they are picked in perfect condition (windfalls will be bruised and useless for storage). In the modern world very few people know how to deal with their apple crops and we find we have lots of friends who are only too happy for us to appear with a party of kids to gather up all their windfalls. Cider-making parties with kids are a pleasant way of enjoying an autumn day.

Crushing

When you have collected a pile of apples – a wheelbarrow-full at least – you're ready to start cider-making. A few strong helpers will be useful if you can find them. You will need chopping equipment, a crusher, a fruit press and a 5-gallon (4-litre) fermentation bin. You then crush the apples. Crushing is an arduous task. Traditionally, this was done by a horse or an ox pulling a huge, round stone round a circular stone trough. I did have one friend who used to put his apples through a horizontal mangle, which reduced them to pulp very effectively.

Firstly, get all the equipment cleaned from storage and set up. Make sure your fruit press is fastened to a solid, heavy object. I have made a small, heavy table for the purpose and screw the press down so strength can be exerted without knocking things about. The apples go from the wheelbarrow into a chopping box, where they can be chopped with a clean spade. Once chopped, the apples can be put through the crusher – watch out for catching fingers here. When your press is full, get squeezing. The first juice is always exciting – and delicious to drink straight from the press.

THREE SIMPLE STEPS

1 **The chopping box** Use a strong wooden box for chopping your apples with a clean, sharp spade. Fill it full enough so the apples cannot move about, but not so full that the pieces spill out.

2 **The crusher** Put your chopped apples into the crusher when they are no more than 1 inch (3 cm) square. Crush the pulp straight into the fruit press – and be careful you don't catch your fingers.

3 **Pressing the juice** Fix your press to a stout table or work surface and squeeze the juice straight into the fermenting vessel. Remove the "cheeses" of squeezed pulp, feed a bit to the pigs, with the rest for your compost heap.

Direct the juice from the press straight into your large fermenting bin and do not worry too much at this stage about bits of apple or grass floating about in the vat. When all the juice has been squeezed from the first batch of apples, you must undo the press to take out the "cheese" of crushed pulp. I keep a separate wheelbarrow to hand for this. You can feed some pulp to pigs and cattle, but most must go on the compost heap. Refill the press and continue the process until your fermenting vessel is nearly full.

Covering and fermenting the apple juice

Your vat of apple juice is full of natural yeasts from the skins and the orchard. It will probably have a fair bit of other bits and pieces in it, too. The next step is to cover with a muslin cloth to keep out the vinegar flies, then leave the vat overnight in a sink. In the first day or so there is likely to be a rapid fermentation and you want this to bubble over the sides. This trick means that the bubbles carry with them large quantities of dirt and muck over the edge and down the plug hole. As soon as any violent fermentation has finished you can decant the whole bin into another clean fermentation bin through a sieve. This removes any remaining bits and pieces of apple, and you can leave behind any heavy sediment that may have formed in the first vessel. You should now have 5 gallons (23 litres) of reasonably clean apple juice ready to complete its fermentation. At this stage I normally add a few teaspoonfuls of commercial brewing yeast and 1–2 pounds (0.5–1 kg) of granulated sugar. The yeast makes sure that you get completion of a good fermentation as you cannot altogether rely on natural yeasts. The amount of added sugar is up to you – more sugar makes a stronger cider that will keep much better than a watery version. Add the sugar by making up a syrup with a couple of pints of boiling water: pour this into your vat. Put a tight-fitting lid on your fermenting vessel and keep it warm; you can use a warming pad, a heated belt or mat (electronically-heated wraps to keep fermentation at a constant temperature), or an airing cupboard. The brew needs to ferment completely so that all the sugars are transformed into alcohol. This may take anything from 10 days to 3 weeks. You can check on progress by looking at the brew to see if there are still bubbles coming up or by tasting it (if it is still sweet, then there is a long way to go). When fermentation is complete, the liquid will begin to clear and a brown scum of yeast will be left around the edges. You are now ready to rack off into bottles for storage. Take the fermenting vessel into a cool place, preferably where you are going to rack it off. Leave it for at least 24 hours so it can settle; then rack it off into bottles just as you would beer. Add a little more sugar to each if you wish to ensure a secondary fermentation, but this often happens on its own. I have stored flatish cider through the winter months (outside), then, when the warmer months come, a secondary fermentation starts and gives an excellent fizzy lift. Cider improves greatly with age and will certainly keep for up to a year.

VINEGAR

Vinegar is wine, beer, or cider in which the alcohol has been turned into acetic acid by a species of bacteria. This bacillus can only operate in the presence of oxygen, so you can prevent your wines, beers, and ciders from turning to vinegar by keeping them protected from the air.

Yeast produces carbon dioxide in large quantities, and this expels the air from the vessel that the beverage has been stored in. But yeast cannot operate in more than a certain strength of alcohol, so fermentation ceases when so much sugar has been converted into alcohol that the yeast is killed or inhibited by its own action. This is the moment when the vinegar-forming bacillus, *Acetobacter*, gets active, and the moment when your beverages need protecting most rigorously from fresh air and bacterial infection.

However, if you want to make vinegar you must take your wine, beer, or cider and expose it to the air as much as possible. If you just leave it in an open barrel it will turn into vinegar in a few weeks. But it is better to speed up the process, to ensure that smells from the surrounding atmosphere do not taint the vinegar and hostile bacteria have little time to attack.

To hasten the process, take a barrelful of beech shavings. Beech is a traditional component for this stage of the proceedings but, really, any shavings will do as long as they do not come from a very resinous tree. First, soak the shavings well in a good vinegar of the type you are trying to make. Then, put a perforated wooden plate in the barrel over the shavings and pour your wine, beer, or vinegar on to this plate.

The liquid will drip slowly through the holes, which must be very small – about the size of pin-holes. Slowly, the liquid drips through the shavings. This ensures that the liquid is well exposed to both air and *Acetobacter*. At the bottom of the barrel the vinegar-to-be is drawn off through a cock. Just leave the liquid in an open cask and it will turn into vinegar within a week.

MAKING VINEGAR

Soak a barrelful of beech shavings in vinegar of the sort you are making. Put a wooden plate, perforated with pin-sized holes, on top of the shavings in the barrel. Pour your alcohol on to the plate. It will drip slowly through the barrel, and be well exposed to air and the vinegar-forming bacillus. After about a week in an open cask, it will turn into vinegar.

"Every householder should have several small dustbins: one for organic waste, one for aluminium, one for non-returnable glass, one for tin cans, and one for plastic. Plastic, by the way, is hugely overused. It hadn't been invented fifty years ago, or most of it hadn't, and the world got on surprisingly well without it. We should refuse to buy goods which are overwrapped in plastic. As far as recycling goes, by far the best solution is one that has been developed in Germany: melting it down and turning it into building panels which are strong and good insulators. Our poor suffering old planet just cannot stand the wastage and pollution of rubbish dumping any more. We owe it to our children and our children's children to put an end to this scandal: we are rifling their inheritance."

JOHN SEYMOUR CHANGING LIFESTYLES 1991

CHAPTER NINE

ENERGY

&

WASTE

Food for the Garden

MAKING COMPOST

If you pile vegetable matter up in a heap, it will rot and turn into compost. But to make good compost, and to make it quickly, you have to do more than this.

You can make the best compost in the world in 12 hours by putting vegetable matter through the guts of an animal. To make it any other way will take you months, whatever you do. But the principle of compost-making is this. The vegetation should be broken down by aerobic organisms. These are bacteria and fungi, which require oxygen to live. The bacteria that break down cellulose in plant matter need available nitrogen to do it. If they get plenty of available nitrogen they break down the vegetable matter very quickly, and in doing so they generate a lot of heat. The heat kills the weed seeds and disease organisms in the compost. If there is a shortage of available nitrogen, it takes the organisms a very long time to break the vegetable matter down. So in order to speed the process up as much as you can, you try to provide the things that the compost-making organisms need: air, moisture, and nitrogen.

You can provide the air by having rows of bricks with gaps between them underneath the compost and, if you like, by leaving a few posts in the heap as you build it, so you can pull them out to leave "chimneys". You can provide the moisture either by letting rain fall on the heap, or by throwing enough water on it to moisten it well.

And you can provide the nitrogen by adding animal manure, urine, fish meal, inorganic nitrogen, blood, blood meal, or anything you can get that has a fairly high nitrogen content.

Dung and fertilizers

The natural, and traditional, way to make compost is to throw your vegetable matter (generally straw) at the feet of yarded cattle, pigs, or other animals. The available nitrogen in the form of the animal's dung and urine "activates" the compost. The urine also provides moisture and enough air gets between the straw. After a month or two, you dig the heap out and stack it carefully out of doors. More air gets into it and makes it rot down further. Then, after a few months, you cart it out and spread it on the land as fertilizer.

But if you don't have any animals your best bet is to build compost heaps by putting down a layer of bricks or concrete blocks with gaps in them, and laying coarse, woody material on these to let the air through. Then put down several layers of vegetable matter, sprinkling a dusting of some substance with a high nitrogen content between them.

Ten inches (25 cm) of vegetable matter and a couple of inches (5 cm) of chicken dung, or a thick sprinkling of a high-nitrogen inorganic fertilizer, would be ideal. Some people alternate lime with the nitrogen.

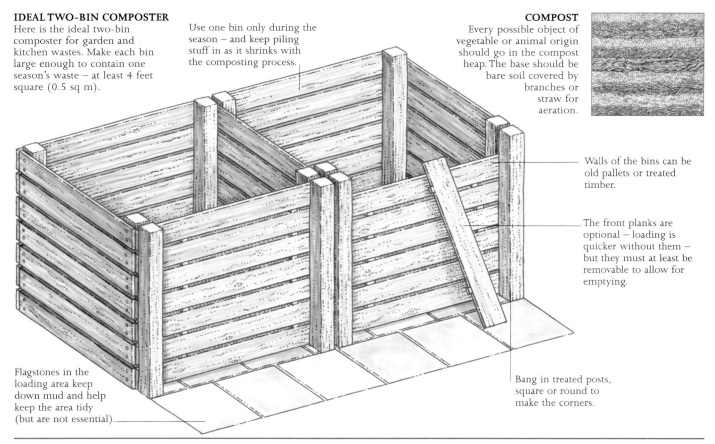

IDEAL TWO-BIN COMPOSTER
Here is the ideal two-bin composter for garden and kitchen wastes. Make each bin large enough to contain one season's waste — at least 4 feet square (0.5 sq m).

Use one bin only during the season — and keep piling stuff in as it shrinks with the composting process.

COMPOST
Every possible object of vegetable or animal origin should go in the compost heap. The base should be bare soil covered by branches or straw for aeration.

Walls of the bins can be old pallets or treated timber.

The front planks are optional — loading is quicker without them — but they must at least be removable to allow for emptying.

Flagstones in the loading area keep down mud and help keep the area tidy (but are not essential).

Bang in treated posts, square or round to make the corners.

Keep the sides vertical using walls of either wood, brick, or concrete, and keep it decently moist but not sopping.

Don't forget that anyone can pile up stuff to make a heap but only a true countryman can make a good stack with vertical sides. You will be surprised how much material from the garden you can deal with in this way. Just make sure you spread out each addition into a thin layer rather than a heap. And mix plenty of rich green stuff (weeds, grass cuttings, and so on) with the more woody wastes so you keep a good mix. Keep piling vegetable wastes in for a whole season. Make sure you allow enough space to do this – 8 x 4 feet (2.5 x 1.2 m) does it for our half an acre of garden. The pile will grow with new material one week and shrink back down the next as the material decomposes; in fact you will be surprised just how much shrinkage takes place.

At the end of the season (October or November) you will stop adding any further material to your season's compost. From now on you will use the second composting area, leaving the first to ferment undisturbed. When it comes to early spring (February or March, most likely) you will be able to take off the top and unrotted material from the outside of last season's heap and put this into the bottom of the second heap. You can then spread your rotted compost on the garden as you choose: some for your runner bean bed, some for the soft fruits, some for new deep beds, as you fancy.

Of course it is a great help if you can obtain farmyard manure from a local farmer if you cannot provide it yourself. You often see farm signs for compost or manure along the roads, but you'll need a trailer of some sort.

The closed urban composter

I like to keep a closed plastic composting bin for the richer kitchen wastes that attract flies and rats. You can buy these with a closing top and a rat-proof base from garden centres and some local authorities are starting to supply suitable plastic bins for this. Of course, if you have a composting toilet, then most of this stuff can go down that route.

Essentially the same principles apply as to the open garden compost heap: a good mixture and plenty of it. I put an aluminium sheet with holes punched through it as a base for my plastic composter and this keeps out rats but lets in worms. Once you have sufficient mixture to generate a good fermentation you will see the material rapidly shrinking. Add urine if you feel it needs extra nitrogen. The composting bin has a hatch in the bottom which can be opened to shovel out the rotted material after several months of fermentation. You may want to add this to your existing garden compost heap if it is not quite ready.

Worm composting

In warm climates worm composters are very popular and easy to use. These depend on small red manure worms which either appear naturally from nowhere or that can be bought from angling shops. Worms need moisture and warmth to thrive. And one neat way of providing this is to use a column of used motor tyres as a container for organic wastes. Add manure worms at the bottom and they will steadily multiply and eat their way upwards. You can then remove the lower tyre (and its compost) as the column fills up. The worms will have moved upwards by then, so you keep the process on-going.

PLASTIC URBAN COMPOSTER
Typical store-bought composter suitable for urban use. Made of heavy grade plastic and should be both fly and rat resistant.

Base must be rat-resistant yet allow worms to enter and moisture to leave; perforated aluminium sheeting is ideal. Sliding door allows rotted material to be removed.

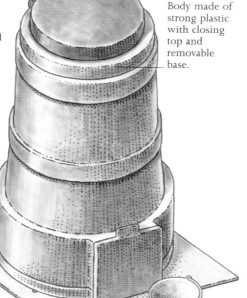

Body made of strong plastic with closing top and removable base.

WORM COMPOSTING
A pile of old car tyres can make a good container for worm composting with manure worms. The worms like warmth and heat. You remove the bottom tyre regularly as the worms move up the stack.

The Dry Toilet

Our dry toilet has been the subject of many jokes, much interest, and a couple of television documentaries. Amazing, really, considering the fact that the idea is not exactly new. What is new, of course, is that such things are no longer a common accessory, at least not for modern urban living. So great is the deep human emotion on this topic that we find media men simply unable to overcome their fear and open the door to the Killowen composting "throne". So if you find all this X-rated material, now is the time to stop reading!

This fear of basic functions seems to be part of the modern human condition, which finds the idea of dealing with waste and death extremely challenging. To illustrate the other extreme, I must tell you about a friend we have from the United States. He is, in fact, one of the founders of the growing bio-regional movement. David regards the flush toilet as one of the greatest sins of modern man. So much so that he takes a small shovel with him on all trips to "civilization" so that he can deposit his own waste carefully into the soil, rather than pollute the water table. Madness? I don't think so. The flush toilet is a remarkably expensive way to pollute fresh drinking water, while at the same time wasting the very nutrients that are essential to maintain fertility in the soil. One pull of the lever and the waste becomes somebody else's problem. We just pay our taxes and allow our children to pick up the real inheritance of all this pollution. But, for the smallholder, there is great satisfaction in being able to deal directly with what he or she consumes and what he or she produces as waste. The composting toilet has a great role in this, and, as a way of explaining this role, I give you the text pinned up on the wall that introduces guests to the "thunderbox".

Our Marvellous Thunderbox Loo

Now the human being is a very strange beast with capabilities good and bad
Not frightened of nature, no not in the least our follies are often quite mad
The toilet that flushes fills our souls with glee
A brainwave by Thomas Crapper
Mixes shit with clean water and pours out to sea
As if the dirt did not matter.

Out of sight, out of mind, muck shoots down the pipes
An incredible fabric of magic
Squandering food for the soil as the water we spoil
It's a tale that is terribly tragic.

But all is not lost for at a marginal cost
Another solution comes easy
The composting loo; yes that's our reposte
And your tummy need not feel too queasy.

The vent goes up high, sending gas to the skies
And the lid fits snug so no entry to flies
Two years it will take, our compost to make
And our river's not sorry the flush to forsake.

No water, no tricks – it's all built with bricks. The shit and kitchen waste too
All go together making food for the soil in our marvellous thunderbox loo.

Here are the short instructions that help users to help the Thunderbox do its job.

Using the Thunderbox

The Thunderbox is a dry composting toilet built to designs well-tested in tropical countries and now approved by some building authorities.

To work effectively its contents should always be a good mixture of organic materials, with a suitable blend of carbon and nitrogen within the mix *and* plenty of air to make sure the fermentation does not become anaerobic.

The toilet does not smell because it is vented above the roof height and the composting process in any case is not particularly smelly.

When using the toilet please:
1 Try not to put in too much pee and use the bucket for pee (no paper please).
2 Do not use too much paper in the Thunderbox.
3 Pour in a box of sawdust after use.
4 Always close the lid when you have finished so that we keep out flies.

Kitchen and garden waste are regularly put into the toilet to give a richer composting mix – so you may see all sorts of vegetable material if you happen to peer down below. We also put in straw and well-rotted manure to help the process.

Design

The design of the Thunderbox derives from the dry toilets commonly used in many tropical parts of the world. So far, our toilet has eaten up wheelbarrow loads of waste – from humans, kitchen, and garden. For a family of six, with occasional guests, it has taken more than 3 years to fill the first chamber. There are thousands of different bugs, bacteria, and manure worms working away down there. Large quantities of water vapour and methane go up the chimney. It is such a pity the methane is wasted, but it would be difficult to process such relatively small quantities of waste in a methane digester.

The essential principle of the Thunderbox is that it is a two-chamber system. One chamber rests and composts while the other is in daily use. This means that the size of the chamber must be sufficient to take all use for at least one whole year to allow sufficient time for complete fermentation before the chamber is emptied after a minimum of 2 years since first use. The design must allow for the front of the chamber to be somewhat permeable so that any liquids can percolate slowly through if things are getting too wet in the mix. In practice, I have never found this to be a problem – possibly because I frequently add woody garden wastes. By using tongue and groove timber fronts we allow percolation and complete replacement of the front (if necessary) after one cycle of use. With a stone or concrete run-off area in front of the chambers it is a

THE THUNDERBOX TOILET

The ultimate eco-friendly toilet is the Thunderbox. Although usually found in hot countries, this need not stop you from building your very own in temperate climes. Different countries also have different regulations about the use and disposal of composting toilet material. Check this with your local authority if in any doubt. Composted toilet material should be placed around fruit trees or bushes, where it does not come into direct contact with foodstuffs.

Thunderbox door
Make sure the door shuts properly – for privacy as well as to keep out flies.

Second chamber
The previous years' waste composts down in a sealed chamber.

Percolation
Tongue and groove timber fronts allow liquids to percolate.

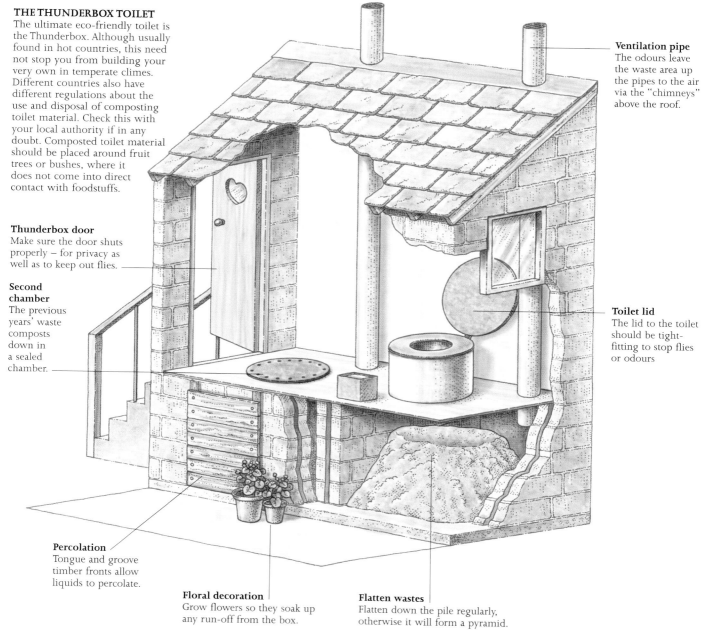

Ventilation pipe
The odours leave the waste area up the pipes to the air via the "chimneys" above the roof.

Toilet lid
The lid to the toilet should be tight-fitting to stop flies or odours

Floral decoration
Grow flowers so they soak up any run-off from the box.

Flatten wastes
Flatten down the pile regularly, otherwise it will form a pyramid.

simple matter to grow a few geraniums or other pot plants that can soak up any small run-off. If there is a great quantity of run-off then action must be taken by adding more sawdust and stirring the mix (with a long pole inserted through the top opening).

It is good practice also to stir the compost regularly, say, every month, to prevent a pyramid building up under the opening. Good ventilation is vital, with the chimneys positioned well above the roof. The warm gases create their own draught without the wind, taking away the water vapour and any odours.

It is also essential that the composting does not get waterlogged. Composting in a confined space benefits from a good mix of ingredients. Floor sweepings, dust, and succulent garden waste provide variety, while leaves and sawdust keep things from getting too wet. But, as always, composting is something of an art form – you do get better with practice.

Close-fitting lid

Keeping flies out of the toilet is the hardest job and some small fruit flies are almost inevitable. If the lid fits closely, keeping out all light, and the only source of daylight is down the chimney, then the flies should fly up to the mesh and drop back exhausted. It is vital to get this part of the design correct.

Managing Waste

Over the last 50 years, rubbish collection and the processing of human waste has become a major challenge for urban civilization. We take from nature, we use (briefly) and we throw away – tons and tons of supposedly useless "waste". Most of this is put into landfills which will, of course, become the mines of tomorrow if not the methane power stations of the 22nd century. Much more than half of this "waste" is paper and kitchen waste and lots more is glass, metal, or clothing that can be recycled. Every self-sufficient holding should take pride in producing as little waste as possible – indeed, zero waste should be the goal.

The first and most elementary step is to avoid buying in waste in the first place. Give up your daily newspapers for a start, you probably have news blasted at you many minutes a day anyway from television and radio. Avoid plastic packaging and take advantage of laws that allow you to take unavoidable packing back to the shop for disposal by them. The second step is to make sure you separate your waste into different containers. I have separate bins for:

1 Compost heap Your main compost heap takes bulk vegetable waste like potato peelings, weeds, leaves, fruit skins, apple pulp, and so on. Your closed composter takes stuff that would attract rats – for example composting disposable nappies, old tea bags, and rotten meat or fish unsuitable for pigs. (*See p.234 for more on the compost heap.*)

2 Pig and poultry bucket This takes waste food and scraps and must be emptied and washed out daily to keep it fresh. Stale cakes and buns, meat fat, and fish skins – these are pure heaven for your pigs who will convert the whole lot into beautiful bacon. The more varied your pig's diet the more tasty the bacon will be.

3 Paper Paper can be burned, composted (after shredding which is tiresome), or recycled if you are in the right area. Put a stop to junk mail and stop taking the Sunday papers and you will avoid excesses of paper.

4 Plastic Plastic is almost impossible to avoid in the modern world beholden to the oil industry. Recycle plastic 2-litre bottles (find out what your local authority will take) and keep clean for bottling your beer and cider.

5 Tins Most local authorities now recycle both aluminium and steel tins so you need to separate these out. Aluminium cans are crushed easily whilst steel cans will flatten if you take off both ends.

6 Glass The crashing of green, brown, and clear glass bottles is now a regular sound at recycling centres as people get rid of waste glass. Quite why we have to smash up perfectly good bottles when millions of kilowatts of energy could be saved by having a few standardized shapes beats me. The fuel companies would hate us but for goodness sake we could even take wine bottles to be refilled as they still do in parts of France; and I do remember a time when milk came in bottles that were re-used without difficulty.

With all the above it's sensible to check with your local authority first to find out what recycling they offer. But the best recycling is that of composting your rubbish to use later as a fantastic fertilizer for your crops.

THE BLACK ART OF SHIFTING MUCK

We are told that the Zen master of housework spends a lifetime perfecting the job of washing up so that the activity ultimately becomes an art form of sublime pleasure and satisfaction. What greater success could there be in life? What magic is it that can transform drudgery into high pleasure?

On the previous pages I've discussed how to make the best compost for your garden, but here I'd like to concentrate on an even better source of food for next year's vegetables. It may seem like a foul task: that of emptying out a winter's build-up of cow manure and straw. Let's face it, digging out and wheelbarrowing 10 tons of smelly manure does not sound immediately appealing. But in the art of self-sufficiency the good student will enjoy great satisfaction from the sight of a beautiful heap of rotting manure. Just think of the hot, sweaty, smelly shovelling as being like riding a bike up a beautiful mountain; the pain of the climb is more than balanced by the prospect of whizzing downhill for many miles on the other side. In this case the pain of the shovelling is more than compensated for by the prospect of a good pint of beer afterwards and fabulous "food" for next year's vegetables.

The fork and wheelbarrow

The first requirement, as all students of Zen will know, is to get to know and love your tools, in this case, the manure fork and the wheelbarrow. The right equipment is vital. Your manure fork will have four prongs or tines made of spring steel and with curved shoulders, like a pitchfork. The tines should have a nice curve on them which not only allows you to carry more manure but also serves as a vital lever to prize up layers of heavy, downtrodden, wet straw. The curved shoulders mean that straw is less likely to stick to the fork when you toss it into the barrow. A heavy, square garden fork is absolutely the wrong tool for this job: it is straight, its tines are too thick, and its shoulders catch constantly on the straw.

Your wheelbarrow has to be strong enough to carry up to 250 pounds (100 kg) of muck at a time. The wheel should be of the inflatable type, as this runs more smoothly. And don't forget you can use different grades of tyre. The more expensive are much less likely to give problems with punctures. A galvanized steel barrow is probably the best of all. You will also want your wheelbarrow to fit close to the cowshed gate so you do not have to take more than a couple of steps each time you put a forkful into the barrow. Minimize walking and save time and energy.

Now get yourself organized so you have a clear run between the cowshed and the compost-manure heap. You would not want this to be more than 25 yards (23 m).

Make sure your manure heap has good strong sides than can take the pressure of a few tons of manure – drive in extra treated posts with your post rammer if necessary. We are talking here about a manure heap which might typically be 10 feet (3 m) across and 6 feet (1.8 m) deep. It will be open at one side and bounded on the other three by timber slats held in place by treated fenceposts. If you have steps or similar to push up, then you will make sure any ramp is well secured and supported before you start. The straw bedding will be heavily compacted by the weight and hooves of the animals. Each barrowload may weigh over 220 pounds (100 kg) so you will need a good run to get up any slopes.

The art of digging

You cannot dig out manure like sand. Instead you must push in your fork tines at a low angle and use the curve of the tines to act as a fulcrum as you lever gently down on the handle of the fork. This will slowly lift up a layer of manure perhaps 3 inches (7.5 cm) thick with a sucking sound as the compaction is released. You can then lift this, with a bit of jiggling, and toss the forkful into your barrow. You can pile up the barrow pretty high, as the manure holds itself together well. One tip worth remembering is not to put your hand over the end of the fork handle. If you do and accidentally bang a wall, it will hurt!

An almost inevitable spin-off from the manure shifting will be, alas, a broken fork handle. Manure is heavy stuff and it is very easy to press a bit hard and break your fork handle. Many modern tools are sold with poor quality handles. You want one made of ash where you can see a good straight grain running down the handle. It should be thicker in the middle, tapering to the top. When you break your fork handle, you can look forward to a 45-minute rainy day job (*see p.252*).

Making a straight-edged stack

Now you have the task of making a good stack of manure to compost. This is an art form in itself for, as every countryman knows, anyone can make a pile, but only a countryman can make a good stack. And a stack is going to be there for all to see as a work of art, or otherwise, for perhaps a year or more! The key to making a good stack is to have the material (straw in this case) sloping down into the stack from the edges. If it is sloping down from the centre to the outside then it's going to slip and you will not get a straight vertical side to your stack. In the old days of farming the stacking of the crop was a vital skilled task, for a stack that slipped or let in the weather was a major disaster. Stacking your manure heap is not quite that important but a neat stack will keep its shape and give you a better compost than a loose heap. To make a straight-edged stack you must start by putting material along the outside edge. You want a wide mat of material with plenty of long strands, ideally running at right angles to the edge of the stack. If you have a forkload of loose material which is in small pieces then discard it for the moment: toss it to the back of the space. By building up a straight edge right across the front of the heap about 9 inches (23 cm) high and then working back from it, you will make sure your material is sloping down away from the edge. Every time you have a good forkload with long strands, use this to strengthen the edge. The long strands will be trapped by the other material above them and hold the edge firm.

You have a major strategic choice to make when digging out your compost. Do you make a solid heap of manure which will compost itself? Or do you keep your good manure in reserve so you can take out a 6-inch (15-cm) layer every couple of weeks to add to your general compost heap? It depends on the amount of manure, the routine of your garden, and the need you have to clean out the cowshed quickly for more stock or whatever. By mixing manure with other material for composting you will achieve an excellent result. There is a lot to be said for taking a little time over shifting 10 tons of muck. Like many tasks in the world of self-sufficiency, a great deal can be achieved by regular application. When eager beavers from the city try to race into a long job they will literally bust a gut if trying to set some arbitrary target. Targets are a big feature of the business world but they often become counter-productive in the world of self-sufficiency.

The only way to tackle a huge job is to make a point of doing a little but doing it regularly. I often make moving muck a pleasant break from other daily chores by fitting in say, 6 barrowloads in one session. That might take about 30 minutes – and move about half a ton of muck. Do this every day for about 3 weeks – it's better than going to the gym (certainly much cheaper) and the job will get done even though each day you can hardly see any change. It is the same logic for digging a large vegetable bed: little and often is much better than taking a huge bite at one time. So that's the story. Good tools and the right attitude can transform drudgery into a sublime pleasure where you can bask every day in the beautiful light from the end of your metaphorical tunnel. And each day after the task is finished you will pass your manure heap and enjoy the memory of your daily sweat and application.

MANURING

If your land has had proper additions of farmyard manure or the dung of animals added direct, or seaweed (which has in it every element), or compost, it is most unlikely to be deficient in anything. By getting your soil analyzed when you take it over, and adding once and for all whatever element the analysis shows the soil to be deficient in (nitrogen, phosphorus, potassium, or calcium), and thereafter farming in a sound organic way, the "heart" (fertility) of your land should increase continually until it is at a very high level. There should be no need to spend any further money at all on "fertilizers". And, very often, if land is virgin, or if it has been properly farmed in the past, you may not even need to get it analyzed.

Saving Energy

THE ALTERNATIVES

Throughout this book I have advocated an integrated approach to the land: the encouragement of organic beneficial interaction of soil, crops, and animals. When considering energy we must adopt this same approach. We should look upon our holding of land as having a certain energy potential that we can use for our own good purposes, and we should aim to make our holding autonomous in this respect, as we have aimed to make it for food.

There is something wrong about burning coal to heat water on a hot, sunny day, or burning oil to warm a house when there is a fast-flowing stream next to it. Or, for that matter, using mains electricity to drive a mill or a power loom, when there is potential wind or water power nearby.

Water power is most available in hilly, rainy countries and wind power in flat lands, but wind power should never be used where water power is available. The simple reason for this is that the wind is fickle, while water is relatively reliable and consistent. Where there is hot sun it it is ridiculous not to use it. It is obviously unproductive to feed cold water into your water boiler when the corrugated iron roof over your shed is so hot you can't hold your hand on it.

A characteristic of natural sources of energy is that they lend themselves much more to small-scale use than to large-scale exploitation. For example, more energy can be got out of a given river more cheaply by tapping it with a hundred small dams and waterwheels right down its length, than by building one enormous dam and driving one set of huge turbines. The wind's energy can be tapped better, and now increasingly is (thankfully for our environment) by a myriad of small windmills, rather than some gigantic wind-equivalent of a power station.

It doesn't need an "Earth Summit" to tell you that every house in a city could have a solar roof, and derive a great part of its energy requirement from it, whereas a solar collector big enough to supply a community is still in the realms of fantasy. Scattered farmsteads can easily make their own methane gas, but to cart muck from a hundred farms to some central station, make gas from it, and then redistribute it would be madly uneconomical. So these "alternative energy devices" commend themselves especially to the self-supporter.

HEAT LOSS

A house built in the traditional way loses vast amounts of heat through the roof, doors, windows, floor, and outside walls. Use a combination of the methods illustrated and you can save as much as two-thirds of your annual domestic energy requirement.

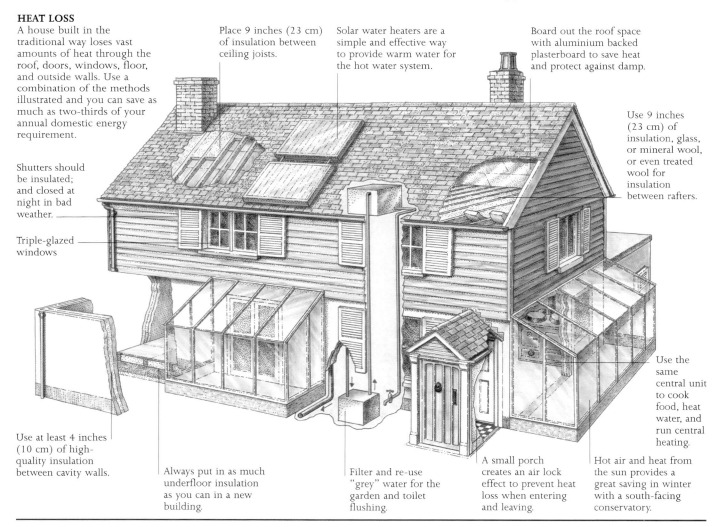

Place 9 inches (23 cm) of insulation between ceiling joists.

Solar water heaters are a simple and effective way to provide warm water for the hot water system.

Board out the roof space with aluminium backed plasterboard to save heat and protect against damp.

Use 9 inches (23 cm) of insulation, glass, or mineral wool, or even treated wool for insulation between rafters.

Shutters should be insulated; and closed at night in bad weather.

Triple-glazed windows

Use at least 4 inches (10 cm) of high-quality insulation between cavity walls.

Always put in as much underfloor insulation as you can in a new building.

Filter and re-use "grey" water for the garden and toilet flushing.

A small porch creates an air lock effect to prevent heat loss when entering and leaving.

Use the same central unit to cook food, heat water, and run central heating.

Hot air and heat from the sun provides a great saving in winter with a south-facing conservatory.

COMBINING NATURAL ENERGY SOURCES

Now it may well be that it is better for the self-supporter to combine several sources of energy instead of concentrating on just one. For example, you could have a big wood-burning furnace (*see p.287*) that does the cooking for a large number of people, and heats water for dairy, kitchen, butchery, bathroom, and laundry. If you preheated the water that went into it with solar panels on the roof, you would need less wood to heat more water.

Then if you had a methane plant to utilize animal and human waste and used the methane to bring the hot water from the furnace to steam-heat for sterilizing dairy equipment, better still. Then you could use a pumping windmill to pump up water from the clear, pure well below your holding, instead of having to use the very slightly polluted water from the hill above.

And what about lighting your buildings using the stream that runs nearby to drive a small turbine? All these things are possible, would be fairly cheap, and would pay for themselves by saving on energy brought in from outside.

Keeping heat in

There is little point devising elaborate systems for getting heat from natural sources until you have plugged the leaks in the systems you have already got.

For keeping heat in a house there is nothing to beat very thick walls of cob, stone, *pise* (rammed earth), or brickwork with small windows and a thatched roof. The thin cavity walls of modern brick or concrete block housing only insulate well if plastic foam or some other insulating material is put between the walls and laid on the joists in the roof. The big "picture windows" beloved of modern architects are terrible heat-losers. Double glazing may help, but it is very expensive. Country folk, working out of doors for most of the day as people were designed to do, want to feel, when they do go indoors, that they are indoors; they get plenty of "views" when they are out in it and are part of the view themselves. Therefore, for country housing, big windows are a mistake.

Huge chimneys, very romantic and fine when there are simply tons of good dry firewood, send most of their heat up to heat the sky. In a world short of fuel they are inexcusable. Long, straggly houses are also great heat-wasters. A compact shape is more desirable. A round building will lose less heat than a square one, because it has a smaller surface area in comparison with its volume. A square building is obviously better than an oblong one. It is always best to have your primary heat source in the middle of the building rather than against an outside wall. Most insulation nowadays is achieved with high technology products, and these are very expensive. What we can do is search for cheaper and more natural materials. Wherever the cork oak will grow it should be grown, for it provides an excellent insulator and in large quantities.

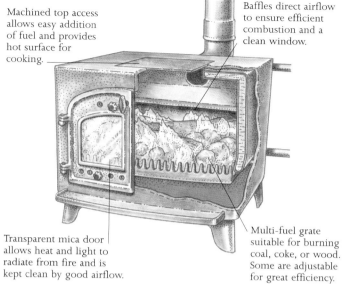

Machined top access allows easy addition of fuel and provides hot surface for cooking.

Baffles direct airflow to ensure efficient combustion and a clean window.

Transparent mica door allows heat and light to radiate from fire and is kept clean by good airflow.

Multi-fuel grate suitable for burning coal, coke, or wood. Some are adjustable for great efficiency.

CLOSED BOILER

Most wood stoves and ranges will burn almost anything combustible. Dead wood and all other inflammable rubbish provide additional free fuel supplies which can be saved and stored during summer and autumn for use in winter. A large old-fashioned cast-iron range can supply the major source of domestic heat as well as being used for cooking meals.

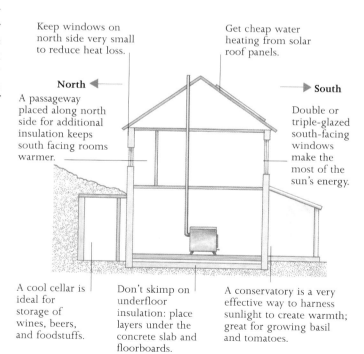

Keep windows on north side very small to reduce heat loss.

Get cheap water heating from solar roof panels.

North ← → **South**

A passageway placed along north side for additional insulation keeps south facing rooms warmer.

Double or triple-glazed south-facing windows make the most of the sun's energy.

A cool cellar is ideal for storage of wines, beers, and foodstuffs.

Don't skimp on underfloor insulation: place layers under the concrete slab and floorboards.

A conservatory is a very effective way to harness sunlight to create warmth; great for growing basil and tomatoes.

USING NATURAL FEATURES FOR ENERGY SAVING

By taking full advantage of the lie of the land and the orientation of your house you can create a comfortable and energy-efficient living space. Building into a south-facing slope is the ideal: the effect can be magnified by careful tree planting to create shelter on the north and east sides. Before cheap fossil fuels, traditional builders took full advantage of all these possibilities. Both thatch and thick cob walls were excellent insulation.

Power from Water

Parakrama Bahu, King of Ceylon, as it was in the 7th century, decreed that not one drop of water that fell on his island should reach the sea: all should be used for agriculture. In wetter climes, where irrigation is not so necessary, the inhabitants would do well to take the same attitude, but change the objective a little: "let not one river or stream or rivulet reach the sea without yielding its energy potential".

Water power is completely free, completely non-polluting, and always self-renewing. Unlike wind it is steady and reliable, although of course there may be seasonal variations, but even these tend to be consistent. Like wind, it is generally at its strongest in the colder months, and is therefore at its greatest strength when we most need it.

The primitive water wheels that have stood the test of centuries are not to be despised, and for many uses are better than more sophisticated devices. For slow-flowing streams with plenty of water, the "undershot wheel" is appropriate. With this you can take in the whole of quite a large stream, and thus exploit a river with a low head but large volume. Your wheel will be slow-turning, but if you use it for a direct drive to slow-turning machinery, a corn mill for example, this is an advantage. It is a common mistake of "alternative energy freaks" to think that all power should first be converted to electricity and then converted back to power again. Energy loss is enormous in so doing.

If you want to generate electricity with your water power you will need something more sophisticated than a waterwheel, for this requires high speeds to which more complex water engines are well suited. For small heads, from as low as a yard (1 m) to up to 20 feet (6 m), the propeller turbine is very good.

WATER WHEEL (OVERSHOT)
The oldest method of using water power, the overshot water wheel is up to 70 percent efficient. The water goes over the top, filling the buckets round its rim. Water wheels like this turn quite slowly but with considerable force, making them best suited for driving mill-stones or other heavy, slow-speed equipment. Depending on flow rate, power from this wheel might be from 5–20 hp (4–16 kW).

WATER WHEEL (UNDERSHOT)
Undershot wheels are less efficient than overshot, but are used when there is insufficient head of water for it to fall over the wheel. They can produce from 2–5 hp (1½–3 kW).

BREAST WHEEL (UNDERSHOT)
An undershot wheel with straight blades is up to 30 percent efficient; fitting curved blades increases this to 60 percent. A breast wheelmakes twice the power from the same source.

WATER POWER

To calculate the available water power of a stream, measure the flow rate of the water, and multiply it successively by the density (62.4 lb/ft³), the head of water, and the efficiency of the turbine or water wheel you will be using – e.g., in a hilly area the flow rate might be ½ ft³ water per sec. This flow falling 40 feet through an 80 percent efficient Pelton wheel would give:
½ x 62.4 x 40 x $^{80}/_{100}$ = 998 $^{ft lb}/_{sec}$
$^{998}/_{550}$ (1 hp) = 1.8 horsepower.
If used for generating electricity, 60 percent of the turbine's power might be converted into electrical power: i.e., in this example 60 percent of 1.8 = 1.08 hp and as 1 hp = 746 watts, this is the equivalent of 805 watts.

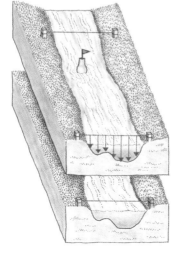

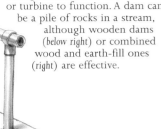

POSITIONING AND TYPES OF DAM

In order to build up a head of water and control its flow, it is often necessary to build a dam or a weir (above) across the main stream, usually at a narrow point or where there are rapids. A head race or leat, dug along a contour above the stream, will create enough head for a water wheel or turbine to function. A dam can be a pile of rocks in a stream, although wooden dams (below right) or combined wood and earth-fill ones (right) are effective.

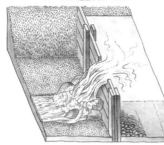

Flow rate

Water power depends primarily on the flow rate and available head of water. The flow rate of a stream therefore needs to be measured, as well as its fall, to predict the available power. A simple method is to find a length of the stream that is straight and has as constant a cross-section as possible. The cross-section of flow is estimated by taking soundings at regular intervals across it and calculating the average depth: area equals average depth times width. This should be repeated at several points to arrive at an average cross-sectional area for the chosen length. A sealed bottle (left) is then timed as it drifts along the middle of the chosen section. Flow rate for the stream will be around 75% of the speed of the bottle times the average cross-sectional area of that length of stream. (For an example of a water-power calculation using flow rate, see the caption, left.)

HYDROELECTRIC POWER

If you are fortunate enough to have a stream available for use, then potentially you have the capability to produce a free and continuous supply of electricity. Don't use a water wheel to run a generator because this turns so slowly that an enormous step-up ratio of gears or belts and pulleys is needed to arrive at the required generator speed. Small turbines turn much faster and need little more than a pair of pulleys and vee belts to connect them to a generator. They are less expensive to build because their smaller size means they need much less steel, and they are also slightly more efficient than water wheels. There are many different types of turbine. The Pelton wheel turbine (top right) is for high head applications where the fall is 40 feet (12 m) or more, and is up to 80 percent efficient. A special nozzle directs the water at high speed against a set of spoon-shaped deflector buckets set around the periphery of the turbine wheel. The Banki turbine (centre right) is for medium head, up to 65 percent efficient and best suited to a fall of 15–40 feet (4.5–12 m). A special nozzle directs water into the periphery of a spool-like wheel with curved blades. The propeller turbine (bottom right), up to 75 percent efficient, operates best on heads of under 20 feet (6 m), down to 6 feet (1.8 m). A propeller in a pipe, it is the best substitute for an old mill water wheel. To obtain reasonable efficiency, the water must be given a spin the opposite of that of the propeller. This is best done by running it through a spiral volute before entering the draft-tube containing the propeller.

Pelton wheel turbine

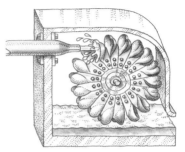

Banki turbine

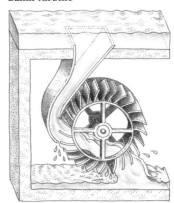

Propeller turbine

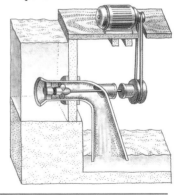

Heat from the Sun

The most practical solar collector is a wood, for woodland can collect the sun's rays from vast areas, and, properly managed, can continually convert them into energy, while to cover a few square yards with a man-made solar collector costs a lot of effort and money. But if collecting and storing the sun's heat can be done relatively easily and cheaply, as it usually can on the roof of an existing house or wall, then, if nothing more, solar energy can be used to reinforce other sources of energy. The drawback is that in cold climates we want heat in the winter and we get it in the summer, but if the winter gap is filled in with wind or water power (which may be at their best in the winter) a consistent system can be evolved.

The practical choices open in temperate climates are:
1 Heating water by letting it trickle over a black-painted corrugated roof under a transparent covering which turns the roof into a heat collector. You will have to buy your transparent covering and a pump to circulate the water. All the same, this will allow you to collect the sun over a large area.

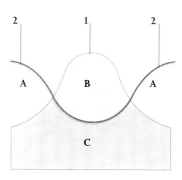

SOLAR ENERGY
Solar energy 1 is most abundant in mid-summer while our heating requirements 2 are greatest in mid-winter. Most solar collectors provide more heat than we need in summer B and less than we need in winter A. The productive use of solar energy C reaches peaks in spring and autumn. Received energy per day per sq m might be 4 or 5 kW-hr in summer, and ½ to 1 kW-hr in winter in a temperate climate.

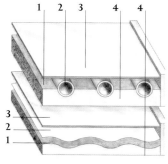

FLAT-PLATE SOLAR COLLECTORS
Most solar energy collectors use a black surface 1 to absorb the sun's radiation and produce heat. You transfer the heat into a hot water tank, or to heat space, by passing water, or air in some cases, through pipes or channels 2 behind the absorbing surface. A glass or plastic covering 3 prevents heat loss from the front of the collector, while insulation 4 prevents it from rear and sides.

HEATING AIR: THE TROMBE WALL
Named after Professor Trombe, this is a clever method for making use of solar energy in winter. The Professor perfected the wall high up in the Pyrénées, where the sun shines quite often in winter, albeit weakly. You use a vertical double-glazed plate glass window 1 which faces south, and allow a black-painted wall 2 behind it to catch and trap the sun's heat. When you require heat inside the house you open ventilators 3, 4 and these allow warm air to circulate between the glass and the wall. An over-hanging roof 5 prevents the high summer sun from striking the glass and also protects the building from getting overheated. An alternative to the Trombe wall is a glass-covered extension to your house – in other words, a conservatory. This will warm the house if properly ventilated.

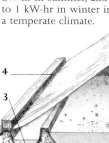

SOLAR STILL
This shallow concrete basin, painted black or tarred 1 contains a few inches of polluted water. A heavy-gauge polythene tent 2 encloses this, and condensation runs down the inside surface of the tent into a pair of collecting gutters 3. The condensation is pure distilled water which you can syphon off. Hold down the plastic sheeting with heavy wooden battens 4 and close the cover ends rather like a ridge tent. You can replenish the polluted water through a hose 5.

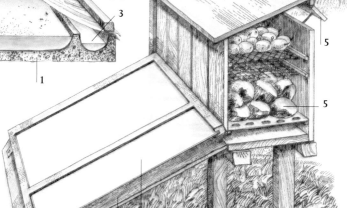

SOLAR DRIER
An inclined, glazed, flat-plate solar air heater admits air through an adjustable flap 1. The air heats up as it crosses over a blackened absorber surface 2, because the heat is trapped by glass panels 3. The heated air rises through a bed of rocks 4 and then through a series of gratings which hold the produce to be dried. A flap 5 under the overhanging roof allows air supply to be adjusted or closed off. The rock bed heats up in the course of the day and continues releasing a measure of warmth to the crop after sunset, thereby preventing condensation from occurring. There is a door in the unit's back to allow crops to be added or removed.

TRICKLE ROOF HEATER
Providing an entire south-facing roof for a solar water heater can be worth it in the long run, by having water trickle down a blackened, corrugated, aluminium roof behind an area of glazing. Insulation behind the aluminium prevents overheating of the roof space and keeps most of the heat in the water. A small pump drives water round the system whenever a sensor on the roof tells a control-box that the roof temperature is higher than that of the water in the copper immersion heater.

2 A second option is heating water with black-painted pipes behind transparent material. This has the advantage that there is no obscuration by misting-up and you don't need a circulatory pump because hot water rises. But it is expensive to cover large areas.

3 Solar stills: these are arrangements for using the heat of the sun directly for distilling water or other liquids.

4 Solar dryers: these can be used for drying fruit, grain vegetables, malt, and many other things.

5 Solar hot air heaters: provide extra heat for a greenhouse.

6 Solar-heated walls such as the Trombe wall: store heat during the day and release it during the night.

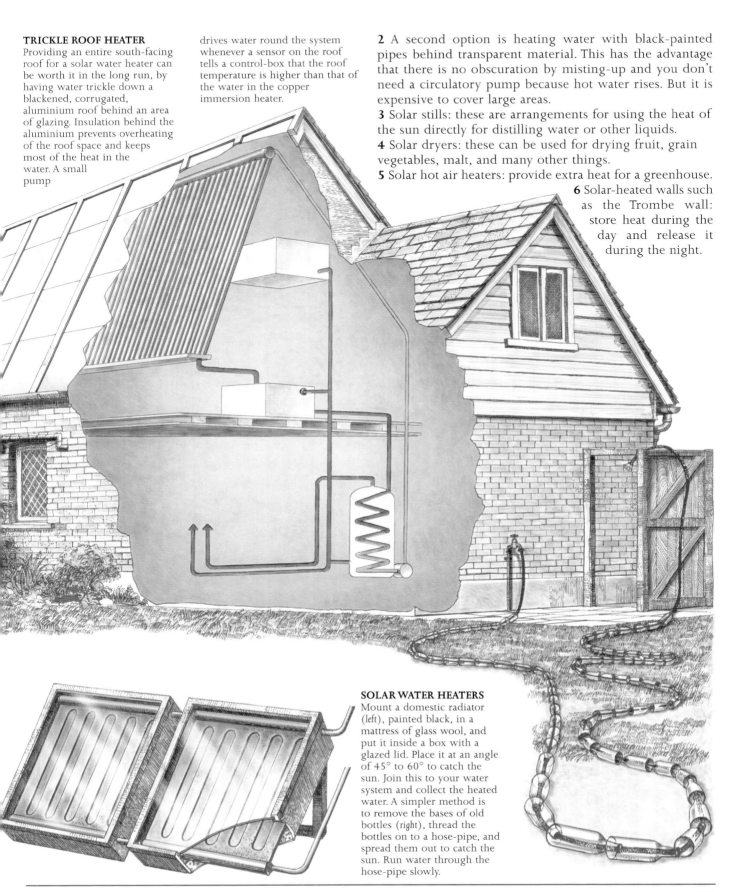

SOLAR WATER HEATERS
Mount a domestic radiator (*left*), painted black, in a mattress of glass wool, and put it inside a box with a glazed lid. Place it at an angle of 45° to 60° to catch the sun. Join this to your water system and collect the heated water. A simpler method is to remove the bases of old bottles (*right*), thread the bottles on to a hose-pipe, and spread them out to catch the sun. Run water through the hose-pipe slowly.

Power from the Wind

The common factory-built steel pumping windmill, seen by the thousand in all lands where water has to be pumped up from deep boreholes, is one of the most effective devices ever conceived by humans. Many an old steel "wind pump" has been turning away, for 30 or 40 years, never failing in its job. Such machines will pump water comfortably from 1,000 feet (300 m) and work in very little wind at all. The tail-vane is arranged on a pivot so that the windmill can turn itself sideways to the wind in a storm.

Wind power has followed the same trend as water power in that low-powered but high-speed devices are now wanted for driving dynamos to produce electricity. But the wind, of course, is completely unpredictable, and so you must either accept that you cannot use a machine in calm weather, or in severe gales, or you must be able to store electricity, and that is very expensive. However, if you can use the power when it is available – say, for grinding corn, or store it, as heat, for example – the total wind energy available over a period of time tends to be fairly constant.

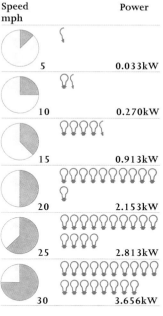

Speed mph	Power
5	0.033kW
10	0.270kW
15	0.913kW
20	2.153kW
25	2.813kW
30	3.656kW

HOW MUCH POWER?

The main problem in harnessing wind is that it carries very little power when it blows lightly, but offers an embarrassing surfeit in a gale. The power of wind is proportional to its velocity cubed: in other words, if the wind speed doubles, its power potential rises eight-fold. This means that a fairly large windmill is needed if useful amounts of power are to be extracted from a light breeze, and that the windmill must be protected from storm damage by having a hinged tail-vane which can swing the wind-rotor out of the wind. Or it can have removable sails or blades which can be made to twist into a "feathered" position where they act as air brakes to slow the rotor. The diagram shows how many 100-watt lightbulbs a 15-feet (4.5-m) diameter electricity-generating windmill can power.

Sail windmill

All-metal

Tubular steel

Triple-blade

This is a variation on a Mediterranean sail windmill. Used for irrigation water pumping by market-gardeners on Crete, it is readily improvised.

This typical all-metal windmill is used for pumping water. A swinging tail-vane turns it out of the wind in a storm. You might be able to renovate an old one.

A water-pumping windmill in which the rotor runs in the lee of the tubular steel tower; weights at the blade roots swing them into a feathered position in gales.

This windmill needs only three aerodynamically profiled blades. The machine trickle-charges a bank of batteries to supply low-powered appliances.

To be self-sufficient in electricity

Wind power is hard to win and store, so you should always use wind-generated electricity sparingly. Never use it for heating appliances. In order to exploit wind power you must have an average wind speed of at least 9 mph (14 km/h), with no lengthy periods of low winds; even so you will need battery storage to cover up to 20 consecutive days of calm. Apart from an electricity-generating windmill, you need a voltage regulator and a cut-out to prevent the battery from overcharging. Total battery storage capacity needs to be 20 x average current needed in amps (watts ÷ volts) x average usage time in hours per day, measured in amp hours. Standard domestic electric appliances requiring 220 volts a.c. can be driven from a bank of 12 volt (d.c.) batteries by an electronic invertor. Alternatively, low-voltage appliances may be used directly. A typical 2 kW, commercially manufactured windmill will often generate at 110 volts d.c. to charge a bank of low-voltage batteries, wired in series. You might get 5,000 kW-hr annually from a 2 kW windmill. One kW is equal to one unit of electricity.

Generate your own electricity

The typical electricity-generating windmill is available in kit form or as a do-it-yourself design. The aluminium or fibreglass blades are pivoted from the hub: centrifugal force works on the balance weights and overcomes a set of springs attached to the hub shaft, so the blades feather automatically if the rotor overspeeds.

A toothed rubber belt drives a car alternator to produce up to 750 watts. Power is transmitted down the inside of the tower, either through a conducting slip ring and brush, or by a cable which can be released when it is twisted, thus providing a breakable connection. Similar arrangements might be improvised though they might suffer in reliability.

ROTOR TYPES
Here are a selection of rotor blades running on electrical generators, pumps, and even a Dutch-style millstone – ideal for the self-supporter. Locally constructed and operated windmills have been highly successful in many rural communities in Denmark as it happens. Sadly, incentives for small-scale windpower are few.

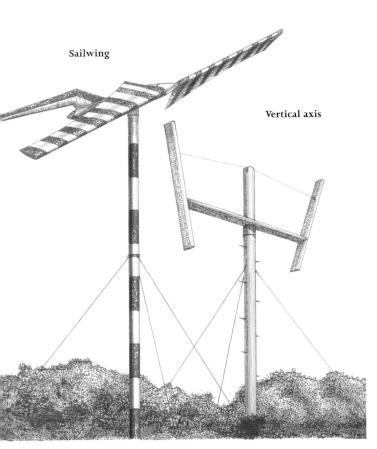

Sailwing

Vertical axis

This is the simple and cheap sailwing, developed at Princeton University, USA. A fabric sleeve is stretched between the two edges of the "wings."

Reading University, England, developed this vertical axis windmill. Aerofoil blades are spring-loaded and fold outwards to prevent over-speeding.

Rotor type		Typical load	rpm	Hp	Torque
Propeller (lift) double and triple		electrical generator	high	0.42	low
Darrieus (lift)		electrical generator	high	0.40	low
Cyclogiro (lift)		electrical generator or pump	moderate	0.45	moderate
Chalk multiblade (lift)		electrical generator or pump	moderate	0.35	moderate
Sailwing (lift)		electrical generator or pump	moderate	0.35	moderate
Fan-type (drag)		electrical generator or pump	low	0.30	high
Savonius (drag)		pump	low	0.15	high
Dutch-type (drag)		pump or millstone	low	0.17	high

SOLAR POWER

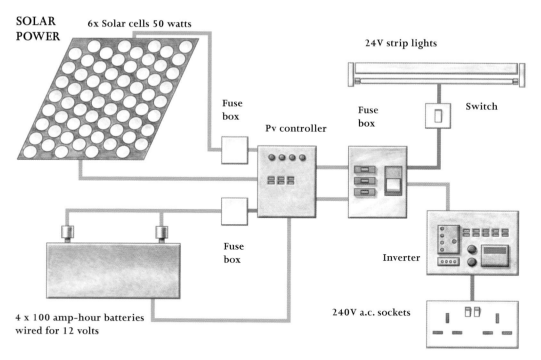

6x Solar cells 50 watts

24V strip lights

Fuse box

Pv controller

Fuse box

Switch

Fuse box

Inverter

4 x 100 amp-hour batteries wired for 12 volts

240V a.c. sockets

PHOTOVOLTAIC SYSTEM
These are the main elements of a typical photovoltaic system. A battery of cells is aligned to the sun. This is often tilted downward somewhat to optimize the performance in winter months when the sun is low. Power flows through to a bank of storage batteries through a control unit, which also links the batteries to domestic appliances. Lighting, audio, and computers can all work well from 24 volts. There is also an inverter for converting the power to 240 volts alternating current for occasional use in larger household appliances. As a guideline, a car battery produces just 12 volts across its terminals, whereas in your house there is 240 volts between the live and neutral wires. High voltage electricity (above 50 volts) is dangerous; low voltage is not. Solar cells normally produce low voltages.

Solar electricity generation

One of the greatest potential sources of electrical energy lies in the direct conversion of sunlight by photovoltaic cells. This is a remarkable piece of physics and today's solar cells have become so effective that a gram of thin-film cell silicone will produce as much energy over its lifetime as a gram of uranium!

In modern photocells purified silicon is used in its crystalline form. There are various grades of cell with different efficiencies and, not surprisingly, different prices. The best cells convert over 15 percent of the incident energy into electricity. They work by creating what amounts to a one-way valve for electrons: when light excites the electrons they end up going more in one way than the other – hence electricity.

Photovoltaic cells are readily available today and will provide up to several hundred watts of electricity for a normal home. Obviously, they only work when the sun shines so you must have some storage capability. Normally, batteries will serve for this. Lead acid batteries are much the most common and well understood form of storage. You must look after them carefully, keeping the electrolyte topped up and not leaving them uncharged.

The output of the photocells is usually limited to 24 volts, so an electronic device called an "inverter" must be used to convert the voltage up to normal domestic levels (240 volts). But the great challenge is in fact to limit your domestic use of electricity to such things as lighting, audio, and computing which all work fine from 24 volts and use very little power. All space heating, water heating, cooking, and heavy machines like washing machines and the hellish beast which is the tumble drier have a heavy appetite for power.

In the northern European climate you are obviously not going to have constant recourse to sunlight. So, as with all alternative energy sources, it is wise to plan your electrical system using several different sources of electricity. If you can combine wind power, water power, and stand-by diesel generation with your solar array and put the whole lot together through a modern control and storage set-up, then you will be well on the way to success.

Understanding electricity

Volts Scientists think of electricity as electrons flowing down a wire, almost like water down a pipe. The higher you lift the top end of the pipe, the greater the pressure and the faster the water flows. It's the same with electricity. If you apply more "volts" you push the electrons harder and more "current" will flow down your wire.

Amps You measure the amount of electricity going down a wire in amps: like gallons of water per second in a water pipe. The more amps there are going down a wire, the hotter it gets and the more energy is being transmitted. All domestic appliances run on what is called alternating current – this changes backwards and forwards 50 times a second and produces power and heat whichever way it goes. Alternating current is easier to manage in the national grid and in motors, but solar cells and batteries produce direct current. This only goes one way and it is the only kind of electricity which batteries can store.

Watts This is how you measure the power produced by electricity. It is an amount you calculate simply by multiplying the voltage by the amps. If you run a machine at 1,000 watts for an hour then you have used one kilowatt-hour of energy. This would apply to something like an electric heater.

Fuel from Waste

The attitude that has grown up in the Western world that all so-called "waste" from the body, human or otherwise, is something to be got rid of at all costs and very quickly, becomes harder to sustain as our planet's fossil fuel comes into shorter supply. If we can take the dung of animals or human beings, extract inflammable gas from it in quantities that make the effort worthwhile, and still have a valuable manure left over to return to the land, we are doing very well.

Methane is a gas which is produced by the anaerobic fermentation of organic matter: in other words, allowing organic matter to decay in the absence of oxygen. It is claimed that after the gas has been produced the resultant sludge is a better manure than it was before, for some of the nitrogen which might have been lost as ammonia is now in a fixed form which will be used by plants. As the methane gas itself is quite as good a fuel as natural gas (in fact, it is the same thing) and is non-toxic and safe, methane production from farm and human wastes seems very worthwhile.

Methane is made with a methane digester, which is fine for animal wastes but there's a limit to the amount of bulky vegetable matter that you can put into it. This precludes filling it with either tons and tons of straw or with the large quantities of valuable manure that result from the traditional practice (well proven) of bedding animals with straw. The spent sludge from the digester is itself an excellent manure, but my own feeling is that, rather than dump it straight from the digester on to the land, it should flow on to straw or other waste vegetable matter. There it will undergo further fermentation, this time aerobic, and at the same time activate the bacteria which will break down the tough cellulose content of the litter.

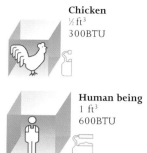

Chicken
½ ft³
300BTU

Human being
1 ft³
600BTU

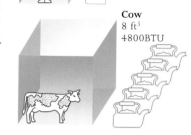

Cow
8 ft³
4800BTU

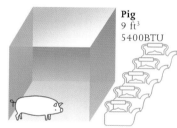

Pig
9 ft³
5400BTU

HOW MUCH GAS?

The diagram above shows the amount of gas produced by the waste of different animals in a day. The gas is sufficient to boil the number of kettles shown.

THE METHANE DIGESTER

The process shown below involves the digestion of organic wastes by bacterial action in a sealed container from which all air is excluded. Animal manure mixed to a slurry with water is added to a holding tank **1** daily. The input is fed into the digester by gravity when a valve **2** is opened. The stirrer **3** has an airtight joint where it enters the digester and prevents scum from building up. The tank is well insulated with straw or similar material **4** as the process only works effectively at temperatures close to blood heat. Each fresh addition causes an equivalent amount of digester sludge to overflow into the slurry collector **6**. The digestion process takes from about 14 to 35 days depending on the temperature of the digester, so the daily input should vary from ¼ to ⅓₅ of a digester volume to achieve the desired "retention time". The gas bubbles up through the slurry into space **5**, and is syphoned along a delivery line **7** to the gas holder **8**. An important safety precaution is a brass or copper fine mesh flame trap at the entry to the delivery pipe to protect the gas holder if air gets into the line and causes a burn-back. The gas produced, called bio-gas, is a mixture of about 60 percent methane (the inflammable fuel component) and 40 percent carbon dioxide, which is inert but harmless. The digested sludge makes a valuable fertilizer, being rich in nitrogen and trace elements.

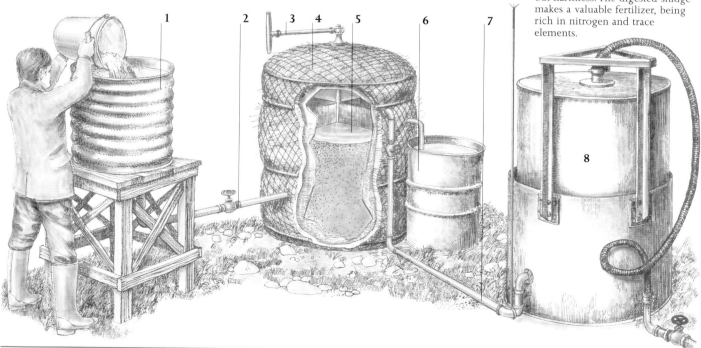

"No machine-made artifact can be beautiful. Beauty in artifacts can only be put there by the hands of the craftsman, and no machine will ever be built that can replace these. Machines might one day be made which will appreciate the beauty of articles made by other machines. People can only be truly pleased by articles made by other people. As far as possible we buy only things that have been made by people, not machines. I do not mean necessarily by the bare hands of people — hands can be magnified by tools and machines — but the hands must be there. Cloth woven on a loom by a person is fine. Cloth woven on a loom by an automatic device is dull. It is grudging...it only serves the one purpose."

JOHN SEYMOUR FAT OF THE LAND 1976

CHAPTER TEN

CRAFTS
&
SKILLS

The Workshop

Some people meditate in their potting sheds; myself, I meditate in the workshop. You may think it is a form of overkill to consider making your own dedicated workshop, but believe me, you will find plenty of time to use it. In real life things are always breaking and wearing out – usually just when you don't want them to. If you have all your tools to hand, in good shape, and a suitable place to use them, your blood pressure will benefit greatly. Make mending tools and equipment a source of satisfaction and you transform an irritating chore into a pleasure.

First things first: a workshop needs to have a good solid worktop. You need ½ inch (1 cm) of good quality plywood; this will just about do, but 2-inch (5-cm) planking would be better. Make a few large holes in the worktop with an augur so you can put in wooden pins to keep work still. You will also need a good vice. You must have good light to work effectively – but you do not want hot sun coming through your window so a big northern skylight would be ideal (you will see these in many old factories). The workshop does not need to be huge – even 8 square feet (6–7 sq m) would do. It should have electric power outlets in convenient places – at working height. The workshop should have a floor that is easy to brush clean, but it must not be slippery: plain concrete or unvarnished boarding is ideal.

It is true that one of the new portable workbenches can be a great help for the smallholder. I have two and both have now reached a battered old age; believe me, these things suffer serious punishment. And that's the bottom line. Do not buy one unless it's robust: the deck should be marine-quality ply and not any form of chipboard. The fittings should be rust-resistant in some shape or form. Anything else will disintegrate pretty rapidly.

Sharpening stone and guides

Tools are worse than useless unless they are sharp. And being sharp in my book means blades with straight edges and no chips. The blades on chisels and planes have been heat tempered on the edges and are very easily chipped if they are dropped, put down carelessly, or used as scrapers or levers. All too often they are borrowed by would-be helpers who use them to scrape off paint or lever up nails – this spells disaster and many hours of hard work to get back a decent edge. My tip to avoid this is to keep good chisels tucked away safely out of sight and leave an old chisel out in a rack or on the workbench, where it will be an effective decoy for visitors or helpers who are searching for a tool!

If you want to have planes and chisels with straight edges, then you need to buy a first-rate carborundum or diamond stone: this will not come cheap but your tools are no good without it. And the way you use the stone will affect the flatness of the surface and hence the straightness of the blade. To keep the surface flat you must use the full width of the stone and turn it around from time to time so both ends get worn equally. You simply cannot do decent work without properly sharpened tools. I have to let you in on a secret here: I always have two sharpening stones. When a stone is worn and old I leave it out on the bench as a decoy (much like my chisel duplication), as well as for sharpening kitchen knives in the hope that any strangers visiting the workshop will not therefore be tempted into looking for my carefully manicured newer stone which I have hidden. That is how seriously you need to take this crusade for properly sharpened tools. When you sharpen chisels or planes, always use a rolling guide to keep the blade at exactly the correct angle. You can see whether you are on the right track by looking at the bright shining area produced by the stone. If your angle is correct this should cover the entire area of the blade. There is no way you can sharpen tools effectively without the guide.

Fixing a broken handle

Follow the steps illustrated below to make the repair. If the old handle will not budge even in the vice after step 2, then I always try cutting off the end straight across at the end of the metal.

FIXING OLD TOOLS

We are talking here about spades, forks, axes, and sledgehammers. You can buy new handles for all of these at the builders merchant and save yourself a lot of money as well as keeping your favourite tools going for another few years. It is a strange thing how familiar tools become so comfortable to use; it's a great help to develop and keep a good relationship like this. When you choose a handle, take care to find a good straight piece of ash – the tighter the grain, the better. Do not be seduced into buying a clean, white piece of some tropical wood – it may look good, but believe me, they always break.

1 To get the steel spade, shovel, or fork off the broken end of the old handle you will need a grinder to grind off the ends of the old rivets securing the broken wooden end.

2 You should then be able to hammer out the rivets using your punch. With a bit of luck you may then be able to dislodge the broken wooden end and free it from the metal tube and blade.

3 If the broken handle still won't budge, put the whole item into a fire – wood burns but the metal doesn't and thus you will have a clean space to insert the new handle.

IDEAL WORKSHOP LAYOUT

Large north-facing skylight illuminates without overheating. Even 8 square feet (6–7 sq m) will make a good workspace and turn irritating chores into satisfying challenges. All tools have a habit of "walking", so keep them in their place and make sure putting tools back is rule number one in your workshop.

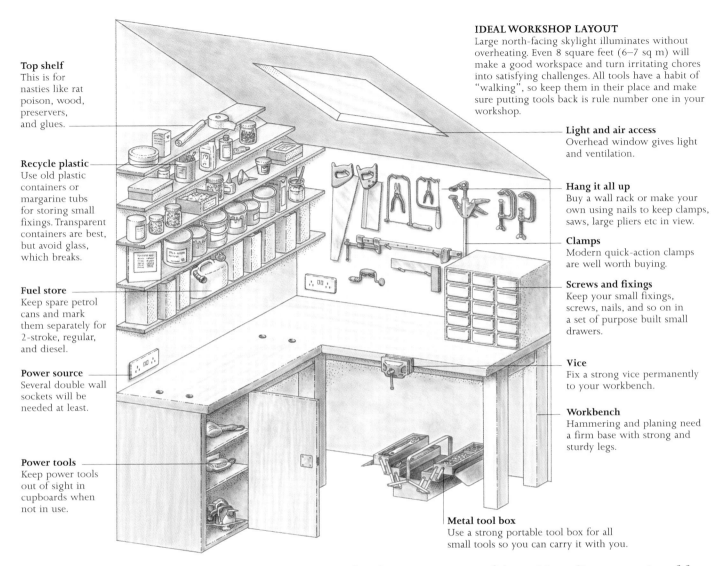

Top shelf
This is for nasties like rat poison, wood, preservers, and glues.

Recycle plastic
Use old plastic containers or margarine tubs for storing small fixings. Transparent containers are best, but avoid glass, which breaks.

Fuel store
Keep spare petrol cans and mark them separately for 2-stroke, regular, and diesel.

Power source
Several double wall sockets will be needed at least.

Power tools
Keep power tools out of sight in cupboards when not in use.

Light and air access
Overhead window gives light and ventilation.

Hang it all up
Buy a wall rack or make your own using nails to keep clamps, saws, large pliers etc in view.

Clamps
Modern quick-action clamps are well worth buying.

Screws and fixings
Keep your small fixings, screws, nails, and so on in a set of purpose built small drawers.

Vice
Fix a strong vice permanently to your workbench.

Workbench
Hammering and planing need a firm base with strong and sturdy legs.

Metal tool box
Use a strong portable tool box for all small tools so you can carry it with you.

You can then use a drill or auger to hollow out the wood – this often takes the pressure off and allows the shaft to come free. Now if all this fails, do not waste knuckle scraping hours fighting with the broken implement – go straight to step 3.

I use a power plane to pare down the new shaft to get a good fit. If you twist the shaft backwards and forwards slightly you will see from the brown marks how well it fits. Plane away the brown marks. Bang the shaft hard down on a concrete floor to force the tool on before you put a couple of big screws through the empty fixing holes. Put the screws right through the tool by using a small pilot hole made with your drill; it is not as easy as you think to hit the hole in the steel at the other side of the handle. Finally grind off the ends of the screws and file or hammer them down to make a firm fitting. Bingo: the tool is as good as new.

Machinery maintenance

Some people like flowers and vegetables, some love animals, and some even enjoy machinery. I must confess that I am not one of them. My earliest memories of farm work are extremely uncomfortable ones of lying under broken pieces of machinery trying to get rusted bolts loose. Machinery designers seem to have an uncanny knack for making their progeny almost impossible to get at to make repairs. As with so many things in life, prevention is much better than cure, and here are a few golden rules:

1 Check the oil regularly on all engines (and gearboxes, back axles, and the like) and change it according to the official maintenance instructions. There is research to suggest that for some strange reason there are far fewer breakdowns of any kind on engines where the oil is checked regularly. Could this be some sort of emotional response to this form of love? I wonder.

2 If you hear any strange noises – stop! Check for loose nuts and check oil levels. Do not continue until you find the cause. Strange noises can be very expensive.

3 Be very wary of lending machinery to "friends": the level of care they have for your loved one may be less than you imagine.

Building

For some reason the self-supporter often seems to spend considerable time building. Sometimes this amounts to repair work to keep existing buildings going against the ravages of damp and weather – or even the battering given by pigs or cattle. More often the self-supporter simply wants to extend his or her accommodation and storage space, whether for humans, animals, or equipment.

It's all in the planning

If you are building anything substantial you ought to contact your local building inspector. You may well need planning permission – which means paperwork. Do not be deterred. In most cases the building inspector can offer you very constructive advice about the pitfalls and peculiarities of local conditions. Far better to get advice early than have to knock down your prized blockwork and start again.

Building is labour-intensive but much of the work is far less skilled than you might imagine. It is a great help if you have a good eye for straight lines and right angles, and you do need to be fairly systematic. The vital first step is to make sensible plans. Sensible in terms of where you are siting your building, how big you are going to make it, what materials you are going to use, and how much of a financial budget you expect to have available. Last but not least, do not take on more than you can manage: start with a small, simple project and move on from there. A great deal of the work in building consists of lugging heavy, uncomfortable material from one place to another, often lifting up a considerable height. This will give you strong arms and a patient disposition; alternatively, you may make some new friends who have plenty of energy and good, strong backs!

There is a wide choice of materials available for building. Bricks, concrete blocks, breeze blocks, wood, and metal are the most common. Local stone may be an option if you are lucky and there is an abundance of suitable building stone. Of course you can also use more outlandish materials such as straw bales or rammed earth, but for most practical self-supporters it is easier to make a trip to the builder's merchant. For a cheap, strong, and long-lasting building my preferred material is concrete block: you can face this with cement, or plaster or wood, if you don't like the look of it.

Before you buy materials, ask around to find out which suppliers have the best quality and most reasonable prices. You will find enormous variations in both. Many suppliers will deliver much more cheaply if you buy direct from the factory. This means buying a lorry-load (and making sure

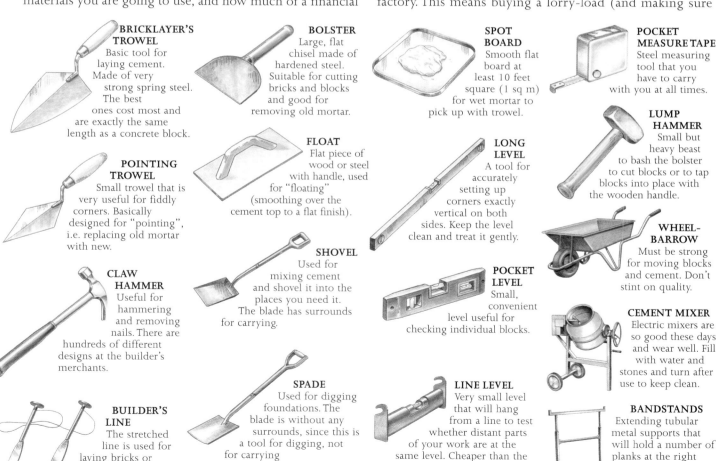

BRICKLAYER'S TROWEL Basic tool for laying cement. Made of very strong spring steel. The best ones cost most and are exactly the same length as a concrete block.

POINTING TROWEL Small trowel that is very useful for fiddly corners. Basically designed for "pointing", i.e. replacing old mortar with new.

CLAW HAMMER Useful for hammering and removing nails. There are hundreds of different designs at the builder's merchants.

BUILDER'S LINE The stretched line is used for laying bricks or blocks level, or in line.

BOLSTER Large, flat chisel made of hardened steel. Suitable for cutting bricks and blocks and good for removing old mortar.

FLOAT Flat piece of wood or steel with handle, used for "floating" (smoothing over the cement top to a flat finish).

SHOVEL Used for mixing cement and shovel it into the places you need it. The blade has surrounds for carrying.

SPADE Used for digging foundations. The blade is without any surrounds, since this is a tool for digging, not for carrying to or away.

SPOT BOARD Smooth flat board at least 10 feet square (1 sq m) for wet mortar to pick up with trowel.

LONG LEVEL A tool for accurately setting up corners exactly vertical on both sides. Keep the level clean and treat it gently.

POCKET LEVEL Small, convenient level useful for checking individual blocks.

LINE LEVEL Very small level that will hang from a line to test whether distant parts of your work are at the same level. Cheaper than the modern laser equivalent.

POCKET MEASURE TAPE Steel measuring tool that you have to carry with you at all times.

LUMP HAMMER Small but heavy beast to bash the bolster to cut blocks or to tap blocks into place with the wooden handle.

WHEEL-BARROW Must be strong for moving blocks and cement. Don't stint on quality.

CEMENT MIXER Electric mixers are so good these days and wear well. Fill with water and stones and turn after use to keep clean.

BANDSTANDS Extending tubular metal supports that will hold a number of planks at the right height for your work. Must be very strong.

your access is suitable for such) which probably amounts to over 1,000 blocks. One thing about blocks and building materials is that this kind of thing does not deteriorate and believe me, you will always find uses for blocks. Make sure you think things through carefully before you tell the lorry driver where to stack the blocks. The modern lorries have very long reach with their hydraulic cranes. Get the blocks as close as you can to where they will be used.

Damp is the great enemy of the self-sufficient builder. If you have damp in your floors or walls you can expect all sorts of trouble from rot. So a good roof, good damp-proof course, proper membrane under any concrete slab, and well-made cavities between walls are essential. Concrete and plaster act like blotting paper and seem to have an uncanny knack for absorbing any moisture going.

Roofing

Materials for roofing also come in all shapes and sizes – from thatch at one extreme to artificial asbestos tiles or corrugated iron at the other. In between you have tiles and natural slates. Each type of roofing material requires a different approach. Your choice will obviously depend on the aesthetics of your design, your budget, and what is easily available locally.

If you are using heavy roofing materials you will need to construct a strong wooden framework. Slates and tiles are extremely heavy – this is one reason they make a good roof that will resist the ravages of the wildest winter storms. Make sure the roof is well ventilated to prevent damp rotting the timber. Use stainless steel, aluminium, or copper nails for fastening roof coverings.

Basic block laying

First load out your blocks in piles close to where they will be needed. Make sure you have a couple of spot boards for dumping mortar ready for use. Mix your cement (usually a 4-to-1 mixture of sand and cement respectively) and add some mortar plasticizer to make the cement flow better. With practice you will get the sloppiness just right: not too wet and not too dry (wetter is probably better than drier to start with). With your large trowel lay a bed of wet mortar in the place where your first block is to go. Place the block carefully in position exactly to the markers you have set up in advance (corner and line). Slide it backwards and forwards to ease it into the mortar. Check with your pocket level that it is level in all directions. Tap it with the lump hammer end if it refuses to settle correctly. Block after block, keep each one positioned accurately to your builder's line.

BUILDING FOUNDATIONS AND A WALL

1 Mark out the foundations with posts and lines. Corners and diagonals must be exact. Dig by hand or mark out with white lime for a small digger to work to. As a rule dig to at least 2 feet (60 cm).

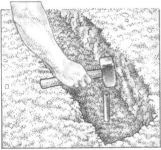

2 Tap your metal bars into the bottom of your trenches until their upper ends are in exactly the right place to give you the correct thickness of concrete.

3 Check your levels carefully across the tops of the metal bars. Use a long, straight piece of wood in conjunction with your long level – and be fussy. Mistakes at this stage will cost you time and frayed nerves later.

5 When the foundations have set solid, you are ready to begin your "footings". Again, make sure you have your corners marked accurately and take great care to get the corners precise before you set up lines from them for the rest of the blocks.

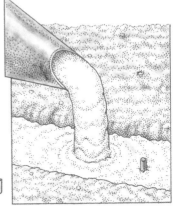

4 Pour your concrete to the tops of the metal bars and spread it out carefully. Tamp it down with a long, straight piece of wood, making sure the levels are correct.

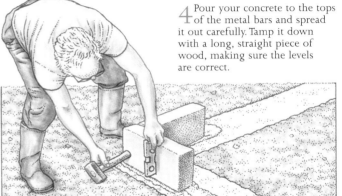

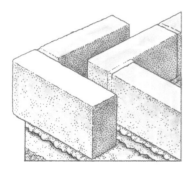

6 Place the inner course of blocks in exactly the right position inside the original blocks. Spacing must be exactly right for a substantial layer of insulation material (usually at least 4 inches /10 cm). Do not forget to put in wall ties to connect the two walls every few feet. And never let mortar drop down between the walls, since this can cause damp if it bridges the gap.

Subsoil   **6-inch concrete slab**   **Insulation "bats"**   **Cavity insulation**

CROSS SECTION
This shows the foundations, footings and layers of insulation. There is insulation under the concrete slab and between the two courses of blocks that form the walls. You also have a strong damp-proof membrane between the slab and the earth underneath.

7 Use reinforcing mesh to strengthen your concrete slab if the area to be covered is large. Support this a couple of inches up with a few pieces of stone or brick.

8 Find or make a long, stiff wooden batten to extend right across the slab so you can "tamp" down the concrete to a perfect flat, level surface. Always pour a little more concrete than you need so you can pull off the excess with the batten.

9 Lay your damp course carefully over the top of the footings on a bed of wet mortar. The damp course must be well above ground level if it is to work effectively to keep out damp.

10 Split blocks to size using a lump hammer and bolster. Cut a shallow groove in each side of the block using a few blows at a time.

11 Use your long level with great care to make sure corners are exactly vertical on both sides. Don't cut corners or the project will go awry.

12 Once the corners are set, use a builder's line pulled tight between the pins at each corner. Check levels regularly using a hanging level at the middle of the line.

Springs & Plumbing

Sinking a well

The easiest way of finding water is to drill a hole with a drilling machine, and if you can get hold of one it is well worth using it. But they are expensive, even to hire, and all they really save is time and energy. If you have got some of each to spare you can dig your well yourself by hand.

Sinking a well in earth or soft rock is very easy, if laborious. You just dig in, keeping the diameter as small as you can, just leaving yourself room to use a shovel. As you get deeper, you send the spoil (dug earth) up to the surface in a bucket hauled up by a friend with a windlass, and you go up the same way. It is almost always necessary to line the well as you dig to stop the earth from falling in. The easiest way to do this is with concrete rings sent down from the top. As you dig down you dig under the lowest concrete ring which causes it to fall and all the other concrete rings on top fall with it. From time to time you put another concrete ring on top. Where timber is cheap you can use a timber lining on the same principle.

Sinking a well through rock is harder in that you have to blast it, but easier in that you probably don't have to line it. In days gone by the rock was shattered by building a fire on it and then quenching this with water. The rapid contraction shattered the surface of the rock. Of course "powder" and modern detonants are the direct route to a well and in the past I would do it myself, but nowadays you need to hire experts.

Whichever way you sink a well, when you come to water, go on sinking. Even if you have to spend half of each day winding up water in the bucket, go on sinking until the water beats you, because if you don't, when there is a drought and the water table sinks, your well will go dry. When you have got your water the best thing you can do is install a steel pumping windmill (*see p.246*). It will pump water from 1,000 feet (300 m) and go on doing it for years, free, and with very little attention.

Plumbing

I have no regrets about saying goodbye to the drama of hot solder and the mysteries of copper piping. Those cunning plastic gizmos that can join pipes, make T-junctions, and fix taps have made plumbing a lot easier (and believe me, your labour is much cheaper than a plumber's).

A tip I will pass on: always put in-line service taps into your new plumbing. These small taps fit into the line of your piping and can be turned on or off with a screwdriver. By installing these you can easily isolate new pipework from the rest of the house so any leakage can be fixed without cutting off the entire water supply. If your garden is anything like mine you will inevitably cut through or damage buried water pipes! So keep a range of plastic joining fittings in your workshop: this will save you a great deal of time, not to mention nerves, as water has to be turned off and kitchen, bathroom, and washing machine grind to a halt.

MANAGING A SPRING
Make good use of a spring as a source of clean, fresh water. Explore the source of the spring carefully so you cover it effectively with the chamber. Avoid mud and soil contamination by ensuring the outlet pipe is properly placed.

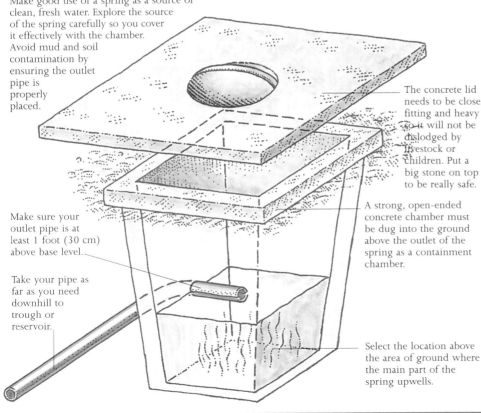

The concrete lid needs to be close fitting and heavy so it will not be dislodged by livestock or children. Put a big stone on top to be really safe.

A strong, open-ended concrete chamber must be dug into the ground above the outlet of the spring as a containment chamber.

Make sure your outlet pipe is at least 1 foot (30 cm) above base level.

Take your pipe as far as you need downhill to trough or reservoir.

Select the location above the area of ground where the main part of the spring upwells.

PLUMBING FITTINGS
You will find a whole range of easily used plastic plumbing fittings at any good builder's merchant. Make sure you know exactly the sizes of pipe you are using – take a piece in if you are unsure.

A typical all-purpose joint where the inserts can be of different sizes – very convenient to keep as a spare.

This is a cheaper plastic joint which must be the correct size to fit piping.

There are many types of push-fit connectors for taps and pipes (left). A typical T-junction (below) has brass "olives" compressed to make the joint watertight. Keep extra new olives for each different size of pipe so you can re-use the joints.

Knots & Ropework

The use of rope and string is probably older than the wheel. Their use is certainly one of the greatest of all human inventions. If string and rope came first, then knots, nets, and weaving came pretty soon afterwards. Ropes and cordage are made in all sorts of clever ways by spinning and twisting strong fibres together. Hemp, linen, cotton, and wool are obvious examples of fibres used in making rope and cord. But today we are more likely to come across the modern plastic equivalents in terylene, nylon, and other so-called synthetics.

Different kinds of rope and cord are suitable for different types of jobs. And there are a thousand different types of knots that humans have developed over the centuries for specific purposes. Learning about knots and cordage is one of the simplest skills, yet surprisingly few people know more than one or two knots and, even then, do not know their proper names. Those of you who have seen the movie *Jurassic Park* will know just how important it can be to tie a proper bowline that makes a loop which will not slip, or jam, or come undone. Our hero's lorry was tied to a tree by some foul Hollywood hitch and was always destined to slip off into the clutches of rampant waiting dinosaurs.

Knots – a life skill

They should teach knot-making to children in school: knots are cheap, they require dexterity, and they are extremely useful. Indeed, the right knot in the right place can be a life saver. Animals frequently pull and twist on any rope used to tie them. The wrong knot will either jam or come undone – in the first case needing a knife to free it, in the second leaving the animal free to cause havoc elsewhere. A stockman's hitch or falconer's knot is tied quickly with one hand, yet allows for quick release if the animal panics or has to be let free quickly. Using the right type of rope or cord can also be vital: some stretch and some do not, some are easy on the hands, some are as hard and sharp as sandpaper, some are stiff, and some flexible.

The right knot is one that does the job it is intended for with minimum effort. An anchor bend, as its name suggests, is for securing an anchor and it will never slip even when it is allowed to tighten and slacken on every tide. It may jam and have to be cut free, but that is a fair price to pay for not losing your anchor. The knot is tied tight up against the anchor ring so that it will not chafe and rub. And so it goes on.

One of the most satisfying moments a teacher can have on any self-sufficiency topic is the look of amazement on a student's face on seeing how years of struggling with rope can be swiftly brought to an end with the right knot. My father always used to tell the tale of how the "thief" knot was used instead of reef knots to tie up important parcels. The reef knot is the simplest of knots: right over left and left over right and the sails are tied up tightly in place as the wind strengthens. It is so simple that the knot is invariably used to tie up parcels. On first glance the thief knot looks just like a reef knot: left over right and right over left – certainly sufficient to deceive a would-be snooper. But there is one crucial difference: the loose ends of string in a reef knot come out both on the same side of the tight string, while the ends on a thief knot come out on opposite sides. The simple logic of the story was that you could always tell if someone had meddled with your mail if it was tied with thief knot!

Just for the record, the reef knot is often used for tying together two pieces of rope, yet it is quite unsuitable for this as it will come undone if the tension of the rope is varied or jerked. The correct knot for tying together two pieces of rope is the sheet bend, or double sheet bend if the ropes are of different thicknesses.

Managing rope and cord

All ropes should be properly coiled when they are not in use. They should be kept free of knots and kinks which will, in time, weaken them. And their ends should not be allowed to fray. Ends can be kept tidy in three ways:

1 Using heat from a match or hot piece of steel if the rope is of synthetic material. Watch out if you are doing this and the rope catches fire: the molten material that drips off can cause very nasty burns, since it is sticky and extremely hot.

2 Applying strong twine in the form of a whipping which binds the loose ends together over at least an inch of the rope.

3 Undertaking a back splice on three or four stranded ropes, which makes the job much more permanent.

Ropes have a nasty habit of getting twisted and knotted. There is one very simple reason for this. When a rope is made into a single coil, it has to be twisted to make it coil neatly. Consequently, if both ends are tied and the rope then unravels, it can only do so with all the twists intact. As soon as tension comes off the rope and it gets loose, then it seems to tie itself up in knots. (You often have this problem with telephone leads or that devilish cable that connects up the vacuum cleaner to the electricity.) Even worse, the problem is often the final straw in the kiting exploits of young children. There is a simple answer to this, which I learned when studying falconry. When training a young falcon, the falconer must tie a long, lightweight string to the falcon so that on its first few flights it cannot simply disappear up the nearest tree. This line is called a creance. One of the essential skills the falconer learns is how to wind up the creance (one-handed, of course, as the falcon sits on a glove on the other). The only way to wind up the creance is in a figure-of-eight on a special wide stick made for the purpose. Miracle of miracles, this figure of eight system means that the falcon can pull out the string *without* making any twists. The twists at each end of the figure-of-eight cancel each other out and this makes running out the rope much much easier. Remember this when you are flying a kite with kids and life will be much simpler!

ESSENTIAL KNOTS

BOWLINE

The bowline is the essential method of making a loop in rope that will not slip and will not jam – an important factor when stringing a bow (for which this knot was originally intended). If you only ever learn one knot, this should be it.

FALCONER'S HITCH

Learn to tie this knot with one hand as you have to hold the animal with the other. One pull on the loose end undoes the knot letting a struggling animal free.

SHEETBEND

The sheetbend rivals the bowline in importance. It is the definitive knot for tying together two pieces of rope under tension. The thin rope can be passed around twice for extra security to make a double sheet bend. For extra security whip the loose end of the thicker rope tight against the part under tension (the "standing" part).

ROLLING HITCH

A rolling hitch is used to fasten a rope to a pole or beam. The tension on the rope should be pulling against the double turn side of the knot which jams the turns tight preventing slippage.

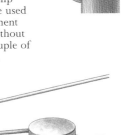

CLOVE HITCH

This is a quick way to take tension on a pole or bollard. But it does slip so cannot be used for a permanent fastening without adding a couple of half hitches.

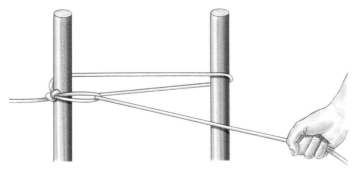

LORRYMAN'S HITCH

Use this knot to tighten up a rope already under tension. The loop acts like a block and tackle to double your force. You can then secure with a quick half hitch against the loop, or a slip half hitch if you want to undo it more easily.

This knot is now largely replaced by the use of woven straps with hooks at each end and a ratchet tightener. These are powerful tools; I always keep a couple in the workshop.

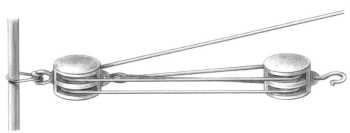

BLOCK AND TACKLE

Modern sailing equipment has produced some extremely light but powerful versions of the block and tackle. It is amazing what force you can exert provided you have something to pull against. Remember, always pull the free end in the same direct as the force you want to apply. Every self-supporter should have a block and tackle available for emergencies. The best ones even have self-jamming cleats to maintain tension when you let go.

SPLICING

A simple splice – that is, the joining up of the ends of two ropes by interweaving strands – is easy enough to understand. Simply weave the cords making up one rope through the cords making up the other rope. Even three or four tucks makes an extremely strong join. Pare the ends of the cords down gradually if you want a neat finish that will not jam in blocks or other guides or fasteners.

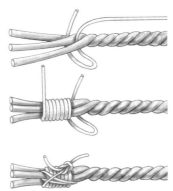

WHIPPING

A very useful technique for making a permanent neat end on ropes using strong twine so they cannot fray and come apart. Simply wind the twine over a loop of itself, then put the loose end through the loop and pull it under the turns you have already made.

HOLDFAST

By driving three posts into the softest mud or soil you can achieve amazing staying power with this layout of holdfast. This is the way to pull a boat, machine, or animal out of the mud if you have a block and tackle but no convenient tree to pull on. It really works!

Each post should be at least 3 feet (about 1 metre) long and driven well into the ground. Use rolling hitches to fasten the ropes. You can use a similar configuration of posts and wire to anchor the ends of stock fencing before applying tension: drill holes right through the centre of the posts to fasten your wires.

Basketry

You can go for a walk in the country with no tools other than a sharp knife and come back with a basket. Traditionally, one-year shoots from willow (called osiers) are used for the strongest baskets, but you can use other materials, such as straw or rushes too. Osiers are grown by a very simple process of pushing stout lengths of willow into the ground. These should be about 1 foot (30 cm) long and at least ½ inch (1.5 cm) thick, planted about 18 inches (45 cm) apart with rows a couple of feet (60 cm) apart. These will sprout roots at one end and rods at the other. The rods are cut in winter when the sap is down. Sort the rods into lengths after harvesting. You can let them dry naturally, you can strip the bark, or you can boil to obtain different finishes. Rods are tied into "bolts" (bundles) for sale.

Basic basketry

Good willow should be smooth and straight, with nice thin tips. It can only be worked when it has been soaked. Choosing rods the correct size for what you need is more critical than you might think. Even small variations in thickness can allow the stronger rods to dominate the work.

The base of your basket, the "slath", can be round, oval, or square. The base is the foundation for your work and must be firm and well-shaped. Choose strong, smooth rods and use "pairing" to form them into the shape required. The uprights are then pushed firmly into the base to begin weaving the sides. To make uprights for a square basket you must cut away the willow at the butt of the stake to make what is called a "slype". Otherwise simply use the traditional bodkin (see below) to insert a stake into the weave on each side of each stick of the base. Push the stake right to the centre of the base if you can. If the stakes are too stiff then you will have to prick them – that is to say, make a small crease with your thumb or the back of a knife at the point about ¼ inch (½ cm) from the base edge where they are to turn up. Now tie the ends of the stakes together as you prepare to weave. The first few rows of your weave (the "upsett") are critical to the shape of the finished basket. The weave used for this is called the "waling". It is vital to get the stakes into the correct positions with this weave and you will probably use a rapping iron to firm them down. You always finish a round with the tip ends.

Borders A good border is critical. We show one border here (another common border is the 3-rod-behind-1). Once you have 3 pairs of rods facing towards you from 3 consecutive spaces you simply take the right-hand rod of the left-hand set in front of the next upright and behind the next. You can then bend the first upright down beside it and continue the sequence.

Handles A bow handle is made by forcing a stout rod of the correct length down into the weave on each side of a basket. Use a bodkin and grease to ease the sharpened rod into position as far down the weave as you can. Now cover the handle bow with thin rods, taking four at a time, pushing them in besides the handle. Wind them around the bow three times as you go to the other side where they can be woven in. Repeat with another four rods from the opposite side. Add more rods until the handle is evenly covered.

To make a twisted rope handle, take four rods long enough to reach across the handle twice with some to spare. Slype the first one and insert it beside one end of the handle. Twist and wind the rod five times around the handle to the other side. Pass the rod through or under the border and bring it back keeping the twist beside the first. Tuck the loose end through the weave to keep it in place and repeat the process with each successive rod. Twist the loose ends of the weavers together and weave them under the border to complete the handle.

BASKET MAKING EQUIPMENT
The bodkin is a heavy iron rod, tapered down to a pointed end to part the weave in order to insert a new rod. The grease horn stores the bodkin when not in use. Use tallow mixed with plumber's hemp for the grease. A curved knife with sharp point is really useful for cutting scalloms. Weights are heavy chunks of metal to keep work in place. The rapping iron is a heavy flat strip of metal used to tap down each row of the weave. The screwblock is a clamp used to hold weaves flat, particularly for making bases. For a measure use a rigid steel or retractable wooden rule. The secateurs are for cutting rods.

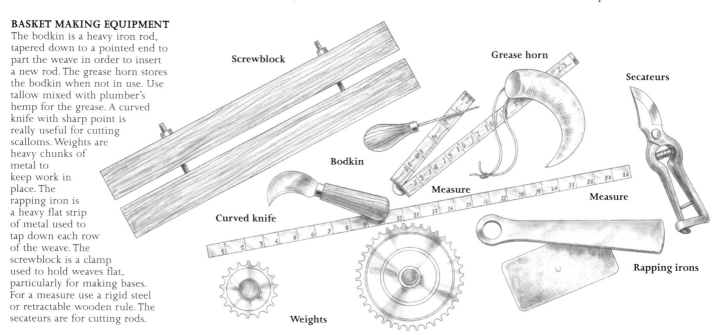

Screwblock • Grease horn • Secateurs • Bodkin • Measure • Curved knife • Measure • Rapping irons • Weights

MAKING A HARD BASKET

You need three different types of rod: eight short, stout rods for the "slath", or base; a number of strong but bendy rods for your side stakes; and some weavers – the long, thin whippy rods that hold the basket together.

Side stakes are generally about 8 inches (20 cm) longer than the intended height of the basket. Weavers can be any length, but they should be at least long enough to go round the basket once. They come in varying thicknesses.

Soak all your rods for an hour before using them, in a container big enough to hold them all. They might need to be weighed down to keep them under water.

The base of the basket is the most critical part of your work as it forms the foundation for all that follows. The size, shape, and number of sticks in your base will determine virtually the entire form of your basket.

The patterns and strength of your basketry comes from using different weaves – but the shape and quality comes from how well you perform each stroke of the weave and being vigilant about the shape and evenness of your work.

STRUCTURE OF A WILLOW ROD

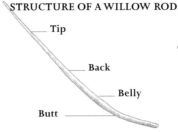

Tip

Back

Belly

Butt

1 Choose the best length of rod for your work: 3 feet (1 m) for small baskets, 4–6 (1.2–6 m) for shopping baskets, and 8 feet (2.4 m) for large items. Sort into four groups of matching thickness.

2 Soak all your rods for an hour before using them. Cut the rods for your slath, and cut slits in two of them. Poke the others through to form a cross.

Alternative base
Instead of piercing or threading, you can start your base with sticks overlaid as shown.

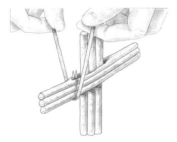

3 Begin the base weave (tie in the slath) by pushing the sharpened butt ends of two weavers into the split centre of your base sticks.

4 Begin pairing (the most common weave for creating a base) and separating out the sticks so as to open them out into the correct configuration.

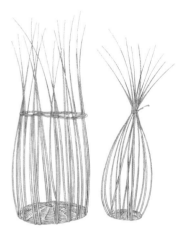

5 When your base is completed by tucking tip ends of the final weavers into the weaving, you are ready to stake up to form the basket sides. Try to make sure all your stakes are the same thickness. A tied stake (*above left*) and hooped stake (*above right*) are shown ready for the upsett.

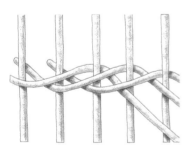

3-rod wale
Insert three weavers into three adjacent spaces. Here the third rod is being woven. The twist in the weave locks the rods into position.

Scalloms
Each scallom goes over the "tail" of the one before and holds it in position. Start weaving easily from any base shape by using scalloms.

Undersides
Make all joints either tip to tip or butt to butt. Slip the ends of the new weavers under the ends that

are finishing – you can overlap with a few strokes if you are changing at the tips.

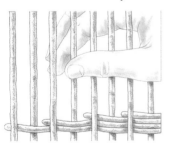

Slewing
Quick and useful for using up oddments but not easy to control. This is a four-rod slew: add a new rod as the bottom is used up.

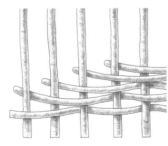

Randing
A single weaver is used to go in and out of the stakes and it creates a spiral effect. Start each one off one space to the right.

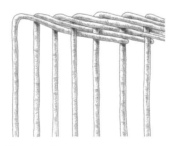

Trac borders
Trac borders are simple to grasp: simply bend over the top of the stakes and weave them through three or four stakes to the right.

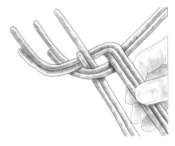

Plaited border
For decorative effect on bowls or trays where there are no handles, work with pairs over each upright stake through to the inside.

Pottery

Clay is very often overlain with earth, and so you may be walking on it, or even living on it, without knowing it is there. Prospect where a cutting has been made, or a well, or anything that exposes the subsoil. If it looks like clay, and when wetted becomes plastic and sticky, it is clay.

Testing clay

Once you have found it you have got to find out if it is any good. It probably isn't. Wet some down to a plastic state and then allow it to dry out. If it has a noticeable scum, usually a whitish stain, on its surface after it has dried, it contains undesirable alkali and probably isn't worth using. Drop a sample of the clay into a beaker containing a 50 percent solution of hydrochloric acid. If it fizzes, forget it. Too much lime. If the clay looks dark brown or black and is very sticky, then there is too much humus. Clay very near the surface may be like this, but there is often better clay underneath.

To test for plasticity, which is important, make some clay into a stick the size of a pencil and see if you can bend it into a ring an inch in diameter without breaking it. If you can, it is good clay.

If there is too much sand in your clay it may be hard to mould or throw on the wheel. If this is so, mix a fatter, less gritty clay with your sandy clay and try that. You can screen sand out of clay, but it is a laborious job and probably not worth it.

Mixing and screening

If you want to mix it with other clay, or screen it, you must mix it with water to a pretty sloppy liquid. Throw the clay into a tank full of water (don't pour the water on the clay) and mix. You can do this by hand, or with a paddle, or in a "blunger" which is a special machine for the job, or in an ordinary washing machine. The semi-liquid clay is then called "slip". The slip can be poured through a screen, to screen it. Use a 60 mesh to the inch screen for ordinary earthenware and a 100 mesh for porcelain or china. If you want to mix two or more clays, make slip of them all and then mix them up in that condition.

The next job is getting the water out of the clay again. An easy method is to let your slip sit in a barrel or tank for a few days until the clay all sinks to the bottom. Then you siphon the water off, much as a wine maker racks his wine. There is a machine called a filterpress which will then extract the rest of the water, but if you haven't got one you can place the slip in bowls of unglazed earthenware and leave these in a draughty place. The absorbent earthenware draws the water out of the clay. The water is then dried off by the air, and after a few days the clay is fit to work.

Preparing clay

If you are very lucky, you may find a clay that you don't have to combine with anything else, or screen, and all you have to do is dig it up and let it weather, or age. All clay is better aged even if only for a fortnight, because bacteria do good things to it. Then you must mix it with water and "pug" it, which means you must tread it well with your feet. Finally, you must "wedge" it. This is the process of pushing the clay away from you on a board, pulling it towards you, rolling it, cutting it up, and recombining it: in fact, giving it a thorough kneading just as you would knead bread.

Shaping pots

There are many ways of shaping a pot. Almost certainly pottery was discovered because baskets used to be plastered with clay to make them hold water. One day a basket got burnt and the clay became hard and durable. This was the first pot made with a mould. There are various techniques for making pots by "pinching", "coiling", and with "slabs". But before you begin to throw pots on a potter's wheel you need to understand clay: what happens to it when it is pulled about, when it dries, when it is fired, and so on. This is best done by shaping some pots by hand before you ever try the wheel. The invention of the potter's wheel was the great breakthrough, and there is really no substitute for it. You "throw" a lump of clay on the wheel, "centring" it plum in the middle of the wheel by pressure from both hands as the wheel revolves. Then you shape it with pressure from your hands, fingers, tools, and so on. Remove the pot from the wheel: usually you cut it off with a piece of wire. Set it aside to dry. Then replace it on the wheel by sticking it with a little water, and "turn" it, that is, spin it round and smooth off the rough edges with a steel cutting tool. Turn it twice: once with the pot the right way up, and again with it upside down.

Making a wheel

In primitive countries they still use wooden cart or wagon wheels as potter's wheels, and if you can get hold of one you can do this, too. You mount the wheel horizontally near the ground, ideally on a short section of its original axe. Make a hole in the side of the wheel towards one edge or in a spoke if it has them. To use it you squat by the side of the wheel, put a stick in the hole and set the wheel turning. Because the wheel is very heavy it goes on turning by its own momentum and your hands are then free to throw a pot or two.

A more sophisticated potter's wheel can he made by casting a reinforced concrete wheel, say 28 inches (71 cm) in diameter and 3½ inches (9 cm) high with a 1-inch (2.5-cm) diameter steel shaft about 30 inches (76 cm) long through it. The bottom of the shaft should protrude a couple of inches and steel reinforcing bars should be welded radiating from the shaft so as to be embedded in the concrete. This is not the wheel you throw on, but one you kick to make the throwing wheel revolve. Then build a table-high wooden frame which has a bearing let into it to house the top bearing of the shaft, and a thrust bearing at the bottom to take the bottom of the shaft. The frame

should also include a seat for you to sit on, and a table to place clay. Fix the concrete wheel and shaft into the frame. And now you must fix on your throwing wheel. Weld (*see* p.274) or braze a wheel-head, say a foot (30 cm) in diameter and ¼–½ inch (0.7cm–1.2cm) thick to the top of a steel hub (a short piece of water pipe will do). Put this on top of the shaft and weld or braze it on. To use your wheel just sit on the seat and kick the concrete wheel round with your foot. Being heavy, it has plenty of momentum.

Firing

Firing is necessary to harden the clay. With most glazed ware there are two firings: the "biscuit firing" which is just the clay and not the glaze, and the "glost firing", which is the biscuit ware dipped in the liquid glaze and fired again. You can fire pots to flower pot hardness in a large bonfire, although you cannot of course glaze them like this. Lay a thick circle of seasoned firewood on the ground, lay your ware in the middle, build a big cone of wood over it, and light. Pull the pots out of the ashes when they are cooled.

Traditional kilns are "updraught" kilns and you can build one yourself if you can lay bricks. "Downdraught" kilns are a more recent development and a little more difficult to build. The kiln is arranged so that the heat from the fire is sucked down through the pots before it is allowed to rise up the chimney. Much higher temperatures can be achieved using this method.

Temperature can be a matter of experience, or can be measured with "pyrometers" or "cones". Cones are little pyramids of different kinds of clay mixture which are placed in the kiln and which tell us the temperature by keeling over when they get to a certain heat. You can buy them very cheaply, but if you plan to use them remember to build some sort of a peep-hole in your kiln so you can see them.

Glazing

Most glaze is a mixture of silica, a "flux", which is generally an oxide of some metal (like rust), and alumina, which is clay. China clay is the most usual form of alumina in glazes. The silica melts and solidifies on cooling to form a coating of glass on the ware. The flux helps the fusion, lowers the melting point of the silica, and provides colour. The alumina gives the glaze viscosity so that it does not all run down the side of the pot when you put it in the kiln.

Anybody can make their own glazes. You must grind the components down fine either with a pestle and mortar or in a ball mill. The latter is a slowly revolving cylinder which you fill with flint pebbles and whatever material you want to grind. You can make a raw glaze from 31 parts washing soda (the flux – sodium is a metal); 10.5 parts whiting; 12 parts flint (the silica); 55.5 parts feldspar. Grind this, mix it, and pass it through a 100 mesh to the inch (2.5 cm) lawn, which is a piece of fine linen. There are hundreds of glazes and the best thing you can do is get a book on the subject and experiment with a few.

BEFORE SHAPING YOUR POT
Let newly dug clay age for at least a fortnight. Then pug it to get the air out. The easiest way is to mix it with water and trample on it.

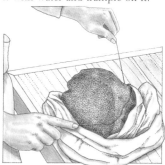

1 Use a wire to cut a workable lump from your store of pug.

2 Wedge the lump to make it a soft, homogeneous mass, free

of air bubbles and foreign bodies like bits of stone and grit. You can wedge in the same way as you would knead dough for bread. Roll the clay towards you with both hands, twist it sideways, and push it down into itself. Press out air bubbles and pick out bits of dirt. If you are mixing two clays, wedge until your clay is one uniform colour.

AFTER SHAPING YOUR POT
Most glazes are applied after the first firing in the kiln, the biscuit firing. The commonest method is to dip your pot in a soup of powdered glaze and water, but it takes practice to avoid finger marks. You can pour glaze so that it flows over the pot. You can spray it, or paint it on with a brush.

A SOLID FUEL UPDRAUGHT KILN
You can get kilns which use electricity, gas, or oil, but a solid fuel kiln can be equally efficient, and you can build it yourself out of ordinary bricks. Updraught kilns are the simplest. You have your fire box at the bottom. If you burn wood you can do it on the ground, but coal and coke should be burned on steel firebars so that

the ash can drop through. Build your pot chamber directly above the fire, by supporting a system of shelves made of firebrick on steel firebars. Include a peephole so you can watch your pots progress. And as long as you build the whole structure firmly, the chimney can be directly over the pot chamber.

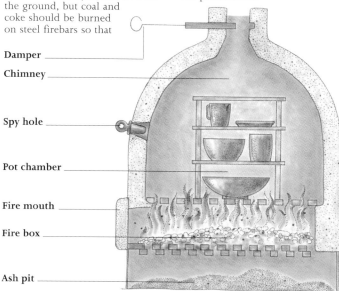

Damper

Chimney

Spy hole

Pot chamber

Fire mouth

Fire box

Ash pit

Spinning Wool & Cotton

WOOL

Wool should be selected (or sheep should be selected) for the job to be done. Different breeds of sheep give wool of varying lengths of "staple" or fibre. Long staple wool is better than short for the hand spinner. Rough, hairy wool is fine for tweeds and rugs: soft, silky wool for soft fabrics like dress material. There are no hard and fast rules, though.

To turn raw wool straight from the sheep into yarn ready for weaving, you usually begin by "teasing" to straighten the wool out and get rid of dust, burrs, and other rubbish. Then you "card" to create "rollags" (*see opposite*), which are rolls of well-combed wool ready for spinning. Spinning (*see opposite*) is done with a spindle, a hedgehog, or a spinning wheel, and whichever it is, the principle is the same: to stretch and twist the straight fibres of wool from your rollags to make lengths of yarn ready for weaving or knitting. The subtle feature of a spinning wheel is that the endless twine which acts as a driving band goes over two pulleys of different sizes. This means that the bobbin and the flyer, which the pulleys drive, revolve at different speeds. The flyer is therefore able to lay the yarn, as it is spun, on the bobbin at the right tension.

Roving

A self-supporting friend of mine wears the most flamboyant garments, very warm and good-looking, and he makes them entirely from wool, with no other tools but five sticks and a needle. He spins them on one stick and weaves them on the other four. Now it is possible to spin wool without carding first. Instead you have to "rove" it, which can be done with the hands alone. Take some teased wool in your left hand, release a little of it between your finger and thumb, and pull out in a continuous rope with your right hand, but not pulling so hard that you break or disengage the rope. This is not easy as it sounds and it takes practice. When you have pulled out all the wool, bend it double and do the whole operation again. Bend it double again (sometimes you might like to triple it) and go on doing this until you are satisfied that it is fairly parallel and well teased-out. This is now a "roving" and you can spin it direct.

Types of yarn

For weaving you generally use single-ply wool. The warp yarn should be fairly tightly spun: the weft yarn less so. If you intend to knit with the yarn, double it.

PREPARING RAW WOOL
You can turn a fleece into spun wool with just a spindle, but it is easier to use cards, and a hedgehog, or spinning wheel.

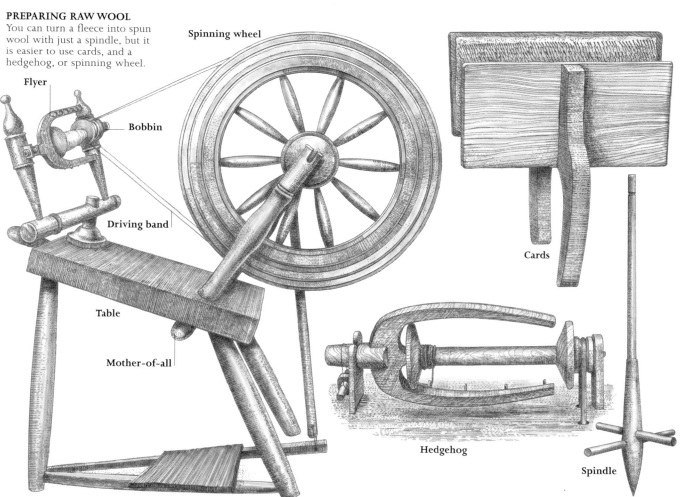

Flyer

Spinning wheel

Bobbin

Driving band

Table

Mother-of-all

Cards

Hedgehog

Spindle

To do this put two full bobbins on a "lazy kate", which is simply a skewer held horizontally at each end (two upright pegs will do just as well), put the ends of the two yarns together, feed them into the spindle on your spinning wheel just as if you were going to spin, put them round the flyer (*see illustration*), tie them to the spindle, and then turn the wheel backwards, or from right to left. This will make two-ply wool. If you want three-ply do the same thing with three bobbins.

COTTON

Cotton is often "willowed" before being carded. In the West this generally means being put in a string hammock and beaten with whippy willow rods. The vibrations fluff out and clean the cotton very effectively. It is then carded just like wool, but it cards much more easily, the cotton staples being much shorter.

Spin it as if it were wool, but keep your hands much closer together, treadle more quickly, and don't hold the cotton back too much with the left thumb and finger or it will kink. Angora hair, if you can get it, is delightful stuff and can be treated just like cotton. It makes amazingly soft yarn, much softer than most wool.

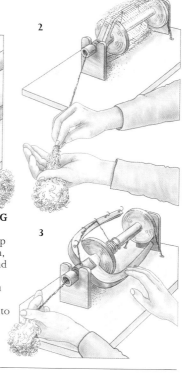

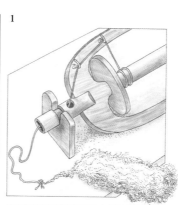

SPINNING WITH A HEDGEHOG

A hedgehog fits on to a treadle. **1** Tie a string round the bobbin, loop over first two hooks, poke through, and tie to your rollag. **2** Treadle, and pull unspun wool from your left hand with your right. **3** When you have a good length of spun wool, stop treadling, move the string on to the next hook, hold the outer bracket still, and treadle. The yarn will be drawn on to the bobbin.

TEASING AND CARDING

1 To tease take raw wool and pull out small pieces. **2** Lay teased locks evenly over your left card. **3** Stroke the left card with the right until the fibres are well combed. **4** Transfer fibres from left card to right. Comb and transfer about five times. **5** Get all the wool on one card and roll it off. Make a rollag, by rolling between the card backs or on a table.

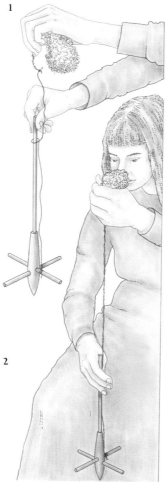

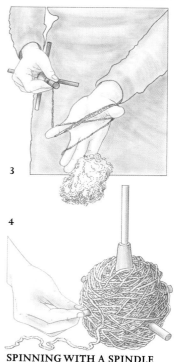

SPINNING WITH A SPINDLE

1 Tie spun yarn to spindle, take a turn round the hand, and tie to your rollag. **2** Spin spindle. Pull unspun wool out between forefinger and thumb of left hand. **3** When spindle reaches the ground, haul it up wrapping newly spun wool round fingers. Transfer spun wool back to spindle. Spin new length of wool. **4** Pull out dowels to release wool.

Dyeing & Weaving

DYEING

Stock dyeing – that is dyeing the fibre in the skein before it is woven – is best for the self-supporter. It is easier thus to get an even distribution of colour.

Natural dyes will generally only dye natural materials: they will not dye nylon and the other synthetics. But the right natural vegetable dyes, used with the correct "mordants", will dye any natural fabric with good and fast colours. (Mordants are chemicals which bite into the fabric and give the dye something to fix on.) Although aniline dyes, which are derived from coal tars and other strange chemical substances, can get close to natural colours, they can never quite match them. But if you want very brilliant colours, then you will probably need artificial dyes.

Some plant-derived dyes don't need a mordant, but most do. The mordants that you should be able to make for yourself or come by very readily are vinegar, caustic soda, and ammonia. To get a greater range of colours you need substances like cream of tartar, alum, chrome (potassium dichromate), tin (stannous chloride), and iron (ferrous sulphate). Alum is the most useful one, and if that is the only one you have, you can still do a lot of dyeing.

To mordant with alum heat 4 gallons (18 litres) of water, dissolve 4 ounces (114 g) of alum and 1 ounce (28 g) of cream of tartar in a little water, and then add it to the 4 gallons (18 litres). Immerse 1 pound (0.5 kg) of clean, scoured (washed), dried wool in the form of a skein and simmer for an hour, stirring occasionally. Lift the wool out and press gently.

To prepare your vegetable dye, cut up your vegetable matter into small pieces, let it stand in cold water overnight, and boil it for an hour. Then add more water if necessary. You will need 4 gallons (18 litres) of dye for a pound (0.5 kg) of wool. Drop wetted, mordanted wool into the dye all at once. The dye should be warm. If it isn't, heat it. Leave the wool in for an hour, stirring occasionally very gently. Then take it out and drain.

A few materials that make strong colour are listed below, but the field is open to endless experiments.

Yellow Bark of ash, elder, brickthorn, apple, pear, and cherry; leaves and shoots of broom and gorse; privet leaves; onion skins (not very fast in sunlight, though); marigold flowers; golden-rod; Lombardy poplar leaves; lily-of-the-valley leaves; bog myrtle leaves; dyers' chamomile; spindle tree seeds; pine cones (reddish yellow); barberry roots and stems (no mordant required).

Green Purging buckthorn berries; heather leaf tips; privet berries (a bluish green); bracken leaves; spindle tree seeds boiled in alum; elder leaves.

Brown Walnut roots, leaves or husks of shells (no mordant required); slow or blackthorn bark (reddish-brown); boiled juniper berries.

Red Spindle tree seed vessels; blood root.

Black Oak bark, which will dye purple if mixed with tin (stannous chloride). Oak galls make ink.

Purple Bilberries are much used for tweeds in the Highlands of Scotland and are a fine dye (no mordant required); willow roots.

Violet Wild marjoram.

Orange Lungs of oak, *Sticta pulmonacea* (no mordant required).

Magenta Lichen makes a magenta on the first dye and other colours as you enter successive dye-lots into the same dye. When the dye seems quite exhausted, freshen it with vinegar and you will get a rosy tan.

BLEACHING

Fabrics can be bleached by soaking them in sour milk and laying them in the sun. A mixture of chlorine and slaked lime also bleaches and is good for flax and cotton. Wool and silk can be bleached with fumes of sulphur. Simply hang the skeins over burning sulphur in an enclosed space.

WEAVING

Weaving on a good hand loom is a magnificent accomplishment, and if you can do it you have made a big step towards true self-sufficiency. Once you have the loom, and are proficient, you can achieve a considerable output of very good cloth. Machine-woven cloth does not compare with hand-woven, nor have machines yet been devised that can even imitate the hands of the weaver.

Fasten four sticks in a square frame shape, tie lots of threads over them all parallel with each other (the "warp"), and haul another thread (the "weft") through the threads of the warp with a needle or sharpened stick, going over one and under the next thread of the warp and so on. Then bring the needle back with another thread on it, going over the ones you went under before. Keep on doing this and in no time you will see your cloth appear.

If you need to make cloth seriously you will soon find yourself inventing ingenious devices to make your task easier and your cloth better. Firstly you will devise a comb (*see illustration*) to poke between each pair of threads in the warp and beat the threads of the weft together so that the weave is not too loose. You will have invented the ancestor of the "reed".

Then you will find that it is tedious to go on threading the weft through with a needle and so you will invent an arrangement of two sets of strings, with loops in their middles, hanging from sticks, and you will thread each thread of the warp through the loop in one of these strings, each alternate thread going to a different set of strings from its neighbours. You will have invented the "heddle". You will lift each set of heddles alternately, on a frame called the "harness" and it will leave a space called the "shed" between the two sets of threads. You will be able to throw your needle through the shed so that you can criss-cross, or weave, the threads without having to pick through each individual warp thread with your needle. Next you will find it a nuisance having to attach a new weft thread to your needle each time.

THE SQUARE WEAVER

The simplest loom is the 5-inch (13-cm) "square weaver". It makes 4-inch (10-cm) squares of cloth which can be sewn together as patchwork. String the warp as shown below and weave the weft with a 5-inch (13-cm) needle. Design your own patterns on graph paper (right): on black squares the weft goes under, on white squares it goes over.

Pattern with plain weave **Simple checks** **Diagonal weave**

Weaving comb

Stick shuttle

Boat shuttle

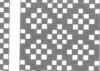

Warping frame

THE FOUR-HARNESS TABLE LOOM

A table loom takes up much less space than a floor loom and does all the same things. It is a little slower because the harnesses are operated with handles instead of pedals.

1 Harness **4** Cloth beam
2 Heddles **5** Shed
3 Reed **6** Warp beam

So you will carve notches at either end of a stick and wind the thread round it in such a way that the stick can turn and release, or pay out, the yarn.

You will have invented the "stick shuttle" (*see illustration*). As you get more inventive you may invent the "boat shuttle" (*see illustration*), into which you can drop a reel of thread, ready wound. You will soon find that, with all your new gadgets, you quickly come to the end of your weaving frame and only have a small piece of cloth, so you will invent a roller at each end of your loom, one for rolling the threads of the warp on, the other for rolling the newly woven cloth on. This time you will have invented the "warp beam" or "warp roller" (*see illustration*), and the "cloth beam" or "cloth roller" (*see illustration*). You will also find lifting the alternate harnesses up to form your shed a nuisance. So connect the harnesses up to some foot pedals with an elaborate arrangement of strings. You will have invented "treadles" with "marches" or "lamms" above them to transmit the motion to the harnesses.

Then, if your life depends on weaving an awful lot of cloth, you will devise a sling device worked by a handle, which will fling the shuttle backwards and forwards through the warp without your having to touch it. By this time, you will have invented the "flying shuttle" and, believe it or not, you will be getting dangerously near the Industrial Revolution.

Now, when you come to thread your new patent loom up with the warp threads you will find that it is so difficult that you nearly go mad, so you invent a revolving spool, a "warping mill" to wind the threads of the warp around, or else a rack, a warping frame (*see illustration*), with pegs that serve the same purpose.

Finally you will realize that by having four harnesses instead of two you can greatly vary the pattern formed by the warp, for you can lift different combinations of warp threads. And by having two or more shuttles, with different coloured weft thread in them, you can alter the pattern in other ways.

But to learn to weave you simply must get somebody who knows how to do it to teach you: you cannot learn it out of a book, although a good book on weaving will help.

FINISHING CLOTH

"Fulling" partially felts cloth and makes it denser and stronger. You do it by beating the cloth in water. Try putting it in the bath and stamping on it hard. If you add "fuller's earth" you will fill up the pores of the cloth.

"Raising" is done by picking the surface of the cloth, traditionally with "teazles" which are the heads of large thistles. You can often find them growing wild, or you can cultivate them yourself. The effect of raising is to give the cloth a fluffy surface.

Spinning Flax

Flax is the most durable of all the fibres available to us. Man-made fibres haven't been invented long enough yet to know whether they will outlast flax: my guess is they won't, for quite good-looking pieces of flax linen have been dug up in Egyptian pyramids and my corlene rope won't last two years.

The crop is harvested before the seed ripens, which is a pity because it means losing the oil the seeds would ultimately produce. It is pulled, not cut, then tied in sheaves and stacked.

Preparing raw flax

Flax must first be "rippled", which means pulling the heads through a row of nails with their heads filed to points. This removes the unripe seeds, which make a marvellous stock-food. Then flax is "retted", which really means rotted. Lay it in stagnant water for 2 or 3 weeks, until the fibrous sheaf separates easily from the central woody portion. You can ret in running water but it takes much longer, or you can spread your flax on grass for about six weeks and let the dew do the job. After retting, dry the flax carefully.

Then you must "scutch", which is the process of breaking the stems of the flax. Do this by beating the flax on a table with a broad wooden blade, or with a special "scutcher".

"Hackling" is the next step, and consists of dragging the flax across a bed of nails to remove the "tow" which is all the short fibres, and leave the "line" which is the long ones. The tow can be used for caulking deck seams on boats, or stuffing mattresses, or it can be carded and spun to make a rather coarse and heavy yarn. The line can be spun to make linen thread.

To spin (you don't card line) you have to dress the line on a "distaff" which is simply a vertical stick, or a small pole, which can be stuck into a hole in a spinning wheel.

Dressing a distaff

Dressing a flax distaff needs considerable skill. Put an apron on (if you don't happen to be wearing your long bombazine skirt), tie a string round your waist leaving the two ends a few inches long, and sit down. Take a handful of line, such as falls naturally away from the larger bundle, and tie round one end of it carefully with the two ends of the string round your waist and secure with a reef knot. Cut the two loose ends of the string. Lay the flax out full length on your lap with the knotted end towards you. Hold the bundle with your left hand at the end furthest from you, pull a few fibres away from the main bundle with your right hand, and lay them on your right knee. Pull some more fibres away and lay them next to the first few. Go on doing this until you have made a thin, fine fan of flax on your lap. Remember that the end closest to you is tied fast.

Now grab the main bundle in your right hand and reverse the process, laying a second fan from left to right

DRESSING LINE ON A DISTAFF
Before it can be spun, line must be dressed on a distaff so that the fibres are separated out. You take a handful of line and tie it at one end with string which you have first wound round your waist. Sit down and carefully spread out a series of fans of fibre on your lap: one on top of the other. Cut the knot, lay the distaff in its hole and tie with ribbon.

on top of the first fan, but be sure to pull from the same part of the main bundle. Go on doing this, alternating hands and directions, until all the flax of the bundle has been laid out, in criss-crossing fan shapes, one on top of the other. As you work try to criss-cross the fibres, otherwise they will not pull out properly when you come to spin.

Now cut the string, take it away and slightly loosen the top end of the bundle where it was tied. Then lay the distaff on one edge of the fan, with its top where the knotted string was. Wind the fan up on the distaff, winding very tightly at the end nearest you, but keeping the flax very loose at the bottom of the distaff. Then put the distaff, with the flax fan round it, upright into its hole and tie the middle of a ribbon tightly round the top. Then criss-cross the two ends of the ribbon downwards round the cone of flax until you reach the bottom. Tie the two ends in a bow.

Spinning flax

Take the yarn that you have already tied into the bobbin of the wheel and catch it in the flax at the bottom end of the distaff. Spin. Have a bowl of water by you and keep wetting your fingers so as to wet the flax. Use your left hand to stop the spin from going up into the distaff, and your right to clear knots and pull out thick threads. If you have done your dressing operation right the line should steadily feed itself through the thumb and finger of your left hand into the spinning thread. Turn your distaff as required, and, when you have cleared the distaff as far as the bow in the ribbon, untie the latter and tie it further up to expose more fibres. Keep doing this until you come to the very top of the distaff and the last few fibres.

Curing & Tanning

Animal skins become hard, like boards, when they have been pulled off the carcass and dried for some time, and then they are good for practically nothing. Early on, mankind found two ways to overcome this disadvantage: mechanical methods which produce rawhide, and chemical methods which produce leather.

To make rawhide you must take hide straight off the animal and begin working it before it gets hard. In this way you will break down the fibres which set to make it hard and it will remain permanently soft. A great deal of working is needed. Innuit women, we are told, do it by chewing the hide. Undoubtedly chewing, and working between the hands, for long enough (probably pretty constantly for about a week) will do the trick.

Curing

I use a method which is part mechanical and part chemical to cure sheep-skins, fox skins, and especially rabbit skins, which come up beautifully. The end product is a cross between rawhide and leather.

Wash your animal skin well in warm water and then rinse it in a weak borax solution. Then soak it in a solution of sulphuric acid made by mixing 1 pound (0.5 kg) of salt with 1 gallon (4.5 litres) of water and pouring in ½ ounce (14 g) of concentrated sulphuric acid. Don't throw the water on to the acid or you may lose your eyes and spoil your beauty.

After 3 days and nights take the skin out and rinse it in water and then in a weak borax solution. If you put it in the washing machine and let it churn about for an hour or two, so much the better (after you have washed the acid out of it, of course). Next hang it up and let it half dry.

Take it down and rub oil or fat into the flesh side and work it. Scrape it and pull it about.

Pulling it with both hands backwards and forwards over the back of a chair is a good method. Leave it hanging over a chair and pull it about every time you go past. Rub more fat in from time to time. It will become quite soft and as good as tanned leather.

Tanning

Tanning with tannin is a purely chemical method, and the end product is leather. It takes half a ton of good oak bark to yield a hundredweight (50 kg) of tannin, and this will cure 2 hundredweight (100 kg) of fresh hides. Wattle, elder, birch, willow, spruce, larch, and hemlock also contain tannin. The bark must be milled: that is, pounded up small and soaked in water. The hides must be steeped in the resulting solution for 4 months, in the case of small hides, to a year in the case of big ones. For really perfect results it is best to soak hides in a weak solution at first, putting them in increasing strengths as the months go by. A fool-proof method is to soak the hides in a weakish solution for, say, a month, and then to lay them in a pit or tank with a thick layer of bark between each skin. Then just cover the pile with water. Leave like this for at least 6 months.

A quick way of tanning a skin is the "bag method". You make a bag out of a skin (or take the skin off whole). Hang the bag up, and fill it with tannin solution. After a week or two the hide should be tanned. To get the hair off skins, lay them in a paste made of lime and water for 3 weeks, or in a lime-sulphide paste for a day. De-lime by washing in a weak vinegar solution.

Sewing leather is as easy as sewing cloth: all you need are a few large needles (sailmaker's needles are fine), an awl for making holes in the leather, and some strong waxed thread. Any thread dragged through a lump of beeswax is waxed thread. (*For stitches, see illustrations.*)

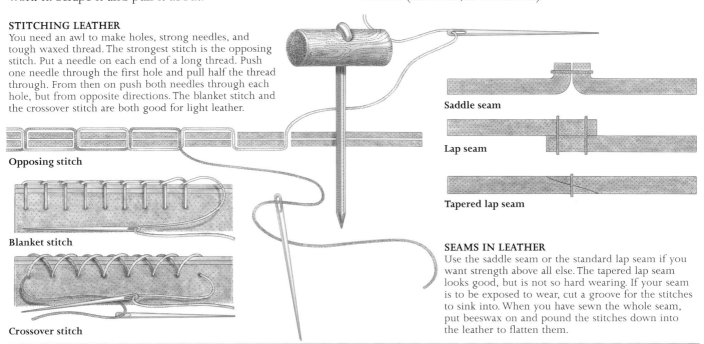

STITCHING LEATHER
You need an awl to make holes, strong needles, and tough waxed thread. The strongest stitch is the opposing stitch. Put a needle on each end of a long thread. Push one needle through the first hole and pull half the thread through. From then on push both needles through each hole, but from opposite directions. The blanket stitch and the crossover stitch are both good for light leather.

Opposing stitch

Blanket stitch

Crossover stitch

Saddle seam

Lap seam

Tapered lap seam

SEAMS IN LEATHER
Use the saddle seam or the standard lap seam if you want strength above all else. The tapered lap seam looks good, but is not so hard wearing. If your seam is to be exposed to wear, cut a groove for the stitches to sink into. When you have sewn the whole seam, put beeswax on and pound the stitches down into the leather to flatten them.

Making Bricks & Tiles

If you can avoid it, don't buy clay to make home-made bricks. Instead, try the different clays on your land and in your locality. You are quite likely to find one that makes a good brick, and save yourself a lot of money.

When you have found it, dig the clay and puddle it. You can do this by laying the clay in a pit, wetting it, and trampling it for an hour or two with your feet. This method works very well, but any way of working the clay well with water will do. Then, when the clay is of the right consistency – that is, solid but malleable – you can make bricks using the method described.

Drying and firing bricks

In countries with a rainless, dry season the easiest way to dry bricks is to lay them out in rows on level sand and just leave them. In rainier climes they must be under cover, and are usually piled up about six courses high, criss-crossed to leave spaces for the air to circulate.

Bricks have to be left to dry for anything from a week to a month, according to the climate, and then they must be fired. To fire bricks you must build a clamp, which is basically a rectangular pile, at least the size of a small room, made of bricks criss-crossed so as to leave cavities between them. There are two ways of using the clamp. The first is to leave fireplaces sufficiently large to contain fair-sized wood fires at roughly 3-feet (1-m) intervals on the two long sides of the clamp. Then you plaster the whole clamp with clay, except for some small chimneys at the top of the leeward side, and light fires in the fireplaces on the windward side. If the wind changes, block up the fireplaces on the new leeward side, open up the fireplaces on the new windward side, and use them.

The fireplaces can be rough arches of already burnt, or half-burnt, bricks, or false arches made by stepping bricks. After firing for a week, let the fires go out and allow the clamp to cool. Open up, pull out the well-fired bricks, and keep the half-fired ones to be fired again.

The other method, which I think is easier and better, does not require fireplaces. Instead, you fill the gaps between the bricks with charcoal (coal, anthracite, or coke will do). The clamp can be smaller, 7 feet (2.2 m) high by any width or length you like. Plaster the whole clamp with mud, except for a hole at the bottom on the windward side and a hole at the top on the leeward side. Light a small wood fire in the hole on the windward side and go away and forget it. The charcoal will quickly catch. After 5 or 6 days, when it has cooled, open the clamp and take your bricks out. You get more completely fired bricks than with the other method.

Tiles

Tiles can be made of the same clay as bricks, but it must be carefully puddled and mixed. They can be flat, or they can be pantiles, which have a convex and concave side, or they can be, as most Mediterranean tiles are, half-cylinders. In Spain and Italy the latter are commonly tapered because, it is said, in Roman times they were moulded on a man's thigh. These can be, and often are, made by throwing a cylinder on a potter's wheel and splitting it in half before drying and firing. Any other tile must be made in a mould.

Fire the tiles in the same clamp as the bricks and build it so the bricks take the weight. Tiles are not strong, so they must have holes for nailing or pegging.

BRICKMAKING: THE ESSENTIAL TOOLS
Most important is the mould, which you should make precisely the right size to allow for the shrinkage of your particular clay. Make it from jointed wood and, if you want it to last, put a steel pin secured with a bolt through each end. Make your bow by bending a length of hazel and stringing a wire across it. The cuckle is an excellent tool for moving large amounts of clay. Your sand tray should be deep enough to immerse the mould in. You need a knife for cleaning the mould and boards for drying bricks.

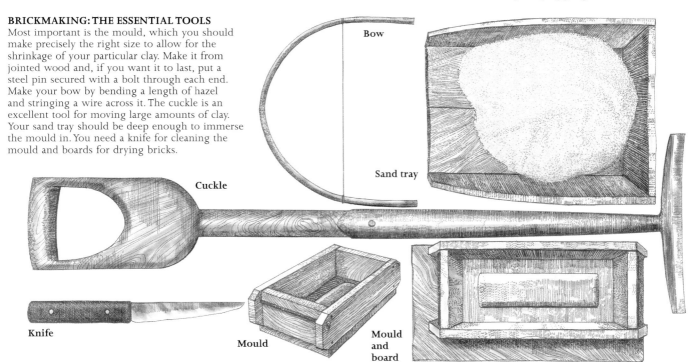

Bow

Sand tray

Cuckle

Knife

Mould

Mould and board

MAKING BRICKS IN A MOULD

Time and practice has determined that bricks should measure 9 inches (23 cm) by 4½ inches (11 cm) by 2¼ inches (5 cm). Depending on your clay and how much it shrinks, your mould must be marginally bigger. Experiment and then make a mould to suit your clay.

1 Clean the inside of your mould by scraping around it with a knife.

2 Coat the inside of the mould with sand as you would a cake tin with flour, by dipping in sand and shaking.

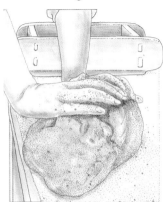

3 Take what you think is the right amount of clay and begin to form a "warp" (a brick-sized lump).

4 Work the warp into the right shape by rolling on a board. Sand the board and your hands to stop the clay sticking to them.

5 Once you have the right shape, gather the clay towards you, rolling the ends in.

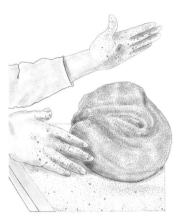

6 Drop the clay with a spinning action so that it thuds onto the bench. This will knock out all the excess air.

7 Throw the warp hard into the mould so the clay spreads out towards the corners.

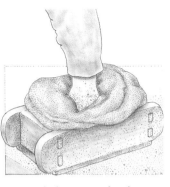

8 Punch down into the clay to push it into the corners and leave a hole in the middle.

9 Ram more clay in the hole and press down very hard into the mould.

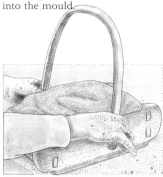

10 Cut off any excess clay by running your bow across the top of the mould.

11 Peel off the severed clay and return it to your pile.

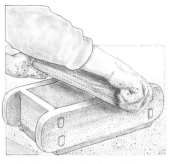

12 Dip a length of wood in water and use it to smooth over, or "strike", the top surface. Then sprinkle the top with sand.

13 Pick up the mould and tap its corners against the bench until you can see gaps on all sides of the clay.

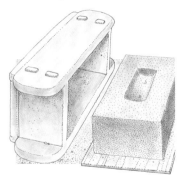

14 Dry the clay brick on a board for about a month.

Working in Stone

Some stone, particularly granite, is awkward for building because it does not split easily in straight lines. Other stone, most sedimentary stone, in fact, has been laid down in layers under water and therefore splits easily along horizontal lines, but they are not necessarily horizontal in the ground. The beds may have been tilted. Other stone, which builders and quarrymen call freestone, splits easily both horizontally and vertically. This is what the builder is looking for and if he can find it, he is a very lucky man. Much of the oolitic limestone from Britain's Jurassic sea is like this, and it gave rise to the superb school of vernacular building which stretches from Dorset to Lincolnshire; the huge quarries of Barnack in Northamptonshire having supplied much of the material for the great gothic churches and cathedrals of Eastern England. The Jurassic limestone of the Isle of Portland is a freestone *par excellence*, as is Purbeck stone.

Freestone can often be split out with wedges instead of explosives. Holes are drilled in a line along the rock, the wedges are driven in, in sequence, further and further until suddenly the rock splits along the line. If you are splitting off a big piece you can use "the plug and feathers". The feathers are two pieces of steel that you put down either side of a hole drilled in the rock. The plug is a wedge that you drive in between them. The advantage is that the feathers exert a more even pressure than the plug alone and so the rock splits evenly when it comes away from the parent rock.

Holes are driven in rock by a rock drill, which is a steel bit with an edge like a chisel sharpened at a very obtuse angle. You either hit the bit in with a hammer, turning it between each blow, or you drive it in with a percussion drill. You can drill the hardest rock in the world like this and in soft rock go quite quickly even with a hand hammer. Put water in the hole for lubrication, and get rid of the ground rock dust by splashing. Wrap a rag round the bit so you don't get splashed in the eye with rock paste.

You can break out, subdivide, and dress to rectangle any rock, even the roughest and most intractable basalt or granite: the harder the rock, the harder the work. You can build with uneven, undressed boulders, and fill the inevitable spaces with – well, just earth, or earth and lime, or in these decadent days concrete made with cement so the rats can't get in. But there will always be places where you need a solid, rectangular stone: doorsteps, lintels, hearth stones, and other similar things.

Slate is a metamorphic rock, which means it is a sedimentary rock that has undergone great heat and pressure. The original layerings or laminations have been obliterated and others have developed more or less at right angles to the first. It cleaves easily along these. Generally there are faults or weaknesses in large masses of slate more or less at right angles to the laminations of the slate. These make it possible to break out large blocks without too much blasting. Slate is the very best roofing material, and thicker slabs are ideal for shelves in larders.

HANDLING MASON'S TOOLS

To prepare stone you need two types of chisel and hammer. The points and the edging-in chisel are given sharp, direct blows with a steel hammer. Claws and other chisels must be given softer blows, so for these you use a wooden mallet.

POINTING
Hold the point at an angle and hit sharply with a steel hammer.

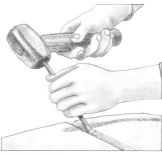

EDGING-IN
To help you control the edging-in chisel, place your thumb across it. Use small, hard strokes, keeping the chisel in position on the stone.

CLEAVING A BLOCK OF STONE
Mark round with a pencil. Drill and chisel deep V-shaped slots in the top and sides. Crowbar the block up and place a steel section below the future

FINISHING A SURFACE

Skim rhythmically with the claw or chisel: **1** place tool on surface; **2** hit firmly with mallet; **3** draw back tool and mallet together and repeat.

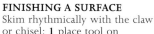

1

2

3

breaking point. Put steel wedges in the slots and hit in sequence with a steel hammer, listening to the ring of the stone. It dulls as the stone cleaves.

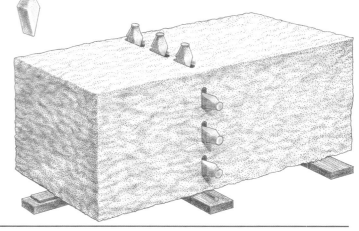

THE MASON'S TOOLS

The heads of the claw and chisel are specially flattened for use with the wooden mallet. The heads of the points and edging-in chisel round over as they are hit repeatedly with the steel hammer.

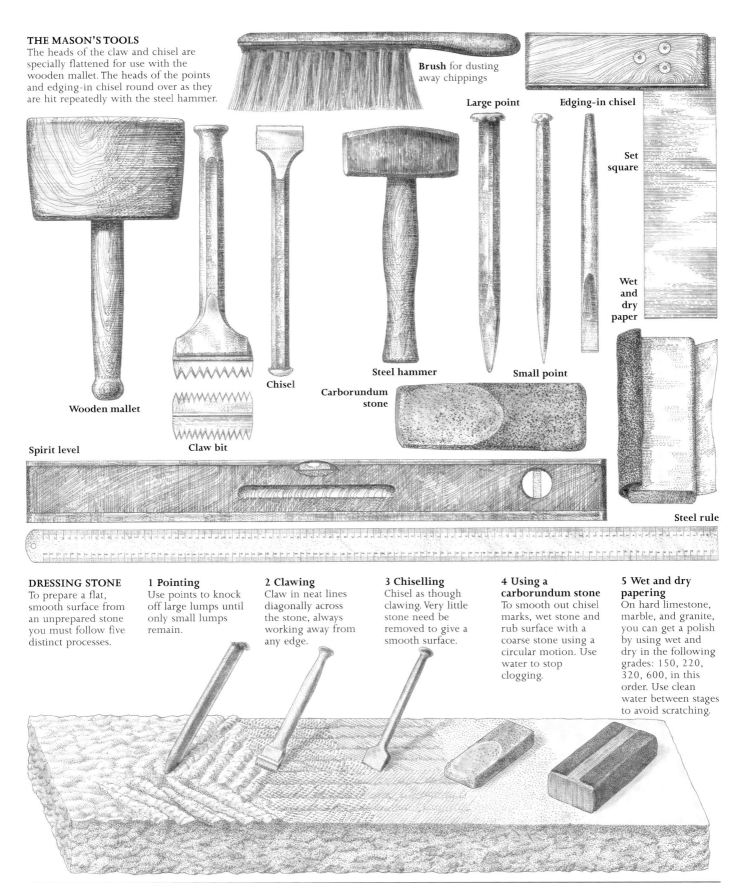

Brush for dusting away chippings

Large point

Edging-in chisel

Set square

Wet and dry paper

Wooden mallet

Chisel

Steel hammer

Small point

Carborundum stone

Claw bit

Spirit level

Steel rule

DRESSING STONE
To prepare a flat, smooth surface from an unprepared stone you must follow five distinct processes.

1 Pointing
Use points to knock off large lumps until only small lumps remain.

2 Clawing
Claw in neat lines diagonally across the stone, always working away from any edge.

3 Chiselling
Chisel as though clawing. Very little stone need be removed to give a smooth surface.

4 Using a carborundum stone
To smooth out chisel marks, wet stone and rub surface with a coarse stone using a circular motion. Use water to stop clogging.

5 Wet and dry papering
On hard limestone, marble, and granite, you can get a polish by using wet and dry in the following grades: 150, 220, 320, 600, in this order. Use clean water between stages to avoid scratching.

Working in Metal

BLACKSMITHING

To learn to be a proper blacksmith should take 7 years, but you can learn to bend, shape, and weld wrought iron in a few hours. To do it well takes practice though, and you will ruin plenty of iron first. If you are planning to work with iron a lot you need equipment: a forge, an anvil, a bench with at least one good vice on it, and suitable hammers, and tongs. But I have done simple forge work by crawling about on my hands and knees in front of an Aga cooker, poking bits of iron into the firebox, and hammering them on the head of a sledgehammer laid on the ground. A little knowledge may be a dangerous thing but it helps sometimes.

Blacksmiths work with ferrous metal and there are many kinds. Wrought iron is the blacksmith's classic material. It is made from pig-iron (the stuff which runs out of the bottom of blast furnaces) by persistent heating and hammering. It has enormous advantages for blacksmithing: you can shape it, split it, weld it, in fact, treat it as if it were clay or plasticine, provided you get it to the right temperature. When cold it is hard (but nothing like as hard as steel), tough, and strong, ideal for much agricultural machinery, chains, shackles, split-links, and the iron components of carts and boats. It doesn't corrode easily.

Cast iron is what it says it is: cast in moulds. It is extremely hard, but brittle. It will not stand hammering and it is no good for edge tools, as the edge would just crumble off, but it does not corrode easily. Malleable iron is only used for a few things, like the fingers of mowing machines, which have to be shaped when cold. Steel comes in many forms and qualities. "Mild steel" is much used by blacksmiths nowadays because they cannot get wrought iron. It is nothing like so good because it is harder to work and it rusts easily.

For forging wrought iron you need an ordinary blacksmith's forge. This is a fire tray, or hearth, with a pipe, called a tuyere, or tue iron, which blows air into the fire. The tuyere commonly passes through a water bath before it reaches the fire so that it keeps cool enough not to burn away, but sometimes it simply passes through a massive piece of cast iron. Cast iron can stand great heat without melting or burning. The fire can be of coal, coke breeze, or charcoal. If you use coal or coke, clinker will form, and hamper your work. Let the clinker solidify and remove it.

Keep the fire as small as possible by wetting the fuel around its centre, and place the work to be heated in the heart of the fire. Draw wet coal in sideways as needed: don't dump "green" coal on top of the fire. The blast can be provided by a bellows worked by hand, by an electric air pump, or by a vacuum cleaner turned the wrong way round so it blows instead of sucks but don't use more blast than you need.

Different jobs require different degrees of heat.

Blood red is for making fairly easy bends in mild steel.
Bright red is for making sharper bends in mild steel, or for punching holes and using the hot chisel in mild steel.

Bright yellow is the heat for most forging jobs in wrought iron, and for drawing down and upsetting (making thinner or making thicker) both wrought iron and mild steel. It is also right for driving holes in or hot chiselling heavy work (iron or steel more than an inch or 2 cm thick).
Slippery heat is just below full welding heat and is used for forging wrought iron and for welding mild steel if it proves difficult to weld it at a higher temperature. It takes speed and skill to weld steel at this heat.
Full welding heat is for welding wrought iron and most kinds of steel. When you reach it, white sparks will be flung off the white hot metal, making it look like a sparkler.
Snowball heat is the temperature for welding very good quality wrought iron, but it is too high for steel. If you go beyond snowball heat you will burn your metal.

Tempering

Tempering is the process of heating and then cooling metal to give it different degrees of hardness and brittleness. The general rule is that the higher you heat it and the quicker you cool it, the harder it will be, but the more brittle. When tempering a steel cutting tool you harden it first, by heating it to somewhere between black and blood red and then plunging it into water. When you have done this you temper it by heating it again, dipping the cutting edge into water so as to cool it, then letting the colours creep down from the rest of it until exactly the right colour reaches the edge and then quenching it again.

Welding

To weld wrought iron or mild steel, first get the metal to the right temperature. Then take the first piece out of the fire, knock the dirt off it, and lay it face upwards on the anvil. Whip the other bit out, knock the dirt off it, lay it face downwards on the first bit and hit it in the middle of the weld with a hammer, hard. Keep on walloping it: on the flat if it is flat work, around the beak of the anvil if it is, say, a chain link. But all this has to be done very fast. If the weld hasn't taken, put it in the fire again. If the centre has taken but not the outsides, fire again, or "take another heat" as blacksmiths say.

To weld anything harder than wrought iron, more modern forms of welding must be used. From the point of view of the self-supporter these are oxy-acetylene and electric arc. Neither of these are as formidable as they sound: every gypsy scrap-dealer uses oxy-acetylene and many a farmer has his own small electric welding set and uses it, too. But for either gas or electric welding, always wear goggles or a mask. It is possible to blind yourself permanently if you gaze at an arc or a gas flame for just a second or two – very easy to do severe damage to your eyes.

Oxy-acetylene

Oxy-acetylene tackle consists of two pressure bottles, one of oxygen and the other of acetylene. The latter gas, in the presence of oxygen, gives off an intensely hot flame, and a

flame, furthermore, that acts as a protection against oxidation for the hot metal before it cools. The two gases are brought together by pipes and then burnt at a nozzle. It is the inner flame that you must use – not the outer. The aim of the welder is to melt rod metal and use it to fuse two metal faces together, and also to fill in any spaces between them. Ideally the edges of the steel plates should be bevelled where they meet, and the space left filled with the rod metal.

There are two methods then of oxy acetylene welding. One is "leftward", or "forward", welding. In this method the rod, which is made of metal of more or less the same type as the work to be welded, is held in the left hand and moved to the left while the torch is held in the right hand and follows the rod. The edges of the pieces of steel are pre-heated. Be careful not to keep the flame in one place too long, or the metal will be distorted. In "rightward", or "backward", welding the torch is moved to the right, and the rod follows it. Less rod metal is used with this method, and it is considered better than leftward welding, particularly for joining larger pieces of steel, anything over ¼ inch (0.6 cm).

Electric arc

Electric arc welding is a simple matter of using a very high voltage to create a spark at the top of a rod. Held between the two surfaces to be welded, the spark melts them and also the tip of the rod. The material to be welded must be earthed. You can buy quite cheap and simple a.c. welding sets that work off the mains, and also portable sets that have a small motor to generate current for them.

Sharpening tools

The principle of sharpening is that, if it is just a freshening up of the edge you want, you use a "whetstone", but if the tool has begun to lose shape then you put it on the "grindstone" first, grind it down to shape, and then hone it on the whetstone afterwards. Whetstones come either as slipstones, which are shaped to be held in the hand, or as oilstones, which are mounted in a wooden box and used on a bench. Both should be oiled with thin oil when used. Grindstones are coarser and are frequently circular and mounted with a handle over a trough of water so that they can be kept wet.

Most sharpening stones that you can buy nowadays are artificial with carborundum embedded in them. They are undoubtedly better than anything except the best Arkansas stone, which is an almost pure quartz, grainless and hard. You must grind your cutting tools at the right angle. This will be a compromise between the acute angle needed for easy cutting and the more obtuse angle needed for strength. Thus a chisel to be whanged with a mallet must have a more obtuse edge than one to be used for delicate carving in the hand. You can buy a guide, or jig, to help set the angle.

USING YOUR ANVIL
If you want to work seriously with metal, get an anvil. Most of your work is done on the anvil's "face"; the flat section at the top. You should use the "table", the short step down, pulling out nails

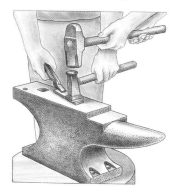

for cutting or chopping, because its surface is softer and will not suffer damage. The "beak", the long pointed bit, is for working anything that needs a curved edge. To flatten metal, or remove marks made by a hammer, you should hold hot metal on the face of your anvil with tongs, hold a flatter over it and whack it with a sledgehammer until you achieve the desired effect.

THE HAMMER AND PRITCHELL
To make holes in any metal, and particularly in horseshoes, you should use a pritchell, which is a square-handled punch. Again use this on your anvil's face and hit it with a heavy blacksmith's Warrington pattern hammer (above), or with a ball pane hammer. If you need to make a large hole, do it over the round hole in the "tail wedge" of your anvil to avoid damaging the table underneath. Some of your most important tools will be your pincers. You need them for bending metal, and just holding

things. The longer the handles, and the smaller the head, the greater the leverage.

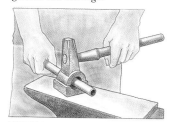

THE TOP AND BOTTOM SWAGE
The swages are for shaping circular rods from hot iron, or for bending rods or pipes. The bottom swage slots into the square "hardie hole" in the anvil's tail wedge.

HOT SETT AND COLD SETT
The hot sett (above, left) with an edge sharpened to about 35 degrees, cuts hot metal. Place it on the metal and wallop with a sledgehammer. The cold sett (above, right), whose edge is about 60 degrees, will cut light iron or mild steel cold.

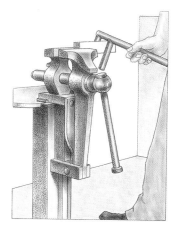

LEG VICE AND SCROLL WRENCH
The best vice for a smithy is a leg vice. It will stand up to heavy hammering because it is made of wrought iron, instead of the more normal cast iron, and some of the load is transferred through the leg to the floor. The front arm is held on a hinge and is opened by a spring. The leg vice is excellent for bending metal, because the leg will stand up to heavy levering. A scroll wrench has rounded jaws for pulling strip metal, especially wrought iron, into curves. Quite intricate designs can be made with it.

Building & Thatching

COB & MUD BUILDING

The cheapest way to construct a solid building is to use mud and thatch, and don't be put off by the way it sounds. Mud is rot-proof and fire-proof, and it keeps sound out and heat in pretty efficiently. Mud for building should be fairly free of organic matter, so dig it from well below the surface: from 2–3 feet (60–90 cm) is best. Save your humus-laden topsoil for growing things.

Your building should be simple, with large areas of unbroken wall, few and small windows, all loads well spread on timber plates, and no outward-thrusting roofs.

An easy but effective method of building with mud is cob building. Cob is simply clayey or chalky mud, mixed with straw and laid in 1-foot (30-cm) layers with a shovel and trowel. Each layer is laid at a different angle from the one below it so that there is a certain amount of binding. The wall should be at least 18 inches (45 cm) thick, 24 inches (60 cm) if your building is to be more than one storey. You cannot build very fast with this method, for each course has to dry out to some extent before the next one is laid on it. The resulting wall is only weatherproof if you keep "its head and its feet" dry. In other words, give your building a good overhanging roof and solid foundations using concrete if possible. And if you can, build a base wall of stone or brick, preferably with a damp-proof course (slate is impervious to water and makes a good one), on top of your foundations up to ground level. The outside, too, should be protected by cement rendering if possible: otherwise with a lime and sand mortar rendering, or at least a thick whitewash. Broken glass is sometimes embedded in the base of a cob wall to deter rats. Window sills must be protected by slate or other stone or concrete.

Rammed earth (otherwise known as adobe) blocks are an improvement on cob because shrinkage takes place in the brick before the wall is built, you can make smoother surfaces, and you can easily build cavity walls. The blocks are made by ramming a mud and straw mixture into wooden moulds. Dry them in the shade, so that they don't dry too quickly and crack. The earth should be, like brick earth, of just the right consistency for the job: that is a benign mixture of clay and sand. The higher the clay content, the more straw you should add, up to 20 percent straw by volume.

African hut

To make an African hut, dig a circular trench, stand straight wall-high poles in it so that they touch one another, and stamp them in, leaving a space for the door. You can have one section shorter than the others if you want a window. Then, on the ground, make a conical roof of what is basically giant thatched basketwork. Get some friends to help you lift the roof and lash it on to the circular wall. Plaster the pole wall with mud, preferably mixed with cow dung. If you rub the earth floor with cow dung and sweep it every day, it will become as hard and clean as concrete.

BUILDING WITH MUD AND THATCH

You can build yourself a warm and solid house or barn mainly out of mud and straw. To make your building last you should build foundations. Cyclopean concrete (large stones embedded in concrete) is fairly cheap and effective. On top of this build a stone or brick wall to just above ground level and top it with a damp course, ideally of slate. Walls can be built of cob (mud mixed with straw). Make your wall at least 18 inches (45 cm) thick and lay in "courses", 12 inches (30 cm) deep. Allow 2 or 3 weeks' drying between courses. To keep rats out you can set broken glass in the wall at ground level. Use slate sills and timber lintels for windows. Render the outside with cement or lime "rough cast", a lime and sand mixture. Apply two coats of pitch at the base to keep it dry. Top the wall with a timber wall plate and to this attach your tie beams, which run right across from wall to wall. Each beam carries a king post, which supports the ridge and the struts, which in turn support the principals. The purlins run the length of the building from principal to principal and carry the rafters, to which you nail battens which will key your thatch. Secure joints with suitable strong bolts.

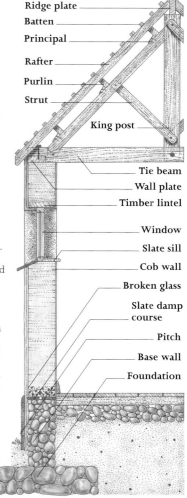

Ridge plate
Batten
Principal
Rafter
Purlin
Strut
King post
Tie beam
Wall plate
Timber lintel
Window
Slate sill
Cob wall
Broken glass
Slate damp course
Pitch
Base wall
Foundation

THATCHING

Phragmites communis, commonly called "Norfolk reed", is the best thatching material there is. A good roof of reed will last 70 years. A roof of "wheat reed" which is simply wheat straw that has not been broken in the threshing may well survive 20 or 30 years. Wheat straw that has been threshed and stored in a stack can be used for thatching ricks. To get it ready for thatching you "pull" it by hauling some down to the foot of the stack and throwing several buckets of water on it. Then you pull the wet straw in handfuls from the bottom of the heap. Because the straw is wet the handfuls come out straight with the straws all parallel to each other. Lay the straws in neat piles about 6 inches (15 cm) in diameter. Tie these with twine or straw rope to make your "yealms". The secret of thatching is that each layer should cover the fastenings that tie down the layer below it, so that no fastenings are visible or exposed to the weather. In practice this means that each layer must cover just over three quarters of the layer below it.

Ricks

Rick thatching is fairly easy and uses comparatively little material. You only need a coat of thatch 2 or 3 inches (5–8 cm) thick to shed the rain.

TOOLS FOR THATCHING

A shearing hook **1** and thatching shears **8** trim thatch, a rake **2** combs it, and a leggat **3** shapes it. A whimbel **4** is for making "bonds", the long twists of straw for tying yealms. Brortches **6** cut from hazel with a spar hook **5** hold the thatch down, and iron hooks **7** secure it to the rafters. Protective knee and hand pads **9** are essential.

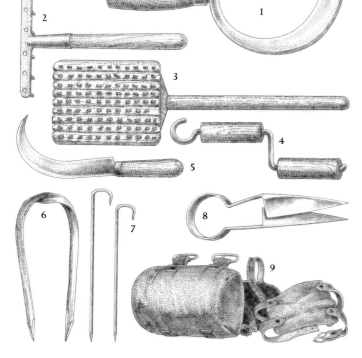

THATCHING A ROOF

Always begin thatching at the eaves on the right hand side. Secure a short row of "yealms", straw bundles, to the roof with "sways", lengths of bendy hazel held to the rafters with iron hooks. Gather the straw up at the upper end of each yealm by pushing in a "brortch". Keep laying rows of yealms, each overlapping the one below, until you reach the ridge. Then move the ladder to the left and thatch another stretch of roof. Carry on like this until you reach the ridge on both sides of the roof.

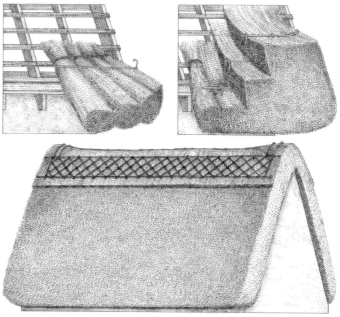

To thatch the ridge lay a row of yealms horizontally along it, and cover them with more yealms folded over the ridge and secured on both sides with sways, brortches, and hooks. You can use hazel sways to decorate the roof.

Lay the straw or reed, ears upwards, in a row along the eaves of the rick. Hold this first row down with one or two lengths of string and hold the string itself down with "brortches". These are 2-foot (60-cm) lengths of hazel or willow (I prefer hazel), twisted in the middle, bent into a hairpin shape and sharpened at both ends. Ram your brortches down over the string and bang them into the rick with a mallet, so that they hold the string down tight. Space the brortches at the intervals that common sense suggests (every thatcher has his own ideas). Now lay your next layer of straw so that it overlaps a little more than three quarters of the first layer and covers the strings. Peg this down too with string and brortches. Go on, layer after layer, until you get to the top.

You then have the problem of ridging. Make bundles of straw, about big enough to clasp in both hands, tie them tightly with string, and lay them along the ridge of the rick. Then lay long straw over these bundles so that it overlaps the top layer of thatch on both sides of the ridge. String and brortch this down on both sides. Or, better still, use hazel or willow rods instead of string here, and brortch them down. Make a pretty criss-cross pattern if you like. Of course with a round rick you don't have a ridge, but a point, and this makes the job much easier.

It is a very simple matter to fashion a conical cap of straw and fasten it down with brortches.

Buildings

You can thatch a building with a comparatively thin layer of straw laid on, much as in rick thatching, pretty well parallel with the slope of the roof. This makes a watertight thatch provided the pitch is steep enough, but in a wet climate it is unlikely to last more than 2 years.

Thick thatching is quite different (*see illustration*). The bundles of reed are laid on much nearer the horizontal, so that the coat of thatch is nearly, but not quite, as thick as the reed is long. Such a roof takes an enormous quantity of material, a lot of time, the right equipment, and a great deal of skill. But if made of true reed, it will last a lifetime. It is completely noise-proof, very warm in winter, and cool in summer: in fact it is, quite simply, the best insulation in the world. If you are building a mudhouse or barn, use rough, unsawn, and unriven poles for the framework of the roof, and they don't have to be seasoned. Thatch is flexible and if the timber moves it doesn't matter. The timber will season naturally in the well-ventilated conditions of a thatched roof, and generally last at least as long as the thatch.

Scything

The traditional scythe consists of a long blade of sharpened spring steel which is fixed by a simple ring and wedge to an ash frame (*see also p.148*) – and a well-swung, sharp scythe is a thing of great beauty. There is little that can beat the wonderfully satisfying scrunch of a really sharp scythe as it cuts the grass and magically peels it away to form a windrow. There are three quite distinct skills involved in scything: the first is how to set and sharpen your scythe; the second is how to perfect a smooth stroke to cut the grass; and the third is how to plan your attack to make best use of the lie of the vegetation.

Setting the scythe

The set of the scythe must be adjusted to suit the size and style of the user. Traditionally this was done by the local blacksmith who could heat up the fixing pin and bend it to just the correct alignment. Today you might have to persuade your local garage to make any adjustments.

Pin prevents blade from flexing

English scythe

Northumberland A-frame scythe

SCYTHES
English sneads (scythe handles) were traditionally steamed and bent (*above*) and the trusty tempered blades were made by the village blacksmith. Modern scythes come with lightweight alloy stales (straight handles) and 2 wooden handles.

SHARPENING A SCYTHE

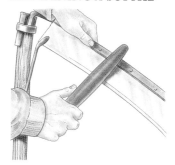 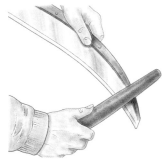 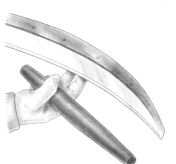 

1 Turn the scythe upside down; grasp the back of the blade firmly in your left hand so the sharp side is facing towards your right hand. Press the sharpening stone firmly against the blade.

2 Sharpen at the shallowest possible angle so the stone is just touching the band on the outside of the blade. Move the stone along the blade in small circles, from tip to handle.

3 The result is a slight burr or sliver of metal pushed down off the blade's edge. Feel for this carefully by pulling your fingers at right angles to the line of the blade across the underside.

4 Feel for the burr very carefully from the outer edge. Now, strop (swipe back and forth) the blade quickly to remove the burr, and do this along both sides of the edge in turn.

USING A SCYTHE

1 Test the sharpness of the edge by resting the blade on the ground as if to cut. Gently press down and move backwards and forwards. It should cut even the shortest grass easily.

2 The scythe is pulled around in a short arc with the back of the blade in firm contact with the ground. Do not use as a chopper, and remember it does not need to move fast to cut.

3 Throughout the stroke you must keep the tip of the scythe up and the heel of the scythe pushed firmly down using the right hand. You are looking for a gentle rolling action.

4 The outside of the blade is blunt, so slip the point between vegetation and a chosen tree and then slide the blade forwards. Continue around the plant until you clear a full circle.

LIE OF THE LAND
Plan your cutting to take advantage of the lie of the grass produced by wind and rain – every area is cut from left to right. You cut away from standing vegetation toward the cut area.

The objective is to have the blade of the scythe running parallel to the ground and lying comfortably on it when held in the scything position. Make sure the blade is very firmly fixed to the handles by getting a good fit between the wedge and the steel ring. You hammer down the wedge by turning the scythe over and resting the handle on the ground. The short stay should be a tight fit, being fastened to the frame by a single screw.

Sharpening your scythe

A good scythe blade is quite literally as sharp as a razor and will cut paper with ease. It is vital to keep the blade razor-sharp at all times – your scythe will not cut if it is blunt and your efforts will almost certainly damage the instrument as well as your self-esteem.

When it comes to sharpening the tool, be very careful to avoid touching the blade. The only time you will have to touch it is when you're feeling for burrs, followed by stropping with the stone, which simply takes off the burr and gives a nice clean edge (*see illustration, left*).

If in doubt I always recommend you try to watch someone with more experience first if you can. You will see the area you are sharpening become bright as the tarnished metal is worn away. But your scythe is nowhere near sharp until the bright area stretches all the way to the edge of the blade.

In tough grass you may have to sharpen the blade every 20 or 30 strokes – but you will waste much less time if you do this frequently, rather than waiting until the blade is really blunt, when you will have a tough time getting it sharp again. It may take 20 minutes or more to bring up a good edge the first time you get to work on a new blade. Sharpening is the hardest skill of all but is the essence of good scything. If you cut yourself, run your hand under cold water immediately, and you will find it heals very quickly and neatly, since the blade must have been good and sharp!

Cutting grass

Set your scythe so the blade runs parallel to the ground comfortably when you hold the handles. Take advantage of the lie of the grass produced by wind and rain. Every area is cut from left to right (*see illustration, left*). It is best to experiment at the beginning with short, firm little strokes backwards and forwards to give you that first feeling of how a sharp blade will sweetly cut the grass. You should gradually extend the length of your stroke, always remembering to keep the back of the blade pressed against the earth. You will only tire yourself and develop bad habits if you try to take too big a swing at first.

You can cut along grand by just scything across about 18 inches (45 cm) of grass in each stroke, and by taking the next stroke just 1 or 2 inches further on. You need to keep a firm grip of both handles and avoid any temptation to "chop". As soon as the scythe loses its edge, get to work with your stone. Always keep a careful watch for stones or posts well to the left of the row you are cutting. The tip of your scythe will extend maybe 3 feet (90 cm) and it is vital not to damage this important part of the scythe.

Carry your sharpening stone with you, preferably in your back pocket. These stones get lost, "walk", or break very easily. I always wear decent-length Wellington boots, even in hot weather. If you cut through a nest of ants, they make very uncomfortable companions inside ordinary shoes. You have been warned! You will quite often come across frogs when you are scything, so try to keep an eye out for them.

Tips for the tip

Beginners frequently have problems because they have not realized how important it is to have the scythe very sharp at the tip. You will always have to give this part of the blade special attention. The tip is the first part of the blade to come into contact with the grass, and if it pushes it over rather than cutting it, your stroke will be much, much harder. When a tip breaks off you must grind down the end of your scythe to make a new tip – a scythe with a broken end simply will not work.

Special jobs for the scythe

It is a pleasant and important task to scythe around young trees in May. Once is usually enough to keep the grass down, provided you cut right to soil level. Using a sharp scythe is a pain-free way of clearing brambles and small scrub. The scythe will slice through those bramble stems quite easily with firm pressure and a good edge. Use your scythe to cut weeds before running a rotavator over the vegetable garden. This way you can rake off the vegetation and dump it straight on to the compost heap in a matter of minutes. For trimming rampant grass edges along turf around the vegetable patch, you cannot beat the scythe for speed and simplicity.

Working in Wood

Being able to work in wood is a vital skill for the self-supporter. It may be making a small box to store apples, repairing stalls for animals, building a hen house, or making a roof. Hardly a day will pass without some activity connected to woodwork. The modern cordless powered screwdrivers and drills are absolutely indispensable. Combined with crosshead posidrive screws, these tools have revolutionized woodwork. No longer is it necessary to drill pilot holes for woodscrews. The new design of screws simply make their own holes. But remember, a screw which has no smooth length to the shank will not pull two pieces of wood tight together. Always keep a decent supply of screws and fastenings, glues, mastics, paints, and thinners. Generally, it is much easier to buy screws by the box, and you will be surprised how fast they get used up. There is a vast range of screws and nails available: nails are usually sold by weight.

QUICKFIT CLAMPS

A quick and easy alternative to the traditional (and heavy) vice, quickfit clamps are a relatively new product and you will find that having a couple always to hand means you have effectively another pair of hands. You can clamp pieces of wood firmly and quickly to your bench (or simply a large plank if you are working on a building site) and you can clamp pieces together easily to assist with fixing. The one-hand quick-release mechanism is very handy when you have several tasks on the go at the same time. I also keep a much larger pair of what are called "sash cramps", which I tend to use to hold together large pieces of work during fixing and glueing. Like all tools, these do not deteriorate over time, so having a couple to hand is usually a good investment.

WOOD SCREWS AND NAILS

Here are the essential screws and nails needed for any woodwork projects. Keep a selection of these handy in your tool box.

Countersunk woodscrew
Traditional workman-like fixing for woodwork. It needs a hole drilled first to avoid splitting the wood plus countersink to recess head. Use with a screw-cup for added strength.

Roundhead screw
Another traditional fixing which requires a hole drilled first and leaves the head proud.

Posidrive screw
Now almost universally used with cordless, powered screwdrivers. The beauty of these screws is that no pilot hole is required. The fixing will also grip tight even if screwed into endgrain wood.

Coach screw
The coach screw is a very strong fixing for outdoor work. It requires a pilot hole and must be tightened with a spanner. Use the galvanized versions for long life.

Lost head nail
The "lost" element refers to the small round head that is easily knocked into the work, especially useful for floorboards. You might will need to use a nail punch to push it below the surface and make it invisible.

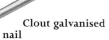

Clout galvanised nail
A nail for external work such as for fixing roofing battens. These nails also provide a cheap way to fasten slates. Clouts come in all shapes and sizes – both length and thickness. Make sure you know exactly which size you need, especially if fixing roofing where you do not want sharp ends going through your felting.

Flat head wire nail
The standard nail for fixing. It also comes in galvanized form for outdoor use. You can't have too many of these handy fixers.

Cut clasp nail
Often used for fastening floorboards, this square section nail is excellent for avoiding splits in the wood.

Roofing nail
A heavy twisted and galvanized nail which will not pull out of roofing in wind or gales. Use for roof sheeting.

Masonry nail
Toughened steel nail made for fastening directly into masonry. Can be difficult to use: watch out for sparks and splinters.

Ribbed flooring nail
A nail used to fix hardboard and chipboard flooring down so it will not loosen and squeak under use.

Staple
Comes in different sizes and is usually galvanized for fixing wire to fencing – such as chicken wire to posts for a run.

Frame fixing
A wallplug type fixing in many sizes which is simply hammered home into the correct sized hole made by a masonry drill.

Wallbolt
The ultimate fixing for fastening into stone or masonry. Tremendous load-bearing capabilities. An expanding masonry bolt requires the correct size of masonry bit to make a good joint.

DOWEL JOINTS

For a neat, strong, and relatively easy joint between two pieces of wood, you can now buy dowel kits from most good hardware shops. The kit has short lengths of different diameter hardwood dowelling rod, a set of wood drills exactly the right size to take each size of dowelling rod, plus a small plug cutter so you can seal and disguise dowelling holes if you need to. To achieve workable dowels, holes of exactly the right size must be drilled in each piece to be joined.

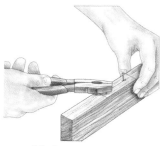

3 Drill holes in the piece to be joined and insert dowels with glue to half their length.

DOWEL JOINT

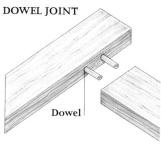

Dowel

Positioning must be exact so the dowels (round pins) match up in correct alignment within the glued joint.

4 Now marry up the fixed dowels to the end grain of the marked piece. Mark centres exactly and use a set square to drop down to centre line.

1 Cut wood to the exact size for a flush fit against the piece to be joined. Use a set square and make sure you cut the correct side of the drawn line.

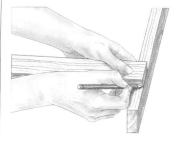

5 Now drill out the two final dowel holes, taking care to keep the drill perfectly perpendicular to the end grain.

2 Mark the centre of the end grain with a sharp scribe.

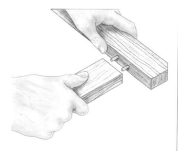

6 Hand-join the two bits of wood and tap the dowels home with a little glue, and your joint is complete. Clamp if necessary until the glue is set.

MORTICE AND TENON JOINTS

The basic joint for the "joiner" who graduates from using dowels is the mortice and tenon joint. The joint, or variants of it, would be useful for tables or gates. With a little practice and care you will soon get the hang of it and produce strong and neat results. Accurate marking out is vitally important, and you must learn to decide whether to cut lines off or cut inside them. Think about this carefully or your tolerances will be too great. Always use a good glue to finish the job.

MORTICE AND TENON JOINT

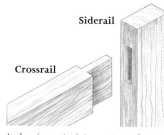

Siderail

Crossrail

A slot (mortice) is cut into the siderail exactly the correct length and width to fit a "tenon" cut out of the endgrain of the crossrail.

1 Mark out the length of the mortice carefully using the crossrail. Use a set square throughout to extend lines across the rail.

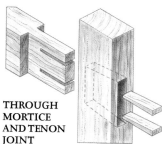

2 Mark off the length of the tenon (again with a set square). Don't make it too long or you may go through the back of the mortice on the siderail.

3 Carefully scribe the marks for the edges of the tenons at each end of the cross rails. Cut these out accurately with a sharp saw and a firm vice. Pare down with a sharp chisel if the saw goes awry.

4 Mark out the correct width for the mortice. With a sharp chisel and wooden mallet, cut it out bit by bit. The mortice is better too small, rather than too large.

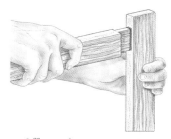

5 Offer up the tenon into the mortice to check for a tight fit. Use your chisel again if you need to perfect the fit. You will see why a smaller mortice is better.

THROUGH MORTICE AND TENON JOINT

Mortice right through the siderail for a really strong joint. Cut slots into the tenon; make correct-size long wedges to hammer in from outside when joint is positioned.

A good quality power screwdriver will last for many years if you treat it well and use the battery sensibly. These batteries must not be recharged when they are only partially discharged. If you do this you will destroy the ability of the battery to store charge. So always run the battery down until it is absolutely flat and you will get many years of use from it.

Choosing your wood

For most of your tasks you will be selecting your wood from the builder's merchant or timber yard. And I mean selecting: do not trust the yard man to choose timber for you or you will certainly end up with all the lengths left by others. Remember, too, that each timber yard will have

its own standards. Some managers will be very fussy about the timber they will accept from suppliers; others simply do not give quality of timber any priority and get whatever is cheapest. Try to use a builder's merchant or timber yard that sets reasonably high standards if you can find one: it will save you a lot of trouble, even if it costs a bit more.

As a quick checklist, here is what to look for in your timber.

1 Wood that is straight. Use your good, old "mark one" eyeball, that is, inspect it thoroughly for warps; you may be horrified by what you might have been fobbed off with.

2 Wood that has no splits.

3 Wood without many knots in the wrong places.

4 Wood that has, if possible, the grain running in a way

CHOOSING WOOD

Learning to understand the grain of wood and how it will affect its future importance is critical. All wood shrinks as it dries out and "seasons". If the grain runs straight through the wood (below, left) then the wood shrinks evenly and will neither warp nor split. If the grain runs across or diagonally, then the wood will warp (below, right). Choose wood which looks like the former rather than the latter.

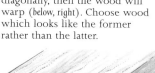

MAKING A BOX

This is a simple but effective method of constructing a strong wooden box which will come in handy for all sorts of useful jobs on your self sufficient holding. Use good quality plywood and well chosen 6-by-1 pine planking. For best results add round beading around the inside and up the insides of the corners.

1 Cut your 6-inch (15-cm) planking to accurate lengths, using a set square to make sure the ends are exactly at 90 degrees to the sides. Next, cut the plywood base to the correct size.

2 When you cut the plywood base, make sure again that you have corners which are accurate right angles. Cut the base too big rather than too small so you can plane down later.

3 Take off any rough edges with sandpaper or a sharp plane. This applies to saw edges as well as the sides of the planking. You'll need to apply some pressure as you sandpaper.

4 Now screw the ends of the box on to the plywood base. Use countersunk screws either galvanised or brass for outdoor use. These are purpose-made for weathering.

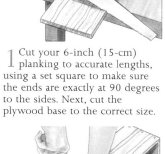

5 The sides can now be screwed into the end grain of the ends which have already been fixed. Posidrive screws work well for this and will pull themselves into the wood.

6 Make sure corners fit exactly and sand off any rough or overlapping edges. You have to work quite hard at this as the corners are quite tricky to make contact with.

7 Complete screwing on the base and use plenty of screws to make the job effective. Battery powered screwdrivers are best for this. Cut handle holes if necessary using a drill and a power jigsaw.

8 Cut hand holes as needed. First use a drill to make a hole large enough to accept the blade of the jigsaw. Then carefully cut out the hand holes, smoothing down with sandpaper afterwards.

that will prevent future warping. This is particularly important for critical items like door posts.

5. Wood that comes in quarter-sawn planks for any kind of quality joinery. When you look at the grain end-on, it should run at right angles to the side of the board all the way across. Remember that the businessmen are trying to get as many planks as they can from each tree, and this is a much more important consideration for them than cutting for stability.

Making a lathe

A simple wooden foot-powered lathe can be made very easily, and although it works slowly it works as well as any other lathe for wood-turning. Plant two wooden uprights in the ground, or if you live indoors attach them to the floor, about 3 feet (90 cm) apart; 6-inch by 4-inch (15-cm x 10-cm) posts are ideal. Nail a block of wood to each post, just at hand height, and on a level with each other. Drill a hole in each block big enough to take the ends of

the "stock", which is the piece of wood to be turned. (You will have to whittle down the ends of this with a knife to make it small enough to fit the holes.) Arrange a simple foot pedal below. This can just be a piece of wood, held at one end by a pin which is supported on two short stakes. Then arrange a bendy, horizontal pole of ash, or other springy wood, above the contraption, so that one end sticks out and can be bent up and down. You can use trees, stakes, or, if indoors, the rafters to support this pole.

Tie a piece of rope to the foot pedal, take one turn with it round the stock, tie the other end to the end of the whippy pole. Nail another piece of wood across the posts next to your stock to rest your chisel on, and you have a lathe. Depress the pedal and the stock turns one way, release it and the pole above your head straightens and turns it the other way. You only make your cut when it's turning the right way, of course. Wood turning is skilled work (*see illustrations below*). If you can, go and watch a skilled craftsman at work.

TURNING A BOWL ON A LATHE

These pictures show a bowl being turned on a simple lathe powered by electricity, but you can turn a bowl in the same way on a treadle lathe, or, rather laboriously, on a chair bodger's

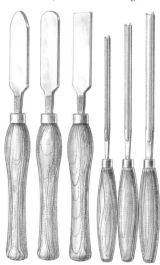

pole lathe. If you use the latter, you must replace the stock with a rod fixed to a chuck to which the bowl can be attached. For the heavy work of removing unwanted wood you need three gouges of differing thicknesses (above right), and for the more delicate shaping and smoothing you need scrapers (above left). Never press hard with any tool, particularly a gouge. If they stick you are in trouble. Keep your tools sharp.

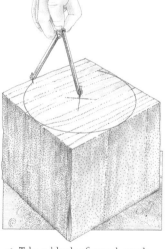

1 Take a block of wood, mark the centre with a cross, and draw a circle with a compass slightly larger – say ¼ inch (6 mm) – than the intended diameter of your bowl.

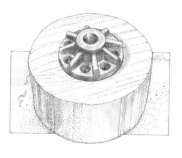

2 Cut roughly around your circle with a saw. Then establish the midpoint for your chuck and screw it on evenly.

Use short, strong screws because the base of your bowl must be thicker than your screws are long. Your work will be ruined if you come down to a screw when shaping the inside.

3 Round off edges with a large gouge.

4 Shape the outside with a smaller gouge. Use the handrest and keep the gouge moving slowly along it.

5 Smooth off the outside with your scrapers. Then, keeping the bowl on the lathe, rub with sandpaper, which will give the wood a gloriously smooth finish.

6 Move the handrest to work on the inside. The unbreakable rule for hollowing wood on a lathe is: begin at the outside and work towards the centre. Start with a gouge, then scrape with rounded scrapers only, and finally sandpaper it smooth. Remove the chuck from the lathe, unscrew it, and fill the holes with plastic wood. Polish it all with beeswax and glue felt on the bottom.

283

Household Items

SOAP

The first lion I ever shot had been eating a friend of mine's donkeys in Africa. It had a thick layer of fat on it, and my friend's mother turned this into soap. She did it by the simple method of boiling the fat with caustic soda. It worked, but was pretty rough stuff.

The chemistry of soap-making is to boil an alkaline with fat, which is an acid. The alkaline, or lye, as soap-boilers call it, can be practically any alkali, and caustic soda will do. But there is a simple way of making your own lye. Knock some holes in the bottom of a barrel, lay some straw in the bottom, fill the barrel with wood-ash, and pour a bucketful of cold water on top of the ash. Pour on a bucketful every 3 or 4 hours on the first, third, and fifth day. The water that drips out of the bottom of the barrel will be lye.

Now, to make soap, take your fat and clarify it by melting it in a slow oven, straining it into cold water and then skimming it off the water. If you haven't shot a lion, practically any fat will do: dripping, lard, chicken fat, goose fat, and so on. Melt the fat again and let it cool to luke-warm. At the same time, warm your lye to luke-warm, too. Then, very slowly pour the lye into the fat (if you pour it too fast it will not mix) and stir it very gently with a wooden spoon. When the mixture begins to drip from your spoon like honey, stop pouring. If you want to make your soap stronger, pour in a solution of borax and water (8 tablespoonfuls of borax to 1 pint /0.5 litre of water) and a dash of ammonia. To half a gallon (2 litres) of soap mixture add 1 pint (0.5 litres) of the borax solution and half a cup of ammonia. Put a board over the mixture, cover it with a carpet, leave until the next day and cut it.

If you want a soap that will make you and your friends smell good, take:

1 lb (0.5 kg) good fat or tallow

1 cup of olive oil

1 cup of peanut oil

½ cup of water with 2 tablespoons lye in it

1 cup of water with perfume in it

To perfume the soap, use the aroma of your choice from a range of essential oils bought from the chemist. Use about 3 tablespoons of it. You can also make your own perfume out of lavender, rosemary, lemon balm, or a score of other flowers or herbs you can grow yourself, in which case you would probably add more. Melt the fat, add the oils and the scent to it, and warm to 90°F (32°C), stirring all the time, of course. Meanwhile, mix the lye and the water and pour it into the fat and oil mixture – and don't stop stirring. When the mixture thickens, pour it into moulds of any fancy shape you like.

Saddle soap

To make saddle soap take:

6 cups of tallow

1 cup of lye

2½ cups of water

Heat the tallow to 130°F (54°C). Dissolve the lye in the water and then let it cool down to 95°F (35°C). Once it reaches this temperature, pour it slowly into the tallow, stirring all the time. Just before it is ready for moulding, pour in 1 cup of glycerine and stir.

SUGAR

From sugar beet

Cut the tops off your sugar beet and press the juice out of them any way you can: with a cider press, a car jack, or an old-fashioned mangle. Boil the juice until all the liquid has evaporated and you are left with unrefined sugar. Refining it is a complicated process involving lime and carbon dioxide. Anyway, it would be madness to refine this further, for unrefined sugar is nourishing and good for all the purposes of sugar, while refined sugar contains 99.9 percent sucrose, absolutely no vitamins, and nothing else that is of any use at all to body and soul.

From sugar cane

Sugar cane must be thoroughly crushed to produce syrup. Cane is tough stuff and full of long fibres, so you either need a lot of strength and a mortar and pestle, or a steel crushing mill. Put the syrup you've managed to extract into a copper boiler, over a fire that you can fuel with the spent cane. Boiling turns the syrup into what in India is called "gor", which is unrefined sugar. As I have said above, it is a waste of time to refine sugar any further and it is much better for you like this.

Maple sugar

To make this you must tap the sugar maple tree in the chilly month of March, by drilling the trunk and driving in a "spile", which is a short tube you can buy or make yourself out of bamboo, willow, sumac, elder, or anything you can hollow out. Hang a container under the spile (an old can will do, or a bucket, or a plastic bag) and cover it over to keep insects out.

As soon as the sap runs, carry it to the "arch" (leaving it too long will spoil it). The "arch" is a boiler placed over a wood fire that needs to be kept blazing by a strong draught. The arch must be out of doors as a great deal of moisture is given off. Don't let the sap get more than a couple of inches deep. Keep the level by pouring in more sap. It is an advantage to have two boilers. Use one for the fresh sap and keep ladling the partially boiled sap into the other from which you "syrup off", meaning take the syrup.

Skim the scum off from time to time and watch constantly to see that the sap doesn't boil over. If it starts climbing up the pan, add some fresh sap, or drop some creamy milk on the climbing froth, or draw a piece of fat across the bubbles. Test the sap's temperature with a thermometer. When it is boiling at 219°F (104°C) it has turned into syrup. Strain it off into jars, cover while hot, and put away to cool. This is maple syrup – delicious with pancakes, ice-cream, in fact, anything.

If you want sugar, go on boiling until the temperature is 242°F (117°C). If you pull a spoon out and the drip forms a thin, spidery thread, that is boiling enough. Remove from the fire, leave to cool for a few minutes, then stir with a wooden spoon. When the syrup begins to crystallize, pour it into moulds and you have sugar.

SALT

If you live near the sea you can make salt by simply boiling and evaporating seawater. You can use driftwood for fuel — and nowadays the oil spillage that coats most driftwood means it gives even more heat. A mobile iron boiler, such as those you boil pig-swill in, is ideal. Never use a copper boiler. The copper and salty seawater will react with one another.

PAINT

Very good paints can be made from a mixture of sour milk, hydrated (slaked) lime, and any coloured earth pigment that you can find. The lime and the sour milk must have neutralized each other, and this can be tested with litmus paper: if the paper turns red, add more lime; if it turns blue, add more sour milk. The pigment you add to this can be any strongly coloured earth, sediment or clay. Dig it out of the ground and boil it in water several times, each time in new water. Strain off the water and dry the sediment in a warm place. Pulverize it as finely as you can and store. Mix this powder with the milk–lime mixture until you get the colour you want.

PAPER

It is possible to make paper from any fibrous plant, or wood, cotton, or linen rags. Nettles, flax, hemp, rushes, coarse grass, and tall fibrous plants like *Tagetes minuta* all make very good paper. Ret (moisten) the plants first by soaking them in stagnant water. Then chop them up as small as you can into, say, ½-in (1-cm) lengths. Put the chopped material into a vat and cover with a caustic soda solution made up of 2 dessert spoonfuls of caustic soda per quart (1 litre) of water. Boil until the material is soft and flabby. Put it in a coarse sieve and drain. Hold the sieve under the tap, or plunge it up and down in a bath of water. This will clear the pulp away. If you want white paper, soak the fibre you now have left in a bleach solution overnight. If left unbleached, the paper will be the colour of the material you are using. Drain the bleach off through a fine-meshed sieve (you don't want to lose fibres).

Next you must beat the material. You can do this best with a mallet or any kind of pounding engine. When you have beaten it thoroughly dry, add some water and continue beating the pulp. A large food-mixer or a large pestle and mortar will do very well for this stage. Put some pulp in a glass of water occasionally and hold it up to the light: if there are still lumps in it, go on beating. If you want to make interesting papers, don't beat too long and your paper will have fragments of vegetation showing in it.

You make the paper on moulds, which can be simple wooden frames covered with cloth. Cover the moulds with a thin layer of pulp by dipping them into it and scooping. As you lift the mould out of the pulp, give it a couple of shakes at right angles to each other. This helps the fibres to "felt" or matt together. If you find that the "waterleaf", which is what your sheet is called, is too thin, turn the mould upside down, place it in the vat, and shake the pulp off back in the water. Then add more fibre to your vat. Turn the mould upside down on a piece of wet felt and press the back of the cloth to make the watersheet adhere to the felt. Take the mould away and lay another piece of wet felt on top of the watersheet. Repeat the operations with another watersheet. Finally, you need a press. Any kind of press will do. Make a "post", which is a pile of alternate felts and watersheets, and put the post in your press. Press for a day or so, then remove the paper from the felt and press just the paper sheets. Handle the paper very carefully at this point, and then lay it out on racks to dry.

RESIN, ROSIN, AND WOOD TAR

Long-leaf pine, maritime pine, Corsican pine, American balm of Gilead, cedars, cypresses, and larches can all be tapped for their resin. To do this: clear a strip of bark, about 4 inches (10 cm) wide and 4 feet (1.2 m) high, off a large tree. This is called a "blaze". With a very sharp axe, take a very thin shaving of true wood off at the base of the blaze. Drive a small metal gutter into the tree at the base of this cut and lead the sap, or resin, into a tin. Every 5 days or so, freshen the cut by taking off another shaving. When you can get no more out of the first cut, make another just above it. Repeat until you have incised the whole blaze (this may take several years). Don't tap between November and February. If you grow conifers for tapping, clean the side branches off the young trees so that the trunks are clean for tapping.

If you distil resin, that is, if you heat it and condense the first vapour that comes off it, you will get turpentine. "Rosin" is the sticky stuff that remains behind and it is good for violin strings, paints, and varnishes. If you heat coniferous wood in a retort, or even just burn it in a hole in a bank, a black liquor will run out of the bottom. This is wood tar, and it is the best thing in the world for painting boats and buildings.

CHARCOAL

Charcoal is made by burning wood in the presence of too little oxygen: you set some wood alight, get it blazing well throughout its mass and then cut off the air. I have tried many ways of doing it and the simplest and best is to dig a large trench, fill it up with wood, and set it alight. When it is blazing fiercely, throw sheets of corrugated iron on to start smothering the fire and then very quickly (you will need perhaps half a dozen helpers) shovel earth on top of the iron to bury it completely. Leave for several days to cool, then open up and shovel the charcoal into bags. You can use the charcoal for cooking fuel or making bricks.

Making a Pond & Fish Farming

MAKING A POND

If you are going to keep ducks, or if you want to try the highly rewarding process of fish farming, you will need a pond. You can just dig a hole, but if the bottom or sides are porous it will probably be necessary to puddle clay and tamp it in so as to form an impervious sheet, or else bury a large sheet of thick plastic.

Simply piling earth up in a bank to form a dam to impound water seldom works. The fill material may be too porous and "piping" will occur, meaning water will seep through and erode a hole. Or the material may contain too much clay and there will be drying, shrinkage, and cracking. If the soil is just right, and well compacted, and an adequate spillway to take off the surplus caused by rainwater is constructed, a simple earth dam may work, but where there is doubt the dam should be made of porous soil with puddled and tamped clay embedded in it. Nowadays plastic sheeting is sometimes used instead. If your pond is for fish farming then good topsoil should be placed in the bottom for plants to grow on.

FISH FARMING

Fish are marvellously efficient producers of high protein human food: far better, in fact, than other livestock. This is because they don't have to build a massive bone structure to support their weight (the water supports it), and they don't have to use energy to maintain their body heat (they are cold-blooded). In the tropics, particularly in paddy-growing areas, they are a major crop. Modern commercial fish farming, in which only one species of fish is fed on expensive high protein in water which is kept weed-free with herbicides, is ecologically unsound and requires absurd inputs of expensive feed or fertilizer. We should all start experimenting with water ecosystems which achieve a proper balance of nature, and in which a variety of fish species can coexist with a cross-section of other marine life, both animal and vegetable.

Strangely enough in the 16th century the matter was far better understood – even in England. At that time a writer named John Taverner wrote that you should make large shallow ponds, 4 feet (1 m) deep and more, and keep them dry one year and full of water the next. When dry, graze them with cattle, and when wet, fill them with carp. The ponds grow lush grass because of the sediments left by the water, and the carp benefit from the fertility left by the cattle. This is the true organic approach to husbandry. You should have at least two ponds so that there is always one full of fish and one dry. Drain the wet pond dry in late autumn, and take the best fish out then to put in your stewpond near the house, where they are ready for eating. Put a lot of young fish in your newly flooded big pond.

Carp

Carnivorous fish, such as trout, are poor converters of food into flesh. Vegetarian fish are far better. This is why the aforementioned Taverner and the monks of old in Europe had carp in their stewponds. Carp will give you a ton of fish per acre per year without any feeding provided they are in a suitable pond. The way the monks farmed them was to let them breed in larger ponds, but then to catch them and confine them to small stew ponds near the house in the autumn.

The stewponds were just deep enough to stay ice-free and the carp were therefore easy to net. As well as being vegetarian, carp are healthy, quick-growing, and they can live in non-flowing water. They need half their food from natural provenance, and can be encouraged by a certain amount of muck or rotting vegetation dumped in the water. This is transformed into the sort of food carp eat by bacterial action, but they will also eat oatmeal, barley, spent malt, and other similar food.

The Hungarian strain of the Chinese Grass Carp has been tried in England, with success. In China these fish grow up to 100 pounds (45 kg) in weight: in England 30 pounds (13.6 kg) is considered a good fish, but they are fine converters of vegetable food. Unfortunately they need temperatures of 122°F (51°C) to breed, and so are propagated in heated tanks and released out of doors, where they flourish.

Tilapia

The best fish of all for fish-farming are the African Tilapia, but because they are tropical fish they need warm water. Nevertheless, putting yourself out for them may well be worthwhile. Research has shown that the average family could provide all its animal protein requirements in a 3,000-gallon (13,640-litre) covered and heated pool full of tilapia. The water should be about 80°F (27°C): less than 55°F (13°C) will kill them.

Tilapia mossambica, which is one of the best of the many species, can be bought from pet shops. The hen fish produce about 25 to 30 young, which live in their mothers' mouths for the first period of their lives, and the hens bring off several broods a year. Much of their food can be supplied free with a little labour by incubating pond water, slightly fertilized with organic manure, in tanks. After 3 weeks or a month, carefully pump this water into the tilapia pond with the organisms that it contains. The incubation tanks should be partially roofed with glass, but access for mosquitoes and other flying insects should be provided.

In temperate climates *Tilapia mossambica* can be kept in heated pools, and they don't require constantly running water. A combination of solar heating and wind or electric heating has proved successful for growing them in North America. They will produce 2 tons of good meat per acre per year. When adult they will feed on algae or any vegetation you like to put into the water (within reason), or they will eat oatmeal. When young they need protein, which can be supplied in such forms as mosquito larvae, maggots, worms, or as fish meat, or blood meal. They are probably the most delicious of all fish to eat.

The All-Purpose Furnace

Firewood is a renewable resource, and the best solar energy collector in the world is a stretch of woodland. Woodland cut for firewood should be coppiced (*see p.136*). In other words, the trees should be cut right down every 10 to 15 years, depending on how fast they grow, and the stumps left to coppice, or shoot again. Cut over systematically in this way, two or three acres of woodland will yield a constant supply of good firewood and other timber. To burn wood effectively and economically requires several things. The wood must be burned on the floor of the furnace, not on a grid. The fire must be enclosed and there must be a means of carefully regulating the draught. A huge, open fire is a romantic thing, but all it does is cheer the heart, freeze the back, and heat the sky. Where wood is in limitless supply it may be justified, but not otherwise.

It is an advantage to burn wood in a dead end, admitting air from the front only. A tunnel with the back walled off is ideal. Logs can be fed into the dead-end tunnel and lit at the end closest to the door, and the fire then slowly smoulders backwards into the tunnel. The draught control should be such that you can load the tunnel right up with dry logs, get a roaring fire going, and then actually put it out by cutting off the air. If you can feed your furnace from outside your house you will avoid a lot of mess inside. And if you can organize things so that your furnace can take long logs, you save an awful lot of work sawing.

Now any decent economical furnace should be capable of doing at least four things: space heating, oven baking, hot-plate cooking, and water heating, and if it can smoke meat and fish as well, so much the better. We built a furnace that would do all these things on my previous farm, and as the farm was called Fachongle, I called it the Fachongle Furnace. But don't try to build one like this

unless you know you can get, for not too much money: firebricks, a cast-iron plate big enough to cover the whole furnace, and a massive cast-iron fire door.

BUILDING A FURNACE

We built a firebrick tunnel, 4 feet (1.2 m) long, inside the house. It is bricked off at the back, but the front falls 4 inches (10 cm) short of the exterior house wall. The house wall there is lined with firebricks. On either side of the tunnel we built a brick wall slightly higher than the top of the tunnel. The bases of these need not be of firebrick, but the tops must be able to withstand the heat. On top of the two outer walls we laid a steel plate. (This has since warped slightly, which is why I advise you to get a cast-iron plate.) This goes from the back of the furnace right to the wall of the building. On top of the steel plate away from the wall of the building we built an oven, and at the very furthest end from the wall we built a chimney. We knocked a hole in the outer wall and in that set a furnace door with firebricks. The furnace is fed through this door, and the heat and smoke has to come back to the front end, curl up through the 4-inch (10-cm) gap, hit the iron plate, curve back, and go under the oven and on up the chimney. We built a back boiler into the back wall of the tunnel, and the pipes from it come back between the tunnel wall and the outer wall, and then come out through two holes in the latter. We partially filled the cavity between tunnel and outer walls with sand to insulate and store heat.

Managing your store of firewood is an essential part of operating any woodburning stove. The wood must be dry, that is to say, it must have been cut for at least 12 months before burning. We constantly add to the woodshed, keeping two piles of wood: one new, and one (which is 12 months old) for use now.

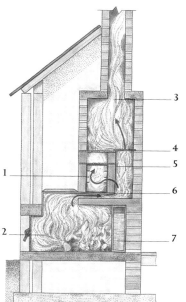

A VERSATILE FURNACE YOU CAN BUILD YOURSELF
We designed and built the Fachongle Furnace with the idea of getting as much benefit as we could from burning wood. The furnace gives us: space heating for a large area, a lot of very hot water, a hot plate and oven for cooking, and a smoke chamber. We burn the wood in a tunnel made from firebricks. The back is closed off and contains the back boiler for water. The front comes almost up to a hole in the house wall where we have built a fire door, so that the furnace is fed from outside. We built brick walls on either side of the tunnel right up to the house wall, and rested a steel plate across them. The front of the plate serves as a hot plate, and over the back we have built an oven. There is a slit in the steel plate so that heat circulates right round the oven. At the very back is the chimney, which widens out above the oven to form the smoke chamber. Heat from the fire comes forward along the tunnel, curls up under the hot plate and oven, and on up the chimney. It is quite likely that your requirements will be different and will necessitate a modified design.

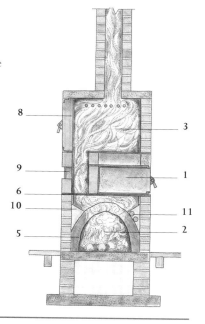

1 Oven	7 Back boiler
2 Fire box	8 Smoking box door
3 Smoking box	9 Access door
4 Damper	10 Sand filling
5 Firebrick	11 Water pipes
6 Flue passage	

"*Well, what have we learned? What has altered since I wrote* The Fat of the Land? *The first big change is that now we are not alone. When we found ourselves becoming self-sufficient in food we were probably the only family in England living in this way. Now there are hundreds doing it — and tens of thousands who would like to. Other changes are the soaring price of land, the drying-up of the empty cottages [to rent], of cheap horse-drawn and other old implements and the even greater intrusion of the State into every corner of private life.*

But the thing that I have learnt is that it was a mistake to try to live like this alone. We have tried to do too much, have worked too hard, have forgotten what it is to sit and listen to music in the evening, or read something for pleasure or to engage for hours in amusing and interesting conversation.

If a number of families could get together and then cooperate in a flexible sort of way — 'A' keeps cows and keeps us all in dairy products, 'B' grows corn and keeps us in flour, animal food, and malt, 'C' keeps pigs, 'D' keeps poultry: something like that — then I think they could lead very good lives indeed."

JOHN SEYMOUR FAT OF THE LAND 1976

CHAPTER ELEVEN

THINGS YOU NEED TO KNOW

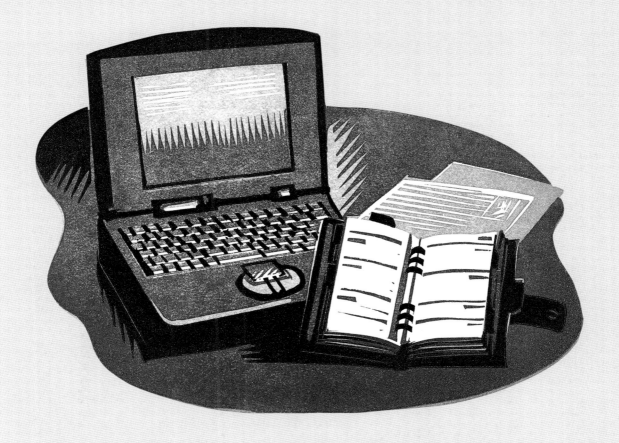

Becoming a Self-Supporter

Making the break

One thing I have noticed is that the settlers who fail and have to go back to the city are generally the ones who tried to adopt, unchanged, the patterns of traditional husbandry in the areas they settled in. In this they are not aiming for self-sufficiency, indeed they will buy much of their food from normal shops, like other farmers, and try to make money to do it by selling just one or two products. This simply cannot work. Traditional farming is up against the ropes, the farmers who inherited their land and their occupation from their families are finished. True it will take a long time before they finally break up and disappear – and it will be an uncomfortable and sad process – but the industrial world that makes the rules has no time for traditional farmers.

I cannot tell you what patterns of activity will work for you in combining a self-sufficient lifestyle with money-earning activities, but I can give you some examples:

A retired couple (the man from engineering) have bought a house and roughly a dozen acres. The wife milks the 6 Jersey cows and sells the milk to be made into cheese. The man keeps a herd of pigs in his woodland and buys some of the whey back to feed them. Besides growing most of their own food and brewing their own beer, they have a small pension. They work hard, have an enjoyable life, and live well.

A divorced lady lives on 5 acres of rough land with her 13-year-old son. She is a successful potter and earns the money she needs from her work. She can just afford to run an old van. She milks one cow and rears calves on another, kills one young steer a year for the deep freeze, keeps two sows, kills two fat pigs a year for bacon and ham, makes all her own butter, cheese and so on, sells some, sells a few of her fat cockerels and ducks, and buys wheat from a local farmer and grinds it by hand to make her own bread. The son is a great help being utterly at home on the farm, a budding carpenter, and a good shot.

A salesman, who retired early, and his wife bought a very remote 25-acre holding. They have done up the house, and fenced and improved the land. They keep a herd of goats and some poultry from which they get all their milk and meat. They run a fine garden, and the man works part time as an engineer for a local firm. The wife does most of the manual work on the holding and loves it.

A music teacher in schools bought a 20-acre farm. He and his wife rear calves, milk a cow, run the farm very efficiently (taking good advice from their neighbours), and grow most of their own food. The man teaches the piano to local kids to earn extra cash. He turns down requests for more work because he has already as much as he wants to do and does not need much money anyway.

A group of people each put £10,000 into a kitty a few years ago and bought a good-sized farm, around 50 acres, with a large house on it. Most of the originals have left now but others have come and they all live in the big house and work the farm co-operatively. Having had enough capital at the start they have good machinery and farm well. They grow strawberries to sell, run a big market garden, make some pretty awful cheese which they eat themselves, run a shop which sells their produce, and have a stall at the local market. They seem to live very well although the house is not neat and tidy. There are lots of happy kids living there.

Now I could go on like this for pages. The lesson to draw from this is that all the successful recruits to self-sufficiency have some skill or business which will provide cash for part of their living. And none of them have debts.

Checklist

It is useful to make yourself some kind of checklist which might include such items as:

Why – Get clear in your mind your major motivations for making a change.

When – There is a marvellous passage in Thoreau's *Walden* in which the writer relates that a certain young man, wishing to become a poet, went off for several years first to make money to support himself. Thoreau remarks "He would better have gone up the garret at once". In other words, his years as a money grubber would have made him completely unfit to write any poetry. His Muse would have been killed, and no amount of money could bring it back. If you have got to buy land, then perhaps the sooner you get your foot in the door, the better.

Where – You may have to travel around extensively before you get a real feel for what you are looking for. Finding your ideal smallholding is a matter of good research, cool judgement, and a great deal of luck. Many of the best smallholders find their ideal spots by some process of serendipity. Maybe they get lost on a trip to friends or their car breaks down or they see a lucky advert. You just have to leave yourself open to this kind of adventure. And do not forget that you will have to live within the local culture as well as on the local land – try to take a dip into this by going to local pubs, markets, and shops. Talk to people and see what it feels like. A local health food shop or library will have many notices of local events and services – that's a useful guide to what life there may have to offer.

Who with – Well, you could be alone, with a spouse and family, with one trusted partner, or in a community, or one of many other permutations. Many people have dreams of living in communes, but I can tell you this is much more difficult than it sounds. Decision-making, money, and sex are a trilogy of troubles that humans find extremely difficult to deal with effectively in groups.

How much – Nearly all new settlers are too ambitious and go for too much land, spending too much money and taking on more than they can manage. We are all familiar with the television sitcoms showing city folk struggling with stroppy animals and unco-operative neighbours as the weather devastates their romantic rural home. What you must remember is that you do not have to do all this at once. It is perfectly possible to start slowly.

Many of our students have done just this – some begin by baking their own bread or making their own beer. Others may start with a few chickens or just a small vegetable garden. All of these are activities which give the would-be self-supporter a new feeling of achievement and a chance to experience the satisfaction of making their own requirements first hand.

What buildings – Do not be put off by poor-quality buildings or lack of buildings. In many ways it is better not to spend hard-earned cash on buildings which you could (in time) renovate or build yourself. What you absolutely do not want to do is to buy yourself a pile of work and trouble, especially in the form of old, damp buildings that will break your heart (and your wallet) to repair. And whatever estate agents may tell you, make sure you conduct some of your own research before committing yourself.

What money-earning activity – This is a matter of the most crucial importance and, as we have seen and will see, there are many ways of approaching it. But do not despair if you cannot see how this will happen – life has a strange way of filling spaces. And you may of course be of that determined breed who say to themselves "well my city profession as a poodle faker is not going to be in great demand so I will learn a new trade". Remember there are nothing like enough good craftsmen in the world – plumbing, electrics, mechanics, carpentry, basket making, and so on are all things you can learn. But many city folk do have skills that find use in the countryside – computer use, accounts, marketing, nursing, and teaching for example.

What crops – To a large extent this will depend on the type of land you have bought, the area you are in, and your own predilections. Once again my advice is not to be too ambitious – make haste slowly, and take care to keep a good diary, and learn from your own mistakes.

Getting it together with others

If you really want to make a change in your lifestyle then you are going to need help from others. And, what's more, the bigger the spread of talents you can find, the better your self-sufficient lifestyle will become. It makes no sense, for example, to imagine you can milk a cow just for a single family. You will have far too much milk and far too much of a tie that means you can never take a day off. You are unlikely to have all the skills – to make clothes, shoes, furniture, buildings, pottery, and the like. But find others who can do these things and your life will become more interesting and much more comfortable.

This logic does not simply apply to those who are seeking the rural idyll, it applies probably even more strongly to those of you who live in towns and want to make some important changes in how you live. Of course some of you will be able to take on allotments or make vegetable gardens; others will keep bees, chickens, or rabbits; some will be experts at bicycle repair; others will

know of local farmers who produce organic food; then there will be those who enjoy making bread, or beer, or wine. And so it goes on. Your challenge is to find such people and, having found them, to energize them and yourself to do things differently. By developing and using these different skills you can make a richer, happier, and healthier lifestyle for you and your family. Yes, and even create more of your own entertainment, too – meeting socially for parties, music, walks, poetry readings, meals, and visits.

All this human interaction and mutual support is what we mean by "getting it together". Strangely, these skills in human relations have been all but destroyed by our modern urban lifestyle. This puts all the emphasis on tight little nuclear families and a separate social life at the workplace. Each family has to paddle its own canoe, so to speak, earning its own money, paying its own bills, and surviving its own crises. Work, school, and survival take up so much time it seems there is no spare time left to develop alternative human contacts. Now is the time you can begin to change all of that.

The first thing you will want to do is to find out whether a self-sufficiency group is already meeting anywhere on a regular basis. Try local papers, shop noticeboards, and the local library for starters. If nothing has been done to bring people together, then it is up to you to make the first move. To some, this will be second nature, but to others it may take quite a bit of courage. There are really a couple of sensible alternatives here: Try an "at home" evening or organize a local pub or other meeting place session. The important thing is to try. It is a great chance to exchange ideas, to laugh about your mistakes, and to find out what else is going on in your area. As you get to know each other better you can plan more joint activities and develop a barter of goods and skills (*see pp.296-297*). Once you have got the ball rolling, make sure you meet regularly, once each month is a good start. Now, here are a few tips and hints about how to have successful meetings:

1 The first thing you may want to do is to search out a local pub or café that seems suitable for meetings. A public place is good because it is neutral ground and does not involve anyone in having to "show off" or tidy, or prepare food at home. And you can go there without involving all the other members of your family in the meeting. A local health food restaurant may be another option, or even a local hotel.

2 One way or another my preference would be a place with real beer, real wood, and a landlord who is happy to have you. Very often pubs have separate function rooms that you can book free of charge – but don't use these unless they have a good atmosphere. Somewhere that is cold and dusty without any "buzz" is not going to be a pleasant meeting.

3 If you find a positive response from your first contact list then take a stab at finding a convenient date for a first meeting. This will be quite exciting if you haven't met face to face before: like a scene from some spy movie perhaps.

So the group meet up using a secret sign – in your case this may be the colour of your hat, the newspaper you are reading or, alternatively, the copy of *The New Complete Book of Self-Sufficiency* that you will have on the table.

4 To those who are confident and outgoing it is a relatively simple matter to meet up with and get to know new faces. But to many it will seem a daunting prospect and you may have some tense silences while people try to figure out what they are supposed to be doing and what you all have in common. With small groups (six or less) it is pretty simple to have a joint conversation where everyone has a chance to take part.

5 If groups get bigger then the conversations will split up or, alternatively, you will need more rules about how each person can make their contribution. Some groups will be very able to "police" themselves so as to make sure everyone feels included. Others will have to work harder and may need a bit of leadership or what is called "facilitation" if they are to work successfully. It is really helpful if you have someone in the group who is used to working with people and has a sensitivity to the mood and the needs of the meeting.

6 If you have "called the meeting" and set things moving, then people will expect you to take a lead in getting things started. Usually people like to introduce themselves and say a little bit about what they do, their family perhaps, and what they are doing at the moment to be self-sufficient. If things are a bit sticky you can invite people to speak either by starting yourself and taking each person in turn around the table, going alphabetically, or writing down names on slips of paper and pulling them out of a hat. This is not as far-fetched as it sounds and is actually more fun and gets better results because people speak more spontaneously when they do not know when their turn will come. Remember, really the only important objectives of the first meeting are to get to know each other and to fix a date for the next meeting. Do not be put off if you find some of the people irritating or bizarre in some way. You have to remember that several of these people are likely to become very good friends as time goes by.

Building a community of support

Many modern people have simply never had the experience of working with others to create a community. You would be wise to work hard making these friendships work. "Community" is an important aspect of self-sufficiency. It may take a real effort and much probing to find out the skills and talents of those who have come together around the table. People who may be hopeless at "smalltalk" in a pub may be wonderful carpenters, musicians, cooks, or gardeners – with skills to share. You need to take the view that all your new contacts have amazing hidden talents and it is up to you to find them. In the meantime you have to make allowances all round for the poor development of social skills which is so prevalent in the modern industrial world.

Your meeting may well be a rip-roaring success. If so, that's great. But do not be surprised if the first meeting seems to be rather hard work. And do not despair. Remember, the modern human has got used to having all the group work, team building, and community activity provided in a package by the workplace. It is no accident that the most popular television shows are the soaps which provide what amount to virtual communities. The viewer can enjoy the ups and downs of these virtual communities at the turn of a switch, and without any personal cost or effort to themselves.

But now you have the challenge of getting the best from your group. Each of you will need to find ways of encouraging and using its varied talents and finding ways of enjoying yourselves together that strengthen and reinforce your friendship. If you can work as a group then your chances of really achieving big steps towards self-sufficiency will be very much greater. Now, be under no illusions – creating a community is a fundamental human skill which has been largely destroyed by modern urban living. It is no accident that follow-up studies which have looked at the progress of over 1,000 eco-villages and organic communities have found that almost all of these initiatives failed. And the prime reasons for failure were not lack of skill in producing food, or lack of people with many diverse skills, or lack of a good local school, or failure to build eco-homes, process waste, or manage livestock. No, the prime reasons for failure were the lack of effective community decision-making procedures, and the inability to manage money. There are two routes you might take to build a community and avoid these pitfalls. One is following a self-selected leader – such communities can work very well in human terms and they can continue for many years, provided there are effective arrangements to replace the leader on death or sickness. Alternatively, the "democratic" choice of leaders is much favoured. But this "model" must work by consensus – general agreement – and the voices of concern listened to very carefully.

Making your contribution

By meeting regularly, as I have said preferably once a month, you can build up confidence between the members. This is a great help in getting people to share their feelings and show their talents. You can start to do things together as simple as arranging to visit some place of interest together, going for a walk perhaps. Or, even better, you can meet up to do a project at someone's home. Here are some examples:

Making cider from loads of apples to be chopped, crushed, and juiced
Clearing waste; brambles and thorns to cut, trees to chop up
Butchering a pig into joints and bacon
Helping move livestock
Hay-making or wine-making
Building projects
Digging and weeding
Fruit harvesting and processing

As your group develops its strength, you may want to experiment with new ways of being together and talk about matters rather more personal and of deeper meaning. You will have to find ways of short-circuiting emotional stresses that creep into all human relationships. You will have to find ways of building up your energies for particularly difficult tasks or local campaigns. For example, I would explore those excellent techniques, many developed over the centuries in the East, such as yoga, meditation, and the like. In the West, groups such as the Quakers or Alcoholics Anonymous, have developed highly effective ways of using groups effectively in supporting the actions of each individual forming part of the group. It is well worth looking into the subject more fully if your group begins to grow bigger.

On a practical and economic level, by creating a group you also give yourselves the opportunity to bulk purchase foodstuffs and equipment. There are excellent sources for bulk-buying organic produce such as flour, fruit, and cheese. You can research the farmers in your area to see who would sell you organically-reared meat such as pork, lamb, or beef. As a group you can afford to buy a whole animal and cut it up for your deep freezes. You can also explore and create recycling options and urge others to do so, to compost more, to experiment with using less energy, even share a car or buy some land together.

Rules for group meetings

1 Appoint one person to look after the way the group works for that meeting.
2 Decide how the group is going to work during the meeting: you could use talking sticks, talking leaves, timed contributions, talking at random, talking around the table, or talking freestyle.
3 Try to set some timescale on the length of the meeting and have a timescale if you need to limit each person's contribution.
4 Draw up an agenda of points people would like to see discussed at the meeting.
5 Decide how the group is going to make decisions – by consensus, by vote, by summing up, or some other method.
6 Appoint someone to conclude the meeting (can be the same person as for item 1): sum up, thanks, next meeting, short silence, hold hands around table, or other ceremony of your choice.

If your group settles down well then you may begin to think about bringing in new members or contacting other similar groups. However, never allow your group to grow too big: 12 people should be a maximum for a basic group. If the number of people coming to meetings increases beyond this, then you need to create separate groups. This does not mean you do not meet in larger groups but it does mean that you do work differently in such groups. The small group will always be the powerhouse supporting your day-to-day self-sufficiency activity.

To make contact with other groups, you can simply ask one of your group to go along to neighbouring meetings. There will be such contact people in each group and their details will be listed under a "self-sufficiency", "self-supporter", or "smallholder" website. And as things develop we expect to build up a skills directory and a bulletin board for each group, which can include "for sale" and "wanted" items as well as events and services. And I would also make sure to also have a directory of the pubs in every region and town that are holding monthly self-sufficiency meetings.

In all of what we do in our groups and in our lives we must constantly remember that the cardinal rule is that nature is bountiful and full of goodness, provided we stick to the rules. To coin a phrase: "It is fun to have fun but you have to know how". As far as I am concerned the rules are very simple: love life, respect the earth, build human community, and work with nature, not against it.

We must also do what we can to celebrate our successes every year, and what better way than to have annual fairs in every region? If we say that each group of 12 groups makes up a region, then they can choose a couple of neighbouring groups each year to organize their regional fair. Ideally this should take place in June, when the days are long and the hay is being made. It will last 3 days and have entertainment, produce, events, displays, book sales, speeches, and plenty of good food and drink.

MEASURING PROGRESS

The households of most modern consumers take in virtually 99.9 percent of their daily requirements from outside suppliers and, equally, they pass on 99.9 percent of their waste products back to outside agencies via the dustbin, waste management bins, the sewer pipe, the chimney, and their car's exhaust pipe. For the self-supporter this is a profoundly unsatisfactory and even dangerous situation. Trying to measure the extent to which we depend on outside agencies (mostly the huge transnational corporations, government agencies, and supermarkets) is something we need to come to terms with. Below are some of the key "measuring units" by which we can judge our contribution to a self-sufficiency model.

Energy

Typically we will buy in oil, petrol, gas, firewood, petrol or diesel, and electricity. We use these and export the waste, mostly carbon dioxide and pollution, to the world's atmosphere. For the ideal self-supporter, household energy requirements would be provided by the solar economy either directly through the sun or indirectly through wind, water, or self-grown firewood. The self-supporter can tackle energy dependency in a multitude of ways. Saving energy is the first step – more insulation, better design of houses, less travel, lower temperatures, and more clothes. Investigating alternative energy sources is a must and never use electricity to heat anything: this is supremely wasteful.

Food

Huge quantities of food are transported thousands of miles and surrounded by acres of packaging material to make them attractive and available to modern households. This is the pride and joy of supermarket shopping. It is almost certain that there is a direct and unpleasant relationship between the look of the food, the quantity of poisons used to achieve this, and the resulting lack of taste for the consumer. We can eat foods from all over the world anytime we want – but in doing so someone has to pay a price sooner or later. The transport costs in terms of pollution and energy are enormous. Worse still, we no longer have the excitement created by having to wait for foods to be in season. Absence really does make the heart grow fonder and this applies just as much to strawberries as it does to sweethearts.

For the most part, modern consumers have no idea where their food comes from, who grew it, what was put on to it to keep off the bugs, and what was done to the soil to make it grow. Nor do they know what has been done to colour and preserve it. Enough said. Smallholders and allotment holders should aim to grow a large proportion of their own food. The results will be very satisfying, very healthy, and very tasty. We can measure progress towards self-produced food by costing our trips to supermarkets.

Household goods

The great snag with many household goods is that they are not designed to last. Sometimes this is because of style (which may rapidly go "out of date") but often there is simple, old-fashioned built-in obsolescence. The self-supporter will avoid these pitfalls at all costs. Make sure you buy goods that are made to last – even if it costs quite a bit more in the short term. The classic choice for household items is, of course, the light bulb. Many other things are similar – there is a choice between a high initial cost which is offset by long life and cheap running costs, or a low-cost, high-operating cost.

Transport

Most of you will know the old argument about using a car. It goes something like this.

Take your average car and look at the annual costs (always a sobering prospect!). You will find that depreciation, maintenance, and fuel costs for an average year's travel of, say, 8,000 miles (1,300 km) breaks down like this: depreciation is about £50 per week, maintenance around £25 per week, and fuel costs £20 per week. Add on a few extra bits and pieces and you have running costs around £100 per week. If your wages are around £8 per hour then you are effectively devoting 12 hours each week to pay for your journeys (the time taken travelling in modern traffic is probably 8 hours). So 20 hours devoted to travel makes your average speed just 8 miles (13 km) per hour, which shows you that you would be much better to take a bike at 10 mph (16 km/h)!

Roll on the day when we can have family-sized, weatherproof, bike-cars powered by pedal power so we do not need to visit the gym. Quite why no one has produced such a vehicle is one of the great mysteries of the modern age. Just imagine, downhill performance would, of course, be exceptional with aerodynamic shaping of the bodywork, lightweight disk brakes, on-board CD, and rechargeable lightweight batteries taking charge from downhill braking and helping with the uphill sections. It doesn't hurt to dream!

SUSTAINABLE LIVING ON LAND

By understanding the nitrogen cycle and nature's law of return we can steadily build up the fertility and production from our land. A healthy soil is the prime agent which we depend upon to convert natural wastes into fertility and food. But a healthy soil needs careful management and a constant supply of nutrients from human and animal wastes. A healthy soil must not be poisoned by strong chemicals or fed artificially with factory-made nutrients. If the soil is to increase in depth and fertility, then it must contain good quantities of vegetable material (humus). Without humus from decaying plant material the soil will either blow away or wash away in winter rains.

Recycling organic waste
Composting and animal manures are prime agents for recycling organic wastes. By learning the art of compost-making we convert weeds, kitchen waste, and garden by-products into new fertility for the soil.

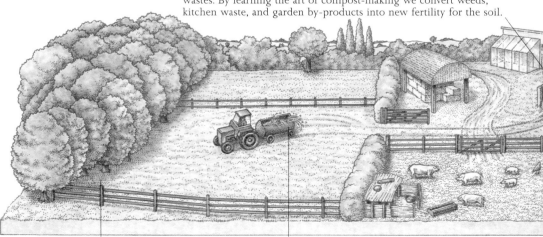

Crop rotation
Different crops need different nutrients and suffer from different diseases. To choose in which order to grow crops in rotation maximizes the available nutrients and minimizes the build up of disease.

Organic farming
For healthy soil and food we must minimize contamination from powerful chemical poisons. An organic farmer will avoid growing large areas of single crops and highly-bred varieties.

Water and "grey" water

For many parts of the world water is the key scarce commodity, yet the modern Western lifestyle uses thousands of gallons of purified fresh water every week in washing machines, dishwashers, toilets, garden sprinklers, and (increasingly rarely) drinking. Indeed, so careless are we that for the most part we do not even meter the quantity of water we use.

There are many ways to reduce dependency on piped and treated water. In the urban situation great savings can be made by collecting rainwater in the old-fashioned water butt. This is, by the way, a really good way to get the water for your greenhouse because if you put the butt into the greenhouse then the water will always be at the correct temperature for the greenhouse plants. We can also be even more creative and get to work with a plumber to use the so-called "grey" water.

Grey water is what comes out of things like washing machines, washbasins, and bathtubs. It may be a bit soapy but the worst of the stuff can easily be removed by suitable filters. Your water authority can even possibly advise on this. The objective is to pump the used "grey" water back up into a special attic storage tank so that it can be re-used for such purposes as flushing toilets, watering the garden, or cleaning the car. Huge savings can be made this way, and everyone benefits. Grey water systems will undoubtedly become the norm within the next 20 years. In the smallholder situation use of water is also likely to be a critical issue. Many smallholders will use their own well water: water from a spring or mountain stream water (I talk more about this on *pp.134 and 257*). Obviously having your own independent water supply is good practice for the smallholder.

Using a grey water process will be very beneficial as will use and collection of rainwater. But, and this is critical, the smallholder will also be able to make enormous savings by avoiding the use of a flush toilet. Mr. Crapper's famous invention is a great way to pour away vital nutrients for the soil whilst, at the same time, creating massive pollution of the rivers and water table. The liquid waste goes off down the sewer pipe to an industrial unit that tries to separate out the organic wastes and chemicals so it can be put back into the natural water cycle without too much danger to the environment (including humans who will have to drink it again before it reaches the sea). If you can use a composting toilet then these problems are avoided (*as described on pp.236–237*).

Waste

When archaeologists of the future try to make sense of this age we live in, they might well name it the "Rubbish Age", as we named ages of the past "Stone Age", and so on. For wherever they are likely to sink their spades into what were once high-tech countries, they will turn up rubbish. We generate more of that than the people of any former age. Soon there will be no more possible landfill sites to dump our rubbish in. We will be forced to stop looking upon it as rubbish, and to look upon it as material which has been used once (or more) and will be used again.

In the conventional urban household all solid waste goes off in bins of one kind or another, usually once each week. For the most part it is still buried (although recycling and composting are slowly changing this). Simple separation of household waste is obviously the first step: glass, paper, metals, organic material, and plastic can all be reprocessed.

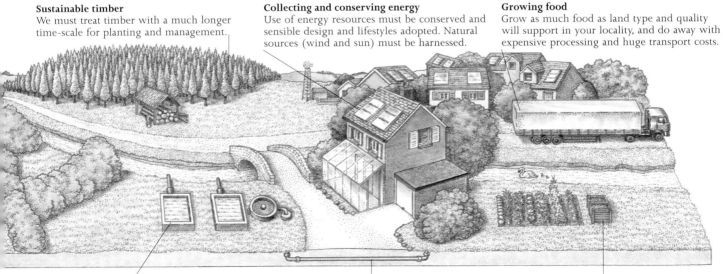

Sustainable timber
We must treat timber with a much longer time-scale for planting and management.

Collecting and conserving energy
Use of energy resources must be conserved and sensible design and lifestyles adopted. Natural sources (wind and sun) must be harnessed.

Growing food
Grow as much food as land type and quality will support in your locality, and do away with expensive processing and huge transport costs.

Nitrate-free water
By careful use of organic methods we prevent serious contamination of drinking water caused by farming's massive dependence on highly soluble cheap chemical nitrates.

Dual water systems
Current use of expensively treated drinking water for flushing wastes, washing cars, or watering must be stopped by creating dual water systems to collect "grey" water for re-use.

Fertility of the land
The challenge of constantly increasing fertility must be one of organic farming's highest objectives. Using chemicals for short-term profit can only damage long-term fertility.

Organic material can be composted in the garden either by bacteria or worms. I can remember an energetic charity worker who I met in Texas getting very excited about some fine-looking red worms she had seen in a local sportshop for fishermen. Just the thing for a worm-composting scheme she was setting up in downtown New York. The right sort of manure worms will work wonders on converting your kitchen waste to soil (and more worms for the fishermen).

Design

Use and layout of the home are really the critical elements in using energy effectively. I have covered this in more detail on pp.240–241

Clothes

God forbid that you are a victim of "fashion". This is a strange disease which has been popular with humans for thousands of years. Essentially it involves constantly wanting to wear the same kind of clothes as well-publicized role models. These models probably started out as emperors or kings but today they are more likely to be pop or film stars, celebrities, or models. The disease reveals that your clothes say a great deal about the person you are. Strangely, this was the main message I brought home after the 1992 Rio Earth Summit. Here we saw the citizens' groups from all over planet Earth wearing ethnic clothing (very beautiful and often handed down from generation to generation), or cheap T-shirts and shorts, whilst the government delegations and the like wore smart suits and ties. Of course they could only do this in their own artificial world of air-conditioned hotels, cars, and conference centres. Out in the tropical heat of Brazil, the suit and tie are about as sensible as wearing an Aran sweater in a sauna. The suit and tie have, of course, become the uniform of the merchants of greed. The tie in particular is some kind of symbol of servitude – slave to the status quo and runner on the wheel of greed. Now there can be few objections to the person who makes their own suit and tie except for the obvious fact that they would have to be pretty crazy to try to do so.

In practice there are few, indeed, in the modern Western world who make their own clothes in any shape or form. Yet for centuries this skill was always a feature of village life and indeed for some civilizations of the past the residue of fabrics and clothes are major indicators of their sophistication. Today most of our clothes are made effectively by machines and cheap labour in disadvantaged non-industrialized countries. Once they have a smart logo and a big advertising budget they become desirable designer items selling at ten times what it costs to make them. The sensible self-supporter will want clothes that are practical, sensible, and made as locally as possible. They will be made from predominantly naturally occurring fabrics – wool, cotton, linen, silk, and leather being obvious choices. Fabrics created from by-products of the oil industry will be particularly avoided, especially plastics and nylon. But the self-supporter's clothes will be stylish as well as practical, for we take pride in our appearance and are proud to hold our heads high as we take more control over our own lives and our impact on the Earth.

Measuring your lifestyle changes

The units shown below are certainly not ends in themselves, just ideas of ways to show how your lifestyle may change over time if you are able to master any or all of the techniques of self-sufficiency. Making a weekly or monthly tally will help focus minds on questions of lifestyle as well as giving a point of reference to compare notes with other fellow travellers on the road to self-sufficiency. Of course the bottom line in measuring progress is the simple question of whether you end up feeling happier about your life and the thousands of activities which make up your particular jigsaw of living.

Family Units
Miles in the car
Hours on the train or bus
Time watching television or videos
Time spent working for someone else
Per capita units (base amount divided by no. of persons)
Weight of rubbish collected
Money spent on energy (electricity, gas, fuel, wood)
Number of toilet flushes/gallons of water used
Money spent on food and drink (includes eating out)
Money spent on new clothes or shoes
Money spent on entertainment e.g. pubs, film, CDs, shows

THINGS TO BARTER WITH
The urban dweller moving away from the cash world of the city may be surprised at how many people in the countryside offer favours and help. This is very much in keeping with the principle of "you scratch my back and I'll scratch yours". And it is not just fruit or vegetable produce or a joint of beef or half a dozen eggs. You can barter with your labour, tools, and skills as well and these are just as useful.

A word on barter and social credit

In our more idealistic moments we may dream of a future where money is a rarely used commodity. Indeed, this may appear an extremely remote prospect. But we should remember that even 150 years ago money was by no means a common medium of exchange in rural areas. In times when villages were more or less self-sufficient in all day-to-day necessities, there was relatively little call for cash. Everyone living in the community made their own contribution in some way with a particular skill or vocation. Living in close proximity everyone was more or less able to "keep score" and avoid running up obligations to other members of the community.

In many rural areas these traditions live on. It is a natural feeling in the countryside to want to be in what might be called "social credit" with your neighbours. When my neighbour's tractor gets stuck outside my front door and I can pull him out of trouble. This is a great bonus for me – he is indebted to me and will certainly repay the favour when the time comes.

Whenever we visit friends (and we do so often, for what could be nicer than an unexpected visit) it is always a good feeling to bring some of our own produce as a gift. Whether it be fresh eggs, elderflower wine, or a joint of beef, the pleasure of giving what we have worked with gives us a good feeling. And these exchanges of things – which are almost part of ourselves – cement friendships and bring their own bounty a hundred times over in returned favours. One of the greatest wisdoms I have learned since moving to Ireland is the importance of giving generously. If you are going to give, then do so with generosity and style – no penny pinching, handing out second-rate produce, or giving half measures. Make the same true of your time, too. If friends visit, then "roll out the barrel" and carve up the side of beef. All the very best parties come through serendipity as one or two good friends or families show up unexpectedly. A good store of home brew and ample supplies of bread and cheese are the ever-ready foundations for such an occasion. And if there is music and singing to be had, then so much the better.

Don't limit your barter by restricting yourself to goods and produce: your labour, tools, and skills can be exchanged too. Every autumn our cider press, for example, "does the rounds" as our ever-increasing circle of cider-making friends take their turn in converting unwanted apples into the sparkling golden pleasure that comes in a bottle. When it comes to more sophisticated tools, always remember that you cannot depend on friends (much as you would like to) to understand their maintenance requirements, so if you want to "lend" these out then it's better in my opinion to lend yourself out as well. Much better being safe than being sorry, for a broken piece of machinery might easily ruin a friendship. Every self-supporter wants to be in social credit with his or her neighbours. Give your products and labour generously and sooner or later you will be generously repaid.

Nuts and bolts of self-employment

Bureaucrats strangled the Roman Empire, and they are making a pretty fair job of strangling our present civilization. Somehow, paperwork has an extraordinary ability to grow and grow, like some cancerous plant that has to be constantly cut away and dealt with. Those who are employed – and that means the majority of today's workers – are spared much of this because the State ensures that the employer fills in the forms. This is one of the stranger boons of being employed. Worse still, the embrace of conventional employment provides a whole range of easy comforts for the dependent employee. I can easily give you a host of examples: subsidized meals, health care, cheap gym membership, a company car, and so on. For those who are contemplating major lifestyle changes that involves moving away from employment and into self-sufficiency and self-employment, all these comfortable aspects of employment will disappear. Not only will they disappear, but we will have to face up to the unpleasant realities of dealing with bureaucracy ourselves. And a very unpleasant and useless task it is, too. Some of my independent minded friends make a feature out of letting their paperwork become a wild jungle of impossibility. Others (very few, I have to say) are models of quiet organization and efficiency. Probably most of us are somewhere in between. Whatever your own predilictions, I do recommend at least keeping your papers properly filed.

Taxes

The greatest difference between being employed and being self-employed is in the way we have to pay our taxes. Long, long ago, governments realized that it is much less painful to extract taxes from ordinary workers *before* they get their hands on the cash. They gave this system various names (in the UK it is "Pay As You Earn" – PAYE in short). The system works fairly automatically with the employer acting as an agent for the government, filling out forms and paying in the necessary taxes after adjusting wages or salary accordingly. With self-employment it is quite different.

A self-employed person only pays taxes when the year's accounts have been prepared. Basically you keep records of all the money you spend trying to run your business, as well as any money people have paid you for what you do or sell. There are plenty of tricky little rules about what kinds of costs you can count as part of your business. Obviously you have to keep the little bits of paper you get as receipts. And, if you have a big enough business, you have even more complications in keeping records for the sales tax people (but this need not concern us here). The essential ground rule for those who are self-employed is to keep hold of all the receipts you collect when you pay out money in the course of your business. You will find it a big help if you keep receipts for cash separate from receipts for cheques. This makes it much easier to check your bank statements – and banks make mistakes, that's for sure.

To keep effective records for paying your own tax, every month you will need to tally up the receipts, entering the amounts either in a book or, if you can manage it, in a suitably user-friendly computer (they can do arithmetic, these machines, as well as driving you mad!). If you are debating about using an accountant (a sort of high priest who professes to understand the tax system), I would say it is generally wise to find one at least when you are starting off on the self-employment road. He or she will tell you how to keep your records and, most important, how to fill in your annual tax return. With the help of your accountant, make a list of the headings which you are going to use for expenditure, such as heating and lighting, telephone, insurance, maintenance, fuel, and so on. Make another list for the different headings of income you may receive: goods sold, services offered, rental or hire payments, and the like. When you write out your cheques make sure you enter all the correct details and then carefully file the relevant invoices.

Every month enter the details for expenditure and income under the appropriate headings and add them up. You can then put the resulting monthly totals into summary form for the appropriate month. Use these totals to compare your figures with the figures sent to you in your bank statement. You must be able to reconcile them every month, and on some occasions it can be a surprisingly tedious piece of detective work to do so.

Keep separate records for cash expenditure and keep these receipts in a separate folder month by month. If you use a personal credit card then treat this the same as cash expenditure because it does not come out of the business account. Keeping track of cash payments can become a nightmare in my experience, so watch out.

You may well object to having to pay accountant's fees after the first few years when all this bookkeeping becomes so routine. Quite right, too. It pays a thousandfold to keep track of your own income and expenditures. Accountants can easily cost more than the taxes they save. I had one eccentric friend who simply kept all his receipts and pay-in slips in a big box, which he then sent to the Inspector of Taxes every couple of years. The Inspector did the necessary paperwork and he was charged the appropriate amount of tax. But he reckoned this was cheaper in the long run than employing an accountant. Whether you could stand the suspense of such an arrangement is another matter and, of course, in many countries self-assessment is now the norm in any case, so you have to do the work whether you like it or not.

Tax and accounts really are most important if you are seriously making the big break. If you get it pretty well right at the start, then you will save yourself a great deal of trouble (and possibly money) later on. The great thing is not to be reticent about putting down every item of expenditure that could possibly be thought of as being necessary for you to earn income from whatever trade (or trades) you are doing.

Claim, claim, claim

Don't be shy to claim expenses: I can remember my first accountant telling me how important it was to put down plenty for taxi fares in your first year, the logic being that taxis are not always going to give you a receipt, so the more you could justify under this head, the better off you would be (since all expenditure counts against income before tax). Of course in claiming generally I am talking about tax avoidance here (quite legitimate) as opposed to tax evasion (which is prosecutable). You should seek professional advice, or ask the tax office itself, if you are in any doubt. Once the taxman's computer has registered a certain pattern for your expenses then this pattern can (indeed probably should) be continued for all time. Over the years a regular few hundred pounds on taxi travel adds up to a small fortune of taxes you do not have to pay.

Pensions and life assurance

Other things to watch out for are pensions, life insurance, and bank accounts. I am specifically excluding any mention of borrowing money here because it is one of the few things I would advise any self-supporter *never* to do. Borrowing can have great attractions in this world of having your cake tomorrow as well as eating it today. But it is a brave man who can predict the future and being harnessed to work for a lending institution is no fun. If you look around in the big towns you will soon see who has the smartest offices and the biggest profits. Are they really the people you want to make even richer?

Pensions are important and almost all governments offer good tax incentives for those who put money aside for this purpose. You will need to check out these rules carefully because your benefits and options may very well vary with what age you are. And if the government is going to help you save for your own future, then so much the better. But be warned, any pension scheme will almost certainly involve handing your hard-earned money over to someone else to look after. Different outfits charge at very different rates for this service as well as offering all sorts of different possible ways of saving. Do your homework and think very carefully before you agree to any saving schemes that involve stock market investments (these are the ones that can go down as well as up).

Life insurance is another of the big scams which tends to be managed by people in suits in tall buildings. There are two main kinds of option – one is essentially a regular savings scheme (linked to a final lump sum or a pension), the other is the completely different insurance deal where you pay a small monthly amount so that if you die before the scheme ends (say, 25 years later) the company will pay your relatives a big lump sum. The former is a compulsory saving, which may or may not be more effective than something you could arrange for yourself (tax breaks aside). The latter is effectively a gamble on your own life expectancy. The "suits" do a few calculations based on your age and a few questions about family medical history.

You might also possibly have to have a medical examination. For your own peace of mind it is a very good thing to have a decent-sized life insurance of this kind, normally called term assurance.

Bank accounts
Bank accounts are a necessary evil, but there is a lot you can do to give yourself peace of mind. Certainly some of the modern telephone and internet banks are very much more cost-effective than old-fashioned high street banks. But all banks spend your money to make a profit for themselves and the way they do this is something you want to find out about. At one end of the spectrum we have agencies like the Credit Unions which only lend your money to other local people. Then there are so-called ethical banks which take a certain amount of care with what they spend your money on, for example, the Co-operative Bank. Finally there are the big money-grubbers who will put your money where it will earn most, irrespective of the activities it is promoting. As a self-supporter I am very much in favour of local banking through Credit Unions, even though they do not operate credit cards or cheque book facilities. They make my money work for the good of others in my area and they have the simplest of all criteria for lending money if you hit a seriously bad spot in your business life.

Insurance
Another subject which is "boring but important" is insurance. If you are going to keep animals, then you have to realize that the law is pretty fierce in the sense that you are likely to find yourself liable for any damage they do. And believe me, in the real world animals do escape, fences break, friends leave gates open, and animals decide they want to explore. It is also possible that animals or equipment may injure someone on your land. In all these cases you certainly should be covered by insurance.

The key point here is that many ordinary domestic insurers will not cover these types of damage. If you find this is the case, then try working through farming organizations. There are always a number of specialist firms who are used to these kind of risks; they are the people you need to talk to.

Craft associations
Many of those who adopt a self-sufficient lifestyle will take up a craft as a sane and satisfying means of earning money. Whether you are a potter or a furniture-maker, this can often be a lonely occupation where marketing your produce becomes very time-consuming and frustrating. One of the best ways of avoiding some of these problems is to join your local craft association. And if you cannot find one, then think very seriously about setting one up. An active craft association will be a great support for you, not only emotionally, but also in the vital task of gaining access to markets. Many craft associations run their own markets and this is an excellent way to sell your work.

IN SEARCH OF THE SELF-SUPPORTING LIFESTYLE
So what, then, does a sensible man or woman really need to be happy? How does the would-be self-supporter achieve this while treading lightly on this much-muddled world of ours? Health, of course, and real positive health, not just absence of disease. This can be achieved only by hard manual work or exercise, fresh air, sunshine, secure shelter, and unadulterated food. All but the last two items here are free (and we have forgotten how much in this world the best things, such as love too, can be free). The inflated price of housing in Western countries (and unsustainable population growth) makes a warm, dry, and comfortable home, surrounded by a vibrant and productive natural world, elusive for many; but it is essential for the self-supporting lifestyle. Other prerequisites are good friends and neighbours and working to keep it that way. Working hard should be balanced by plenty of play with ample opportunity to enjoy good food, good beer, late nights with music, songs, and dancing. (Never let go of fun and conviviality.) It's about having control of your own time. It's about working with the great power of the natural world – just like a surfer works with the waves, or skier with the snow – always aware that nature is bountiful but must treated with great respect. It's also about leaving the world a better place with healthier and happier animals, plants, people, and soil.

Making your own entertainment
Certainly, when I talk of self-sufficiency my ideas do not stop at the simple production of food and drink. We should also want to develop other talents and contribute whatever we can to the whole social fabric around us. The more we give to it, the more it will give back to us. And if we can have great entertainment, which is self-generated and not subject to some huge entertainment industry, so much the better. I have been to wilder nights in the pub when the Gaelic music and ambience have been fantastic. Musicians, singers, and dancers have enjoyed themselves mightily; and as for quality, well the audience are not paying directly for it, and it has been a great thrill to have been involved in the whole performance. Entertainment is only limited by the imagination, and you could encounter drama groups, singing groups, birdwatching groups, music groups, poetry reading, arts and crafts, cardplaying, and so on. Some of these groups will be more or less formal, others will exist only for their members (check your local newspapers and noticeboards). As for participating, do not be shy nor too modest. We all have talent and a capacity to respond to practice. It's taking the first step that can be difficult and, as the old adage goes, it's the taking part that counts. And do not forget that much of what we see in the media is recorded, mixed, and re-recorded until it is perfect. But in real life and in real human interaction it is not plastic perfection that we seek. On the contrary, it is honesty and commitment. When people are doing their best and putting their full soul into it, then that is all we can possibly ask.

Contacts & References

On the following pages you will find organizations, societies, associations, cooperatives, and the like who, in my opinion, have something to say to the self-supporter, however small. I have also listed several reference books that I have found genuinely helpful, and I pass these on to you in the hope that you may find them of use too. Websites are also provided (of course, when I started out on the road to self-sufficiency the world wide web was something used only really by the military, universities, or institutes). Bear in mind that web addresses, like telephone numbers, change with alarming rapidity, but they were accurate at the time of going to press. Certainly, the home computer can really help the self-supporter, especially for getting in contact with like-minded souls. Good luck and good searching!

Every effort has been made to check the accuracy of telephone numbers, and email and website addresses, at the time of going to press. However, the content, as well as the addresses, are subject to frequent changes, and the publisher can accept no responsibility for any inconvenience or distress arising from such changes

The John Seymour School for Self-Sufficiency

The Self-Sufficient Smallholding, Killowen,
New Ross, County Wexford
Ireland
Tel/Fax 00353 51388945

Each year the "School" takes up to 8 students at a time on week-long courses which provide hands-on experience of life on a mixed smallholding. There is accommodation on site and the courses include occasional visits to neighbours, local pubs, and the local town. The Killowen smallholding is only 3.5 acres with about 1 acre taken up by the house and garden. Courses cover a wide range of practical skills which vary depending on the seasons and include: Organic Vegetable Growing; Planning a Smallholding; Top Fruit and Soft Fruit Growing; Pig and Cow Husbandry; Poultry Management; Dairy Work; Beekeeping; Basket-making; Wine, Beer and Juice-making; Jam and Bread-making; Scything and Grass Management; Composting; Blockwork and Basic Building Skills; Ropework and Knots (see also p.302).

COMMUNITIES & ASSOCIATIONS

Eco Village Network

The Create Centre
Smeaton Road
Bristol, BS1 6XN
Email: evnuk@gaia.org
An organisation and magazine promoting the eco-village concept

Earth Village Network

Postbus 1179
1000BD Amsterdam
Netherlands
also at Edinburgh Crescent, Leamington Spa, CV31 3LL
Gives help and advice on forming sustainable permaculture communities

Eco-Design Association

The British School
Siad Road, Stroud, Gloucestershire GL5 1QW
Source of information on ecological architecture

Permaculture Academy

8 Helen Road, Oxford OX2 0DE
Promotes the philosophy of working with nature through permaculture

Sustainable Land Trust

7 Chamberlain Street
London NW1 8XB
Organization dedicated to sound management of the land

Smallholder's Society

Web: www.smallholder.org.uk
For information email roger-walker@tesco.net
Set up in 1999 and now expanding rapidly, provides a comprehensive listing of smallholder activity and information, including meetings and local associations

SEED International (Sustainability Education and Ecological Design)

Crystal Waters Permaculture Village
MS 16, Maleny
QLD 4552 Australia
Tel: +61 7 5494 4833
Web: www.permaculture.au.com

ENERGY

Centre for Renewable Energy and Sustainable Technology

Centre for Alternative Technology
Machynlleth
Powys, SY20 9AZ
Tel: 01654 705959
Web: www.cat.org.uk
Extremely active and wide ranging centre for all aspects of alternative technology; produces many fact sheets, runs courses

Intermediate Technology

The Schumacher Centre for Technology and Development
Bourton Hall, Bourton-Dunsmore
Rugby, Worcestershire CV23 9QZ
Tel. 01926 634400
Web: www.itdg.org.uk
This organization exists to promote all types of intermediate and alternative technology in line with the philosophy of E. F. Schumacher

Wind and Solar Energy

Alternative Energy Systems Co.
1469 Rolling Hills Rd., Conroe, Texas, USA 77303
Tel: +41 936 264 4873
Web: www.poweriseverything.com
Has a very large selection of subjects and related information

Alternative Technology Association, CERES

8 Lee St, Brunswick East VIC 3057 Australia
Tel: +61 3 9388 9311
Web: www.ata.org.au

Energy Development Co-operative Ltd

The Old Brewery, Oulton Broad Industrial Estate,
Harbour Road, Oulton Broad
Suffolk NR32 3LZ
Tel: 0870 745 1119
Web: www.unlimited-power.co.uk
Comprehensive supplier of alternative energy equipment:

British Wind Energy Association
Renewable Energy House, 1 Aztec Row,
Berners Road, London N1 0PW
Tel: 020 7689 1960
Web: www.bwea.com
Trade association promoting wind energy

MAGAZINES
Smallholder Magazine
3 Falmouth Business Park, Bicklandwater Road,
Falmouth, Cornwall TR11 4SZ
Tel: 01326 213303
Web: www.smallholder.co.uk
Monthly magazine for smallholders, with news and sources

Country Smallholder Magazine
Fair Oak Close, Exeter Airport Business Park,
Clyst Honiton, Exeter EX5 2UL
Tel: 01392 447711
Another monthly magazine for smallholders, with news and info

Earthgarden Magazine
PO Box 2, Trentham VIC 3458 Australia
Web: www.earthgarden.com.au

Green Futures
Circa, 13–17 Sturton Street, Cambridge CB1 2SN
Tel: 01223 564334
Web: www.greenfutures.org.uk
A forum on all issues re progress towards sustainable development

Smallholder Bookshop
High Street, Stoke Ferry, Nr Kings Lynn,
Norfolk PE33 9SF
Tel: 01366 500466
Web: www.smallholderbooks.co.uk
Useful mail order bookshop for smallholders

FURTHER HELP
Department for the Environment, Food and Rural Affairs Development and Advisory Service
Ergon House, 17 Smith Square, London SW1P 3JR
helpline@defra.gsi.gov.uk
Tel: 0845 9335577
Web: www.defra.gov.uk
Useful first stop for the would-be self-supporter. Regional offices in UK provide advice and pamphlets (many free) on most agricultural subjects

Agricultural Development and Advisory Service
Whitehall Place, London SW1A 2HH
Tel: 0845 7660085
Web: www.adas.co.uk
For advice and information on general aspects of farming (not free)

The Forestry Commission
Silvan House, 231 Corstorphine Road, Edinburgh EH12 7AT
Tel: 0870 1214180   Web: www.forestry.gov.uk
For advice and pamphlets on woodland

ASI, Importers and Wholesale Distributors
Alliance House, Snape Maltings, Saxmundham
Suffolk IP17 1SW
For a wide range of guns (to purchase you must have a licence and go through a dealer)

Organic Federation of Australia
PO Box Q455
QVB Post Office
Sydney NSW 1230 Australia
Tel: +61 2 9299 8016
Web: www.ofa.org.au

Wright Rain Irrigation
No4 Millstream Industrial Estate, Christchurch Road,
Ringwood, Hampshire, BH24 3SB
Tel: 01425 472251
For irrigation equipment

Henry Doubleday Association
Ryton Organic Gardens
Wodston Lane, Ryton-on-Dunsmore, Coventry, CV8 3LG
Tel: 024 7630 3517
Web: www.hdra.org.uk
Email: enquiry@hdra.org.uk
For general information on soil and organic gardening

W. Gadsby and Son Ltd
Huntworth Business Park, Bridgewater, Somerset TA6 6TS
Tel: 01278 437123
For basketware

Essex Kilns Ltd
Furnace Manufacturers, Woodrolfe Road,
Tollesbury, Maldon,
Essex CM9 8SE
Tel: 01672 869342

Harris Looms
Wotton Road, Ashford, Kent TN23 6JY
Tel: 01233 622686

Jacobs, Young and Westbury
Bridge Road, Haywards Heath, Sussex RH16 1UA
Tel: 01444 412411
For rushes and loom cord

Frank Herring and Sons
27 High West Street, Dorchester, Dorset DT1 1UP
Tel: 01305 264449
For spinning wheels

Alec Tiranti Ltd
27 Warren Street, London W1T 5NB
Tel: 020 7636 8565
Web: www.tiranti.co.uk
For all stone masonry tools, as well as clay

FURTHER HELP

Basket Makers Association UK
Membership Secretary: Sally Goymer, 37 Mendip Road, Cheltenham, Gloucestershire GL52 5EB
Web: www.basketassoc.org; *also for supplies see Willow Basketmakers at www.basketmakers.org*

British Goat Society
34–36 Fore Street, Bovey Tracey, Newton Abbot, Cornwall TQ13 9AD
Tel: 01626 833168
Web: www.allgoats.com
A large website with comprehensive coverage of different breeds and suppliers

Pig Breeders
SM PO Box 233, Sheffield S35 0BP
Tel: 0114 286 4638
Web: www.thepigsite.com
Comprehensive information about different pig breeds and where to obtain them – mostly slanted towards the commercial operator

Linda & Derek Walker/Oxmoor Smallholders Supplies
Harlthorpe, East Yorkshire YO8 6DW
Tel: 01757 288186

Smallholder Supplies
Fenland Smallholder Supplies
Anvil House, Dudleston Heath, Ellesmere, Shropshire
Tel: 01691 690750
Web: www.goats.co.uk

Bridport Gundy Marine
The Court, Bridport, Dorset DT6 3QU
Tel: 01308 422222
Lines and nets supplier

Art of Brewing
Comprehensive source of all brewing supplies – can be viewed online
Tel: 020 8549 5266
Web: www.art-of-brewing.co.uk

Buck and Ryan Ltd
101 Tottenham Court Road, London W1T 4DY
Tel: 020 7636 8565
For metal-working tools

General Woodworking Supplies
76/80 Stoke Newington High Street, London N16 7PA
Tel: 020 7254 6052
For wide range of British and imported woods

SPR Centre
Greenfield Farm, Fontwell Ave. Eastergate, Chichester, Sussex PO20 6RU
Tel: 01243 542815
Web: www.sprcentre.co.uk
A supplier of a large range of animal foodstuffs for smallholders

Ascott Smallholding Supplies
Anvil House, Dudleston Heath, Ellesmere, Shropshire, SY12 9LJ
Tel: 0870 7740750
Web: www.ascott-shop.co.uk
Mail-order supplier of a large range of animal foodstuffs, precision seeders, and the like for smallholders

Moorlands Cheesemakers
Brewhamfield Farm, North Brewham, Bruton, Somerset BA10 0QQ
Tel: 01749 850108
Cheese-making supplies and equipment

Solartwin
Freepost, NWW7 888A, Chester CH1 2ZU
Tel: 0845 1300137
Solar power suppliers

Solar Sales
97 Kew St, Welshpool WA 6986 Australia
Tel: +61 8 9362 2111
Web: www.solarsales.com.au

Cheeselinks
15 Minns Rd
Little River VIC 3211 Australia
Tel: +61 3 5283 1396
Web: www.cheeselinks.com.au

Chr. Hansen UK Ltd
Rennet Manufacturers, 2 Tealgate, Hungerford, Berkshire RG17 0YT
Tel: 01488 689800
For starters and pure cultures for cheese-making

E. H. Thorne Beehives Ltd
Beehive Works, Wragby, Market Raisen, Lincolnshire LN8 5LA
Tel: 01673 858555
Web: www.thorne.co.uk
For bee equipment

Burntstone Ceramics Ltd
19 Redgate, Walkington, Beverley, HUM 8TS
Tel: 01482 868 706
Web: www.burntstone.co.uk
A supplier of furnaces and kilns

Organic Seeds
PO Box 398, Ipswich, Suffolk IP9 2HU
Tel: 01473 310118
Web: www.organicseeds.co.uk
Organic seed supplier

Countrywide
Email: enquiries@countrywidefarmers.co.uk
Livestock care and feed, plant production, and energy needs for smallholders

BOOKS

The Living Soil
Lady Eve Balfour; Faber and Faber 1933
One of the definitive books on the health-giving powers of organic management of soil and garden. Lady Balfour was effectively one of the founders of the reborn "organic" movement

The Growth Illusion
Richard Douthwaite
A painstakingly researched exploration showing how propaganda about the importance of "economic growth" is largely fallacious

Short Circuit
Richard Douthwaite; Green Books 1966
A useful and comprehensive collection of alternative economic systems which bring back control to a more local level. There is also a powerful statement explaining the importance of adopting more of such systems

Soil and Civilisation
Edward S. Hyams; HarperCollins 1976
An accurate and perceptive description of the vital link between soil and the development of modern civilisation. Sooner or later the fate of every urban culture is decided by the wisdom with which they manage their soil

Eco Villages and Sustainable Communities
Gaia Trust research publication commissioned from Context Institute, 1991. Gaia Trust, Skyumvej 101, 7752 Snedsted, Denmark Context Institute, PO Box 946, Langley, WA 98260 USA www.context.org
Extremely important and comprehensive review of the theory and practice of eco-villages worldwide. The report reaches a sobering conclusion that the huge majority of eco-villages fail not because of physical or technical deficiencies but because of financial and social problems

The Guide to Co-operative Living (annual)
Diggers and Dreamers c/o Edge of Time Ltd, BCM Edge, London WC1N 3XX
Useful practical handbook (and other catalogues and information about communities) for people seriously intent on exploring community living. There is a large index of communities from England, Scotland, and Wales

Ecotopia
Ernest Callenbach; Pluto Press 1978
A thought provoking novel which really explores what living in a self-sufficient eco-culture would be like. Full of practical examples of how to do things differently

Blueprint for A Small Planet
John Seymour and Herbert Girardet; Dorling Kindersley, 1987
Comprehensive and common sense review of actions that could be taken in all the routing activities of daily life to save the future of the planet. Full of neat diagrams and plenty of frightening facts about the consequences of our present Western approach to ordinary living

The Forgotten Arts & Crafts
John Seymour; Dorling Kindersley, 1984, 1987, 2001
This very beautiful and thorough book gives detailed and well-illustrated insights into the skills of traditional life. It is an invaluable and extremely interesting reference book — the result of many months of painstaking research and visits to some of the last of the true traditional craftsmen

Home Farm
Paul Heiney; Dorling Kindersley 1998
As the title suggests, lots of practical advice on pig, sheep, horse, and chicken-keeping and a lot more traditional skills (with photographic step-by-steps) beyond the garden

Textbook of Fish Culture - Breeding & Cultivation of Fish
Marcel Huet; Fishing News Books Ltd 1971
A definitive book on fish culture by the Director of the Belgian Research Station for Water and Forests

The Fat of the Land
John Seymour; Metatonia Press 1991
This is the original personal story of how John and Sally Seymour developed their desire for self-sufficiency in rural Suffolk. Packed with anecdotes and real-life dramas, the book makes entertaining reading as well as giving a warning about some of the less expected consequences of searching for self-sufficiency

Organic Gardening
Maria Rodale; Rodale Press, Emmaus, PA 18098 USA, 1998
An inspirational book celebrating the beauty of organic growing and discussing many of its leading ideas

HDRA Encyclopedia of Organic Gardening
Pauline Pears (Editor-in-chief); Dorling Kindersley 2001
About as complete as you can get and with the Henry Doubleday Research Association seal of approval (founded in 1954, it's Europe's largest and most respected organic gardening association)

The Organic Garden Book
Geoff Hamilton; Dorling Kindersley 1987
For growing better tasting fruit and vegetables untainted by chemicals from a much missed man

Flora Britannica
Richard Mabey; Sinclair-Stevenson 1987
A vivid testimony as to how nature and self-sufficiency have collaborated over the centuries

The Complete Encyclopedia of Home Freezing
Jeni Wright (editor); Octopus
This is a useful A-Z guide to freezing techniques and recipes for the freezer

The Complete Book of Preserving
Marye Cameron-Smith; Marshall Cavendish Books
Gives all the technical know-how you need on how to preserve a huge range of fruits and vegetables

The Gaia Atlas of Planetary Management
Norman Myers; Pan Books 1985
This is an ideal book to give younger people a good idea about what is happening on planet Earth. Lots of exemplary pictures, flow charts and information

Glossary

acrospire first stages of the shoot coming from germinating grain

aftermath short grass stubble left after hay-making

anaerobic fungal and bacterial action taking place in absence of oxygen

annual plant which germinates, grows, seeds, and dies in one year

ark small portable house or shed providing shelter for animals

awl sharp spike used to make holes in wood or leather

babbing technique for catching eels by tangling their teeth in threads passed through a bundle of worms

biennial plants which grow in the first year and seed in the second

blanch to dip fruit/vegetables in boiling water briefly before freezing

bloom surface colouring or sheen on plants or animals

blunger robust mixing device for merging different clays

bolt large bundle of reeds or willows

brash cut-off lower branches of young growing trees to make better timber

brassica family of vigorous vegetables like cabbage, sprouts, turnips

break growing a different crop for a season to reduce the likelihood of disease in the crop which is grown more regularly – e.g., a root break between cereal crops

britchin the part of the harness which goes around the backside of the horse to prevent a cart from over-running downhill

broadcast to sow seed by scattering by hand rather than in rows

broiler chicken bred to be fattened up quickly into meat for the table

brood the eggs and young grubs of bees

brortches lengths of flexible hazel rods used to pin down thatching

budding useful way of propagating fruit varieties by grafting a single bud to an existing compatible host tree

bulling the behaviour of a cow when "in heat" and ready for the bull

cake the solid crushed residue of fruit which is left after the juice has been pressed out

cappings the beeswax which is removed from the top of honey cells each time honey is taken from the hive

card the process used to straighten out and align fibres in wool or flax before spinning

catch-crop quick-growing crop sown late in season after another crop has been harvested

caul the white, fatty membrane which supports the intestines

cheese mass of chopped or pulped fruit that is placed into a muslin bag prior to being put in the juicing press

chitlings the stomach and intestines of a pig

churn robust chamber or vessel used to create butter

clamp the pile of earth and straw which is built over a heap of vegetables – usually potatoes – to preserve them through winter

clean the term used to refer to land/soil which is free of weeds

cleave to split wood down the grain

come what happens when cream turns to butter

coping the act of splitting a small stone with a cold chisel

coppice the technique of cutting back growth on mature trees to stimulate fast growth of smaller logs suitable for firewood

cordon a particular way of training fruit trees so they are convenient to pick

couch one term for the persistent rhizome grass weed which is commonly found in light soil

coulter the sharp, vertical knife positioned to cut turf in front of the plough share

creep feed feeder fenced in by a small opening which only allows access to small, young animals

crush device to contain and hold still large animals for veterinary work

curd the solid lumps which form in milk as it is attacked by bacteria or enzymes (like rennet in cheese-making)

curing preserving and softening leather by use of manipulation and chemicals, or the preservation of meat using salt and/or smoke

cutting small piece of wood/plant removed from a mature plant with the objective of being rooted to create a new plant

decant carefully pour off a clear liquid from a sediment

draw remove the guts of an animal after slaughter

drill sow seed in straight lines with a machine designed for this purpose

espalier way of training a fruit tree along horizontal wires, making it easier to manage and pick fruit

fan way of training a fruit tree into a fan shape for ease of management

fell the thin layer of translucent white skin which covers the outside of butchered meat after an animal has been carefully skinned

fisting the forceful removal of skin from a newly killed carcass by using the fist

fleece the wool of a sheep

flitch the usual term for a side of bacon

flocculate what happens when fine particles of material (usually soil) come together to form larger particles which are easier to manage

fold the process of turning grazing animals on to a piece of land where specific fodder crops have been grown, usually with temporary fencing of some kind

fold pritch a heavy, sharp steel rod used to make holes in the ground which can be used to put up hurdles being used to "fold" animals

friable soil which is pleasant to handle and breaks up easily into crumbly pieces is said to be "friable"

frost pocket low-lying place where frost collects because of either the lie of the land or the existence of hedges or trees

fulling the process of creating felt from cloth by bashing it about in water

gambrel strong twin hook used to hoist up freshly killed carcass

grafting the process of fastening one type of growing plant to a host root-stock so as to combine the virtues of both plants

hackling dragging retted flax over spikes to pull out the short fibres and leave the long ones

harden off gradually expose a plant to outdoor conditions after it has been kept in a greenhouse

hardy frost-resistant

harrow to cultivate an area of land using a lightweight metal grid with downward-extending spikes

heavy the description of land which contains a high proportion of clay, making it prone to stay wet and create difficult conditions for cultivation

heddle row of loops on a wooden batten which lift threads to allow shuttle to pass through when weaving

heeling in putting plants in a trench so as to cover their roots whilst they are temporarily in storage

holding bed a part of the garden used to grow seedling plants before they are transplanted to final positions

hulking see "paunch"

humus the largely decomposed masses of vegetative matter which make up a vital part of good soil structure

kive the large vessel used to contain and heat up beer during the brewing process

legume the family of plants which create their own fertility through nitrogen-fixing bacteria in root nodules (clover, beans, peas)

ley special grass mixtures sown as a temporary break for good grazing between cereal crops

light the description of soil which contains a high proportion of sand and is therefore quick to dry out and easy to work

linseed the seeds of the flax plant, which produce an excellent oil when crushed

lye alkaline chemical used to make dyestuff

mash the mixture of malt, hops, and water which is heated up and processed to make beer

mordant chemicals which bite into fabric as a key for dye

mouldboard the large curved part of a plough which turns and shapes the furrow

mulch a layer of reasonably inert material which is spread on to soil to keep down weeds and prevent loss of moisture

must the remains of fruit which has been squeezed out in the process of making wine

pasteurize to eliminate harmful bacteria by heating, to preserve juices

paunching the immediate removal of stomach and guts from freshly caught game such as deer or rabbit

perennial plants which grow on from year to year without reseeding

poach what happens when heavy animals trample soil into a mudbath during wet, wintery weather

pH term used to measure acidity – acid or alkaline

pitching the act of putting yeast into a liquid which is being fermented for wine or beer

polled cattle that have no horns are said to be "polled"

propagation various techniques which are used to create small new plants from existing mature plants

prove the process by which active yeast infuses carbon dioxide gas into dough to make it rise in bread-making

puddle to pour copious quantities of water on to newly transplanted seedlings

rack to syphon clear liquid from above sediment during the process of making beer or wine

rand type of neat, strong weave in basket-making

render thin covering of cement (usually 4 parts sand to 1 cement) used to finish off walls and floors

rett the process of rotting down the substance of fibrous plants to be used for making cloth – particularly flax

rick metal or wood construction used to hold hay for feeding to animals

ripple to pull heads of flax through spikes to remove all the seeds

rive the technique of splitting rather than sawing a length of wood

roots crops which yield a root which makes good food

rotation technique of changing crops grown on a piece of land every year to create a pattern which minimizes disease and boosts fertility

row crop crops which are grown in rows for ease of hoeing, weeding, and harvesting (as opposed to broadcast or deep-bed plantings)

runner vegetative shoot produced by "walking" plants to propagate themselves

scion the piece of living wood which is grafted on to a rootstock to create a productive tree

screen to sieve soil (or any other material) to remove large pieces

scutch see "couch grass", but also the action used to break up flax before the fibres are removed

served when a female stock animal has been inseminated by a male, she is said to have been "served"

sets what you call small onions which are planted as "seeds"

share the pointed sharp end of the plough

shed the open space between threads created during the weaving process to allow easy passage of the shuttle

shuttle the heavy, two-ended needle which carries thread each way through the "shed" during the weaving process

singling the act of thinning massed seedlings with a hoe in the early stages of establishing a row crop

skep small, round straw container used to contain a swarm of bees

slath term used to describe the base of a willow basket

slewing fast weave used in basket-making

slip watery mixture of clay and water

snood piece of stiff wire used to keep hooks away from the line when sea fishing

sparge the action of pouring hot water over spent malt and hops to remove the last of the flavour

spear grass another name for the weed couch grass

spiling driving in stakes to hold out the sides of a well or hole being dug in sandy soil

spit depth of digging equal to the length of the blade of the spade

squab fledgling, young pigeon

squodge what you do to butter to squeeze out the water

stack a neat pile of material quite different from a "heap"

standard the shape of a young tree where a single trunk extends at least 6 feet (1.8 m) from the roots to the first branches

starter a small brew of bacteria which is put into milk to make yoghurt or cheese

stook bundle of newly cut cereal before the grain has been thrashed out

store adjective used to describe beef animals reared on a low diet and needing fattening

stratify what you do when exposing tree seeds (nuts) to frost in order to make them germinate

strike what happens to sheep when fly maggots attack their flesh

swaithe the row of cut grass/ hay/cereal left after the scythe or mower

tailings small grain which is too small to grind into flour

tease to pull out wool into straight lengths using a board full of small spikes

ted to fluff up cut grass to enable it to dry into hay

tempering changing the spring and hardness of steel by alternative heating and cooling

tender what you call plants that are sensitive to frost damage

throw the action of a potter making a shape with clay on a wheel

tine spike fastened to implements that are used to till the earth

tow the pieces of fibre which are too short to spin into yarn, often used for caulking the gaps in the hull of a wooden boat

truss how you tie up a freshly dressed bird to prepare it for roasting

tug the part of the harness which takes the strain when a horse pulls a plough or cart

tuyere tube which blows air into a blacksmith's forge

volunteer plants which grow on in the following year from tubers or seeds left behind from the previous harvest (typically potatoes)

waling very strong and stable weave used in basket-making

walk-up what happens when game is disturbed from cover by a hunter walking on foot

wedge the shape created in threads during the process of passing the shuttle in weaving

whey what is left of the milk after the curd has been taken to make cheese

wicken yet another name for the grass weed called "couch grass"

windrow the row of cut grass left fluffed up to dry after the mower or scythe in hay-making

winnow to separate grain from chaff using wind or blown air

wort the mixture of water, malt and hops when it is ready for fermenting

Index

Acknowledgements

Acknowledgments to this Edition

Messrs Kinsey and Harrison would like to thank the following for their help in the production of this book. Patricia Hymans for indexing. Mel Hobbs and Peter Rayment at Brightside; Martyn's Garden Services, Little Chalfont; David and Anne Sears; Louise Waller, Carla Masson and Christine Heilman at DK.

Digital Colour Enhancement

All existing and new line illustrations were digitally coloured by Simon Roulstone.

Linocut Illustrations

Jeremy Sancha produced the linocuts on pages 2/3, 4/5, 10/11, 12/13, 15, 18/19, 24/25, 39, 41, 93, 94/95, 99, 171, 182, 194, 201, 205, 219, 223, 232, 251, 288.

Line Illustrations

Sally Launder pages 6, 52, 108/109, 180/181, Kathleen McDougall pages 32/33, 36/37, 208/209, 259, 296.
John Woodcock pages 45, 198, 206/207, 230, 252, 260/261, 278/279, 281, 282.
David Ashby pages 26/27, 28/29, 142/143, 184, 185/186, 197, 234/235, 237, 240/241, 248, 253, 255, 256/257, 294/295.
Peter Bull Associates pages 144, 247, 254, 280.
Simon Roulstone pages 22, 70/71.

Acknowledgments to 1st Edition

I would like to thank the many people who have helped me with information and advice, particularly Sally Seymour without whom this book would never have been written. The students on my farm, Fachongle Isaf, also assisted in many ways, especially Oliver Harding and David Lee who helped with the drawings and diagrams.

John Seymour

Dorling Kindersley would also like to express their gratitude to Sally Seymour and the many people associated with Fachongle Isaf. In addition they would like to thank the following for their special contribution to the book:

Susan Campbell
Peter Fraenkel
Mr Woodsford of W. Fenn Ltd.
Cleals of Fishguard
Peter Minter of Bulmer Brick & Tile Co.
Mr Fred Patton of Cummins Farm, Aldham
Rachel Scott
Fred'k Ford
Ramona Ann Gale
John Norris Wood
Richard Kindersley
Barbara Fraser
Michael Thompson and the staff of Photoprint Plates
Barry Steggle, John Rule, Murray Wallis, Mel Hobbs, Peter Rayment and the staff at Diagraphic

MEASUREMENTS & OVEN TEMPERATURES

Weights and measures in this book, (including temperatures) have approximate metric and Imperial conversions (rounded up or down); this applies to the recipes, too. When referring to these measurements, including recipe temperatures (see below), do not mix metric with Imperial.

Gas Mark			
	1	275°F	140°C
	2	300°F	150°C
	3	325°F	170°C
	4	350°F	180°C
	5	375°F	190°C
	6	400°F	200°C
	7	425°F	220°C
	8	450°F	230°C
	9	475°F	240°C

Artists

Dorling Kindersley would also like to thank Eric Thomas, Jim Robins, Robert Micklewright, and David Ashby for their major contributions to the illustrations in this book. Also Norman Barber, Helen Cowcher, Michael Craig, Brian Craker, Roy Grubb, Richard Jacobs, Ivan Lapper, Richard Lewis, Dave Nash, Richard Orr, Edward Kinsey and Alastair Campbell at QED, Christine Roberts, Rodney Shackell, Kathleen Smith, Harry Titcombe, Justin Todd, Roger Twinn, Ann Winterbottom, Elsie Wrigley.

The Authors

JOHN SEYMOUR

John Seymour was born in Essex, England, in 1914 but at the age of 20 he left England for Africa to fulfill one of his dreams which was to be a cowboy! John travelled all over Africa; he managed farms, worked in a copper mine, and became skipper of a fishing boat. When the Second World War broke out, John joined the King's African Rifles and fought in the gruelling Burma campaign – of the 40 officers who started the campaign he was one of only three who lasted to the finish. After the war he came back to England. He had left penniless and returned penniless. He then lived in a trolley bus for many years until he moved on to a Dutch sailing barge. By this time he had begun to write – mostly travel books (one about Africa, one about India, and books about sailing). He also started doing radio talks for the BBC.

When John and his then wife, Sally, had children they moved to a remote cottage without water or electricity in Suffolk. Very quickly they became self-sufficient and John wrote his first Self-Sufficiency book "The Fat of the Land" (extracts of which are reprinted in this edition with kind permission of the author). It is still in print today. After a time they decided they needed more land so they moved to Wales – still with very little money and with a 100 percent mortgage on a 70-acre farm. Once in Wales John wrote "The Complete Book of Self-Sufficiency" and Sally drew the illustrations. Then in 1981 John decided to "retire" and hand his farm over to his children and move over to Ireland. Once again John and his young companion, Angela Ashe, started from scratch on their three-acre Killowen

smallholding. The place was a wilderness; there was no running water nor electricity. John wrote a book about these early days called "Blessed Isle, One Man's Ireland".

Throughout his life John has written over 40 books and has made many films and radio programmes. Most of his later writing has been devoted to country matters, self-sufficiency and the environment. For the last 8 years John, Angela, and now William Sutherland, have been running courses in self-sufficiency from their home at Killowen in southern Ireland. The courses are taken up by students from all over the world who come to Killowen to meet John and to experience and learn about his lifestyle and philosophies at first hand.

John continues to take a front line interest in environmental campaigning and during 1998 became one of the now famous Arthurstown 7 who were arrested for destroying part of the agri-business Monsanto's experimental "genetically-mutilated" sugar beet experiment. More recently he has been a regular contributor to RTE's Nationwide TV programme, commenting on topical issues. Many of the books written by John Seymour are available from Metanoia Press, Killowen, New Ross, Co Wexford, Ireland.

WILL SUTHERLAND

Born in 1945, Will grew up on his father's traditional mixed farm in Northumberland. After taking a degree in Theoretical Physics and Law at Cambridge University, he went on to receive a Diploma in Farm Management and an MSc in Agricultural Economics before undertaking a year's apprenticeship on the family farm. Pursuing his fascination for insights into the reasons why countless human civilizations have failed to stand the test of time, Will then joined the Whitehall civil service becoming a Private Secretary and eventually a Treasury "Mandarin". In 1979 Will's public sector career was interrupted by twin desires to design and build his own organic golf course (The Millbrook Course in Bedfordshire) and to establish a worldwide organization for windsurfing in which he had been UK National Champion on a number of occasions. Having established both, Will returned to the public sector in 1983 becoming Head of the Policy Unit at Westminster City Council and later a Senior Management Consultant with Arthur Young (now Ernst and Young) in the City (of London's financial sector).

In 1989 Will left full-time management consultancy to edit and publish an alternative political magazine called Ideas for Tomorrow Today. Whilst organizing a conference prior to the Rio Earth Summit, Will met John Seymour in London. After taking part in the Citizens' Forum at Rio de Janeiro, Brazil, in 1992, Will devoted the next 18 months to editing and publishing the NGO (Non-Governmental Organization) Alternative Treaties which are now published in many different languages. Will moved to Ireland in 1993 to work with John Seymour and Angela Ashe, who were jointly running a school for self-sufficiency. Later Will married Angela and they now have 3 children.